Steel and its
heat treatment
Second Edition

Steel and its
heat treatment
Second edition

Karl-Erik Thelning
Head of Research and Development
Smedjebacken-Boxholm Stål AB, Sweden

Butterworths
London Boston Durban Singapore Sydney Toronto Wellington

First published 1975
Second edition 1984

© Jointly owned by Butterworth & Co. and K-E Thelning 1984

British Library Cataloguing in Publication Data
Thelning, Karl-Erik
  Steel and its heat treatment.—2nd ed.
  1. Steel—Heat treatment
  I. Title   II. Bofors handbook. *English*
  672.3′6   TN751

  ISBN 0-408-01424-5

Library of Congress Cataloging in Publication Data
Thelning, Karl-Erik
  Steel and its heat treatment.—2nd ed.
  1. Steel—Heat treatment
  I. Title   II. Bofors handbook. *English*
  672.3′6   TN751

  ISBN 0-408-01424-5

Typeset by Phoenix Photosetting, Chatham
Printed in Great Britain by
Mackays of Chatham Ltd, Kent

Preface

Steel and its heat treatment has been thoroughly revised and updated so that the second edition may incorporate the many developments that have taken place in the subject since publication of the first edition in 1975. As a result the coverage has been extended to include the following items.

Chapter 1: The fundamentals of TTT-diagrams are explained in detail. A description of various hardening mechanisms is given.
Chapter 3: Injection metallurgy and continuous casting are discussed; as is the influence of sulphur on the properties of steel.
Chapter 4: Existing CCT-diagrams are subjected to a critical review and a new generation of CCT-diagrams are presented. The mechanisms controlling hardenability are discussed and this forms the basis for a new concept of the cooling sequence during hardening. Examples of various cooling sequences and their effect on the resulting hardening are given.
Chapter 5: Various annealing processes, strain ageing and temper brittleness are discussed in more detail than in the first edition. Solution diagrams on heating are explained and discussed. Various methods of testing cooling media are given along with different interpretations of the three stages of the cooling curve.
Chapter 6: The various steel grades are arranged in accordance with ISO's system, with reference to national standards. A recently developed Swedish hot-work steel is presented. A large section is devoted to boron constructional steels, micro-alloyed steels and dual-phase steels. A scientifically interesting case-hardening test series, which is also of practical use, illustrates the prime importance for a case-hardening steel to have the right hardenability in the carburized case.
   The literature has been critically surveyed and all important references are listed.

The newly-written sections are based in the main on information and research released since 1975 which was the year when the author joined Boxholms AB. The steel division of this company merged in 1982 with Smedjebackens Valsverk (Smedjebacken Rolling Mill) to form the new company, Smedjebacken-Boxholm Stål AB (Smedjebacken-Boxholm Steel Company Limited).

The author is indebted to Dr Allan Hede, the laboratory director of Bofors AB, Professor Rune Lagneborg of the Swedish Institute for Metals Research, Professor Torsten Ericson of the Institute of Technology in Linkoping and Professor Tom Bell of the University of Birmingham, who have scrutinised the new material and made valuable contributions to the work.

As in the preceding edition the English translation has been carried out by Mr Cecil N. Black, BSc (Hons).*

The author wishes to thank all the persons and institutions mentioned as well as all others who have assisted in and contributed to the publication of this work. Last, but no means least, my thanks and appreciation to my wife Iris who has patiently and loyally supported me in all the work connected with this book.

Boxholm 1984   Karl-Erik Thelning

---

* The Swedish edition was edited by Maskinaktiebolaget Karlebo in cooperation with AB Bofors and Smedjebacken-Boxholm Stål AB

International designations and symbols

International system of units

On the 1st of January 1971 the metric system was officially introduced into the UK. For the majority of technicians this involved an adjustment from inches to millimetres. For several years, work had been in progress to devise a common standard international system of units. Such a system, SI (Système International d'Unités), was adopted in 1960.

The following units should be used in ISO standards prepared under the jurisdiction of ISO/TC 164:

1. Stress — N/mm^2
2. Hardness—The hardness designations in current use are retained but the hardness values are regarded as dimensionless numbers. The actual testing load shall be specified as N. Example: HV 5 (Testing load 49.03 N).
3. Impact — J.

During an interim period several European countries have used the symbol kp/mm^2 (kilopond) or kgf/mm^2 (kilogram-force) to signify the unit of stress.

The mechanical strength of steel was previously designated ton/in^2 (TSI) in the UK and kg/mm^2 on the Continent.

The first edition of this book was written and published during the period of transition covering the introduction of SI units both in the UK and other countries that have adopted it. In several diagrams in this book stress is designated kp/mm^2 and in some tables, kgf/mm^2 in accordance with some editions of ISO's recommendations. (BS 970:1970 gives kgf/mm^2 as the designation for stress.) In older ISO documents the symbols for the units of the yield and the ultimate tensile strength are given as kgf/mm^2 and $tonf/in^2$. In such documents $tonf/in^2$ has been replaced by N/mm^2. Also when the impact strength is given as $kgfm/cm^2$ the values have been supplemented with J.

In order to simplify the transition to SI units several diagrams have been drawn with double scales, e.g. inches–millimetres, kp/mm^2–TSI–N/mm^2 and even Celsius (°C)–Fahrenheit (°F).

In some tables two systems of units are used. For the conversion of inches to millimetres the factor 25·0 has often been used since the small error introduced thereby is of no practical consequence. For other, more precise applications, such as for Jominy diagrams, the exact conversion factor has been used.

In other ways, too, an international outlook is favoured, viz. the symbols of hardness units, e.g. HB and HRC. In line with this principle the symbol HV is used instead of DPH or VPN. Conversion tables and nomograms are found in Chapter 8.

In connexion with the change over to SI units, according to the ISO standard a number of designations for mechanical testing have been changed. What is characteristic of this transition is that certain designations in Greek letters have been replaced by Latin ones. The designations generally used for steel are indicated below, partly old ones, partly according to the new standard. An example for the use of SI units is given at the same time.

Old standard

$\sigma_{0.2}$	σ_B	σ_5	ψ	HB	KV	KCU
kp/mm^2	kp/mm^2	%	%		kpm	kpm/cm^2
54	81	19	61	249	7,0	9,2

New standard—SI

$R_{p0.2}$	R_m	A_5	Z	HB	KV	KU
N/mm^2	N/mm^2	%	%		J	J
530	790	19	61	249	69	45

In older ISO materials standards cited in this book the proof stress is designated by R_e. The designation $R_{p0.2}$ was adopted in 1973. The designation $R_{p0.2}$ is used for hardened and tempered steels. The designations R_{eL} and R_{eH} are used for unhardened steels with a clearly defined yield stress range (*see Figure 2.8*).

International steel designations

Under the auspices of ISO extensive work has been in progress for several years on the standardization of steel grades, in particular with respect to composition and mechanical properties. ISO recommendations covering a large number of steel grades have already been published. Among them may be mentioned the group 'heat-treated steels, alloy steels, free-cutting steels and tool steels'.

The tables covering 'Surveys of various types of steel' contain the standards as published by AISI, BS, DIN and SS along with such ISO standards as have been issued.

In the text, tool steels are designated mainly by the type letter and numeral as used in the USA and the UK for standardized tool steels, e.g. H 13, O 1. These designations are so well known by steel consumers all over the world that no qualifying institutional designations are necessary. Steels for which there are no AISI or BS specifications are designated according to DIN or SS standards.

Depending on which steel types are being discussed in the text, constructional steels are designated according to B S standards as well as A I S I, D I N or S S standards, respectively. In several instances use is made of simplified designations, e.g. 42 CrMo 4. Such designations are in general use on the Continent and indicate in a straightforward manner the approximate chemical composition of the steel.

Previously, the Swedish Standard was designated as S I S. All standards that have been revised or issued since 1978–01–01 are designated as S S. For the sake of uniformity all Swedish Standards are designated as S S in this book.

It is the author's aim and hope that this book will help in promoting the introduction of the S I units.

Contents

1

Fundamental
metallographic concepts

Metallography reveals the structure of metals and leads to a better understanding of the relationship between the structure and properties of steel. With the aid of modern developments such as the electron microscope and the scanning electron microscope it is now possible to obtain a much deeper insight into the structure of steel than was possible only some twenty years ago.

In order to understand the process occurring during the heat treatment of steel, it is necessary to have some knowledge of the phase equilibriae and phase transformations which occur in steel as well as of its microstructure. Therefore, a brief summary of these topics is given in this chapter which forms the groundwork for subsequent discussion.

1.1 The transformations and crystal structures of iron

On heating a piece of pure iron from room temperature to its melting point it undergoes a number of crystalline transformations and exhibits two different allotropic modifications. When iron changes from one modification to another heat is involved. This is called the latent heat of transformation. If the sample is heated at a steady rate the rise in temperature will be interrupted when the transformation starts and the temperature will remain constant until the transformation is completed. On cooling molten iron to room temperature the transformations take place in reverse order and at approximately the same temperatures as on heating. During these transformations heat is liberated which results in an arrest in the rate of cooling, the arrest lasting as long as the transformation is taking place.

The two alloptropic modifications are termed ferrite and austenite and their ranges of stability and transformation temperatures on heating and cooling are shown in *Figure 1.1*. The letter A is from the French arrêter, meaning to delay, c from chauffer, meaning to heat, and r from refroidir, meaning to cool. Ferrite is stable below 911 °C as well as between 1392 °C and its melting point, under the names α-iron and δ-iron respectively. Austenite, designated γ-iron, is stable between 911 °C and 1392 °C. Iron is

1

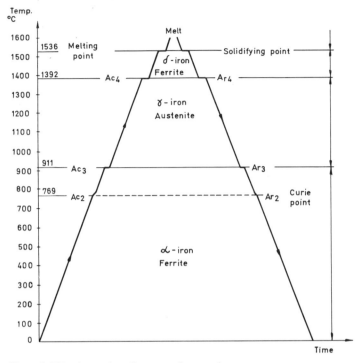

Figure 1.1 Heating and cooling curve for pure iron

ferromagnetic at room temperature; its magnetism decreases with increasing temperature and vanishes completely at 769 °C, the Curie point.

The atoms in metals are arranged in a regular three-dimensional pattern called a crystal structure. In the case of iron it may be pictured as cubes stacked side by side and on top of one another. The corners of the cubes are the atoms and each corner atom is shared by eight cubes or unit cells. Besides the corner atoms the iron unit cell contains additional atoms, the number of positions of which depend on the modification being studied.

Ferrite, besides having an atom at each corner of the unit cell, has another atom at the intersection of the cube body diagonals, i.e. a body-centred cubic lattice (BCC). The length of the unit cube edge or lattice parameter is 2·87 Å at 20 °C (Å = Ångström = 10^{-10} m). Austenite has a face-centred cubic lattice (FCC), the parameter of which is 3·57 Å (extrapolated to 20 °C). the structure of the unit cells of α-iron and γ-iron respectively may be envisaged as shown in *Figure 1.2*. The γ-iron unit cell has a larger lattice parameter than the α-iron cell but the former contains more atoms and has a greater density, being a 8·22 g/cm^3 for γ-iron at 20 °C and 7·93 g/cm^3 for α-iron.

1.2 The iron-carbon equilibrium diagram

The most important alloying element in steel is carbon. Its presence is largely responsible for the wide range of properties that can be obtained

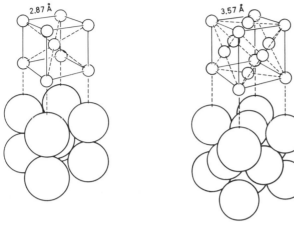

Ferrite   Austenite

Figure 1.2 The crystal structure of ferrite and austenite

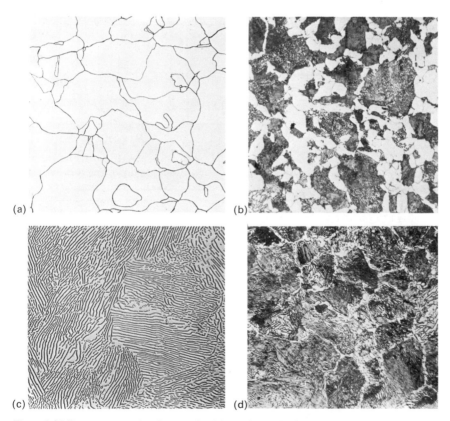

Figure 1.3 Microstructure of carbon steels with varying carbon content.
(a) Ferrite 0·0% C. 500× ; (b) Ferrite + pearlite 0·40% C. 500 × ;
(c) Pearlite 0·80% C. 1000 × ; (d) Pearlite + grain boundary cementite
1·4% C. 500 ×

and which make this metal such a highly useful commodity in everyday life. At room temperature the solubility of carbon in α-iron is very low and therefore the carbon atoms are to be found only very infrequently in between the individual iron atoms. Instead the carbon is combined with iron carbide, also called cementite, Fe_3C. The iron carbide may be present as lamellae alternating with lamellae of ferrite, which together form a constituent called pearlite, the mean carbon content of which is 0·80%. The proportion of pearlite in the structure increases with the carbon content of the steel up to 0·80%. Carbon in excess of this amount separates as grain-boundary carbides. A steel containing 0·80% carbon is said to be eutectoid (*see Figures 1.3a–d*).

When iron is alloyed with carbon the transformation will take place within a temperature range which is dependent on the carbon content as shown in the iron-carbon equilibrium phase diagram. *Figure 1.4* illustrates

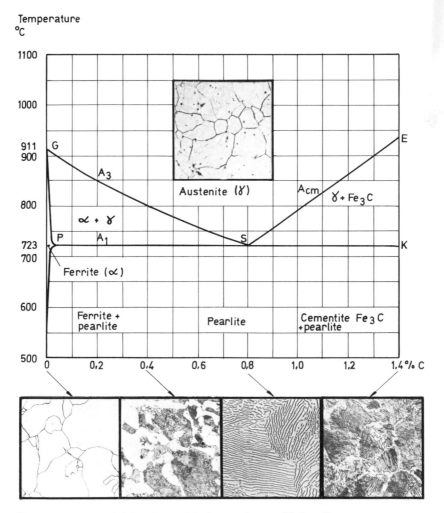

Figure 1.4 The lower left-hand part of the iron–carbon equilibrium diagram

various microstructures appropriate to that part of the iron–carbon diagram which applies to steel heat treatment. For the sake of completeness the phase diagram is reproduced as far as 6% carbon in *Figure 1.5* where it can be observed that the solubility of carbon is much greater in austenite than in ferrite.

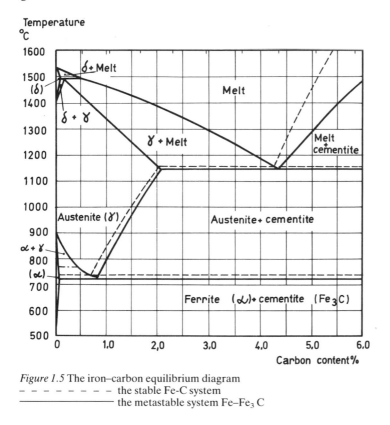

Figure 1.5 The iron–carbon equilibrium diagram
– – – – – – – – the stable Fe-C system
————————— the metastable system Fe–Fe$_3$C

1.2.1 Heating

It was mentioned earlier that α-iron transforms to γ-iron on being heated to 911 °C. This can be seen by looking at the vertical axis of the left-hand part of the diagram in *Figure 1.4*. In a steel containing 0·80% carbon, i.e. a eutectoid steel, the transformation to austenite takes place at about 723 °C. The temperature at which α-iron, γ-iron and cementite are at equilibrium is designated A_1 (line PK in diagram). Steels with less carbon, which are called hypo-eutectoid steels, begin to transform from pearlite to austenite at the same temperature, viz. 723 °C. In the equilibrium region between PS and GS there is austenite, formed from pearlite, and unchanged ferrite. The transformation is not complete until a temperature A_3 given by the line GS is reached. Above this line there is only one stable phase, viz. austenite. If the carbon content is more than 0.80% the steel is said to be hyper-eutectoid. In these steels, too, the pearlite transforms into austenite at 723 °C but the cementite (iron carbide) does not go into solution completely until the temperature rises above the equilibrium line SE, designated A_{cm}.

Let us for a moment return to the course of events taking place during heating. At 723 °C we find that the transformation to austenite begins to take place in steels having more than 0·025% carbon. This means that the atomic configuration changes from ferrite to austenite in which the carbon atoms are more soluble. The positions of carbon atoms in austenite are illustrated in *Figure 1.6*. At temperatures above G-S-E there is austenite only, all the carbon having been dissolved and evenly distributed throughout the austenite.

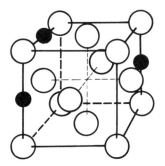

Figures 1.6 Examples of positions of carbon atoms in γ-iron

1.2.2   Cooling

As the temperature of a fully austenitized eutectoid plain carbon steel is slowly lowered below 723 °C the transformation from γ-iron to α-iron begins to take place and as a consequence the carbon is forced out of the lattice and forms cementite. On complete cooling to room temperature the steel has once again its pearlitic structure.

1.3 Time-temperature transformation

The iron-carbon equilibrium diagram is unquestionably of fundamental importance to heat-treatment processing. However, it only describes the situation when equilibrium has been established between the components carbon and iron. In the great majority of heat treatments the time parameter is one of the determinative factors, the influence of which is shown by so-called time–temperature–transformation diagrams. From these diagrams it is possible to follow the effect of both time and temperature on the progress of transformation. For the sake of convenience the time axis is drawn to a logarithmic scale.

1.3.1   Heating

The influence of time is best explained by means of the diagrammatic illustrations in *Figures 1.7a–f*. *Figure 1.7a* shows the familiar iron–carbon diagram in which it is seen that a 0.80% carbon steel, on being heated,

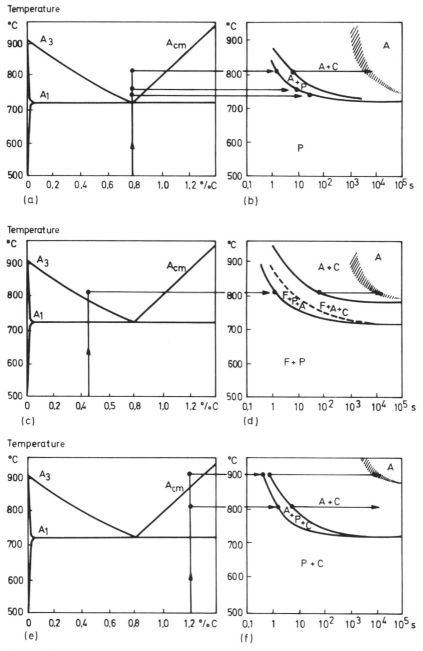

Figure 1.7 Structural transformations on heating steels containing:
(a) 0·80% C; (b) 0·80% C; (c) 0·45% C; (d) 0·45% C; (e) 1·2% C;
(f) 1·2% C. Schematic representation (after Rose and Strassburg[1])
A = austenite, B = bainite, C = cementite, F = ferrite, P = pearlite

8

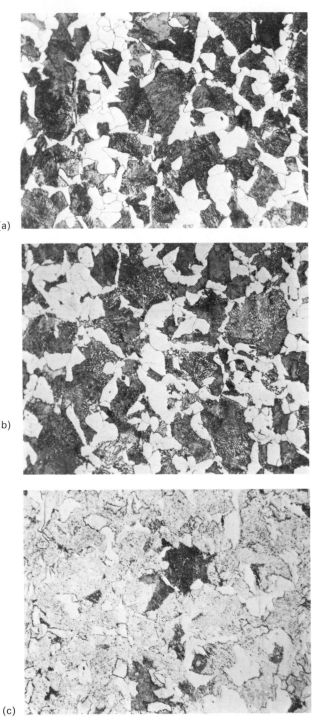

(a)

(b)

(c)

Figure 1.8 Microstructure of a 0·45% C steel after various heat
treatments. (a) Untreated, hardness 220 HV; (b) Heated to 725°C
in salt bath. Holding time 5 min. Quenched in water. Hardness
215 HV; (c) Heated to 735°C in salt bath. Holding time 5 min.
Quenched in water. Hardness 376 HV; (d) Heated to 750°C in salt

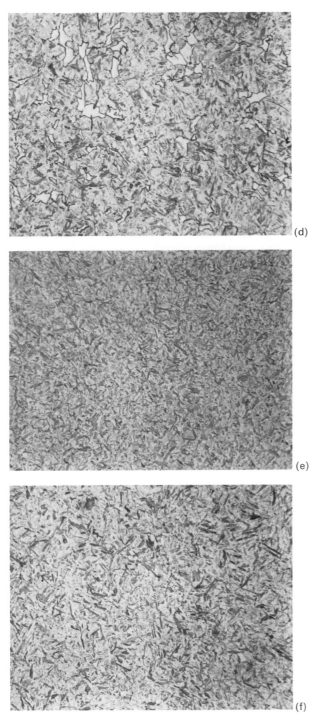

(d)

(e)

(f)

bath. Holding time 5 min. Quenched in water. Hardness 662 HV;
(e) Heated to 775 °C in salt bath. Holding time 5 min. Quenched in
water. Hardness 738 HV; (f) Heated to 825 °C in salt bath. Holding
time 5 min. Quenched in water. Hardness 744 HV

transforms to austenite at 723 °C. However, the diagram tells us nothing about how long this transformation will take. From *Figure 1.7b,* which applies to a 0.80% carbon steel only, it can be predicted that when the temperature is maintained at 730 °C after a rapid heating to this temperature, the transformation will be initiated in about 30 s (logarithmic time scale). If instead the steel is rapidly heated to 750 °C the transformation will begin in 10 s and if heated to 810 °C, in slightly over 1 s, i.e. practically at once on reaching this temperature. The transformation of pearlite to austenite and cementite is completed in about 6 s at 810 °C. If the steel is to be fully austenitic it must be held at this temperature for about 2 h ($7 \cdot 10^3$).

Figures 1.7c and *d* which apply to a 0·45% plain carbon steel show in the same way that at 810 °C, for instance, the transformation from pearlite to austenite starts almost immediately. In about 5 s the pearlite has been transformed and the structure consists of ferrite, austenite and cementite. About 1 min later the carbon has diffused to the ferrite which has thereby been transformed to austenite. Residual particles of cementite remain, however, and it takes about 5 h to dissolve them completely.

On heating a hyper-eutectoid steel containing 1·2% carbon to 810 °C a structure consisting of austenite and cementite is obtained in about 5 s (*Figures 1.7e* and *f*). It is not possible for the cementite to be completely dissolved at this temperature; this is apparent from a perusal of the equilibrium diagram. In order to effect complete solution of the cementite the temperature must be increased to 860 °C at least.

In order to study the rate of solution the steel is heated to a predetermined temperature, and after holding it there for a certain time it is quenched in water thereby 'freezing' the pre-existing structure. By this treatment, however, the austenite is transformed to martensite (for a discussion of this *see* p. 14).

The result obtained from such an experiment, applied to a 0.45% carbon steel, is shown in *Figure 1.8.* In all cases the holding time at temperature was 5 min. *Figure 1.8a* shows the original structure which consists of about 50% of ferrite and 50% of pearlite. On being heated to 725 °C for 5 min some of the cementite lamellae are converted to spheroids. This is called spheroidization. A decrease in hardness is a result of this process. There is no transformation (*see also Figure 1.7*). When it is heated to 735 °C (*Figure 1.8c*) the main part of the pearlite is transformed to austenite which, on being quenched, forms martensite. Some ferrite and pearlite remain untransformed. A holding time of 5 min at 750 °C (*Figure 1.8d*) is sufficient to transform all the pearlite but about 5% of the ferrite remains unchanged. However, 5 min at 775 °C is not sufficient to transform all the ferrite, as is apparent in *Figure 1.8e* which shows traces of this constituent. At 825 °C, on the other hand, all the ferrite has been transformed in 5 min but cementite, in the shape of small round particles, is still in evidence, which is in agreement with *Figure 1.7*.

Varying the rate of heating to the hardening temperature will have an effect on the rate of transformation and dissolution of the constituents. *Figure 1.9* shows a continuous-heating diagram for a steel of almost eutectoid composition and consisting originally of ferrite and pearlite[1]. The third curve from the right represents a heating rate of about 3 °C per

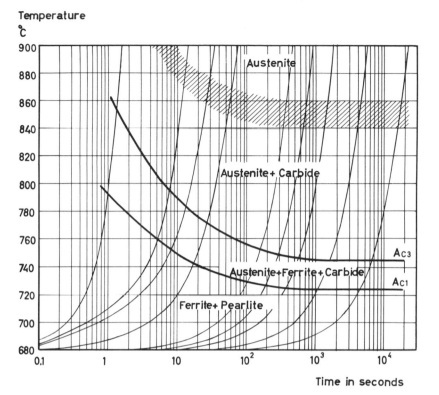

Figure 1.9 Transformation diagram for continuous heating. Dissolution of ferrite and lamellar pearlite in a 0·70% C steel (after Rose and Strassburg[1])

minute. The temperature of transformation increases as the rate of heating increases.

A complete set of continuous-heating curves for various grades of steel would be of great help in the field of practical heat treatment. Since such diagrams are generally not available we must be content with simple estimations of heating and holding times. However, a commendable achievement in this field has been accomplished by the issuing of *Atlas zur Wärmebehandlung der Stähle*, Volume 3 in 1973 and Volume 4 in 1976. This latter volume contains both continuous and isothermal solution diagrams for a large number of commercial steel grades. Some of these diagrams will be discussed more fully in Chapter 5, Heat treatment—general.

1.3.2   Cooling

The general appearance of the structure created during cooling is dependent on the temperature of transformation and on the time taken for the transformation to start. As in the case of heating, the iron–carbon equilibrium diagram can tell us nothing about this. In a manner similar to that described above, the transformation of steel at a certain temperature

12

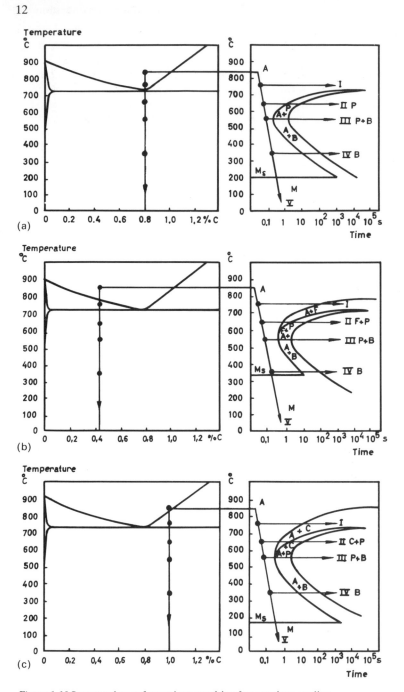

Figure 1.10 Structural transformations resulting from various cooling programmes for steels containing (a) 0·80% C; (b) 0·45% C; (c) 1·0% C. A = austenite, B = bainite, C = cementite, F = ferrite, P = pearlite, M = martensite, M_s = start of martensite formation

may be investigated by cooling it from the austenitic state to the temperature concerned, allowing the transformation to take place and then quenching to room temperature. The structure thus obtained is then studied under the microscope. Cooling techniques of this kind are shown in the right-hand part of *Figure 1.10*, and in the text following the formation of the various structural components is explained. The same structure may also result from continuous cooling. This will be discussed later in the book.

1.3.3   Formation of pearlite

When a eutectoid steel is cooled from an austenitizing temperature of, say, 850 °C to 750 °C and held at this temperature, no transformation will take place. If the temperature is lowered to 650 °C pearlite will start to form after 1 s and the transformation will be completed in 10 s (*see* curve II in *Figure 1.10a*). As the temperature of pearlite formation is lowered the pearlite lamellae become increasingly finer and the whole structure becomes harder. If we allow the transformation of the hypo-eutectoid steel in *Figure 1.10b* to take place at 750 °C, only ferrite separates and a state of equilibrium is established between ferrite and austenite (curve I). If the transformation takes place at 650 °C ferrite separates first followed after a short interval by pearlite. Similarly, in the case of the hyper-eutectoid steel, *Figure 1.10c,* cementite separates first followed by pearlite.

   Pearlite formation is initiated at the austenite grain boundaries or at some other disarray in the austenite grains. The process has been studied in detail by Hillert[2] who found that pearlite formation can be initiated on either ferrite or cementite and that pearlite growth proceeds by branching. Platelets of cementite and ferrite grow in juxtaposition since carbon transport from the austenite to the edges of the cementite platelets results in a simultaneous carbon impoverishment of the edges of the ferrite platelets. *Figure 1.11* shows how pearlite grows according to this model. On the assumption that the pearlite originates from cementite, *Figure 1.12* shows schematically how pearlite is formed. This model is illustrative but is probably not a scientifically accurate one.

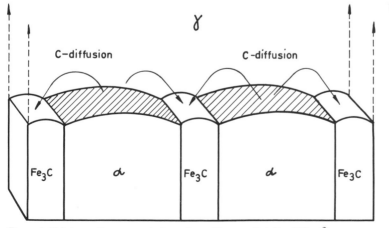

Figure 1.11 Schematic representation of pearlite growth (after Hillert[2])

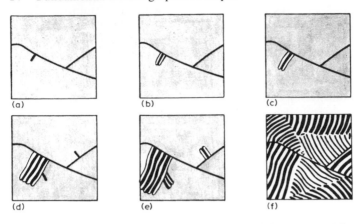

Figure 1.12 Schematic representation of pearlite growth (previous model)

1.3.4   Formation of bainite

At temperatures below about 550 °C another constituent, bainite, starts to separate along with the pearlite. Its formation is assumed to be initiated on ferrite nuclei which grow as platelets from the grain boundaries. The carbon content of the surrounding austenite increases continuously and when it has reached a limiting value platelets of cementite form in juxtaposition with platelets of ferrite.

As the temperature falls bainite begins to form inside the grains as well; at the same time the mode of formation changes. In the metallurgical microscope it may be difficult to differentiate bainite from other constituents since bainite alters its appearance according to its temperature of formation and the composition of the steel. *Figure 1.13* shows bainite in a chrome–manganese steel.

Depending on the temperature of formation of bainite it is classified as upper or lower bainite. The mode of formation of the various types of bainite as well as their properties have been described by Pickering[3]. For the present purpose it is sufficient to indicate that upper bainite is relatively brittle and lower bainite is tough.

1.3.5   Formation of martensite

Referring to *Figures 1.10a–c*, if cooling takes place as represented by curve V, i.e. very rapidly, austenite will start to transform to ferrite on reaching line M_s. As the cooling continues below M_s there is very little carbon migration while the austenite is transforming. Thus, the carbon atoms remain in solid solution in the α-rion. Since the space available for the carbon atoms is less in α-iron than in γ-iron the carbon atoms will expand the lattice. The resulting state of stress increases the hardness of the steel. We say that the steel has been *hardened*. The new constituent, called martensite, is a supersaturated solution of carbon in α-iron.

The mechanism of martensite formation is the subject of considerable

Figure 1.13 Bainite 500 ×

controversy and several theories have been advanced. *Figure 1.14* illustrates a simple model of how we can picture the transformation of γ-iron during martensite formation. The carbon atoms situated on the edges of the martensite unit cube cause the unit cell to increase in one direction, which results in a tetragonal lattice. Even in the case of high-carbon steels only a very small fraction of the number of possible lattice sites are occupied by carbon atoms. The volume of the martensite increases with increasing carbon content.

From the transformation diagrams it will be seen that the formation of pearlite and bainite progresses with time whereas that of martensite does not. Each temperature below M_s corresponds to a definite proportion of martensite but the amount actually formed depends on the grade of steel, the conditions of the austenitizing treatment and the rate of cooling on

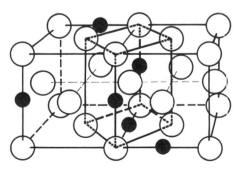

Figure 1.14 Simple model for the transformation of austenite (γ) to martensite (α)

quenching for hardening. As the temperature falls it is possible to follow visually the martensite transformation in a so-called hot-stage microscope. *Figures 1.15a–f* show the progress of martensite formation. It starts at 220 °C and at 175 °C the main part of the austenite has been transformed to martensite. The steel used is a high-alloy steel, designated MAR 2, in which the martensite is composed of nickel and iron.

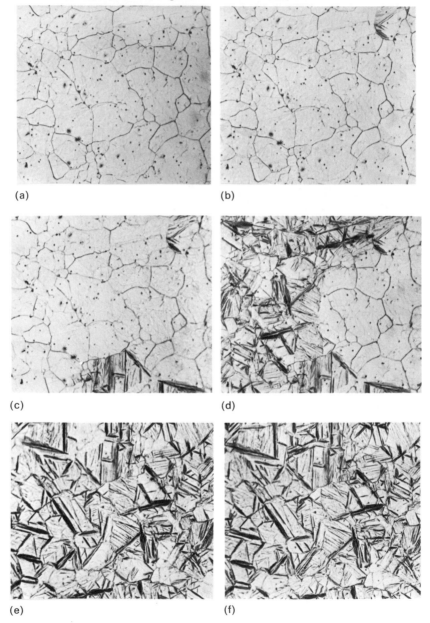

Figure 1.15 Transformations of austenite to martensite at successively lower temperatures. (a) 280; (b) 220; (c) 200; (d) 195; (e) 180 and (f) 175 °C

Marder and Krauss[4] have studied the structure of martensite in iron–carbon alloys and have found two different types called massive martensite or lath martensite, and acicular martensite or plate martensite. The range composition in plain carbon steels in which only lath martensite exists extends up to 0·6% carbon. This type of martensite consists of packets of parallel laths (*Figure 1.16a*), which can only be resolved in the electron microscope. The more familiar plate martensite is found only in steels containing 0·6% carbon and more. The structure of quenched high carbon steels consists of irrationally arranged plates in a matrix of austenite (*Figure 1.16b*). In the range 0·6–1% carbon both types of martensite may occur simultaneously.

For an unalloyed steel the temperature of the start and finish of the

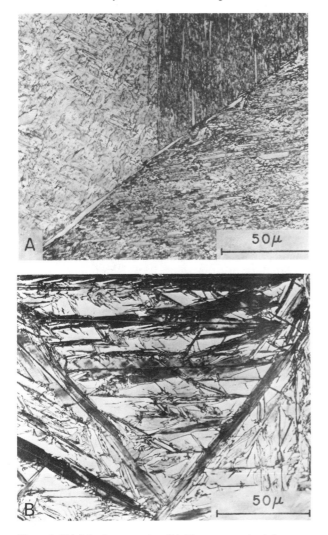

Figure 1.16 (a) Lath martensite; (b) Plate martensite (after Marder and Krauss[4])

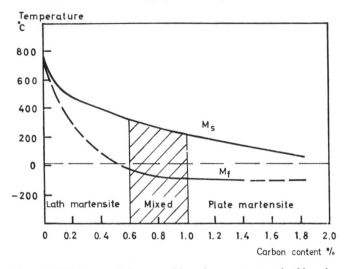

Figure 1.17 Influence of the austenitic carbon content on the M_s and M_f temperatures and the type of martensite formed in unalloyed steel (after Marder and Krauss[4])

martensite formation, M_s and M_f respectively, depends on the carbon content as shown in *Figure 1.17*. According to Krauss and Marder[5], however, the formation of either lath martensite or plate martensite is not only dependent on the carbon content but also on the temperature at which it forms. Alloying elements that lower the M_s temperature, such as Mn and Ni, will therefore promote the formation of plate martensite.

Van de Sanden[6] has investigated samples of commercial carbon steels containing carbon from 0·32% to 0·79% by austenitizing them at 1000 °C for 30 minutes and then quenching them in water. He found that the 0·32% steel sample contained mainly lath martensite, but on examining it with a transmission electron microscope (TEM) he found some areas with plate martensite. With increasing carbon content the amount of plate martensite increased also and at 0·63% C the structure consisted mainly of plate martensite. The 0·79% C steel contained plate martensite only.

Van de Sanden found that the transition zone of lath to plate martensite was displaced towards carbon contents lower than those given by Krauss and Marder. This fact is probably due to the high hardening temperature (1000 °C) which causes complete solution of the carbon, thereby also resulting in a lower M_s temperature.

1.3.6   Retained austenite

Most of the austenite in a eutectoid steel will transform to martensite during the quenching to room temperature. The untransformed part is called retained austenite. *Figure 1.18* shows how the amount of retained austenite in unalloyed steel varies with the carbon content. If the temperature is lowered below room temperature the transformation to martensite continues (*see Figure 1.17*). This method of increasing the amount of martensite is called subzero treatment (*see* Chapter 5 for details

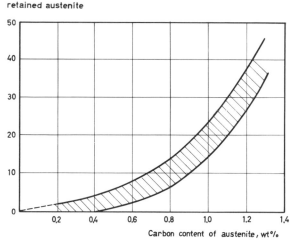

Figure 1.18 Variation in the amount of retained austenite carbon content on hardening

of this treatment).

In the process called martempering (*see* Chapter 5) the cooling is interrupted just above M_s or in some cases just below it, and then the steel is allowed to cool to room temperature. This interruption in the cooling stabilizes the austenite somewhat, which causes the martensite formation to start at a lower temperature, thereby resulting in a higher proportion of retained austenite at room temperature. Hence a martempered or air-hardened steel has in general a larger amount of retained austenite than an oil-hardened one. The reason for the stabilizing effect is assumed to be due to the dissolution at the arrest temperature of the martensite nuclei formed during cooling from the austenitizing temperature. The dependence of the amount of retained austenite on the alloy content and hardening temperature will be discussed in several of the following chapters.

1.3.7   TTT diagrams and CCT diagrams

The formation of the structural constituents or phases discussed above has been studied for several decades in considerable detail and at present the majority of steel grades have been chartered by means of TTT diagrams (Time–Temperature–Transformation). Such diagrams are constructed from data obtained from the transformation of steel specimens at constant temperature, as described in *Figure 1.10* along with the accompanying text. They are also called isothermal diagrams (IT diagrams). The following background information concerning the transformations may facilitate the understanding of the diagrams. The phase change taking place at a certain temperature is dependent on nuclei formation and growth of the phases concerned. The mechanism may be described in simple terms as follows.

Nuclei formation is favoured by supercooling. Hence nucleation takes place rapidly (takes a short time) at low temperatures and slowly (takes a long time) at high temperatures. This is because the driving force behind nuclei formation increases with falling temperature. The rate of growth of a stable nucleus is governed by diffusion and is therefore rapid at high temperatures, but growth cannot take place until nucleation has been initiated. At temperatures at A_{c3} and above there is no driving force and consequently no nucleation and growth.

To illustrate the above-described simplified model of the course of an isothermal transformation in a carbon steel of eutectoid composition the diagram in *Figure 1.19* can be used as an example. The so-called nose of

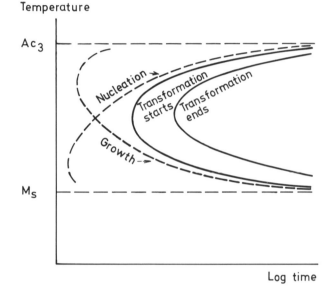

Figure 1.19 Schematic representation of nucleation and growth in a eutectoid carbon steel

the diagram is made up of a pearlite nose and a bainite nose. For a hypo-eutectoid steel there is an additional ferrite nose, and for a hyper-eutectoid steel a carbide nose as shown in *Figure 1.10b* and *c*. Each phase is dependent for its nucleation and growth on special circumstances which are influenced by the alloying elements in the steel.

As has already been stated, TTT diagrams show only those transformations which occur at constant temperature. In spite of this limitation TTT diagrams are frequently incorrectly used to account for continuous-cooling processes. Unfortunately this practice has to some degree delayed a general appreciation of CCT diagrams.

CCT stands for Continuous–Cooling–Transformation, which means transformation taking place during continuous cooling. Such diagrams complement TTT diagrams but it is the CCT diagrams in the main that play a very important role in the heat treatment of steel. Hence they are frequently used further on in this book.

The times for the initiation of transformation on the one hand by the isothermal process and on the other hand by the continuous-cooling process, differ very markedly, as may be seen by comparing such diagrams constructed for the same steel. *See Figure 1.20.*

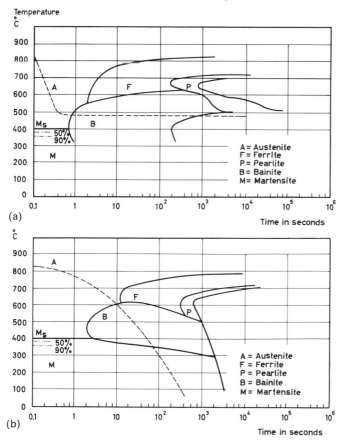

Figure 1.20 (a) Isothermal diagram; (b) Continuous-cooling diagram for BS 708A37 (En 19B)

1.4 Decomposition of martensite and retained austenite on tempering

The martensite formed during hardening is generally too brittle for the steel to be put to practical use without first tempering it. Tempering usually results in an increase in toughness and a simultaneous reduction in hardness. For a better understanding of the mechanism of the tempering process the three stages that a hardened carbon steel passes through when subjected to a continuous rise in temperature are now briefly discussed.

1.   80–160 °C.   Precipitation of a carbon-rich phase called ε-carbide. As a consequence the carbon in the martensite is reduced to approximately 0·3%.

2.   230–280 °C.   Decomposition of retained austenite to bainite.
3a. 160–400 °C.   Formation and growth of cementite (Fe_3C) at the expense of the ε-carbide.
3b. 400–700 °C.   Continued growth and spheroidization of the cementite.

Somewhat different temperature ranges have also been reported and these differences are thought to be due mainly to different rates of heating. *Figure 1.21* shows a tempering curve for a 1% carbon steel with the different tempering stages marked on it.

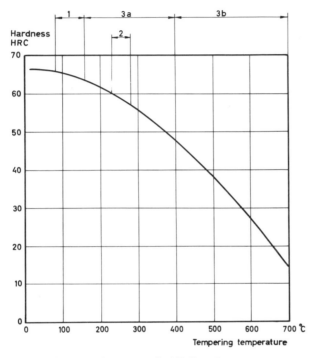

Figure 1.21 Tempering stages of a 1% C steel

For high-alloy chromium steels, hot-work steels and high-speed steels the range of decomposition of retained austenite is displaced towards higher temperatures. The product of decomposition, i.e. bainite or martensite, depends on the tempering temperature and time (*Figure 1.22*). Bainite formation occurs isothermally, i.e. at constant temperature during the tempering process, whereas martensite forms as the steel is cooling from the tempering temperature.

In high-alloy steels a precipitation of finely dispersed complex carbides occurs at about 500 °C. This is called the fourth stage of tempering. The martensite formed from the retained austenite and the, possibly coherent (see p. 45), carbide precipitate together create a hardness peak which is a characteristic feature of high-speed and other high-alloy steels. *Figure 1.23* shows the variations in hardness of some current steels that have been tempered for 2 h at different temperatures.

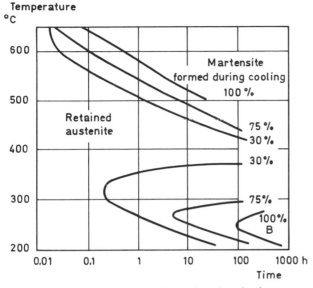

Figure 1.22 TTT diagram for transformation of retained austenite. Steel A2

The interconnection between hardness, temperature and time of tempering may be mathematically illustrated by means of the following expression, formulated by Hollomon and Jaffe[7].

$$P = T(k + \log t)$$

where P = a parameter linked to the tempering process
T = temperature in K (absolute temperature K = °C + 273)
k = a constant
t = time in hours

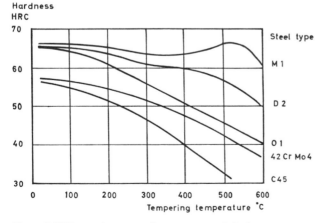

Figure 1.23 Tempering curves for some current steels

Each value of *P* corresponds to a hardness value that can be read off from a so-called master parameter curve which must first be drawn for each grade of steel. The constant *k* is about 20 for the majority of alloy steels. Such a curve for steel H 13 is shown in *Figure 1.24*. The practical application of these master curves is treated in Section 5.4.

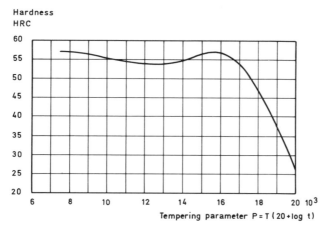

Figure 1.24 Master parameter curve for steel H 13

1.5 Diffusion

1.5.1   The nature of diffusion

Diffusion is the process of migration of individual atoms within materials and takes place in gases, liquids and solids, and consequently occurs in steels and other metallic systems. Diffusion is of immense importance to most heat treatment processes and therefore some appreciation of the mechanisms and laws of diffusion will facilitate the understanding of many heat treatments including carburizing and decarburizing, nitriding and annealing.

The foreign or alloy atoms that occupy sites of the same type as the host atoms in the iron lattice in a substitutional solid solution move about with the aid of empty sites called vacancies. Since steel contains a large number of vacant lattice sites there is a continuous migration of atoms via these vacancies. Carbon and nitrogen atoms are small compared with iron atoms, and can therefore be situated at sites between the iron atoms so forming an interstitial solid solution. These small atoms can diffuse without the aid of vacancies and their rate of diffusion is much higher than that of substitutional atoms (*see Figures 1.42 and 1.43*).

Mathematical studies of diffusion by Einstein have shown that the average distance of diffusion $\bar{x}$ cm is equal to $\sqrt{(2Dt)}$ where *D* is the coefficient of diffusion in cm^2/s and *t* the time in seconds. This equation may be applied in practice to obtain an estimate of how far diffusion can proceed in a certain time when the atoms migrate in one direction. For such purposes the equation may be written:

$$x = \sqrt{(2Dt)} \qquad (1.1)$$

If we put $\sqrt{(2D)} = k$ we obtain

$$x = k\sqrt{t} \qquad (1.2)$$

This equation may be regarded as fundamental to most diffusion processes. If we take an assumed case and let $k = 0.1$ we obtain the graph shown in *Figure 1.25*. Notice the difference between the diffusion distance for example of one hour's diffusion at the beginning of the process and for the same period of time after 10 h.

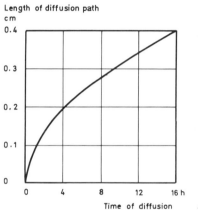

Length of diffusion path
cm

Time of diffusion

Figure 1.25 Dependence of diffusion on time

1.5.2   Factors that influence the rate of diffusion

Diffusion takes place due to a concentration differential (actually an activity differential) which produces the driving force required for the diffusion process. The basic expression for the dependence of diffusion on the concentration gradient is given by Fick's First Law:

$$J = -D \cdot \frac{\delta c}{\delta x} \qquad (1.3)$$

where J = amount of the substance under investigation that passes in unit
time through unit area in a plane normal to the x-axis (g/cm^2 s)
D = diffusion coefficient (cm^2/s)
c = concentration of the diffusing substance (g/cm^3)
x = a coordinate (cm)

The diffusion coefficient is very strongly influenced by temperature. On a rough estimate D is doubled for every temperature increase of twenty degrees. The value of D for different temperatures is given by:

$$D = D_0 \exp\left(-\frac{Q}{RT}\right) \qquad (1.4)$$

where D_0 = frequency factor (cm²/s)
Q = energy of activation (cal/mol/K)
T = temperature in kelvins (K)
R = gas constant (1·987) cal/mol)

Table 1.1 gives approximate values of D_0 and Q for some substances. Different investigators give somewhat different values[8, 9, 10, 11]

Table 1.1

Diffusing element	Diffusing through	D_0 cm²/s	Q cal/mol
Carbon	α-iron	0·007 9	18 100
Carbon	γ-iron	0·21	33 800
Nickel	γ-iron	0·5	66 100
Manganese	γ-iron	0·35	67 000
Chromium	α-iron	30 000	82 000
Chromium	γ-iron	18 000	97 000

1.5.3   Calculation of diffusion distance

From Fick's Second Law

$$\frac{\delta c}{\delta t} = D \frac{\delta^2 c}{\delta x^2} \tag{1.5}$$

we can calculate the diffusion distance. The methods of solving various types of differential equations are obtainable in works of reference[9].

There follows some examples that should be of interest to those involved in heat treatment. They may also be re-read alongside the various processes as they are discussed later in this book.

Carburization

For the sake of simplicity it is assumed that the surface of the steel acquires a carbon concentration corresponding to the carbon potential of the carburizing medium and that the carbon concentration in the outermost layer remains constant during the whole process. The carbon content of the outer layer is usually about 0·8–1·0%. To calculate the carbon concentration figure at a certain depth below the surface after a certain carburizing time the following formula may be used:

$$C - C_0 = (C_1 - C_0)\left[1 - \text{erf}\left(\frac{x}{2\sqrt{(Dt)}}\right)\right] \tag{1.6}$$

where C = required carbon concentration at a depth x below the surface (*see Figure 1.26*).
C_0 = basic carbon content of the steel
C_1 = carbon content at surface of steel
x = depth below surface (cm)
D = diffusion coefficient (cm²/s)
t = time (s)
erf = a so-called error function (*see Table 1.2*)

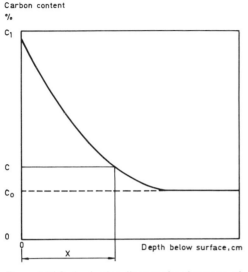

Figure 1.26 Carburization diagram showing some of the quantities contained in Equation 1.6

Table 1.2

y	erf (y)	y	erf (y)
0	0·000	0·8	0·742
0·1	0·112	0·9	0·797
0·2	0·223	1·0	0·843
0·3	0·329	1·2	0·910
0·4	0·428	1·4	0·952
0·5	0·521	1·6	0·976
0·6	0·604	2·0	0·995
0·7	0·678	2·4	0·999

Example 1

How long will it take to carburize a 0·20% C steel at 900 °C so that the carbon content will be 0·40% at a depth of 1 mm? (0·40% C is roughly the carbon content of a steel that on being hardened will have a hardness of 550 HV. This hardness figure is generally taken to define the limit of the depth of case hardening.) The carbon concentration at the surface is assumed to be 1·0%.

Inserting the given carbon values in Equation (1.6) we obtain

$$0·4 - 0·2 = (1·0 - 0·2)\left[1 - \text{erf}\left(\frac{x}{2\sqrt{(Dt)}}\right)\right]$$

$$\text{erf}\left(\frac{x}{2\sqrt{(Dt)}}\right) = 0·75$$

$$\frac{x}{2\sqrt{(Dt)}} = 0·814 \quad \text{(from *Table 1.2*)}$$

$$x = 1·63\sqrt{(Dt)} \tag{1.7}$$

D is calculated from Equation (1.4) and taking the appropriate values from *Table 1.1*.

$$D = 0.21 \exp\left(\frac{33\,800}{1.987 \times 1173}\right)$$

$$D = 1.1 \times 10^{-7}$$

Since the depth x is given (0·1 cm) we can calculate t from Equation (1.7)

$$0.1 = 1.63\sqrt{(1.1 \times 10^{-7} \times t)}$$

$$t \approx 34\,200$$

Converting to hours we find that the carburizing time is 9·5 h. If we apply Einstein's equation and take the 2 outside the root sign we obtain

$$x = 1.414\sqrt{(Dt)} \qquad \text{c.f. Equation (1.7)}$$

In this case the carburizing time is 12·6 h. We see therefore that Einstein's equation could have been used to give a rough estimate of the carburizing time.

Decarburization[12]

Provided that the graph representing decarburization has the appearance shown in *Figure 1.27* it is possible, by using Equation (1.8) below, to calculate the depth at which there will be a given carbon concentration.

$$C = C_0 \cdot \text{erf}\left(\frac{x}{2\sqrt{(Dt)}}\right) \tag{1.8}$$

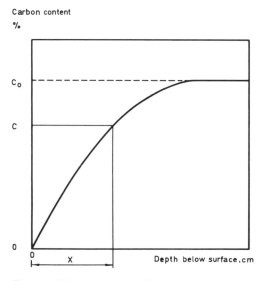

Figure 1.27 Decarburization diagram

Example 2

A plain carbon steel, 0·85% C, has been heated in air for 1 h at 900 °C and as a result the carbon content of the outermost layer has been reduced practically to zero. What depth of stock must be removed from the surface if the permissible carbon concentration in the outer layer is 0·80%? The value of *D* is taken from the foregoing example.

From Equation (1.7) we obtain

$$0\cdot80 = 0\cdot85 \times \text{erf} \frac{x}{2\sqrt{(1\cdot1 \times 10^{-7} \times 3600)}}$$

$$0\cdot941 = \text{erf} \frac{x}{2\sqrt{(3\cdot96 \times 10^{-4})}}$$

$$1\cdot35 = \frac{x}{2\sqrt{(3\cdot96 \times 10^{-4})}}$$

$$x = 0\cdot054 \text{ cm}$$

Homogenizing annealing

Segregations always occur in steel to a greater or lesser extent. By means of a homogenizing annealing, which is carried out at a relatively high temperature, differences in concentration of the segregating elements can be reduced. This is illustrated schematically in *Figure 1.28*. Provided that

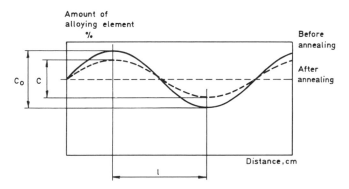

Figure 1.28 Reducing concentration differential by means of homogenizing annealing

the variation in the concentration of the element in question is sinusoidal it is possible to calculate data for the homogenizing annealing according to the equation:

$$C = C_0 \exp\left(-\frac{Dt\pi^2}{l^2}\right) \tag{1.9}$$

where C = concentration differential after annealing
$\quad\quad C_0$ = concentration differential before annealing

Example 3

The concentration differential for manganese is found to be 0·6%. This differential is to be reduced to 0·4% by a homogenizing annealing at 1100 °C. What is the holding time required if the diffusion path $l = 0.01$ cm? D^{Mn} at 1100 °C = 2×10^{-11} cm²/s.

According to Equation (1.9) we obtain

$$0.4 = 0.6 \exp \left(\frac{2 \times 10^{-11} \times t \times \pi^2}{0.01^2} \right)$$

$$t = 2.06 \times 10^5 \text{ s} = 5.72 \text{ h}$$

1.6 Dislocations

In a perfect iron crystal the atoms are arranged in a regular pattern, which we have previously discussed. In order to produce shear in a perfect crystal all the atoms in a slip plane must move simultaneously. *Figure 1.29* shows a section through the slip plane, illustrating how this slip can take place. A nearly perfect crystal of iron (whisker) has a strength more than 100 times the strength of ordinary mild steel. In general, however, the lattice is far from perfect. This is due to the fact that a crystal of an ordinary steel contains imperfections of various kinds. We have mentioned earlier the

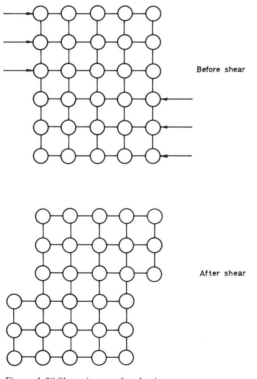

Before shear

After shear

Figure 1.29 Shear in a perfect lattice

presence of vacancies, i.e. empty positions in the lattice by means of which atoms can move about. The crystals also contain linear faults termed dislocations. The symbol ⊥ in *Figure 1.30* marks the end of what is called an edge dislocation which extends further into the lattice, i.e. at right angles to the plane of the paper.

The concept of dislocations helps us to understand and explain many characteristic features of steel, for instance why it is not so strong as it would be if it consisted of a perfect crystal. The following example illustrates this point.

If the shear forces are applied as shown in *Figure 1.30a* a distance between atoms E3 and E4 will increase and the cohesive force between

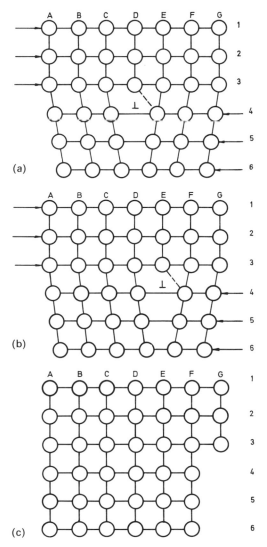

Figure 1.30 Slip in an atomic lattice by means of dislocations

them will decrease. As the shear force approaches a critical value atom E4 will move into the sphere of influence of atom D3 as seen in *Figure 1.30b* where E4 has now jumped into position D4. By still applying the shear force we find that the atoms will jump across, one after the other, into new positions and the lattice will assume the configuration shown in *Figure 1.30c*. As a consequence of this intermittent movement of the atoms and compared with the above-described case of the perfect crystal a much smaller shear force is now required. Hence it is apparent that dislocations facilitate the plastic deformation of the material. But they have other functions.

During plastic deformation new dislocations are created and as a result it becomes gradually more difficult for the gliding dislocations to move through the crystal grains, i.e. the steel work hardens. In a material with precipitate particles, which are too hard to be sheared, work hardening is particularly marked and depends on the rings around particles left by the gliding dislocations, which have passed through a row of particles as indicated in *Figure 1.31*. The actual existence of dislocations has been verified with the aid of the electron microscope. Even though it is not possible to see the dislocations themselves the structural anomalies they give rise to are very obvious (*Figure 1.32*).

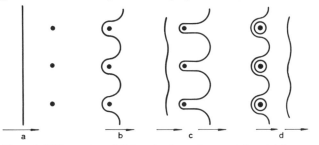

Figure 1.31 Formation of dislocation loops in a crystal containing hard particles

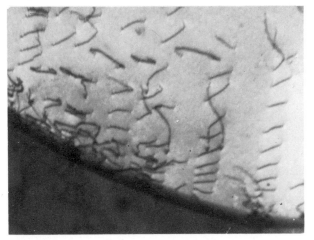

Figure 1.32 Dislocations piling up at a grain boundary in an 18/8 steel

Dislocations markedly affect the properties of metals. The above presentation is merely an introduction and the interested reader will find a wealth of pertinent literature [13, 14, 15]

1.7 Grain size

When a tool or machine component has failed it is customary to examine the fractured surface in order to see whether it is fine or coarse grained. If the latter, faulty heat treatment may have been the cause (generally too high a hardening temperature) or the steel may have been unsuited to the treatment in question. A great number of the properties of steel are influenced by grain size, such as machinability, impact strength, hardenability, etc.

1.7.1   Grain boundaries

Steel contains many grains separated by grain boundaries, of which there are three distinguishable types

Low-angle grain boundaries

Such grain boundaries may be regarded as a series of dislocations lying in the same plane in an otherwise perfect crystal. The crystal lattices of the two adjacent grains give rise to an angle, the size if which is proportional to the dislocation density at the surface boundary (*Figure 1.33*).

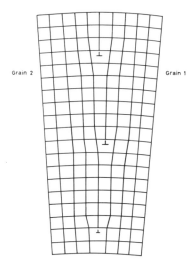

Grain 2   Grain 1

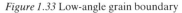

Figure 1.33 Low-angle grain boundary

High-angle grain boundaries

The angle between adjacent crystal lattices is so large in this case and the structure at the grain boundary so disordered that no individual dislocations can be distinguished.

Twinning boundaries

The boundary between twins is an atomic plane that conforms exactly to both crystal lattices, each of which may be regarded as the mirror image of the other (*Figure 1.34*).

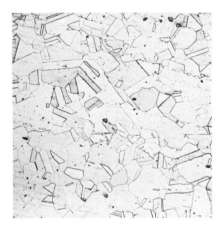

Figure 1.34 Twinning boundaries in austenitic stainless steel

Another term used is phase boundary, i.e. the boundary between two grains of different phases or crystal structures. Such boundaries are comparable to high-angle grain boundaries. It can sometimes happen that two crystal structures have specific similarities along certain planes, with a regular and mutual alignment as a result. Such crystal lattices are said to be coherent. Since the grain boundaries contain considerably more vacancies and dislocations than the grains themselves, diffusion proceeds much more easily along the boundaries than through the grains. This is particularly the case at low temperatures. In addition preferential concentration of certain foreign atoms takes place at grain boundaries. Likewise, precipitation of particles (e.g. carbides in steel) from a supersaturated solution often occurs at grain boundaries, thus making these easy to observe in a metallurgical microscope.

1.7.2   Methods of determining grain size

When determining the grain size of hardened steel the austenitic grain size is implied, i.e. the size of the austenite grains before the steel is cooled and before the austenite has transformed to other structural constituents. The original austenite grains can generally be revealed after cooling the steel to room temperature.

When hypo-eutectoid carbon steels or low-alloyed steels are cooled slowly ferrite separates first from the austenite, then pearlite. The final grain size is therefore smaller than that of the austenite grains.

The grain size can be determined by various methods; the five most frequently used techniques are described below. It is important that the name of the method required be specified when grain size estimations are stipulated.

For ferritic and ferritic–pearlitic steels the concept of grain size has acquired increased significance owing to the introduction of high-strength, low-alloy (HSLA) steels. For such steels grain size is usually determined by one of the methods numbered 1 to 4.

1 ASTM standard classified grain size

The grain size is determined at a magnification of 100 × . The ASTM Grain Size Number corresponds to a certain number of grains/in² according to *Table 1.3*.

Table 1.3

ASTM	0	1	2	3	4	5	6	7	8	9	10
Grains/in²	0·5	1	2	4	8	16	32	64	128	256	512

The relationship between the Grain Size Number and the number of grains/in² is given by the expression:

$$n = 2^{N-1} \tag{1.10}$$

where N = ASTM Grain Size Number
n = number of grains/in² at a magnification of 100 ×

The evaluation of the ASTM Grain Size Number is carried out with the aid of standard grain size charts or with special eyepieces (*see Figure 1.35*).

2 ISO-index G

The number of grains, viewed at a certain magnification, is counted inside an area of 5000 mm², corresponding to a circle of 79·8 mm diameter, on the ground-glass screen of the microscope or on a micrograph. The magnification is chosen so that there are at least 50 grains within this area. The number of grains n considered to be contained in this area is deduced from

$$n = n_1 + n_2$$

where n_1 = number of grains wholly inside the circle and
n_2 = the number of grains intercepted by the circle

From the value of n we calculate m, the number of grains per mm² from

$$m = 2n(g/100)^2$$

where g = the linear magnification.

The ISO-index is obtained from the expression

$$G = \frac{\log m}{\log 2} - 3 \tag{1.11}$$

According to ASTM E 112–63 the correlation between G and ASTM is

$$G = ASTM - 0·046 \tag{1.12}$$

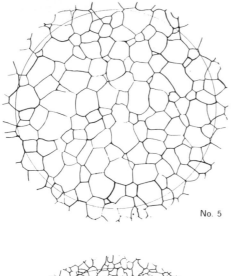

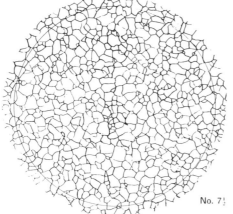

Figure 1.35 Grain size charts for ASTM Grain
Size Nos. 5 and 7·5, 70 × (original
magnification = 100 ×)

Since only whole or half units are recorded we can, for all practical
purposes, put G equal to ASTM. Hence ISO-index G can be evaluated
from the charts or eyepieces that are applicable to the ASTM Grain Size
Number.

3 Mean grain diameter

Grain size is increasingly being expressed as the mean grain diameter
which is usually determined by the intercept procedure. According to this
procedure the grains intercepted by one or more straight lines are counted.
The total length of the reference line must be such that at least 50 grains
are intercepted. The two grains at each end of the line are counted as one.

Table 1.4 below shows the relationship between the mean grain diameter and the ISO-index G or ASTM.

Table 1.4 Relationship between the mean grain diameter and the ISO-index G or ASTM

Mean grain diameter d μm	ISO-index G (ASTM-index)	Mean grain diameter d μm	ISO-index G (ASTM-index)
500	−1	45·0	
		44·2	6
450		40·0	
420	−0·5	37·1	6·5
400		35·0	
354	0	31·3	7
350		30·0	
		26·3	7·5
300		25·0	
297	0·5	22·1	8
250	1	20·0	
210	1·5		
200		18·6	8·5
180		18·0	
177	2	16·0	
160		15·6	9
149	2·5	14·0	
140		13·1	9·5
125	3	12·0	
120		11·0	10
105	3·5	10·0	
100			
		9·29	10·5
90·0		9·00	
88·4	4		
80·0		8·00	
74·3	4·5	7·81	11
70·0		7·00	
62·5	5	6·57	11·5
60·0		6·00	
52·6	5·5	5·52	12
50·0		5·00	
		4·65	12·5
		4·00	
		3·91	13
		3·28	13·5
		3·00	
		2·76	14
		2·00	

4 Intercept or Snyder–Graff method

In this method, which is applied mainly to high-speed steels, a count is taken at a magnification of 1000 × of the number of grains intercepted by a line 0·005 in in length. This number is a direct indication of the Intercept Value I. The estimation can be carried out either on a ground-glass screen or by means of a graduated eyepiece. Ten counts are normally made, the average being taken as the Intercept Value. If we express the mean grain size on the Intercept Scale (I) in the same way as the ASTM Grain Size Number, i.e. as the number of grains $(n)/\text{in}^2$ on the micrograph at $100 \times$ magnification, we obtain:

$$n = (2{\cdot}I)^2 \tag{1.13}$$

Solving for N in Equation (1.10) we obtain:

$$N = 2\frac{\log I}{\log 2} + 3 \tag{1.14}$$

This correlation is illustrated graphically in *Figure 1.36*. Note that 9·5 ASTM is equal to 9·5 I.

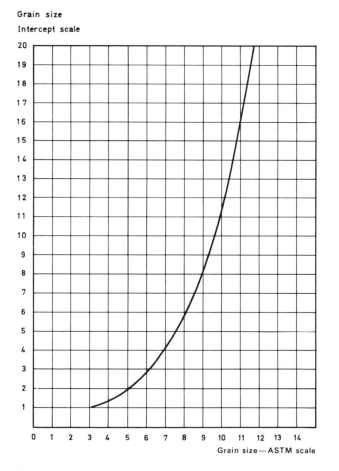

Figure 1.36 Correlation between grain size values as determined according to Intercept and ASTM method

5 Fracture test

For this test we use as a scale a series of hardened and fractured test-pieces, numbered from 1 to 10 (No. 1 being the coarsest and No. 10 the finest fracture). There are two such series, viz. Shepherd's series and Jernkontoret's series (*Figure 1.37*). For most practical purposes these two scales can be regarded as identical. It has been shown that when a hardened steel is fractured the fracture surface usually follows the austenitic grain boundaries that existed prior to hardening. Hence a study of this surface should give an indication of the austenitic grain size. The correlation between the ASTM Grain Size Number and Jernkontoret's fracture number (JK) or Shepherd's fracture number is shown in *Figure 1.38*.

Figure 1.37 The JK fracture series. Natural size

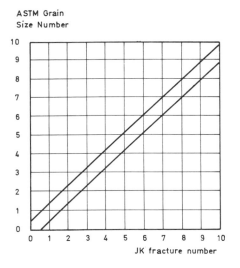

Figure 1.38 Correlation between grain size values as determined according to ASTM and JK fracture number

Carbide particle size

By means of linear analysis in the metallurgical microscope it is possible to estimate the carbide volume fraction. By counting the number of carbide grains in a section it is then possible to compute the mean carbide particle size (in μm^2).

1.7.3 Examples of grain size determinations

Case-hardening steel

The grain size is usually determined by the McQuaid–Ehn method. The steel is carburized at 925 °C for 8 h followed by cooling while it is still packed in the carburization medium. This test gives an indication of the susceptibility of the steel to grain growth during carburization. As the steel slowly cools after the relatively long carburizing period, carbides precipitate at the grain boundaries whereby the estimation of the grain size is facilitated (*see Figure 1.39*). The McQuaid–Ehn method is described in ISO/R 643–1967 Section 3.1 and in SS 11 11 02.

(a)   (b)

Figure 1.39 Micrographs showing coarse-grain (a) and fine-grain (b) case-hardened steel, 100 ×. McQuaid—Ehn test

When the determination of the McQuaid–Ehn grain size is being carried out on steels that have been rather ineffectively fine-grain treated it may happen that the nitrogen take up from the carburizing medium will help to give a deceptively favourable grain-size figure. Another method may then be used. The sample is austenitized for 30 minutes at the testing temperature stipulated in the standard specification for the steel under consideration, and then cooled to room temperature (method No. 1 according to SS 11 11 02).

High-speed steel

The determination of grain size in hardened high-speed steel by means of microscopy is best carried out on the quenched and untempered sample since tempering tends to obscure the grain boundaries. However, at temperatures above 600 °C the grain boundaries become sharp once more (*see Figure 1.40*). In this instance the sample has been etched in 5% nitric acid. More distinct contours would result if a grain-boundary etchant containing picric acid had been used.

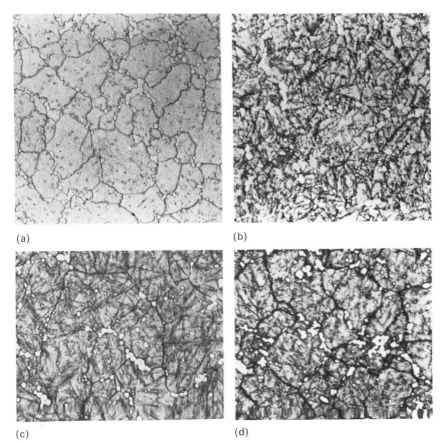

(a)   (b)

(c)   (d)

Figure 1.40 Tempering high-speed steel, M 42, in order to develop the grain boundaries. Oil hardened from 1230 °C, tempered 2 × 1 h, etchant = 5% HNO_3 in ethyl alcohol. 700 × ; (a) Tempered at 225 °C. Etchant 6 min; (b) Tempered at 550 °C. Etchant 10 s; (c) Tempered at 600 °C. Etchant 8 s; (d) Tempered at 640 °C. Etchant 15 s

Constructional steel and tool steel

The grain boundaries of quenched and tempered Cr–Ni or Cr–Ni–Mo grades of steel may be difficult to discern. By tempering at 525 °C for 3 h followed by slow cooling it is possible to bring out the grain boundaries more distinctly. It is then necessary to use a grain-boundary etch (*see Figure 1.41*). A saturated solution of picric acid in distilled water with an addition of 1% of a wetting agent is usual.

When assessing the quality level of a steel or its heat treatment the grain-size rating is very useful. Fine-grain case-hardening steels usually have a stipulated grain-size rating of 5–8 ASTM. According to the most recent SS standard for case-hardening steels the mean grain diameter shall be 30 μm (corresponding to ISO-index G = 7). The largest grain shall be maximum 60 μm (ISO-index G = 5). Quenched and tempered tool steel

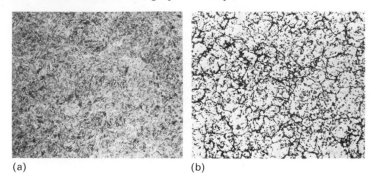

(a)   (b)

Figure 1.41 Steel S S 2550. Microstructure after (a) hardening and
tempering at 200°C, etchant 1% HNO₃; (b) hardening and tempering at
525°C, etchant picric acid with Teepol (Teepol is a wetting agent), 350 ×

should have a grain size not coarser than 7 ASTM which is about the same
as the JK fracture number 7. A high-speed steel quenched from its correct
hardening temperature shows in normal cases a grain size of 9–10 ASTM
or JK. Modern molybdenum-alloyed high-speed steels may show ratings
higher than 10JK. Since the JK scale stops at 10 and ratings higher than
10ASTM are difficult to assess the Intercept method is usually applied to
such steels.

1.8 Hardening mechanisms in steel

1.8.1 Solution hardening

Up till now the hardening mechanism that has been discussed in this
chapter, i.e. the formation of martensite, is by far one of the most
important ones for steel. This mechanism is included in a group called
solution hardening which is one of the four hardening mechanisms to be
discussed below in general terms. Owing to the process of diffusion foreign
atoms are able to migrate in among the iron aroms and form a solid
solution. This may take place in two ways, as follows:

1. *Substitutional solution.* The foreign atoms are unsystematically
   dispersed among the host atoms. They may be larger or smaller than
   the latter, as illustrated in *Figure 1.42*.

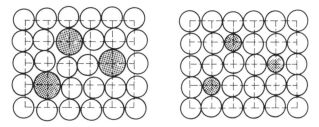

Figure 1.42 Substitutional solid solution. (a) Solute atoms larger
than host atoms; (b) Solute atoms smaller than host atoms

This size difference will create around the solute atoms a strain field which will interact with the strain field set up by the dislocations. When the latter tend to move, the solute atoms will exert an obstructing effect on the dislocations. This will result in an increase in the yield point and hardness of the material. *Figure 3.5* shows how various substitutionally dissolved alloying elements influence the hardness of the ferrite.

2. *Interstitial solution.* The atoms of C, N, H and B, which are all small compared with iron atoms, may be dispersed in among the latter in the manner shown in *Figure 1.43*. Owing to the large size differences between the interstices in the iron lattice and the interstitially dissolving atoms, considerable local strain fields may be set up which will result in substantial increases in hardness.

At room temperature iron usually exists as α-iron and its capacity to dissolve the above-named atoms is very slight. γ-iron, which is stable at high temperatures, has on the other hand, a high solubility potential for these atoms. By rapidly cooling γ-iron specimens containing carbon atoms in solution a supersaturated solution of carbon in α-iron is obtained. This gives rise to the formation of martensite of high hardness, which has been described earlier in this chapter and which will be discussed several times in later sections of this book.

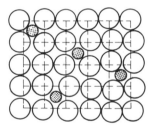

Figure 1.43 Interstitial solid solution

1.8.2   Grain size effect hardening

It has been found that in practice a fine-grained material has a higher yield point and hardness than a coarse-grained one. This observation may be explained by applying the dislocation theories and is summarized in the Petch–Hall equation

$$R_e = R_i + K \cdot \frac{1}{\sqrt{d}}$$

where R_e = yield point
R_i = the stress required to make the dislocations move in the grains
K = a constant
d = the grain diameter

An actual example of this relationship for a 0·11% C steel[16] is given in *Figure 1.44*.

In practice the fine-grain size is obtained by alloying a carbon steel or a carbon–manganese steel with small amounts of Al, Nb, Ti or V in

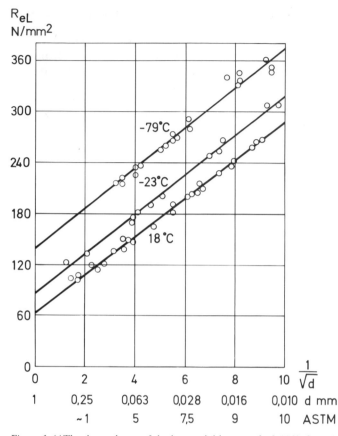

Figure 1.44 The dependence of the lower yield stress of a 0·11% C steel on the grain size (after Petch[16])

conjunction with controlled rolling or with normalizing. As a result the steels will show an increase in both yield strength and low-temperature toughness.

Some of the elements added to promote the fine-grained structure enhance further the yield point by means of another mechanism to be discussed in the following section.

1.8.3   Dispersed-phase hardening

The hardness of a matrix may be increased by creating a separation of particles in the matrix. When fairly large-sized particles or phases separate out the hardening effect may be compared with that produced when a single-phase material is rendered fine-grained, i.e. grain-boundary hardening. Pearlite is an example of this effect: the finer the lamellae, the higher the hardness.

When very small particles separate out and lie closely together the migrating dislocations will be stopped by the particles, but eventually the former will press on past them, either by cutting through them or by forming closed loops round them.

The particles give rise to a considerable increase in hardness. This hardening process is called precipitation hardening or dispersion hardening. The closer the particles lie together, i.e. the smaller and more finely-dispersed they are, the greater will be the hardness increase.

When particles are precipitated their crystal lattice may coincide very closely with that of the matrix, as shown in *Figure 1.45*. This is called coherent precipitation or precipitation of coherent particles. As a consequence, severe strain fields may be created around them which still more impedes the migrations of the dislocations and thereby causes increased hardness.

The precipitation of particles may occur as the steel is cooling from a high temperature or after a special solution treatment. In certain cases a subsequent heating may be required to precipitate particles of optimum size. Examples of such treatment are the fourth tempering stage of high-speed steel and the process of ageing.

1.8.4   Work hardening

When a metal is deformed plastically, such as by cold drawing, dislocations are produced in numbers proportionate to the degree of deformation. A complicated network of interlocking dislocations is thereby created. The finer this network mesh becomes, the harder the material will be. Most evident is the increase in yield strength and the simultaneous reduction in

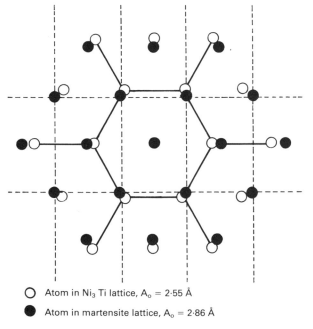

$\bigcirc$ Atom in Ni_3 Ti lattice, A_o = 2·55 Å

$\bullet$ Atom in martensite lattice, A_o = 2·86 Å

Figure 1.45 Coherency between the (110) crystallographic plane of martensite and the (0001) plane of Ni_3 Ti in a maraging steel

both toughness—as measured by the impact test—and ductility—as measured by the elongation and reduction of area. The changes in these properties —as a function of the area reduction—as measured in a carbon steel—are shown in *Figure 1.46*.

By heating cold-worked steel in the temperature range 100 °C to 300 °C carbides and nitrides are precipitated on the dislocations which are thus impeded still more. Thereby an increased hardness and reduced toughness result. This process is called strain-age hardening. (*See also* Section 5.7).

If a precipitation-hardened or solution-treated material is cold-worked and then heated to a suitable temperature the yield point can be increased considerably.

1.8.5   Other hardening mechanisms

The hardening process of boron steels is subject to a very special hardening mechanism which may be described as a combination of several mechanisms based on the supposition that the interstitially-dissolved boron atoms tend to become concentrated at the grain boundaries where they are able to influence the formation of ferrite nuclei in various ways. Some of these mechanisms will be discussed in Chapter 6 in conjunction with other particulars concerning boron steels.

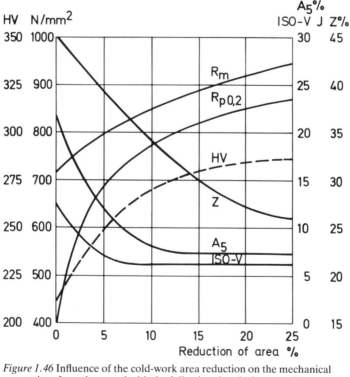

Figure 1.46 Influence of the cold-work area reduction on the mechanical properties of a carbon steel with the following chemical composition:

% C	Si	Mn	P	S	Cr	Ni	Mo	N
0·48	0·21	0·60	0·025	0·028	0·10	0·15	0·03	0·010

References

1. ROSE, A., and STRASBURG, V., 'Demonstration of the Formation of austenite in Hypoeutectoid Steels by means of TTT Heating Diagrams', *Stahl und Eisen,* **76**, No. 15, 976–983 (1956) (in German)
2. HILLERT, M., 'Formation of Pearlite', *Decomposition of Austenite by Diffusional Processes,* Interscience, New York, London (1962)
3. PICKERING, F. B., 'The Structure and Properties of Bainite in Steels', *Symposium for Transformation and Hardenability in Steels,* Climax Molybdenum, 109–130 (1967)
4. MARDER, A. R., and KRAUSS, G., 'The Morphology of Martensite in Iron–Carbon Alloys', *Trans. Amer. Soc. Metal,* **60**, 651–660 (1967)
5. KRAUSS, G., and MARDER, A. R., 'The Morphology of Martensite in Iron Alloys', *Metallurgical Transactions,* **2**, 2343–2357 (1971)
6. VAN DE SANDEN, J., 'Martensite Morphology of Low Alloy Commercial Steels', *Practical Metallography,* **17**, 238–248 (1980)
7. HOLLOMON, J. H., and JAFFE, L. D., 'Time-temperature Relations in Tempering Steel', *Trans. AIME,* **162**, 223–249 (1945)
8. SHEWMON, P. G., *Diffusion in Solids,* McGraw-Hill Inc. (1963)
9. JOST, W., *Diffusion in Solids, Liquids, Gases,* Academic Press Inc., New York (1960)
10. BESTER, H., and LANGE, K. W., 'Estimation of Average Values for the Diffusion of Carbon, Oxygen, Hydrogen, Nitrogen and Sulphur in Solid and Liquid Iron', *Arch. Eisenh.,* **43**, 207–213 (1972) (in German)
11. FRIDBERG, J., TÖRNDAHL, L–E, and HILLERT, M., 'Diffusion in Iron', *Jernkont. Ann.,* **153**, 263–276 (1969), 63 references
12. *Decarburization,* Publication 133, The Iron and Steel Institute (1970)
13. FRIEDEL, J., *Dislocations,* Addison-Wesley, Boston (1964)
14. REED-HILL, R. E., *Physical Metallurgy Principles,* Van Nostrand, London (1964)
15. 'The 2nd International Conference on High Voltage Electro-microscopy', *Jernkontorets Annaler,* **155**, No. 8 (1971) (in English)
16. PETCH, N. J., In *Fracture,* Proc. of the Swampscott Conference (1959)

Materials testing

The properties conferred on steel as a result of heat treatment need to be tested and verified. There are many testing methods available but only those that are standardized and widely used are considered here. (Before reading this chapter further consideration of the section 'International Designations and Symbols' at the beginning of the book is recommended.)

2.1 The hardness test

After heat treatment a steel component is usually hardness tested, and the value obtained is a good indication of the effectiveness of the treatment. The hardness test is carried out by pressing a ball or point with a predetermined force into the surface of the specimen. The hardness figure is a function of the size of the indentation for the Brinell (HB) and Vickers (HV), tests and of the depth of the penetration for the Rockwell (HRC) test. These three methods are the most commonly used tests and each has its special range of application and between them they cover almost the whole of the hardness field that is of interest to the steel producer and user.

2.1.1   The Brinell test

In the Brinell test a ball of hardened steel or sintered carbide is pressed into the surface of the specimen to be tested (*Figure 2.1*). Depending on the material to be tested and the ball diameter, a load up to 3000 kp may be used. For steel the following discrete ball diameters and loads have been standardized:

Diameter of ball mm	2·5	2	10
Load kp	187·5	270	3000

The diameter of the impression is measured and the Brinell hardness, which is the quotient of the load divided by the spherical area of the impression, is read off from a table. The unit, kgf/mm^2 or kp/mm^2, is not recorded after the hardness number.

The piece to be tested should have a thickness of at least 8 times the depth of the impression. The minimum distance between the centres of two adjacent impressions or from any edge should be 4 and 2½ times the impression diameter, respectively. The finer the surface finish the more accurate are the results but a reasonable rough-ground surface is usually acceptable. The upper and lower surfaces do not have to be absolutely parallel since the anvil of the testing machine is usually adjustable to compensate for this. The Brinell hardness test is used for materials of a hardness up to about 550 HB, i.e. unhardened or fully hardened and tempered steel. For a hardness greater than 450 HB it is advisable to use a sintered-carbide ball which may be used up to about 600 HB. For plain and low-alloy carbon steels the Brinell hardness number is approximately proportional to the ultimate tensile strength, UTS (UTS in kgf/mm^2 or kp/mm^2 = HB/3). The Brinell hardness test is often used to supplement other mechanical tests since it is much quicker and cheaper than the tensile test and may be considered as a non-destructive test.

Figure 2.1 Hardness testing with universal testing machine

2.1.2   The Vickers test

For the Vickers test a pyramid-shaped diamond indentor is used, the apex angle being 136°. Normally the indentation load varies between 1 kp and 30 kp, but both higher and lower loads may be used. The average length of the diagonals of the impression is measured and the Vickers Pyramid Number is read off from a table. Like the HB the HV is measured as the quotient of the load divided by the pyramidal area of the impression (kp/mm^2). For steel, the thickness of the specimen or the layer to be measured must be at least 1·5 times the impression diagonal. The minimum distance between the centres of adjacent impressions or between the centre and any edge must be 2½ times the impression diagonal. The specimen is firmly clamped horizontally, usually in a vice placed on the anvil of the machine. The Vickers hardness test is used for both hard and soft materials. Due to the small depth of penetration this test is particularly well suited to testing the superficial hardness of a material. When measuring the hardening depth of case-hardened, nitrided, induction or flame hardened steel, the ISO stipulates that the applied load shall be 1 kp, on a section cut at right angles to the surface of the steel.

The magnitude of the load used (in kp) is recorded after the hardness number, e.g. 850 HV 10. By using different loads it is possible to determine whether a steel surface has been unintentionally carburized or decarburized during the course of the heat treatment. If a low-load hardness is substantially lower than a high-load hardness—in the latter case the indentor penetrates more deeply into the material—decarburization might be suspected. The Vickers test is also suitable for hardness testing on a taper-ground section—which is used when hardness variations across a superficial zone or layer need to be measured. Vickers microhardness testing machines capable of applying loads of only a few grams are also available.

The Vickers test can be used for both very low and very high hardnesses. For example, the hardness of pure iron is about 60 HV and the hardness of certain carbides occurring in steel and sintered carbides is about 2000 HV.

2.1.3   The Knoop test

When the hardness of very thin layers is of primary interest the Knoop hardness test is particularly suitable. *Figure 2.2* shows a comparison between the Vickers and Knoop methods. Under the same load the

Vickers (HV)   Knoop (HK)

Diamond pyramid (square)   Diamond pyramid (rhomb)

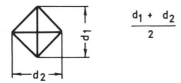
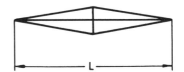

$$\frac{d_1 + d_2}{2}$$

Figure 2.2 Comparison between Vickers and Knoop impressions (after Dengel and Rossow[1])

Vickers:
angle between opposite faces = $136°$
angle between opposite edges = $148° 6' 20''$
depth of penetration = $t \approx d/7$

$$HV = \frac{1.854 \times P}{d^2}$$

Knoop:
angle between long edges: = $172° 30'$
angle between short edges: = $130°$
depth of penetration = $t \approx L/30$

$$HK = \frac{14 \cdot 23 \times P}{L^2}$$

Vickers indentor penetrates more deeply into the material of the specimen than the Knoop indentor. The same depth of penetration is obtained with, say, HV 0·3 and HK 0·718. However, the hardness values may differ slightly from the one method to the other.

2.1.4   The Rockwell test

There are several Rockwell tests, the most usual being Rockwell C (HRC), Rockwell N Superficial (usually HR 30N) and Rockwell B (HRB). Of all hardness tests based on the measurements of the penetration depth or size, the Rockwell test is the quickest since a direct reading is obtained on a dial indicator. The accuracy is relatively high, viz. ± 1 unit.

At the start of the test a pre-load or minor load (F_0) is applied to the indentor. The dial indicator is set at zero and the major load is applied, the time at full load being 5–10 s. After removing the major load only—the minor load still acting—the hardness is read on the dial direct. The hardness is a function of the increase in penetration depth caused by the application and subsequent removal of the major load (*see Figure 2.3*).

The accuracy of the testing machine should be verified with test-pieces of known hardness before each testing operation. For this purpose standard hardness blocks are recommended. The hardness of these blocks should lie as near as possible to that of the workpiece being tested. If the testing

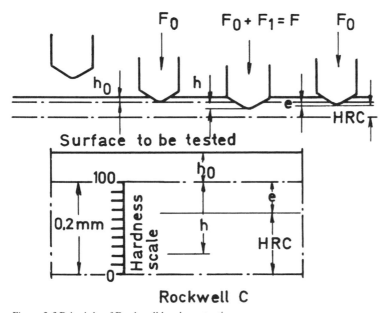

Figure 2.3 Principle of Rockwell hardness testing
F_0 = minor load, N
F_1 = supplementary load, N
$F = F_0 = F_1$ = total or major load, N
$h = h_0$ = depth of impression under total load
$e = h_0$ = depth of impression after removing supplementary load only The unit for e is 0·002 mm

machine has not been in continuous use, consistent readings are obtained only after the third and subsequent indentations.

Since Rockwell C (HRC) is the most frequently used Rockwell test it will be treated in more detail. The description covering this test is also valid, where applicable, for the other Rockwell tests.

For the Rockwell C (HRC) test the Brale indentor is used, i.e. a diamond cone with an apex angle of 120°, the apex being somewhat rounded ($r = 0.2$ mm). This indentor is pressed into the surface to be tested with a minor load of 10 kp, and the dial indicator is set at zero. After application of the major load of 150 kp and its subsequent removal the hardness is read off direct on the dial, the minor load still being applied.

A difference in hardness of 1 HRC corresponds to a difference of 0.002 mm in penetration depth. This calls for a high degree of precision when carrying out the test and therefore test surfaces should be ground to a 280 mesh finish, the opposite surface should be smooth and parallel. The anvil and specimen should be wiped clean before testing. At least three indentations should be made, the first one being disregarded since the possible presence of dirt between the specimen and the anvil can influence the hardness reading. At the first indentation these dirt particles are compressed and will have little influence on subsequent readings. Rockwell C (HRC) impressions should not be closer than 3 mm and at least 3 mm from an edge.

The Rockwell impression is relatively deep. If carried out on thin parts incorrect readings will result; therefore the thickness of the sample should be not less than 10 times the penetration depth. With a hardness of, say, 20 HRC the specimen must be at least 1.6 mm thick, with 70 HRC at least 0.6 mm.

Round, small-diameter parts give low values if hardness tested on the cylindrical surface. This error may be corrected with the aid of *Table 2.1*.

Rockwell testing should be avoided on slanting surfaces, partly because misleading results are obtained and partly because the machine may be damaged by the resulting non-axial or off-centre loading. It is advisable to compensate for the slant by inserting cuneiform gauge blocks between the specimen and the anvil.

Table 2.1 Correction values for cylindrical surfaces

HRC	Bending radius of tested surface, mm								
	3	5	6·5	8	9·5	11	12·5	16	19
20				2·5	2·0	1·5	1·5	1·0	1·0
25			3·0	2·5	2·0	1·5	1·0	1·0	1·0
30			2·5	2·0	1·5	1·5	1·0	1·0	0·5
35		3·0	2·0	1·5	1·5	1·0	1·0	0·5	0·5
40		2·5	2·0	1·5	1·0	1·0	1·0	0·5	0·5
45	3·0	2·0	1·5	1·0	1·0	0·5	0·1	0·5	0·5
50	2·5	2·0	1·5	1·0	1·0	0·5	0·5	0·5	0·5
55	2·0	1·5	1·0	1·0	0·5	0·5	0·5	0·5	0
60	1·5	1·0	1·0	0·5	0·5	0·5	0·5	0	0
65	1·5	1·0	1·0	0·5	0·5	0·5	0·5	0	0
70	1·0	1·0	0·5	0·5	0·5	0·5	0·5	0	0

When case-hardened, nitrided or thin layers generally are being hardness tested, the softer underlying core will be deformed and the values obtained will not represent the true hardness of the superficial layer. The hard case must be at least 0·5 mm thick if a true Rockwell C hardness figure is to be obtained. For thinner layers it is necessary to employ a test method that uses a smaller load, e.g. HR 30 N or HV with 1–30 kp. The Rockwell C test may, however, be applied to case-hardened parts having a case less than 0·5 mm thick. The hardness measured will be lower than the true hardness but the test may be used on mass-produced parts once the relevant hardness limits have been determined, since the values obtained will enable parts with too shallow or too large case depths to be picked out.

The Rockwell C test is normally used on steels in the hardness range 30 to 70 HRC since softer materials will give unreliable values.

The Rockwell N test employs the same diamond cone indentor as Rockwell C. The minor load is 3 kp and the total major load is 15, 30 or 45 kp, HR 15 N, HR 30 N or HR 45 N respectively. The test is used for thin sections or for case-hardened or nitrided steel where the superficial layer is less than 0·5 mm in thickness. HR 30 N is used within the hardness range 40 to 90 HR 30 N.

The Rockwell B test is carried out in principle in the same way as the other Rockwell tests. The indentor is a ¹⁄₁₆-in diameter steel ball. The total load is 100 kp. Rockwell B is used within the hardness range 50 to 110 HRB for unhardened and hardened and tempered steel.

2.1.5   The scleroscope test

This is carried out by allowing a diamond-tipped hardened steel cylinder to fall freely through a tube on to the surface to be tested. On the rebound up the tube the cylinder is caught at its highest point and held there. The height of the rebound is a measure of the Shore C or Shore D hardness. This method does not give very reliable results since the height of the rebound is dependent on the mass of the tested object. It is used for grading the hardness of rolls for cold-rolling. When used in this connexion it has been found that the hardness measurements made on rolls less than 150 mm in diameter give very erratic results whereas larger-diameter rolls give relatively consistent results. Schmitz and Schlüter[2] have demonstrated the difficulties in using the scleroscope to obtain reproducible results. As seen in *Figure 2.4* a scatter of some 20 Shore units is to be expected with each individual apparatus, and a similar variation exists between the testers themselves (*see Figure 2.5*).

At the Bofors Company the figures in *Table 2.3* are used to obtain an approximate correlation between HV and Shore C or Shore D. Here it should be stressed that the Vickers test is the only reliable one for measuring the hardness of rolls for cold-rolling. The Shore test is used only as a guide during the manufacture of these rolls.

2.1.6   Conversion tables for various scales of hardness

All hardness tests are not applicable in all circumstances and conversion tables are very necessary. Such tables have been prepared by institutions

Hardness of test specimens

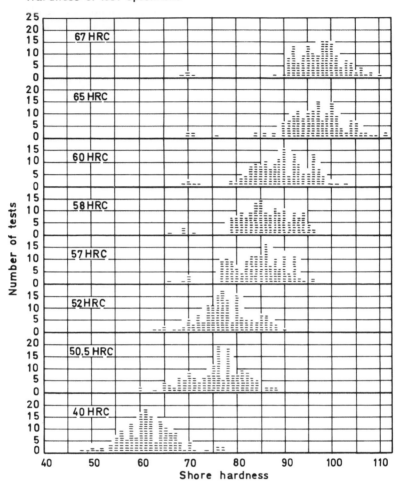

Figure 2.4 Distribution diagram for Shore hardness testing of eight steel test-pieces of different hardness (after Schmitz and Schlüter[2])

such as Jernkontoret, ASTM and EURONORM. There are some relatively minor differences between these tables. The information in *Table 2.2* agrees in the main with the suggestions for revising EURONORM 8. The Brinell test stipulates a load of $F = 30 \times D^2$ and the use of a steel ball for hardnesses up to 440 HB. For higher hardnesses a sintered-carbide ball is used. The amount of scatter must be taken into account when converting, always bearing in mind that the higher the hardness the more the scatter. This point is exemplified in *Figure 2.6*. The hardness shall be tested according to the method stipulated in the instructions or on the drawing. Should this not be possible the method used must always be stated before carrying out any conversion.

When tensile-strength testing is stipulated, hardness testing must not be substituted for it unless by agreement between customer and supplier. If

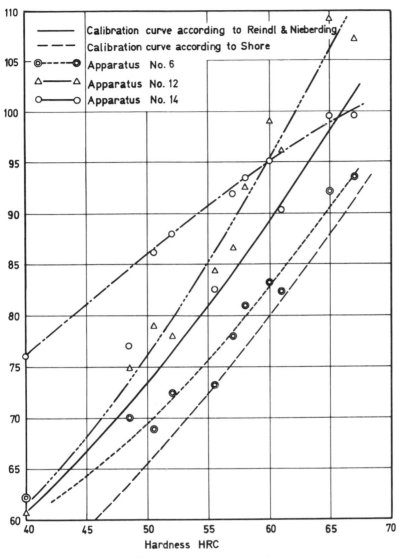

Figure 2.5 Compilation of calibration curves and test results for separate
Shore hardness testing apparatus and their comparison with Rockwell tests
(after Schmitz and Schlüter[2])

the standards required both tensile strength and hardness the values for the
tensile tests are the ruling ones.

2.2 The tensile test

The maximum design stresses that a material may be subjected to on static
loading are determined by means of the tensile test. The following
properties of the material are usually determined by this test:

Table 2.2 Conversion table for hardness—ultimate tensile strength of steel. (Not valid for austenitic or cold-worked steel)

Hardness			Tensile strength		Hardness			Tensile strength		Hardness		
HV	HRB HRC	HB	kp/mm²	N/mm²	HV	HRC	HB	kp/mm²	N/mm²	HV	HRC	HR 30N
100	57·0	95	33	325	400	40·8	380·0	130	1275	700	60·1	77·6
110	62·0	105	36	355	410	41·8	389·5	134	1315	710	60·5	78·0
120	67·0	114	39	380	420	42·7	399·0	137	1345	720	61·0	78·4
130	71·0	124	43	420	430	43·6	408·5	141	1385	730	61·4	78·7
140	75·1	133	46	450	440	44·5	418·0	144	1410	740	61·8	79·1
150	78·8	143	49	480	450	45·3	423	147	1440	750	62·1	79·4
160	82·1	152	52	510	460	46·1	432	150	1470	760	62·5	79·7
170	85·0	162	55	540	470	46·9	442	153	1500	770	62·9	80·0
180	87·3	171	58	570	480	47·7	450	156	1530	780	63·3	80·4
190	89·6	181	62	600	490	48·4	456	160	1570	790	63·6	80·7
200	91·8	190	65	635	500	49·1	466	164	1610	800	64·0	81·1
210	93·7	200	68	670	510	49·8	475	168	1640	810	64·3	81·4
220	95·5	209	71	695	520	50·5	483	172	1680	820	64·7	81·7
230	97·0	219	74	725	530	51·1	492	176	1725	830	65·0	81·9
240	HRC	228	77	755	540	51·7	500	180	1765	840	65·3	82·2
250	22·2	238	80	785	550	52·3	509	184	1805	850	65·6	82·5
260	24·0	247	84	825	560	53·0	517	188	1845	860	65·9	82·7
270	25·6	257	87	855	570	53·6	526	193	1890	870	66·1	82·9
280	27·1	266	90	880	580	54·1	535	198	1940	880	66·4	83·1
290	28·5	276	94	920	590	54·7	543	203	1990	890	66·7	83·3
300	29·8	285	97	950	600	55·2	552			900	67·0	83·6
310	31·0	295	101	990	610	55·7	560			910	67·2	83·8
320	32·2	304	104	1020	620	56·3	569			920	67·5	84·0
330	33·3	314	107	1050	630	56·8	577			930	67·8	84·2
340	34·4	323	110	1080	640	57·3	586			940	68·0	84·4
350	35·5	333	114	1115	650	57·8				950	68·4	84·6
360	36·6	342	117	1150	660	58·3				960	68·7	84·8
370	37·7	352	120	1175	670	58·8				970	69·0	85·0
380	38·8	361	123	1205	680	59·2				980	69·3	85·2
390	39·8	370	127	1245	690	59·7				1000	69·9	85·6

	SI designations	Old designations
Yield point (0·2% proof stress)	$R_{p0·2}$ N/mm²*	(σ_s† kp/mm² or kgf/mm²)
Lower yield point	R_{eL} N/mm²	
Upper yeild point	R_{eH} N/mm²	
Tensile strength	R_m N/mm²	(σ_B kp/mm² or kgf/mm²)
Elongation	A_5 %	(σ_5)
Reduction of area	Z %	(ψ)

In addition, Young's modulus of elasticity E, N/mm² (kp/mm²) may be determined.

---

* In old ISO documents cited in this book the proof stress is designated by R_e. The designation $R_{p0·2}$ was adopted in 1973.
† The designations within brackets were used before the SI was introduced.

Table 2.3 Conversion of diamond pyramid hardness (HV) to Rockwell C (HRC) and different Shore values

Rockwell C (RC)	Diamond pyramid hardness (HV)	Shore			
		Roll calibration		Standard calibration	
		C	D	C	D
45·0	450	68	73	65	67
49·0	500	72	78	69	72
52·5	550	76	82	73	76
55·0	600	79	86	76	80
58·0	650	83	90	80	84
60·5	700	87	93	84	88
61·0	720	89	95	86	90
62·0	740	91	96	88	91
62·5	760	92	98	89	92
63·5	780	94	99	91	94
64·0	800	95	100	92	95
64·5	820	97	101	94	97
65·0	840	98	103	95	98
66·0	860	100	104	97	100
66·5	880	102	105	99	101
67·0	900	103	106	101	103
67·5	920	105	—	—	—

NOTE
The conversion HV-Shore C (roll calibration is related to the ASTM Standard A-427–61, which is applicable to forged and hardened alloy steel rolls.

The relationship between the different Shore values is based upon the tables on pp. 102 and 104 of *Metal Progress Data Sheet*, June 1964 (Hardness Conversion for Forged Steel Rolls).

When measuring the hardness with a scleroscope, a varying range of ± 3·7 Shore (± 43 HV), expected on 95% of hardness measurements must be considered, according to ASTM A 427–61. The varying range when measuring the HV on test rolls of uniform hardness, is normally only ± 7 HV according to test results obtained by Bofors.

When measuring rolls, each scleroscope must be calibrated on rolls of known HV in order that the conversion from Shore to HV shall be as correct as possible.

For correct conversion of kp/mm^2 to N/mm^2 *see Table 8.18.*

Figure 2.7 shows a test bar for the tensile testing of steel. The bar is gripped in a tensile testing machine and subjected to a successively increasing load until fracture occurs; the extension of the test bar can be traced on the diagrams in *Figures 2.8a* and *b*.

The yield stress is the stress at which a relatively large extension of the test bar takes place while it is being subjected to a constant load (*Figure 2.8a*. For steels that have a clearly defined yield stress range, such as unalloyed steels, either the upper yield stress R_{eH} or the lower yield stress R_{eL} is determined, depending on the standard or stipulated requirements. In the case of alloy or hardened steels this point is not clearly defined or it may be completely absent (*Figure 2.8b*). For such steels the yield stress is taken to mean the proof stress, i.e. the stress that, after the load has been removed, produces a permanent elongation of, say 0·2% designated $R_{p0·2}$ ($\sigma_{0·2}$). If a structural part has been stressed beyond its yield point the material suffers plastic deformation. Hence the value of the yield stress is taken as the basis for more accurate calculations of mechanical strength.

The ultimate tensile strength or just tensile strength R_m (σ_B) is the maximal numerical value of the stress which, computed from the original area of the test bar, causes fracture of the specimen.

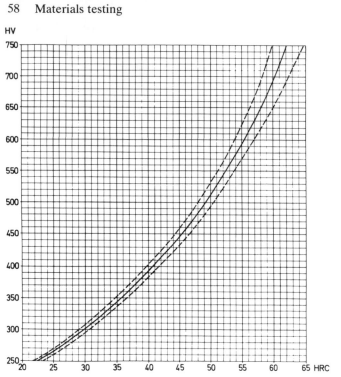

Figure 2.6 Curves indicating maximum predictable amount of scatter when converting from HV to HRC or vice versa. When converting, say, 400 HV to HRC there may be a scatter of ±1 HRC; when converting, say, 700 HV to HRC the scatter may be ±2 HRC. An experienced hardness-testing operator is able to halve the above-indicated scatter

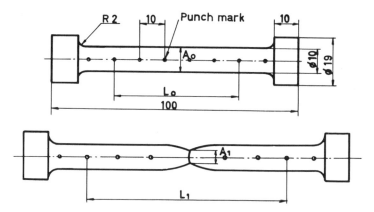

Figure 2.7 Tensile bar test, type SS 10A50, for steel, before and after testing. The punch marks serve as check points for measuring the extension. For an elongation A_5 (δ_5) the extension between 6 punch marks is measured ($= 5 \times d$)

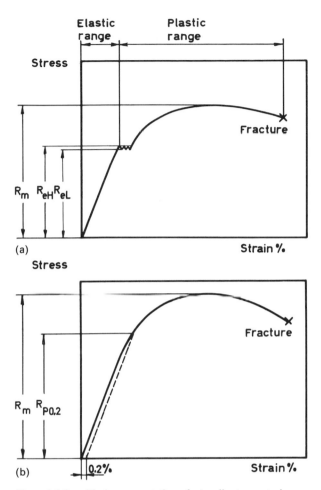

Figure 2.8 Graphical representation of a tensile stress–strain diagram for a steel (a) with a definite yield range and (b) with no marked yield point

The elongation A (δ) is the percentage increase in length of the test bar after completion of the tensile test. It is measured on a gauge length that is generally 5 times the diameter of the test bar and is then designated A_5 (δ_5). On a gauge length of 10 times the diameter, the elongation is designated A_{10} (δ_{10}), etc. The amount of elongation is a measure of the ductility of the material and, taken along with the figure for the reduction of area, it gives an indication of the amount of plastic deformation to which the material can be subjected before it fractures. The elongation is calculated from:

$$A_i\ (\delta_i) = \frac{L_1 - L_0}{L_0} \times 100 \qquad (see\ Figure\ 2.7)$$

The Reduction of Area Z (ψ) is the percentage area reduction at the point of fracture. It is calculated from:

$$Z(\psi) = \frac{A_0 - A_1}{A_0} \times 100$$

The Modulus of Elasticity (Young's Modulus), E, is obtained by dividing the stress at any point below and up to the elastic limit by the corresponding strain. The E-modulus is roughly the same for all steels, viz. about 206 000 N/mm^2 (21 000 kp/mm^2) and is a measure of the slope of the straight-line part of the diagram. The E-modulus may also be computed from data obtained by measuring the resonance frequency of a test bar.

The E-modulus is an indication of the stiffness of the material within the elastic range. There is a wide-spread belief that hard steels are stiffer than soft ones, even if the load is insufficient to cause a permanent set in the material. However, this is a misconception, since the E-moduli of different steels subjected to different heat treatments usually agree to within 2–3%. If a structural member or machine part is to be stiff and strong it must be adequately dimensioned to ensure sufficient rigidity. Thus the possibility of influencing the rigidity by a judicious choice of steel is very small. On the other hand it is quite obvious that a hard material can be subjected to heavier loads before a permanent set occurs.

2.2.1   Comparison between mechanical properties obtained according to different specifications

Tensile and yield strengths obtained according to different specifications are often directly comparable. Only a recalculation of the units is required. Values of the elongation as obtained according to various standard specifications or measured on different gauge lengths are not directly commensurable after recalculation (*see Table 2.4*).

2.3 The impact test

The impact test gives and indication of the amount of energy absorbed by the material at fracture. The test consists in allowing a tup on a rigid pendulum arm to swing from a fixed height and strike a standardized speciment set at the lowest point of the swing (*Figure 2.9*). The angle of displacement of the pendulum after fracturing the specimen is measured and from this is calculated the energy required to fracture the test piece.

The prevalent European specimens for impact testing incorporate a U or V notch and are designated Charpy U or Charpy V respectively (*see Figure 2.10*). These test pieces are standardized according to ISO and are designated ISO-U and ISO-V respectively. In the Charpy U-notch test the value was originally reported as kgfm/cm^2 and was designated KCU. According to SI the result is now reported as J and the test is designated ISO-U or KU. The value obtained from the Charpy V-notch test was formerly reported as kgfm, but again J is now used and the test is designated ISO-V or KV.

In Germany the DVM test piece is also used; in France, the Mesnager test piece and in English-speaking countries the Izod test piece also exist.

Table 2.4 Tensile testing.
Relationship between values of elongation obtained with various gauge lengths,
according to EURONORM

Standard test bar SS 10A50 $L_0 = 5\,d$	Gauge length				
	$L_0 = 10\,d$	$L_0 = 7 \cdot 25\,d$	$L_0 = 3 \cdot 54\,d$	$L_0 = 8\,d$	$L_0 = 4\,d$
			Elongation %		
8	5·1	6·4	10·2	6·1	9·2
9	5·8	7·2	11·4	6·8	10·3
10	6·6	8·1	12·6	7·6	11·4
11	7·3	9·0	13·7	8·4	12·5
12	8·0	9·9	14·9	9·2	13·7
13	8·8	10·7	16·0	10·0	14·8
14	9·6	11·5	17·1	10·8	15·9
15	10·4	12·4	18·3	11·7	17·0
16	11·2	13·2	19·4	12·5	18·1
17	12·0	14·1	20·5	13·3	19·2
18	12·8	15·0	21·7	14·2	20·3
19	13·7	16·0	22·9	15·1	21·4
20	14·5	16·9	24·0	16·0	22·5
21	15·4	17·7	25·1	16·9	23·6
22	16·2	18·6	26·1	17·8	24·7
23	17·0	19·5	27·1	18·6	25·7
24	17·9	20·4	28·2	19·5	26·7
25	18·8	21·3	29·2	20·3	27·7
26	19·6	22·2	30·3	21·2	28·7
27	20·4	23·1	31·3	22·0	29·7
28	21·2	24·0	32·3	22·9	30·6
29	22·1	24·8	33·4	23·8	31·6
30	23·0	25·7	34·4	24·7	32·6
31	23·9	26·6	35·4	25·6	33·6
32	24·8	27·5	36·4	26·5	34·5
33	25·7	28·5	37·4	27·4	35·5
34	26·6	29·4	38·4	28·3	36·5
35	27·5	30·3	39·4	29·2	37·5
36	28·4	31·2	40·4	30·1	38·5
37	29·3	32·1	41·4	31·0	39·4
38	30·2	33·0	42·4	31·9	40·4
39	31·1	33·9	43·4	32·8	41·3
40	32·1	34·8	44·4	33·7	42·3

L_0 = gauge length in millimetres
d = diameter of test bar, in millimetres, measured across parallel gauge length. For non-cylindrical test bars d is taken as 1.13√(cross-sectional area)

In the latter test, one end of the specimen is held fast in a vice and the pendulum hammer strikes the free end. The dimensions of the various test pieces are given in *Table 2.5*.

The impact strength is temperature-dependent and falls off as the testing temperature is reduced. The values obtained from the U-notched specimens usually show continuously falling impact strength, whereas the values from the V-notched specimens arrange themselves in a high-level zone, a transition zone with a fairly steep incline and a low-level zone. See *Figure 2.11*. In the high-level zone the V-notched specimen shows a tough

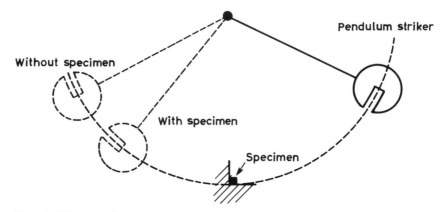

Figure 2.9 Sketching showing the principle of impact testing

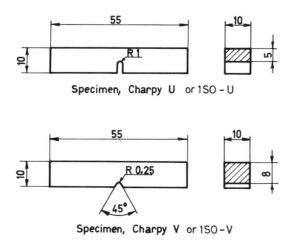

Specimen, Charpy U or 1SO – U

Specimen, Charpy V or 1SO –V

Figure 2.10 Impact test specimens for ISO-U and ISO-V

Table 2.5 Dimensions of various specimens

Type of test	Dimensions of test pieces mm			Dimensions of notch mm				Support distance mm
	Length	Width	Height	Depth	Dia	Aperture angle	Radius	
Charpy, ISO-U	55	10	10	5	2	—	—	40
DVM	55	10	10	3	2	—	—	40
DVM F	55	8	10	4	8	—	—	40
DVM K	44	6	6	2	1·5	—	—	30
Mesnager	55	10	10	2	2	—	—	40
Charpy, ISO-V	55	10	10	2	—	45°	0·25	40
Izod	75	10	10	2	—	45°	0·25	—

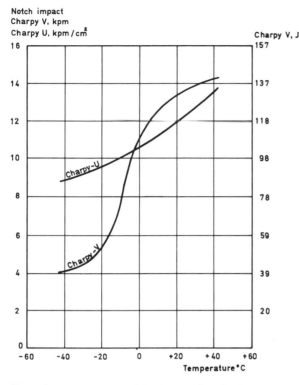

Figure 2.11 Steel C 35 deoxidized and grain refined. The influence of the temperature of test on the notched impact values of Charpy U and Charpy V. Bar diameter 30 mm. Heat treated (hardened and tempered) to 170 HB

or fibrous fracture whereas in the low-level zone the fracture is predominantly brittle and crystalline.

The testing temperature that gives rise to a 50% brittle fracture is called 50% FATT (Fracture Appearance Transition Temperature) or simply Transition Temperature. This is sometimes defined as the temperature at which the impact strength is half the value reported in the high-level zone.

The U-notched test piece is used for testing brittle steels, e.g. tool steels and certain structural steels that have been heat treated to a high hardness. The V-notched test-piece is used either for structural steels heat treated to a somewhat lower hardness or for general engineering steels containing < 0·25% C, unheat-treated or normalized. For this latter steel a minimum impact strength of 27 J (2·8 kgf m) is the usual requirement, the testing temperature being stipulated also.

The conventional Charpy V-notch test pieces have a cross-sectional area of 10 × 10 mm, which implies that the smallest bar dimension that can be tested is about 12 × 12 mm or 16 mm diameter. Since the middle of the 1970s a smaller test piece, measuring 5 × 10 mm across has been used for testing smaller sections, the notch being machined on the 5 mm flat. Test pieces measuring 7·5 × 10 mm are also in use. A working group within

ISO/TC17/WG15 has suggested that the values obtained from the 7·5 mm and 5 mm test pieces should not be less than 80% and 70% respectively of the lowest stipulated value for the 10 mm test piece.

The 1980 Edition of the German Standards DIN 271001 states that the minimum figure for the impact value of the ISO-V test piece of breadth between 5 and 10 mm shall be directly proportional to the width of the test piece.

Robiller[3] has studied this matter in detail, using 20 ferritic steels and test pieces varying in width from 5 to 10 mm. He found that the impact strength, measured in the high-level zone, is almost wholly in direct proportion to the test-piece width, i.e. a test piece measuring 5 mm across, tested in the high-level zone, will yield a value that is 0·5 times the impact strength of a 10 mm test piece. This finding applies equally to longitudinal and transverse specimens. These differences are evened out to a large extent when the testing is done in the transition and low-level zones. *See Figure 2.12* which is representative of both longitudinal and transverse specimens.

Whereas the values of ultimate tensile strength and yield point may be applied direct to expressions dealing with strength calculations, impact values can only be used for judging, from experience, the suitability (or otherwise) of a material for a specific purpose. There are no exact

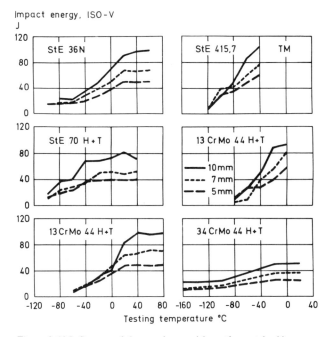

Figure 2.12 Influence of the specimen with on the notched bar impact energy–temperature curve for different steels with tensile strength 483–1093 N/mm². Transverse specimens with ISO-V-notch (after Robiller[3]). N = normalized, TM = thermomechanical treatment, H + T = hardened and tempered

conversion factors for the impact values obtained by the various methods. Published conversion tables only give an indication of the mutual relationship between the various values. This means that the impact values can be guaranteed valid only for the test method actually used (cf. *Table 2.6*).

Table 2.6 Impact testing.
Relationship between different impact values obtained with various methods and specimens, according to EURONORM

Charpy U = ISO-U				Charpy V = ISO-V		
(KCU) kpm/cm^2	(KU) J	DVM kpm/cm^2	Mesnager kpm/cm^2	(KV) kpm	(KV) J	Izod ft lb
1	5	1·2	1·4	0·3	3	2·5
2	10	2·4	2·8	0·7	7	6·4
3	15	3·6	4·2	1·2	12	10·8
4	20	4·8	5·6	1·8	18	16·0
5	25	6·0	7·0	2·5	25	21·5
6	29	7·2	8·4	3·3	32	27·8
7	34	8·4	9·8	4·2	41	34·1
8	39	9·6	11·2	5·2	51	40·4
9	44	10·8	12·6	6·4	63	46·7
10	49	12·0	14·0	7·5	74	53·0
11	54	13·2	15·4	8·7	85	59·3
12	59	14·4	16·8	10·1	99	65·6
13	64	15·6	18·2	11·3	111	71·9
14	69	16·8	19·6	12·6	124	78·2
15	74	18·0	21·0	14·2	139	84·5
16	78	19·2	22·4	15·5	152	90·8
17	83	20·4	23·8	16·2	159	97·1
18	88	21·6	25·2	18·4	181	103·4
19	93	22·8	26·6			109·7
20	98	24·0	28·0			116·0
21		25·2	29·4			122·3
22		26·4	30·8			128·6
23		27·6	32·2			134·9
24		28·8	33·6			141·2

2.4 The torsion impact test

By means of the Carpenter torsion impact machine the torsion impact may be determined on specimens illustrated in *Figure 2.13*. To conduct the test the specimen is clamped to a flywheel which rotates at a certain speed. The other end of the specimen is suddenly gripped and its rotation

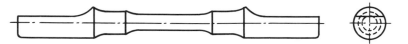

Figure 2.13 Test specimen for the Carpenter torsion impact test

instantaneously stopped, and the specimen fractures as a result of the torsional impact blow. The energy required to break it is recorded in ft/lb. The diagram in *Figure 2.14* has been constructed from the results obtained from hardened and tempered grade O1 steel. This test is very useful in selecting a suitable steel and its heat treatment for components subjected to simultaneous impact and torsion. The test can also be used to give an indication of the tempering temperature range for the maximum impact resistance of the steel. However, the method is not recommended when a choice is to be made among steels differing greatly in composition and characteristics.

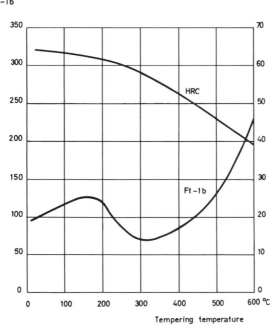

Figure 2.14 Torsion impact and hardness as a function of the tempering temperature. Steel O 1

2.5 The fatigue test

2.5.1   Fatigue in general

Machine component failures are in 90% of all cases due to fatigue. By fatigue is meant crack formation caused by fluctuating loads, the maximum stress always being less than the ultimate tensile strength of the material. The fatigue crack usually starts at some stress raiser in the material such as a slag inclusion or an excessively sharp fillet radius. At first the crack propagates slowly, but its speed of propagation increases successively.

In many service failures the fracture surface exhibits markings resembling the annual growth rings in a tree trunk. These are usually referred to in the literature as 'clam shell' markings or arrest lines. They have been attributed to periods of crack extension followed by crack arrest as a consequence of variation in the loading pattern of the component. These arrest lines indicate the initiation area and growth direction of the fatigue crack. After some time the crack has grown so much that the stress in the remaining part of the section exceeds the ultimate tensile strength of the material and the final fracture occurs. This usually takes place in a brittle manner and with little or no deformation. *Figure 2.15* shows a typical fatigue fracture.

A study of a fatigue on a microscopic level, i.e. in an electron microscope reveals the stepwise propagation of the fatigue crack with one step per loading cycle. This stepwise propagation gives rise to a fine line pattern in the fracture surface known as 'fatigue striations' (*see Figure 2.16*).

2.5.2   Test procedure

In a fatigue test stress may be regarded as resolved into a constant static and a fluctuating stress which varies with time. Hence the applied stress varies between an upper and a lower limit, σ_{max} and σ_{min} respectively. The constant stress is called the mean fatigue stress (σ_m) and is defined as:

$$\sigma_m = \frac{\sigma_{max} + \sigma_{min}}{2}$$

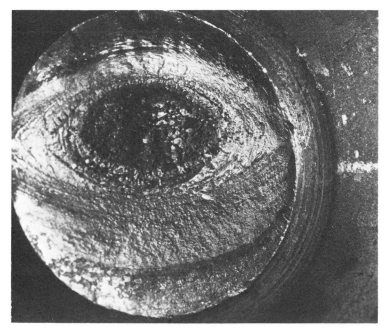

Figure 2.15 Illustration of fatigue fracture, with arrest lines and final structure

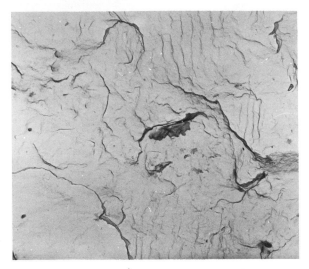

Figure 2.16 Fatigue fracture showing striations, 8700 ×

The fluctuating stress may be defined by its amplitude σ_a, *Figure 2.17*, as follows:

$$\sigma_a = \frac{\sigma_{max} - \sigma_{min}}{2}$$

The applied fatigue stress may be axial (tensile or compressive), flexural, shear or torsion stress. If σ_{max} and σ_{min} have opposite signs during each load cycle they are referred to as alternating stresses; if the same sign, pulsating stresses (*see Figure 2.18*). In addition there is a special type of fatigue test, the rotating-cantilever test. The specimen is loaded so that it is deflected as it rotates, *Figure 2.19*. The bending load rotates round the test bar, as it were, giving rise to alternating bending stresses in it. The results of fatigue tests with alternating bending stresses and reversed flexural stresses respectively, are numerically equivalent.

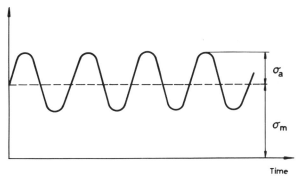

Fracture 2.17 Mean fatigue stress σ_m and stress amplitude σ_a

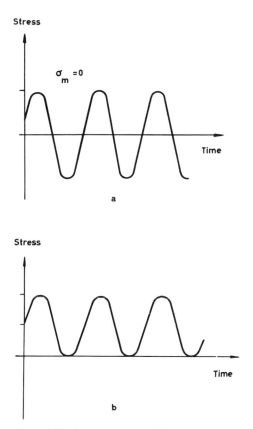

Figure 2.18 Alternating stress (a) and pulsating stress (b)

The results of fatigue tests are usually plotted in Wöhler diagrams, *Figure 2.20*. In these the amplitude of the stress σ_a is plotted as a function of the number N of load cycles to fracture: the latter scale usually being logarithmic. For amplitudes less than the endurance limit, σ_D, the specimen will not fracture.

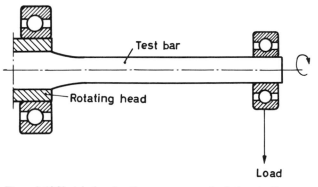

Figure 2.19 Sketch showing the arrangement for fatigue testing using principle of rotating cantilever beam

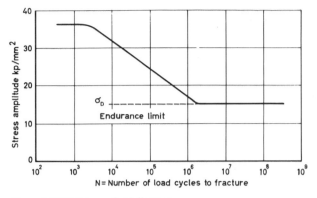

Figure 2.20 Wöhler or S–N diagram

2.5.3   Different types of fatigue fracture

By studying the appearance of the fractured surface of a fatigue failure, it is possible to determine the extent of overloading as well as the relative degree of stress concentration, *see Figure 2.21*. If there are no stress concentrations, the fatigue zones are convex-shaped, viewed from the point of fracture initiation. As the stress concentrations increase there is a gradual and continuous change to concave-shaped zones. The size of the final fracture area increases with the load. In contrast to pulsating flexural stresses, alternating and rotating flexural stresses can initiate fatigue fracture from two sides as well as from one, depending on the size of the stress raisers.

A case illustrating fracture type 2e (*Figure 2.21*) is shown in *Figure 2.22*, where the fracture occurred in a loading hook, the rather poor surface finish of which gave rise to high stress concentrations. *Figure 2.23* illustrates a fracture of type 3e, i.e. caused by rotating flexural stresses with high stress concentrations and a small amount of overloading.

Type of fatigue stress / Stress category	No stress concentration		Low stress concentration		High stress concentration	
	Light over-loading	Heavy over-loading	Light over-loading	Heavy over-loading	Light over-loading	Heavy over-loading
1. Pulsating flexural stress	a	b	c	d	e	f
2. Alternating flexural stress						
3. Rotating bending stress						

Figure 2.21 Illustrations of different fatigue fractures caused by various types of flexural fatigue (after G. Jacoby)

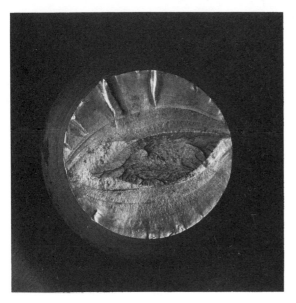

Figure 2.22 Fatigue fracture in a loading hook, 1 ×

Figure 2.23 Fatigue fracture in a rotating shaft, 1 ×

2.5.4   Goodman diagram

The endurance limit σ_D for different mean fatigue stresses σ_m can be shown in a fatigue diagram or Goodman diagram, *Figure 2.24*. The abscissa is graduated in σ_m units and the ordinate in $(\sigma_m \pm \sigma_D)$ units. The diagram is constructed by plotting values of σ_m against the corresponding values of σ_D on either side of the 45° reference line. In this way two curves are obtained as functions of σ_m: one for $(\sigma_m + \sigma_D)$ and the other for $(\sigma_m - \sigma_D)$. These two curves enclose an area. A structural member subjected to stresses, the values of which are confined inside this area, is exposed to only a slight risk of failure by fatigue. The endurance limit σ_m decreases as the mean fatigue stress σ_m increases.

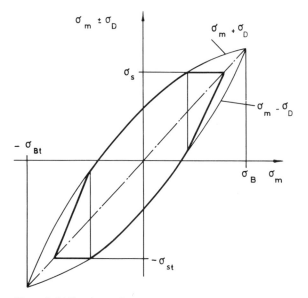

Figure 2.24 Goodman diagram

Usually it is not expected that there should be any deformation of the material during the fatigue test and therefore the maximum allowable stress σ_{max} has its upper limit defined by the yield stress in tension σ_s, and its lower limit by the yield stress in compression σ_{st}. Hence the diagram is truncated at both ends as shown in *Figure 2.24*.

The difference between the negative and positive mean fatigue stresses is mostly so small that only the positive part of the diagram need be constructed. The limiting curves may often be approximated by straight lines. This implies that the diagram can be constructed simply from two separate fatigue tests.

2.5.5   Endurance limit—ultimate tensile strength

For the majority of hardened and tempered steels the relationship between the endurance limit σ_{Dr} and the ultimate tensile strength is $\sigma_{Dr} = 0{\cdot}5 \times R_m$ for values of R_m up to about 1280 N/mm^2 = 400 HV

Endurance limit
N/mm² kp/mm² 1000 PSI

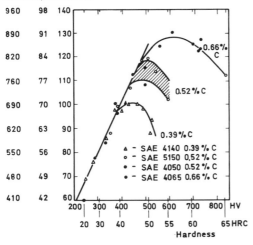

Figure 2.25 Endurance limit as a function of the
hardness of some different steels (from Scott[4])

which can be seen to hold for the SAE steel 4140 (*see Figure 2.25*).
According to this figure the critical limit increases with increasing carbon
content. A steel that has high values of both endurance limit and tensile
strength is grade H 13 which will be discussed further on in this book. The
values obtained in a fatigue test depend on the type of test used. The
approximate relationships given below between the endurance limit
obtained from various fatigue tests are valid in most cases.

Endurance limit for rotating flexural stresses $\sigma_{Dr} = 0.5\,(R_m)$
Endurance limit for reversed flexural stresses $\sigma_{Db} = \sigma_{Dr}$
Endurance limit for alternating stresses $\sigma_D = 0.7\,\sigma_{Dr}$
  (tension-compression)
Endurance limit for torsional stresses $\tau_D = 0.6\,\sigma_{Dr}$

These relationships are illustrated in the modified Goodman diagram,
Figure 2.26.

2.5.6 Surface finish

The surface finish of a structural part subjected to fatigue stresses is of
paramount importance. In general a highly polished surface gives the
highest endurance limit. If the finish is not up to this standard the figure for
the endurance limit obtained from a polished specimen must be reduced
accordingly. For each individual case the reduction is determined by means
of a surface factor K. This factor is a numerical expression of the influence
of the surface finish on the endurance limit.

The reduced value of the endurance limit σ_D is given by:

$$\sigma_{Dred} = K\sigma_D$$

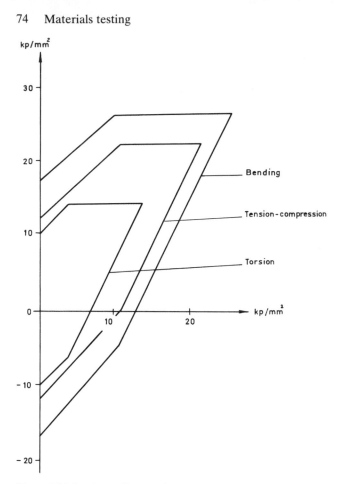

Figure 2.26 Goodman diagrams for bending, tension–compression and torsion

where σ_{Dred} is the reduced endurance limit for the surface finish concerned. *Figure 2.27* shows the surface factor K as a function of the ultimate tensile strength of the material. This diagram is valid, by and large, for the majority of constructional materials with the notable exception of grey cast iron, the surface finish of which has little influence on its endurance limit.

The following example is worth studying.

A steel shaft, $R_m = 70$ kp/mm² (680 N/mm²), fails due to fatigue. There is a skin of mill scale on the surface of the shaft. The endurance limit for a highly polished finish is

$$\sigma_{Dr} = 0.5 \ R_m$$

i.e. $\sigma_{Dr} = 0.5 \times 70 = 35$ kp/mm² (340 N/mm²)

The value of K is 55%, therefore

$$\sigma_{Dred} = 0.55 \times 35 = 19 \text{ kp/mm}^2 \ (190 \text{ N/mm}^2)$$

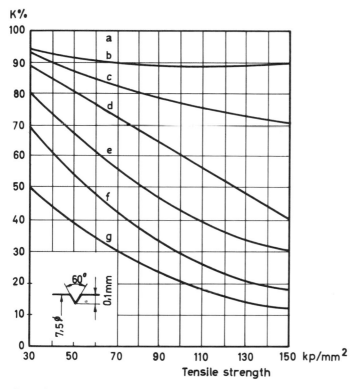

Figure 2.27 Influence of the surface finish on the endurance limit.
(a) Brightly polished surface; (b) Ground surface; (c) Rough-machined
surface; (d) Sharp annular fillet (= V-groove); (e) Surface with skin of
mill scale; (f) Surface exposed to corrosion in fresh water; (g) Surface
exposed to corrosion in sea water

In order to avoid failure by fatigue, a steel having an UTS of
$R_m = 110$ kp/mm^2 (1080 N/mm^2) is chosen for this steel

$$\sigma_{Dred} = 0.5 \times 110 = 0.4 = \underline{22 \text{ kp/mm}^2 \text{ (220 N/mm}^2)}$$

In this case the increase in the endurance limit is insignificant. By choosing
a rough-machined surface it is seen than the first steel gives a better result:

$$\sigma_{Dred} = 0.5 \times 70 \times 0.83 = \underline{29 \text{ kp/mm}^2 \text{ (280 N/mm}^2)}$$

Another example of the influence of notches or stress raisers on the
endurance limit is shown in *Figure 2.28* which refers to grade H 13 with
$R_m = 180$ kp/mm^2 (1770 N/mm^2). This steel has been tested with different
stress concentration factors (K_t) which are computed from special
formulae and diagrams. $K_t = 1.0$ represents a smooth test bar. The larger
the value of K_t, the greater is the stress concentration effect.

2.5.7   Influence of change of section

All deviations from the straight and simple test-bar shape cause local

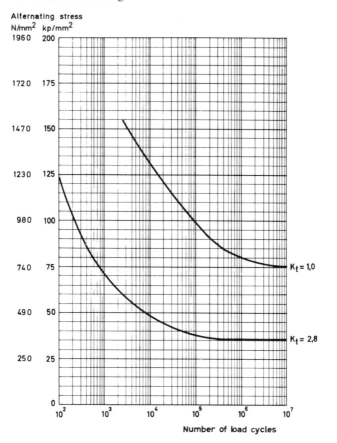

Figure 2.28 The influence of the stress concentration factor (K_t)
on the endurance limit in tension–compression for steel H 13,
heat treated to an ultimate strength of 180 kp/mm² (1770 N/mm²)

concentrations. The maximum stress may be calculated by making use of a
notch sensitivity factor α, the numerical value of which depends on the
type of stressing and the shape of the specimen. *Figure 2.29* shows how α is
estimated from given values of D, d and e. σ_{max} is then calculated from

$$\sigma_{max} = \alpha \times \sigma_{nom}$$

The above simplified passage on the influence of change of section is
included solely for the purpose of drawing the reader's attention to the
great importance of this subject. Detailed instructions on how to cope with
calculations involving the influence of change of section are to be found in
reference books for design engineers.

2.5.8   Ways of increasing the endurance limit

There are many ways of increasing the endurance limit of steel. Most of
them involve the introduction of compressive stresses into the surface of

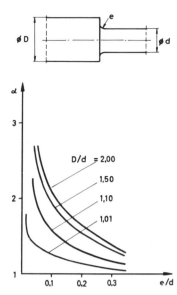

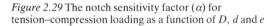

Figure 2.29 The notch sensitivity factor (α) for tension–compression loading as a function of D, d and e

the steel by means of special heat treatments. These points will be discussed later in the book under the appropriate heat treatments.

2.6 The creep test

If a steel is stressed at elevated temperature it is subject to a continuous and progressive deformation with time. This deformation is called creep. If creep is permitted to continue for a sufficiently long time, fracture will occur ultimately. The fracture itself starts in a characteristic manner by microscopic intercrystalline cracking, the final fracture being typically of a brittle nature.

It may be uneconomical to let a structure carry such a small load that creep can be ignored. Since the amount of creep depends on time, and provided the creep strength is known, a structure may be so designed that it will last for a specific time. The resistance to creep of a material is specified as the creep rupture strength and/or the creep strength.

The creep rupture strength

This is a measure of the tensile stress that will cause the steel to fracture after a specified time at a specified temperature. The time is usually specified in periods of 100, 1000, 10 000 or 100 000 h and the rupture stress is reported as, say, σ_{cB} 10 000/650 where c signifies creep, B stands for fracture, 10 000 the number of hours and 650 the temperature in °C.

The creep strength

This is a measure of the tensile stress that causes the steel to show a specified plastic elongation in a specified time at a specified temperature.

The plastic elongation, or strain, is usually specified in terms of strains of 0·2%, 0·5% or 1·0% after 100, 1000, 10 000 or 100 000 h. The creep strength is reported as, say, σ_c 1/10 000/650 which indicates that the creep stress reported will produce 1% permanent strain in 10 000 h at 650°C.

The creep test

When carrying out the creep test, the specimen is held between two pull rods and heated to the desired temperature, whereupon the load is applied (*Figure 2.30*). This load is kept constant during the whole test. The extension of the specimen is plotted as a function of the time and a creep or

Figure 2.30 The creep laboratory section at Bofors

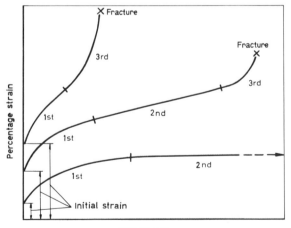

Figure 2.31 Idealized creep curves. 1st = first creep stage (decreasing rate of creep); 2nd = second creep stage (steady-rate creep); 3rd = third creep stage (accelerating creep)

time extension curve is obtained, the idealized shape of which is shown in *Figure 2.31*. The creep curve is characterized by the following stages:

(a) An initial extension
(b) 1st stage. Decreasing rate of creep
(c) 2nd stage. Steady-rate creep
(d) 3rd stage. Accelerating creep

The results obtained from the creep tests are usually brought together and summarized in a creep diagram. As an example, the following information can be obtained from *Figure 2.32*. An applied stress of 290 N/mm² (30 kp/mm²) at 550 °C produces in 5000 h a permanent strain of 0·5%.

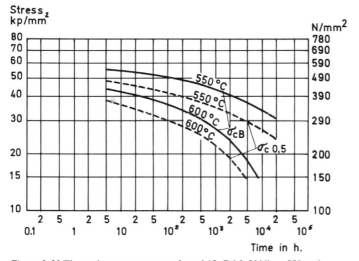

Figure 2.32 The resistance to creep of steel 12 CrMoV Nb at 550 and 600°C. (Creep rupture strength and creep strength for 0·5% plastic strain.)

2.7 Brittle and ductile fractures

When a material breaks the fracture can be related to either of two fundamental mechanisms, viz. cleavage or shear (or sometimes a combination of both). On the scale of atomic dimensions it is easy to differentiate between the two main fracture types. In the case of cleavage, fracture occurs at right angles to the direction of the tensile stress. This results in a disruption of the atomic forces in the tensile direction. Fracture by shear originates in slip and proceeds under the influence of shear stresses, *Figure 2.33*.

Fracture by cleavage occurs as a result of the application of a well-defined stress and in most cases with appreciable deformation. Shear fracture, on the other hand, develops gradually and is completed at considerably greater stress and strain than those operating at the instant the fracture was initiated. A cleavage fracture follows certain definite crystallographic planes in the individual grains and hence it has a

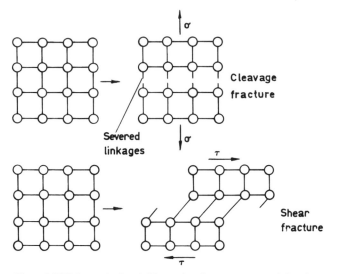

Figure 2.33 Schematic sketch illustrating fractures at atomic levels

crystalline appearance, which is described as brittle. This type of fracture is called transcrystalline which means that the crack is propagated right through the body of the grains.

A shear fracture, which is always preceded by a large plastic deformation, has a fibrous or ropy appearance and is termed ductile. On examination of the fractured surface in the electron microscope it is found that a cleavage fracture shows the characteristic river pattern, *Figure 2.34*, in which the flow lines follow the direction of crack propagation. A shear fracture shows little hollows, termed dimples, *Figure 2.35*.

Figure 2.34 River pattern in a cleavage fracture, 2500 ×

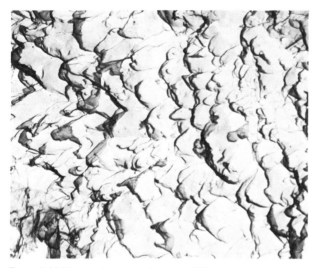

Figure 2.35 Dimples in a shear fracture, 3900 ×

Metals crystallizing in the face-centred cubic system, e.g. austenitic stainless steel, always fail by shear. Metals with a body-centred cubic crystal structure can fail by either cleavage or shear. In the case of martensitic steels the condition of the steel such as its composition, cleanness and heat treatment, all influence the character of the fracture. In addition the *temperature* has a decisive influence, which is clearly illustrated in *Figure 2.36*. The fracture stress σ_f is assumed to vary only slightly with temperature; the yield stress, on the other hand, increases sharply with decreasing temperature. In the figure the σ_f-curve indicates the stress required to propagate initiated cleavage cracks. Although it was stated above that there is practically no deformation associated with the formation of cleavage cracks, a little plastic deformation is necessary for crack initiation, i.e. the yield point must be reached. A consequence of this is that when the temperature falls below T_{D1}, the point of intersection of

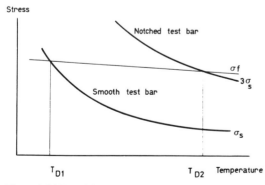

Figure 2.36 Transition from ductile to brittle fracture, according to the Ludwig-Dawidenkow-Orowan hypothesis

the two curves, a cleavage fracture cannot occur until the stress has increased to σ_s, the reason being that the concurrence of conditions for both crack initiation and propagation is a prerequisite of ultimate fracture by cleavage. Thus at temperatures below T_{D1} the cleavage stress coincides with the σ_s-curve. At temperatures above T_{D1} the material begins to flow before cleavage fracture occurs since such fracture is able to set in only when the resulting strain–hardening has enabled the stress to reach σ_f. This type of cleavage fracture takes place only within a rather narrow temperature range above T_{D1}. At a sufficiently elevated temperature the shear fracture mechanism comes into operation and causes fracture before the stress has reached the value of σ_f.

During the course of tensile testing with a notched test specimen there arises in the immediate vicinity of the notch tip a triaxial system of stresses which creates what is called constrained plastic deformation. An increase

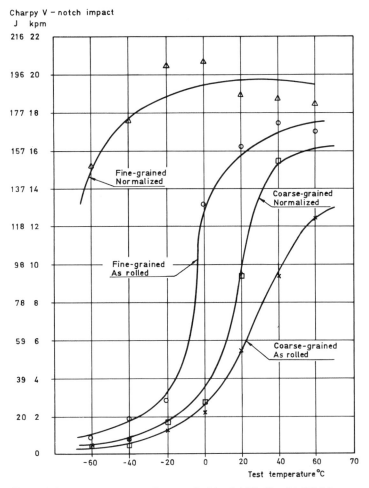

Figure 2.37 Impact-test curves for a steel with a 0·18% C and 1·5% Mn after different treatments. Approximately 60 kp/mm² (590 N/mm²) UTS for all conditions

in tensile stress, applied parallel to the axis of the test-piece, is now required in order that the shear force shall overcome the strength of the material and allow the plastic deformation to proceed. A notched tensile specimen may thus show higher values of yield strength and ultimate tensile strength than a smooth one. As a result of strain–hardening, stresses up to $3 \times \sigma_s$ may develop at sharp notches. According to *Figure 2.36* this would imply that the transition from ductile to brittle fracture takes place at a higher temperature when the specimen is notched.

If the steel is very brittle a notched specimen will not give higher values of yield strength and ultimate tensile strength. Hence the toughness of a steel may be judged by the results obtained from a notched specimen. If a smooth and a notched test-piece give the same results the steel may be classed as brittle. If the value obtained from the notched specimen is about 1·5 times that of the value from the smooth specimen the steel can be regarded as tough.

The geometry of the notch plays a decisive part in the type of fracture the material will develop whether in impact or in tensile testing. This fact is utilized in the Charpy V test-piece, as has already been touched upon (*see Figure 2.11*). In tests conducted at temperatures below the transition temperature the steel breaks with a brittle fracture in the Charpy V test, but is ductile in the Charpy U test. The steel-making process used for the steel as well as the heat treatment has an influence on the impact value, the transition temperature and the type of fracture, which is illustrated in *Figure 2.37*.

The above, much simplified account of fracture phenomena is included in order to explain, if only superficially, why the same steel, used in the same construction, may sometimes fail by brittle and sometimes by ductile fracture. A brittle fracture may have as its cause an unsuitable grade of steel, a faulty design or poor finish, e.g. inferior machining or may result from unsatisfactory heat treatment. It also emphasizes the necessity of using the appropriate heat treatment.

2.8 Fracture toughness

During the last decades the mechanism of fracture has been studied very closely and the newly developed methods of testing are beginning to be applied. The term *fracture toughness* is coming into use more and more frequently in the literature. Since the heat treatment very often has a decisive influence on the fracture toughness, mention of this material property will be found in subsequent chapters. A short note on the concept of fracture toughness is therefore called for.

The fracture toughness of a material is a measure of its capacity to withstand crack propagation from stress raisers when it is subjected to tensile stresses. In practice such stress raisers may be surface defects from a machining operation, incipient fatigue cracks or a faulty structure derived from improper heat treatment. With regard to calculations involving linear elastic fracture mechanics, describing the situation prevailing at the crack tip as it extends, there are two terms to differentiate between, viz. *plane stress* and *plane strain*.

The material property designated K_C is used when dealing with *plane stress*, the conditions for which exist in thin-walled structural members. The numerical value of K_C is dependent on the dimensions of the specimen, on the notch geometry and on the temperature of the test.

The material property designated K_{IC} is regarded as applicable only when dealing with *plane strain*, the conditions for which are inherent in heavy section structural members. The numerical value of K_{IC} is independent of the geometry both of the sample specimen and the notch but is temperature-dependent. However, if the calculated result is to be valid the thickness requirement (t) is that

$$t \geqslant 2 \cdot 5 \left[\frac{K_{IC}}{R_e} \right]^2 \text{ where } R_e = \text{yield point}$$

This thickness requirement implies that in practice the test can be performed only on high-strength materials or on very large test pieces.

When dimensioning for non-linear fracture mechanics, use is made of the J-integral. This concept does not describe the situation at the crack tip as the latter extends but is applicable only at the very moment of initiation itself. By making use of the J-integral the condition for the initiation of crack extension is expressed as

$$J = J_{IC}$$

where J is a value calculated from given data for a structure and J_{IC} is a material constant. For a linear elastic material the J-integral states a fracture condition that is equivalent to the Griffith–Irwin fracture condition but the J-integral may be rendered meaningful even for large plastic deformations and is applied to formulate conditions for crack propagation in any body.

The thickness requirement (t) as applied to the J-criterium has been tentatively put at

$$t \geqslant \alpha \frac{J_{IC}}{R_e} \text{ where } \alpha \text{ lies between 20 and 50}$$

Hence, according to the above J-criterium the thickness requirement for a test piece is considerably smaller, viz. only about 5% of that required for plane strain. In practice this implies that smaller test pieces may be used for steels of various strength levels when the J-criterium is being employed.

In order to evaluate the fracture toughness, standards covering test specimens, methods of measurement and assessment of results have been set up by a committee within ASTM[5]. In the following only the CT (Compact Tension) test bar (*see Figure 2.38*), will be touched upon.

A fatigue-testing machine is first used to initiate a fatigue crack in a previously machined notch and this crack is allowed to grow to a certain depth in the test bar. The final testing is carried out at the desired temperature, using slowly applied loading—tensile or bending as the case may be—until the test bar fractures (*see Figure 2.39*). The typical appearance of a fractured specimen is shown in *Figure 2.38*. The left-hand part of the test-piece shows the machined notch which terminates in a

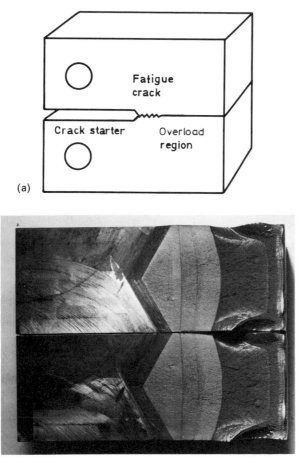

Figure 2.38 Fracture CT specimen. (a) The type of specimen used;
(b) The fractured surfaces

sharp tip. Proceeding from this there is the starter crack which ends in the final ductile fracture. If a fracture test result is to be recognized and accepted it must satisfy certain specific requirements. During the course of the test an X–Y recorder is used to register load and COD (Crack Opening Displacement). In this way a curve is obtained which resembles, as a rule, one of the three representative curves shown in *Figure 2.40*.

Evaluation is carried out by using the load–COD curve obtained during the test. The critical load P_Q in *Figure 2.40* represents the load acting just at the instant when the material in front of the crack tip gives way under the applied force. By applying the formula appropriate to the type of specimen used the fracture toughness of the material can be calculated from this value. The unit of critical stress intensity, K_{IC}, is kp mm$^{-3/2}$, N mm$^{-3/2}$ or MN mm$^{-3/2}$. For conversions *see Table 8.24*. In American literature the unit is given as PSI $\times$ $\sqrt{}$ (in) or KSI $\times$ $\sqrt{}$ (in).

The different modes of COD cases, shown in *Figure 2.40*, illustrate an important limitation of the fracture-toughness concept. From the methods

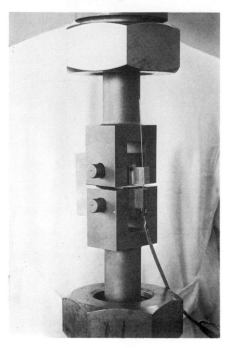

Figure 2.39 CT specimen under test with COD gauge attached

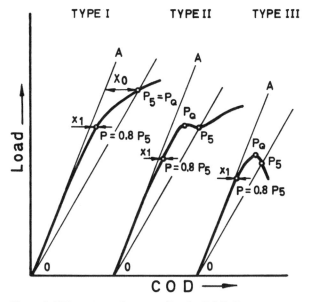

Figure 2.40 Representative type of load—COD diagrams. Type I, the load increases more or less steadily to fracture, which has a ductile appearance. Type II, the curve shows a pop-in which corresponds approximately to the yield point. Type III, the curve resembles that of the tensile stress–strain curve for a hardened and tempered steel. Regarded on a macroscale, the fracture has a brittle appearance

used to estimate the fracture toughness from the load–COD plot, the fracture toughness values may indicate entirely different aspects of the progress of the fracture mode. In Types I and II in *Figure 2.40* the crack propagation is stable well beyond the point at which the fracture toughness is measured; in this case the value of the fracture toughness is thus a measure of the initiation of stable crack propagation. In Type III the crack is unstable and starts to extend in close proximity to the point where the fracture toughness is measured. Hence in this case the fracture toughness is a measure of unstable crack propagation, i.e. what is usually called brittle fracture. The fracture toughness increases with decreasing tensile strength. There is, however, no unequivocal conversion factor that can be applied to convert ultimate tensile strength or yield strength to fracture toughness since different grades of steel at the same strength level present different fracture toughness values. This point is exemplified in *Figure 2.41*.

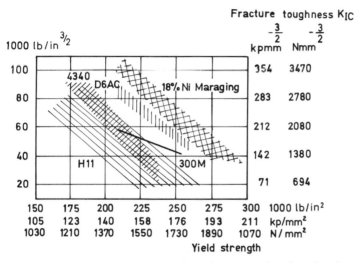

Figure 2.41 Fracture toughness values for various types of steel as a function of the yield strength (after Steigerwald[10])

However, attempts have been made to find a relationship between the K_{IC} values and the Charpy KV notch-toughness values[6, 7, 8]. Results based on empirical correlations will be discussed in Chapter 6.

Fracture toughness data for hard steels and white cast iron are rather scarce. This is mainly due to the difficulty of precracking brittle materials in a controlled way. It is generally accepted that during fatigue of brittle materials, unstable crack propagation will occur immediately on crack initiation. Since a fracture toughness test is the only ductility test that gives relevant data for engineering design, the Jernkontoret Research Organization considered it as interesting to make a fracture mechanics approach to hard steels and cast irons.

A fatigue method permitting stable crack propagation from a machined notch was developed and an alternative method of precracking brittle materials requiring a considerably shorter time than fatigue precracking

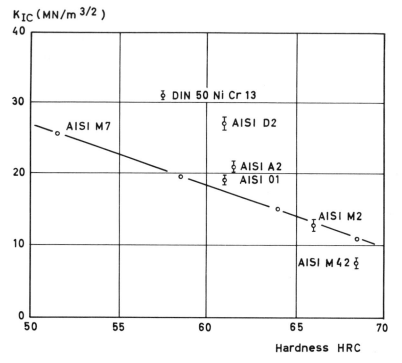

K_{IC} (MN/m$^{3/2}$)

Figure 2.42 Fracture toughness versus hardness for some tool steels and high-speed steels (after Eriksson[9])

(10 min as compared with 8 h) was presented[9]. The test results for some well-known tool steels and high-speed steels are shown in *Figure 2.42*. It is interesting to note that the fracture toughness for D 2 is higher than for A 2 and O 1. This is, however, in accordance with certain practical results obtained.

2.8.1   The implication of fracture toughness

A possession of some knowledge of the fracture toughness of a material enables the designer to dimension the members of the construction so that catastrophic crack growth and other design faults are avoided. It is also possible to calculate the maximum acceptable size of possible defects already inadvertently present in existing constructions. For this calculation it is necessary to know not only the value of the fracture toughness but also the design loading, the yield point of the material and the size and shape of the defects in the materials, such as cracks and similar inhomogeneities.

For further information on fracture toughness the reader is referred to the specialized literature, some of which is quoted in references 5–13.

References

1 DENGEL, D. and ROSSOW., 'Comparison between the Vickers and Knoop Hardness Values', *Härterei–Techn. Mitt.*, **26**, No. 1, 21–27 (1971) (in German)
2. SCHMITZ, H. and SCHLÜTER, W., 'Attempts at Establishing the Reproducibility of Rebound Hardness Measurements', *Stahl u. Eisen*, **75**, No. 7, 411–416 (1955) (in German)
3. ROBILLER, G., 'Influence of the Width of the Specimen on the Results of the Notched Bar Impact Bending Test', *Stahl u. Eisen*, **100**, No. 19, 1132–1138 (1980) (in German)
4. SCOTT, H., *Trans. ASM*, **48**, 145 (1956)
5. *Fracture Testing of Materials*, ASTM Committee E–24
6. BARSOM, I. M., and ROLFE, S. T., 'Conditions between K_{IC} and Charpy V-Notch Test Results in the Transition Temperature Range', *Impact Testing of Metals*, ASTM STP, 466, 281–302 (1970)
7. SANDBERG, O., and ÅKERMAN, J., *The Influence of Tempering Temperature, Microstructure and Carbon Contents on Toughness of Quenched and Tempered CrMo-Steels*, Swedish Institute for Metals Research, IM–1487 (1980)
8. SINHA, T. K., and SANDBERG, O., *Fracture Toughness of a Quenched and Tempered Boron Steel*, Swedish Institute for Metals Research, IM–1526, (January 1981)
9. ERIKSSON, K., 'Fracture Toughness of Hard High-Speed Steels, Tool Steel and White Cast Irons', *Scandinavian J. Metal'*, **2**, 197–203 (1973)
10. STEIGERWALD, A., 'What You Should Know About Fracture Toughness', *Metal Progress*, 96–101, (November 1967)
11. KRAINER, E., 'Defining Fracture Toughness. Evaluation and Implication as a Material Property', *Berg und Hüttenmännische Monatshefte*, **7** (1970) (in German)
12. *Fracture Toughness*, Publication 121, Iron and Steel Institute (1970)
13. *Fracture Toughness of High-Strength Materials, Theory and Practice*, Publication 120, Iron and Steel Institute (1970)

Alloying elements in steel and new steelmaking processes

It is a long-standing tradition to discuss the various alloying elements in terms of the properties they confer on steel. For example, the rule was that chromium (Cr) makes steel hard whereas nickel (Ni) and manganese (Mn) make it tough. In saying this one had certain types of steel in mind and transferred the properties of a particular steel to the alloying element that was thought to have the greatest influence on the steel under consideration. This method of reasoning can give false impressions and the following examples will illustrate this point.

When we say that Cr makes steel hard and wear-resisting we probably associate this with the 2% C, 12% Cr tool steel grade, which on hardening does in fact become very hard and hard-wearing. But if, on the other hand, we choose a steel containing 0·10% C and 12% Cr, the hardness obtained on hardening is very modest. It is quite true that Mn makes for toughness in steel if we have in mind the 13% manganese steel, the so-called Hadfield steel. In concentrations between 1% and 5%, however, Mn can produce a variable effect on the properties of the steel it is alloyed with. The toughness may either increase or decrease. The effects produced by the various alloying elements on different types of steel is described in detail in books of reference[1, 2, 3, 4]. The present chapter will deal with only some of the basic aspects of the effects of alloying elements on steel.

In Section 3.3 some new steelmaking processes are discussed. In several respects these processes exert a decisive influence on the properties of the steel largely owing to their effect on the role played by oxygen, aluminium and sulphur.

3.1 Solids

A property of great importance is the ability of alloying elements to promote the formation of a certain phase or to stabilize it. These elements are grouped as austenite-forming, ferrite-forming, carbide-forming and nitride-forming elements.

3.1.1 Austenite-forming elements

The elements C, Ni and Mn are the most important ones in this group. Sufficiently large amounts of Ni or Mn render a steel austenitic even at room temperature. An example of this is the so-called Hadfield steel which contains 13% Mn, 1·2% Cr and 1% C. In this steel both the Mn and C take part in stabilizing the austenite. Another example is austenitic stainless steel containing 18% Cr and 8% Ni.

The equilibrium diagram for iron–nickel, *Figure 3.1*, shows how the range of stability of austenite increases with increasing Ni-content. An alloy containing 10% Ni becomes wholly austenitic if heated to 700 °C. On cooling, transformation from γ to α takes place in the temperature range 700–300 °C.

3.1.2 Ferrite-forming elements

The most important elements in this group are Cr, Si, Mo, W and Al. The range of stability of ferrite in iron–chromium alloys is shown in *Figure 3.2*. Fe–Cr alloys in the solid state containing more than 13% Cr are ferritic at all temperatures up to incipient melting. Another instance of a ferritic steel is one that is used as transformer sheet material. This is a low-carbon steel containing about 3% Si.

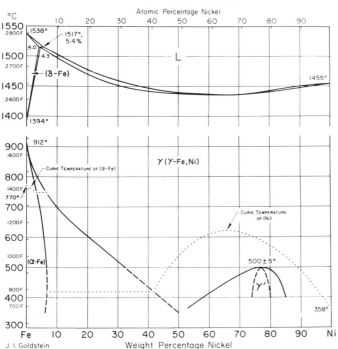

Figure 3.1 Fe–Ni equilibrium diagram (reproduced by permission from *Metals Handbook*, **8**, American Society for Metals (1973))

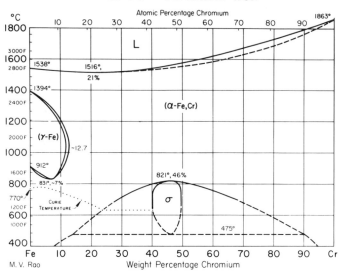

Figure 3.2 Cr–Fe equilibrium diagram (reproduced by permission from *Metals Handbook*, **8**, American Society for Metals (1973))

3.1.3   Multi-alloyed steels

The great majority of steels contain at least three components. The constitution of such steels can be deduced from what are called ternary phase diagrams (3 components). The interpretation of these diagrams is relatively difficult and they are of limited value to people dealing with practical heat treatment since they represent equilibrium conditions only. Furthermore, since most alloys contain more than three components it is necessary to look for other ways of assessing the effect produced by the alloying elements on the structural transformations occurring during heat treatment.

One approach which is quite good, is the use of Schaeffler diagrams (*see Figure 3.3*). Here the austenite formers are set out along the ordinate and the ferrite formers along the abscissa. The original diagram contained only Ni and Cr but the modified diagram includes other elements and gives them coefficients that reduce them to the equivalents of Ni or Cr respectively. The diagram holds good for the rates of cooling which result from welding.

Some examples taken from the diagram

A 12% Cr steel containing 0·3% C is martensitic; the 0·3% C gives the steel a nickel equivalent of 9.

An 18/8 steel (18% Cr, 8% Ni) is austenitic if it contains 0·05% C and 2% Mn. The Ni content of such steels is usually kept between 9% and 10%.

Hadfield steel with 13% Mn (mentioned above) is austenitic due to its high

Nickel equivalent =
% Ni + 30 x % C + 0,5 x % Mn + 11,5 x % N

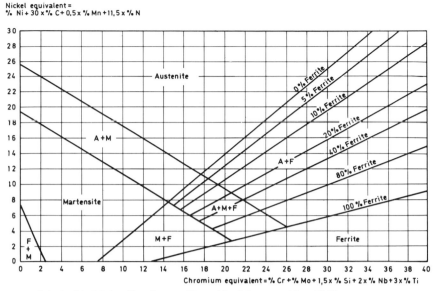

Figure 3.3 Modified Schaeffler diagram (from *Metal Progress*, Nov. 1949)

carbon content. Should this be reduced to about 0·20% the steel becomes martensitic

3.1.4   Carbide-forming elements

Several ferrite formers also function as carbide formers. The majority of carbide formers are also ferrite formers with respect to Fe. The affinity of the elements in the line below for carbon increases from left to right.

Cr, W, Mo, V, Ti, Nb, Ta, Zr

Some carbides may be referred to as *special carbides,* i.e. non-iron-containing carbides, such as Cr_7C_3, W_2C, VC, Mo_2C. *Double* or *complex carbides* contain both Fe and a carbide-forming element, for example Fe_4W_2C.

High-speed and hot-work tool steels normally contain three types of carbides which are usually designated M_6C, $M_{23}C_6$ and MC. The letter M represents collectively all the metal atoms. Thus M_6C represents Fe_4W_2C or Fe_4Mo_2C; $M_{23}C_6$ represents $Cr_{23}C_6$ and MC represents VC or V_4C_3.

3.1.5   Carbide stabilizers

The stability of the carbides is dependent on the presence of other elements in the steel. How stable the carbides are depends on how the element is partitioned between the cementite and the matrix. The ratio of the percentage, by weight, of the element contained in each of the two phases is called the partition coefficient K. The following values are given for K:

Al	Cu	P	Si	Co	Ni	W	Mo	Mn	Cr	V	Ti	Nb	Ta
0	0	0	0	0·2	0·3	2	8		11·4	28	Increasing		

Note that Mn, which by itself is a very weak carbide former, is a relatively potent carbide stabilizer. In practice, Cr is the alloying element most commonly used as a carbide stabilizer. Malleable cast iron (i.e. white cast iron that is rendered soft by a graphitizing heat treatment called malleablizing) must not contain any Cr. Steel containing only Si or Ni is susceptible to graphitization, but this is most simply prevented by alloying with Cr.

3.1.6   Nitride-forming elements

All carbide formers are also nitride former. Nitrogen may be introduced into the surface of the steel by nitriding. By measuring the hardness of various alloy steels so treated it is possible to investigate the tendency of the different alloying elements to form hard nitrides or to increase the hardness of the steel by a mechanism known as precipitation hardening. The results obtained by such investigations are shown in *Figure 3.4*, from which it can be seen that very high hardnesses result from alloying a steel with Al or Ti in amounts of about 1·5%. On nitriding the base material in *Figure 3.4* a hardness of about 400 HV is obtained and according to the

Hardness HV

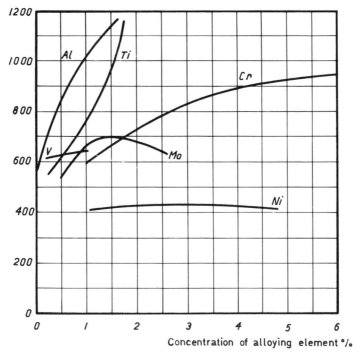

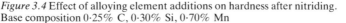

Figure 3.4 Effect of alloying element additions on hardness after nitriding. Base composition 0·25% C, 0·30% Si, 0·70% Mn

diagram the hardness is unchanged if the steel is alloyed with Ni since this element is not a nitride former and hence does not contribute to any hardness increase.

3.1.7 Effect on ferrite hardness

All alloying elements that form solid solutions in ferrite affect its hardness. The hardness increase caused by substitutional solution is shown in *Figure 3.5*. Si and Mn, the most frequently occurring alloying elements, have a relatively potent effect on the hardness of ferrite, while Cr gives the smallest hardness increase. For this reason Cr is a most convenient alloying element in a steel that is to be processed by cold working in which good hardenability is required.

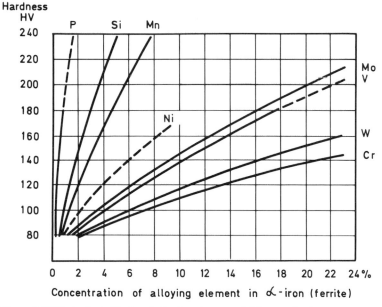

Figure 3.5 Effect of substitutional alloying element additions on ferrite hardness (after Bain[4])

3.1.8 Effect on grain growth

The elements Al, Nb, Ti and V in small amounts from 0·03% to 0·10% are important factors in inhibiting grain growth at the austenitizing temperature. This is because these elements are present as highly dispersed carbides, nitrides or carbo-nitrides (Al only as nitride) and that a high temperature is required to make them go into solution. *Figure 3.6* shows that in a steel containing about 0·05% Nb or Ti and 0·20%C the niobium and titanium carbides are not dissolved until the temperature exceeds 1200 °C[5]. For V and N contents of 0·10% and 0·010% respectively the vanadium nitrides remain undissolved at temperatures up to and somewhat above 1000 °C. Should the temperature rise so high that the phases

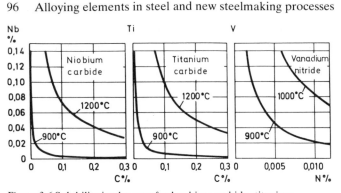

Figure 3.6 Solubility isotherms of columbium carbide, titanium carbide and vanadium nitride in steel at different temperatures (after Irving, Pickering and Gladman[5])

inhibiting grain growth pass into solution there will be a pronounced increase in grain size.

The above-mentioned elements have found great use as microconstituents in the high-strength structural steels, known as the HSLA-steels (High Strength Low Alloy). These steels will be discussed in detail in Chapter 6.

In the production of the fine-grained case-hardening steels, the desired effect is obtained by the addition of the appropriate amounts of Al to the molten steel. The normal practice is that the oxygen concentration is first reduced to a suitable level and then Al is added in amounts corresponding to the nitrogen content of the steel. As the steel cools a dispersion of AlN particles is produced and as a result the steel is rendered resistant to grain growth at temperatures normally employed for heat treatment. The diagram in *Figure 3.7* shows that a steel containing, say, 0·007% N and

Nitrogen content %

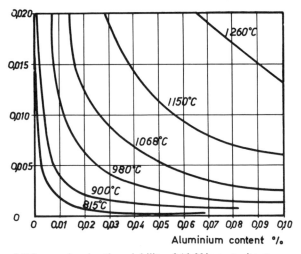

Figure 3.7 Curves showing the solubility of Al–N in austenite at temperatures below 1260°C (after Bain[4])

0·03% Al should remain fine-grained up to about 1000 °C. The critical or coarsening temperature, at which grain growth sets in, is dependent on time as well as on temperature. The diagrams in *Figures 3.8a* and *3.8b*, which apply to the case-hardening steel En 352, illustrate this point.

During the case-hardening operation a carburizing temperature not exceeding 950 °C is customary. Hence a fine-grain treatment with Al and N will suffice for the carburizing temperatures used. However, it is suggested in some quarters that higher temperatures be used, e.g. 1000 °C, in order to speed up the process. In such instances Ti has been found to be a useful alloying element. In high-speed steel and other highly alloyed tool steels the carbides of W and Mo inhibit grain growth.

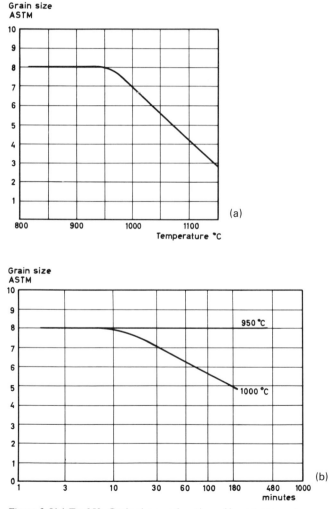

Figure 3.8(a) En 352. Grain size as a function of heat-treatment temperature. Holding time: 30 min; (b) En 352. Grain size as a function of heat-treatment time. Heating at 950 and 1000°C

3.1.9   Effect on the eutectoid point

A_1 is lowered by the austenite-formers and raised by the ferrite-formers. A chrome steel of eutectoid composition, e.g. one containing 12% Cr and 0·4% C, will require a higher austenitizing temperature than a eutectoid carbon steel, whereas a 3% Ni steel will already begin to austenitize below 700 °C, *see Figure 3.9*. This state of affairs is clearly of great practical importance when these steels are being used at temperatures around A_1.

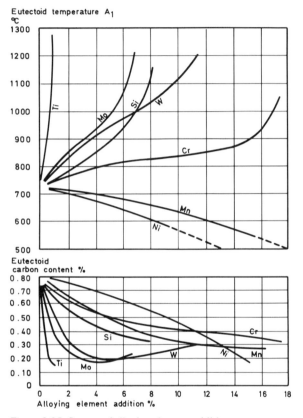

Figure 3.9 Influence of alloying element addition on eutectoid temperature and eutectoid carbon content (after Bain[4])

The eutectoid point in the iron–carbon equilibrium diagram occurs at 0·80% C and 723 °C (A_1). All alloying elements reduce the carbon concentration of this point. For example, a steel containing 5% Cr has its eutectoid point at 0·5% C. The influence of Cr and Mn is illustrated in *Figures 3.10* and *3.11* respectively.

3.1.10   Effect on the temperature of martensite formation

All alloying elements with the possible exception of Co, lower M_s, the temperature of the start of the martensite formation, as well as M_f, the

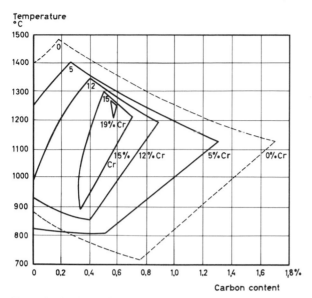

Figure 3.10 Effect of Cr and C on the austenite field (after Bain[4]). The dashed line traces the austenite field of the base alloy. The displacement of the eutectoid point, for instance, indicates that an alloy containing only iron and carbon has not been used

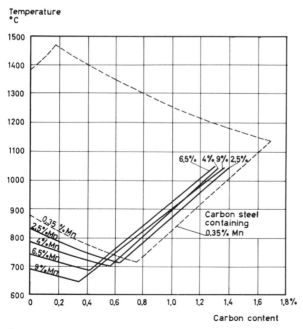

Figure 3.11 Effect of Mn and C on the austenite field (after Bain[4])

finish of the martensite formation, i.e. at 100% martensite. For the majority of steel containing more than 0·50%, C, M_f lies below room temperature. This implies that after hardening these steels practically always contain some residual austenite. M_s may be calculated from the equation given below on inserting the percentage concentration of each alloying element in the appropriate term[6].

$$M_s\,(°C) = 561 - 474\,C - 33\,Mn - 17\,Ni - 17\,Cr - 21\,Mo$$

For high-alloy and medium-alloy steels Stuhlmann[7] has suggested the following equation:

$$M_s\,(°C) = 550 - 350\,C - 40\,Mn - 20\,Cr - 10\,Mo - 17\,Ni - 8\,W - 35\,V \\ - 10\,Cu + 15\,Co + 30\,Al$$

The equations are valid only if all the alloying elements are completely dissolved in the austenite.

Of all the alloying elements it is seen that carbon has the strongest influence on the M_s temperature (*see Figure 1.17*). In *Figures 3.12* and *3.13* an attempt is made to show in a diagram the experimental results of the effect of Mn and Ni on the M_s temperature of various types of steel.

3.1.11   Effect on the formation of pearlite and bainite during the isothermal transformation

All alloying elements except Co delay the formation of ferrite and cementite, i.e. the transformation curves in the TTT diagram are displaced to the right. It is only that part of the element in solution that affects the transformations. It is very difficult to formulate any general rules regarding the influence exerted by the various alloying elements. However, it has definitely been found that some elements affect the bainite transformation more than the pearlite transformation, while other elements act in the opposite manner.

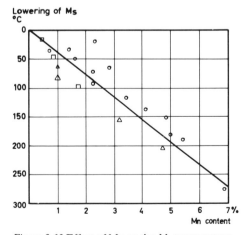

Figure 3.12 Effect of Mn on the M_s-temperature (after Russel and McGuire, Payson and Savage, Zyuzin, Grange and Stewart)

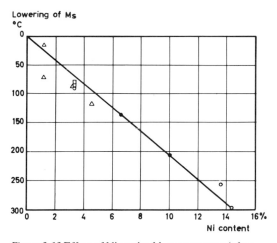

Figure 3.13 Effect of Ni on the M_s-temperature (after Chiswik and Greninger, Payson and Savage, Zyuzin, Grange and Stewart)

Certain elements will, paradoxically, accelerate the transformations if their concentration increases beyond a certain limiting value, this limit being affected by other alloying elements present. For alloyed case-hardening and tool steels the time taken to initiate the pearlite–bainite transformation is reduced as the carbon content exceeds about 0·80% on account of the carbide formation which reduces the amounts of alloying elements in the matrix. For tool steels and constructional steels Si-concentrations of 1·5% and above have been found to promote pearlite formation. This point is further discussed in Chapter 4.

As a general principle it may be stated that by increasing the concentration of *one* alloying element by some few per cent, the basic carbon content being kept about 0·50% only a relatively small retardation of the transformation rates is noticed. For plain carbon steels a successive increase in C from 0·50% to 1% produces but a negligible effect. It is only in conjunction with several alloying elements that a more noticeable effect is produced.

Plain carbon steels

The diagram in *Figure 3.14*, applicable to steel W 1 (1% C) will serve as a basis for this discussion. The shortest transformation time for this steel is less than 1/8th second. Note that the time scale is logarithmic; hence there is no zero time. As has been mentioned previously, both pearlite and bainite form simultaneously in this steel at about 550 °C. Since the curves overlap it is customary to draw only one curve. With increasing contents of certain alloying elements, however, the noses of the pearlite and bainite curves will separate.

The structures shown in *Figure 3.14* are obtained by austenitizing samples of steel W 1 at 780 °C for 10 min and quenching in a salt bath at various temperatures. After holding them for predetermined times at

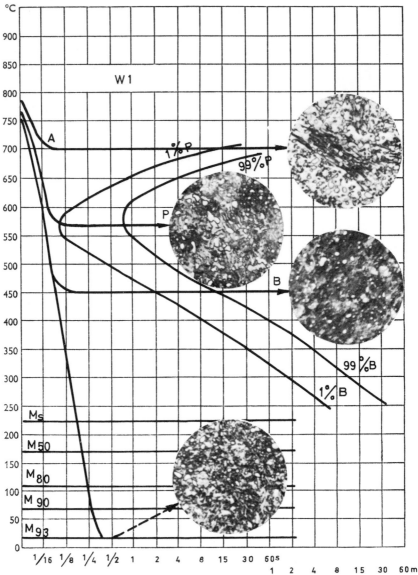

Figure 3.14 TTT diagram for isothermal transformation of steel W 1 (1% C steel). A = austenite, B = bainite, M_s = start of martensite transformation, M_{50} = 50% M, P = pearlite

various temperatures they are finally quenched in water. Before the salt-bath quenching the steel contains undissolved carbides but in view of the composition of the austenite the steel may be regarded as a eutectoid one. The diagram should be studied with the aid of the explanatory text below.

1. Quenching in a liquid bath at 700 °C; holding time 4 min. During this interval the C has separated out, partly as pearlite lamellae and partly as spheroidized cementite. Hardness 225 HV.
2. Quenching to 575 °C; holding time 4 s. A very fine, closely spaced pearlite as well as some bainite has formed. Note that the amount of spheroidized cementite is much less than in the preceding case. Hardness 380 HV.
3. Quenching to 450 °C; holding time 60 s. The structure consists mainly of bainite. Hardness 410 HV.
4. Quenching to 20 °C (room temperature). The matrix consists of, roughly, 93% martensite and 7% retained austenite. There is some 5% cementite as well which has not been included in the matrix figure. Hardness 850 HV.

Alloy steels

In *Figures 3.15a–d* it is seen that Cr produces approximately the same effect on the pearlite nose as on the bainite nose. At Cr concentrations above around 3% the noses separate completely. *Figures 3.16a–d* show that Ni exerts only a modest influence on the transformation and that this element displaces the whole diagram to the right, practically unchanged. The influence exerted by Mo is shown in *Figure 3.17,* from which it is seen that the pearlite nose is displaced farther than the bainite nose. Si, Mn and W in concentrations up to some 2% have only a very slight effect on prolonging the transformation times.

The transformations are affected to a much greater extent by the simultaneous presence of several alloying elements, even if the sum of their concentrations often is less than that of two of the elements discussed above. In fact, a multiplicative effect is obtained. In such instances the influence of C up to 0·5% is considerable and this is illustrated by the diagrams in *Figures 3.18a–d* and *3.19a–d.* With regard to the Cr–Mo steel, an increase in the C content tends to produce a marked retardation of the pearlite transformation, whereas for the Cr–Ni steel it is the bainite transformation that is retarded with increasing C content. The so-called ternary-alloy steels, besides containing C, Si and Mn, have three additional alloying elements which exert an even more potent influence on the transformation rates. This may be seen in *Figures 3.20* and *3.21.* Examples of TTT diagrams for high-alloy steels are shown in *Figures 3.22, 3.23* and *3.24.*

Boron in very small amounts has a surprisingly potent effect on the transformation in steel. A detailed discussion on boron steels appears in Section 6.3.

By varying the hardening temperature, larger or smaller amounts of the alloying elements are made to pass into solution. As a result it is possible to vary in a large measure the time taken to incipient transformation. The grain size is also an important factor. These topics, along with a detailed discussion on continuous-cooling-transformation (CCT) diagrams, will be treated at considerable length in Chapter 4.

104

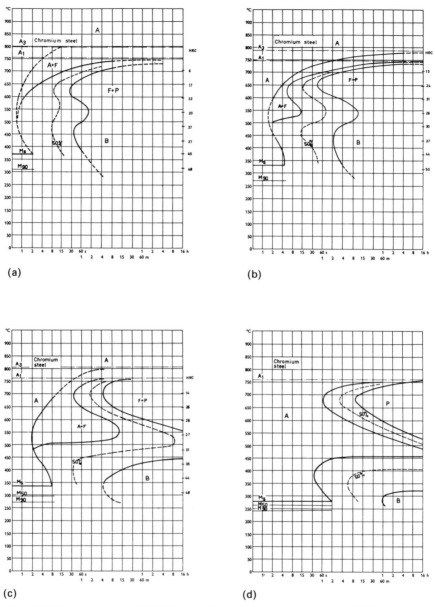

Figure 3.15 Cr-alloyed steels, TTT diagrams for isothermal transformation (USS).
(a) Nominal analysis 0·4% C, 0·5% Cr. Solution temperature 870°C;
(b) Nominal analysis 0·4% C, 0·9% Cr. Solution temperature 840°C;
(c) Nominal analysis 0·35% C, 2·0% Cr. Solution temperature 870°C;
(d) Nominal analysis 0·5% C, 3·1% Cr. Solution temperature 900°C

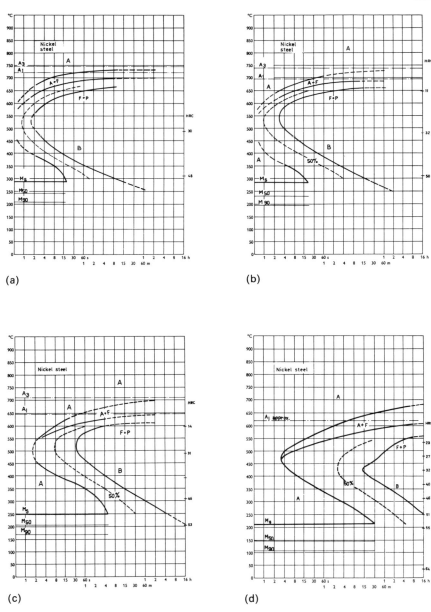

Figure 3.16 Ni-alloyed steels TTT diagrams for isothermal transformation (USS).
(a) Nominal analysis 0·6% C, 1·0% Ni. Solution temperature 800°C;
(b) Nominal analysis 0·6% C, 2·0% Ni. Solution temperature 800°C;
(c) Nominal analysis 0·6% C, 4·0% Ni. Solution temperature 800°C;
(d) Nominal analysis 0·6% C, 5% Ni. Solution temperature 925°C

106

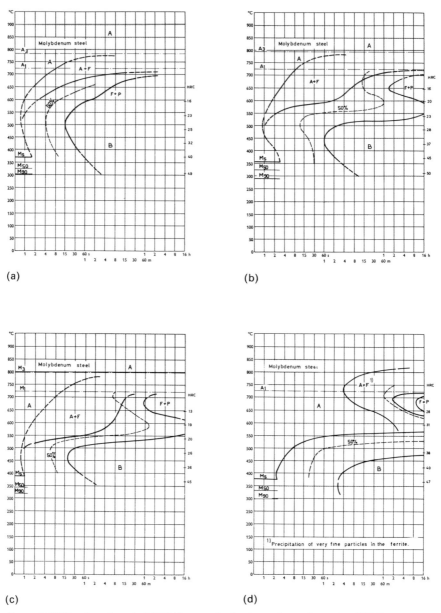

Figure 3.17 Mo-alloyed steels. TTT diagrams for isothermal transformation (USS).
(a) Nominal analysis 0·4% C, 0·2% Mo. Solution temperature 870°C;
(b) Nominal analysis 0·4% C, 0·5% Mo. Solution temperature 870°C;
(c) Nominal analysis 0·4% C, 0·8% Mo. Solution temperature 870°C;
(d) Nominal analysis 0·3% C, 2·0% Mo. Solution temperature 1040°C

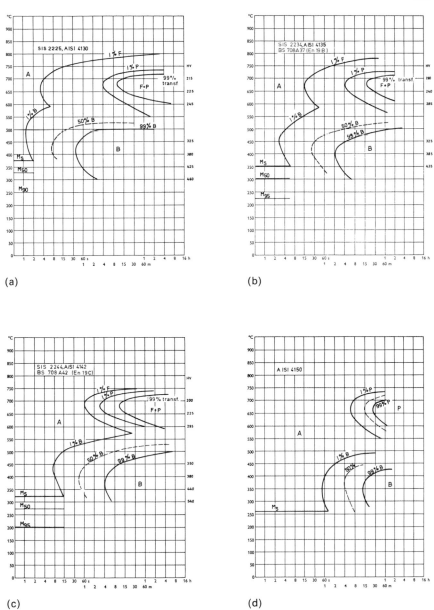

Figure 3.18 Cr–Mo-alloyed steels with different carbon content. Nominal analysis 0·8% Mn, 1·1% Cr, 0·2% Mo. Solution temperature 850°C. TTT diagram for isothermal transformation. (a) Carbon content 0·25%; (b) Carbon content 0·35%; (c) Carbon content 0·40%; (d) Carbon content 0·50%. (This heat has its content of alloying elements close to the upper limit of the composition range.)

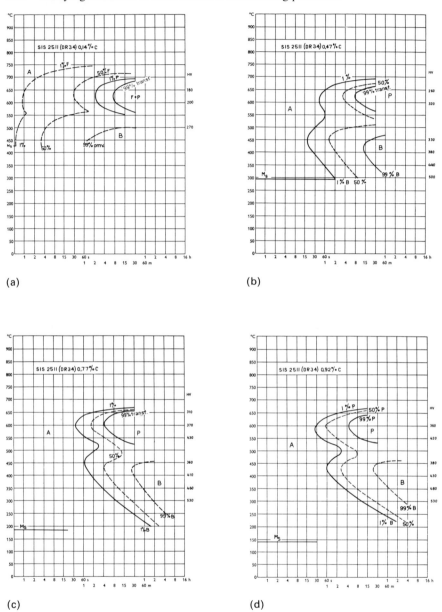

Figure 3.19 Cr–Ni-alloyed case-hardening steel with different carbon contents. Base analysis. (a) 0·14% C, 0·8% Mn, 0·8% Cr, 1% Ni. SS 2511 = B S 637M17 (En 352); (b) 0·47% C; (c) 0·77% C; (d) 0·92% C

3.1.12   Effect on resistance to tempering

Apart from some exceptions, all alloying elements enhance the ability of the steel to withstand loss of hardness on tempering after hardening. There is no simple relationship between resistance to tempering and alloy content

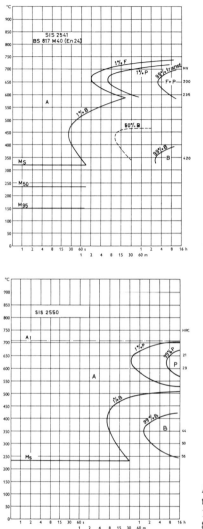

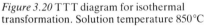

Figure 3.20 TTT diagram for isothermal transformation. Solution temperature 850°C

Figure 3.21 TTT diagram for isothermal transformation. Nominal analysis: 0·55% C, 1% Cr, 3% Ni, 0·3% Mo. Solution temperature 830°C

since there are so many factors involved. Jaffe and Gordon[8] have formulated a method of calculation which is applicable to low-alloy steels. According to this method the tempering temperature required to give a desired hardness in the steel may be calculated from its chemical composition. The method assumes that, after the hardening operation, the steel has a mainly martensitic structure. The original expression had the following form:

$$T = 30 (H_c - H_a)$$

where T = temperature °F
H_c = Rockwell hardness calculated from the analysis
H_a = Rockwell hardness desired after tempering

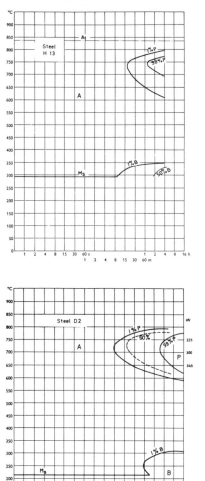

Figure 3.22 TTT diagram for isothermal transformation. Solution temperature 1025 °C

Figure 3.23 TTT diagram for isothermal transformation. Solution temperature 990 °C

Converting to degrees Celsius the expression becomes:

$$T = 16 \cdot 67 \, (H_c - H_a) - 17 \cdot 8$$

The various increments of H_c are obtained from the diagrams in *Figures 3.25* and *3.26*. The expression holds good for a tempering time of 4 h. The grain size of the steel also plays a part in the sense that a fine grain increases the resistance of the steel to softening according to the following scale:

Grain size, ASTM	4	6	8	10
Increment of H_c	0·6	0·9	1·2	1·5

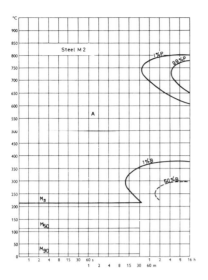

Figure 3.24 TTT diagram for isothermal transformation. Solution temperature 1230°C

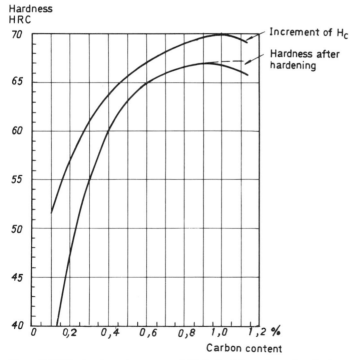

Figure 3.25 The hardness increment of H_c derived from the carbon in the steel: also the hardness as a function of the carbon content after hardening, subzero treatment and tempering at 100°C (after Jaffe and Gordon[8])

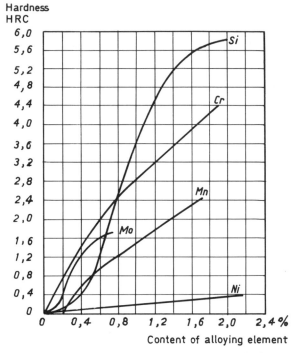

Figure 3.26 Increments of H_c derived from various alloying
elements (after Jaffe and Gordon[8])

Illustrative example

Which tempering temperature should be used when a hardness of 40 HRC
is required for steel SS 2541? Two heats with the following analyses are
concerned:

Heat No.	% C	Si	Mn	Cr	Ni	Mo
A	0·33	0·25	0·60	1·30	1·20	0·15
B	0·37	0·35	0·80	1·50	1·50	0·25

The grain size is ASTM 6 in both cases.
  According to the diagrams the following values of H_c are obtained:

Heat No. A	H_c	Heat No. B	H_c
0·33 C	62	0·37 C	63
0·25 Si	0·15	0·35 Si	0·30
0·60 Mn	0·95	0·80 Mn	1·20
1·30 Cr	3·30	1·50 Cr	3·70
1·20 Ni	0·20	1·50 Ni	0·30
0·15 Mo	0·20	0·25 Mo	0·60
6 ASTM	0·90	6 ASTM	0·90
Total	67·70	Total	70·00

On inserting these values in the expression we obtain:

A $T = 16·67 (67·7 - 40) - 17·8$ $T = 444 °C$
B $T = 16·67 (70   - 40) - 17·8$ $T = 482 °C$

These results are only approximate. For a more accurate calculation it is necessary to make a few adjustments. Those who require more details are referred to the original sources.

Grange, Hribal and Porter[9] have studied the influence of a number of alloying elements on the martensite hardness of steel containing 0·12%–0·97% C after tempering for 1 h at temperatures between 204 and 704 °C.

Pure iron–carbon alloys were hardened from 927 °C in brine and then tempered at various temperatures for 1 h. *Figure 3.27* shows the hardness after tempering as a function of the carbon content. The influence of the other alloying elements tested may be inferred from the diagrams in *Figure 3.28*, as follows. Choose a steel of some given chemical composition after being hardened to a martensitic structure and tempered. First read off from *Figure 3.27* the hardness due to the carbon content. To this value add the hardness increment ($ΔHV$) due to the element in question. This increment may be read off from the diagram representing the tempering temperature concerned.

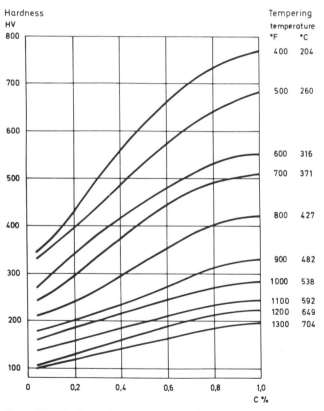

Figure 3.27 Hardness of tempered martensite of varying carbon content (after Grange, Hribal and Porter[9])

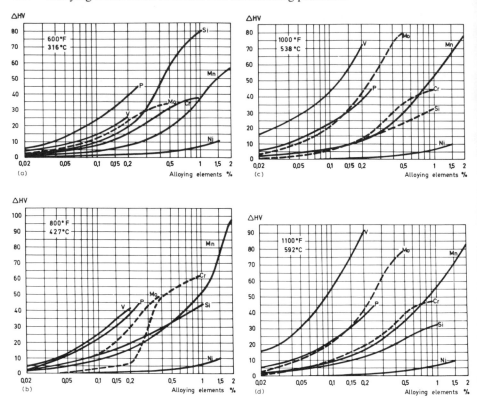

Figure 3.28 Effect of alloying elements on the hardness of martensite, tempered between 316 and 592 °C (600 and 1100 °F) (after Grange, Hribal and Porter[9])

According to this reckoning, steel SS 2541, heat A, tempered at 427 °C, should have a hardness of 427 HV = 43 HRC. This value agrees well with the calculation according to Jaffe and Gordon, which gives 40 HRC for the same steel after tempering at 444 °C. The original article[9] contains diagrams representing 10 tempering temperatures.

The computations described above are not suitable for calculating the tempered hardness of steels containing appreciable amounts of such alloying elements as give secondary hardening. As a result of systematic research[10] it has been possible to determine the influence exerted by such elements when present in fairly large amounts. In the case of Cr, for instance, it has been found that certain steels containing 2·5% of this element are rendered more resistant to tempering than those containing 5%. In other steels their resistance to tempering increases as Cr increases from 2·5 to 5%. *Figure 3.29* shows the influence of Mo, W and V on the resistance to tempering in steels containing 0·3% C and 2·5% Cr. The resistance to tempering is here represented as that temperature which, when applied for 2 h, tempers the steel to an ultimate tensile strength of 150 kp/mm^2 (1470 N/mm^2).

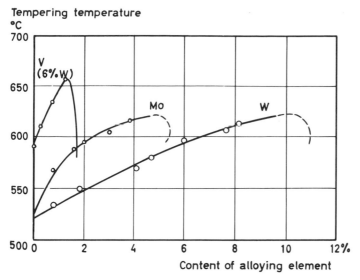

Tempering temperature °C

Figure 3 29 Influence of Mo, W and V on the resistance to tempering for a steel containing 0·3% C and 2·5% Cr. The resistance to tempering is that tempering temperature which, when applied for 2 h, tempers the steel to a UTS of 150 kp/mm^2 (1470 N/mm^2) (after Bungardt, Mülders and Lennartz[10])

3.2 Gases

Since the gases that normally occur in steel are harmful to it, the steelmaker aims at reducing the content of these gases below the critical threshold set for each type of steel. The concentrations are very low and are reported as thousandsths or ten-thousandths of one per cent. One ten-thousandth of one per cent is the same as one gram per tonne or one ppm (part per million). The unit ml/100 g is also used. Note that 1 ml hydrogen/100 g = 0·9 ppm.

3.2.1   Hydrogen

Rate of diffusion and solubility

Of all the elements hydrogen has the highest rate of diffusion in steel. The diffusion coefficients of hydrogen in α-iron and γ-iron, respectively, are given as:

$$D_\alpha = 2\cdot2 \times 10^{-3} \exp\left(-\frac{2900}{RT}\right) \text{cm}^2/\text{s}$$

$$D_\gamma = 1\cdot1 \times 10^{-2} \exp\left(-\frac{9950}{RT}\right) \text{cm}^2/\text{s}$$

The solubility of this element in the different modifications of iron, including molten iron, is shown in *Figure 3.30*. This difference between the solubilities in γ-iron and α-iron is a feature that is in a large measure responsible for certain specific defects in steel.

ml H/100g

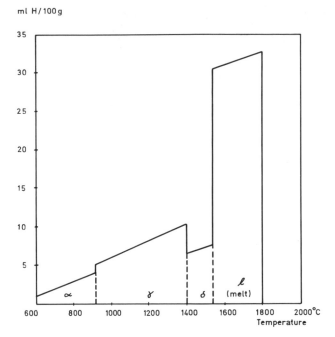

Figure 3.30 Solubility of hydrogen in pure iron (after Sieverts)

The hydrogen content immediately before tapping, of plain carbon or low-alloy steels made according to normal practice, is approximately as follows:

Acid open hearth	3–5 ppm
Electric-arc, double slag	4–8 ppm
Basic open hearth	5–9 ppm

As the steel cools the solubility of hydrogen decreases and the gas may separate in the molecular state, giving rise to pressure of several 1000 atmospheres which causes cracks in the steel. These defects are commonly known as flakes. *Figure 3.31* shows a fractured surface containing typical flakes. Flakes and pores can generally be discovered in the finished steel by means of ultrasonic inspection. The occurrence of flakes and thus the risk of failure increases with the dimensions of the steel.

Austenitic stainless steels of the 18/8 type are not as susceptible to the detrimental influence of hydrogen as are the heat-treatable steels. Contents of up to 12 ppm can be allowed in the former steels but plain-carbon and low-alloy steels are liable to develop flakes when the hydrogen concentration is as low as 3 ppm. For small sections up to 5 ppm may be tolerated. The figures cited apply to molten steel; in the finished steel the concentrations are considerably lower. A small amount of the hydrogen accumulates at the slag inclusions in the steel and thus plays no part in the formation of flakes. Hence a steel with a low slag-inclusion content cannot accommodate the same volume of hydrogen as a steel with a larger amount of slag inclusions. Therefore for such low-inclusion steels

Figure 3.31 Fractured surface showing flakes. Natural size

the hydrogen must be kept at a few ppm lower level in order to avoid flake formation.

Hydrogen removal

On account of the favourable rate of diffusion of hydrogen it is possible, by means of an annealing treatment, to remove that proportion of hydrogen that gives rise to the defects. *Figure 3.32* shows the amount of hydrogen evolved for various times up to 100 min and at different temperatures during the vacuum extraction of the steel bars. The amount of hydrogen evolution is a maximum at an annealing temperature of 600 °C. In actual practice the hydrogen anneal is carried out between 600 and 650 °C.

The amount of hydrogen given off during annealing may be calculated with the aid of the diagram in *Figure 3.33* in which U is set off as a function of T. The plotted values, obtained from a rotor forging during the process of annealing, show satisfactory agreement with the theoretically calculated curve[11].

$$U = \frac{\text{Remaining hydrogen content}}{\text{Original hydrogen content}}$$

$$T = \frac{Dt}{r^2}$$

where D = diffusion coefficient (cm^2/s)
       t = time of treatment   (seconds)
       r = diameter of sample   (cm)

The harmful effect of hydrogen may be diminished by allowing the steel to cool slowly after hot working or by subjecting it to an isothermal anneal.

Evolution of hydrogen
ml H / 100 g

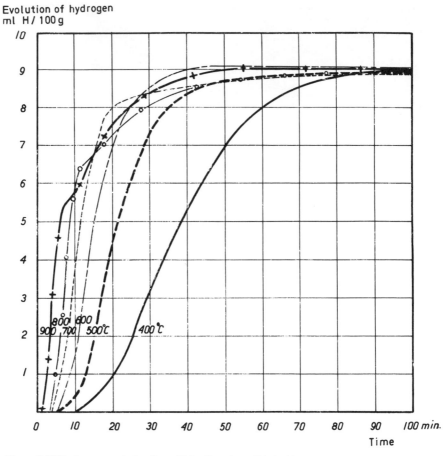

Figure 3.32 Hydrogen evolution from ½ in diameter mild steel bars
subjected to vacuum extraction (after Hobson[11])

3.2.2 Nitrogen

Solubility and rate of diffusion

The solubility of nitrogen in iron at various temperatures is shown in
Figure 3.34. The diffusion coefficient of nitrogen in α-iron is given as:

$$D_\alpha = 6.6 \times 10^{-3} \exp\left(-\frac{18\,600}{RT}\right) \text{cm}^2/\text{s}$$

It is seen from this expression that the rate of diffusion of N is somewhat
higher than that of C at temperatures below 400 °C. For γ-iron the
diffusion coefficient at 950 °C is given as $6.5 \times 10^{-8} \text{ cm}^2/\text{s}$. The more
intimate the contact between the molten steel and the atmosphere, the
greater is the opportunity for N to be absorbed by the steel. Hence it is not
surprising that steel grades produced by the Bessemer or Thomas processes

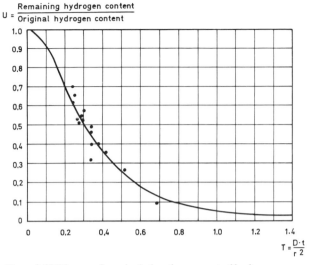

Figure 3.33 Diagram for calculating the amount of hydrogen
evolution from large forgings (after Hobson[11])

are the ones that contain the highest amounts of nitrogen. The simplest
way of reducing the nitrogen content is to prevent contact between the
steel and the atmosphere. LD and Kaldo furnaces which keep nitrogen in
the steel at a low level are operated with oxygen gas to oxidize the carbon
in the pig iron. The more modern bottom-blown converters, e.g. OBM
and AOD, are operated with oxygen and an oxygen–argon mixture,
respectively.

For the sake of completeness some typical values of nitrogen in steels,
produced by different processes, are given in *Table. 3.1.*

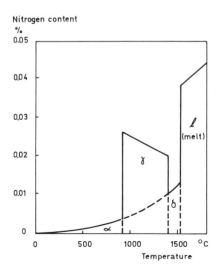

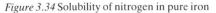

Figure 3.34 Solubility of nitrogen in pure iron

Table 3.1

Steelmaking processes	% N
Bessemer	0·014
Thomas	0·012
Electric arc, one slag	0·007
Electric arc, two slags	0·010
Basic open hearth	0·006
Acid open hearth	0·004
L D	0·004
Kaldo	0·003

Influence of nitrogen on the properties of steel

Nitrogen has both a beneficial and a harmful influence on the properties of steel. As an example of a beneficial influence may be mentioned the properties obtained on nitriding which usually takes place in ammonia gas at 510 °C. The nitrogen combines with the iron and certain alloying elements to form hard, wear-resisting nitrides. The extreme surface hardness is however attributable rather to the state of fine dispersion of the nitride particles than to the inherent hardness of the nitrides themselves. The practical details of the nitriding process are described in Chapter 6.

Another example of the beneficial effect of nitrogen is that it raises the yield point of, *inter alia*, stainless 18/8 steels. *See Figure 3.35*[12]. Low-carbon hardenable chromium steels containing more than about 0·010% N show an enhanced resistance to tempering which, however, is acquired at the expense of toughness[13].

Along with Al, V, Nb and Ti, nitrogen can function both as a grain refiner and as a precipitation-hardening agent. These effects are utilized in the micro-alloyed or HSLA steels. If a yield point of more than 500 N/mm^2 is required the steel must be alloyed with nitrogen up to about 0·015%.

The harmful influence of N begins to make itself felt when the concentration exceeds 0·003%, uncombined nitrogen[14]. Higher contents promote so-called strain-ageing which may lead to harmful effects. One result of strain-ageing is that the yield point of the material is raised if, prior to the test, it had been deformed beyond its original yield point and then aged at some temperature between room temperature and 300 °C. Strain-ageing is caused by an accumulation of N atoms at the dislocations. During the course of the first deformation the dislocations move away from the N atoms, but on subjecting the steel to ageing these atoms find their way again to the lattice defects and once in their new positions they are a contributory cause of impeding the movement of the dislocations on renewed deformation. Carbon produces an effect similar to that of nitrogen but it is most likely that nitrogen is responsible for the main part of the strain-ageing effect at low temperatures. Nitrogen produces this effect at about 100 °C and carbon at a somewhat higher temperature, e.g. 250 °C. Strain-ageing makes the steel brittle and the transition-temperature range, as manifested in impact testing, is displaced to higher temperatures, *Figure 3.36*. An impact test carried out in the

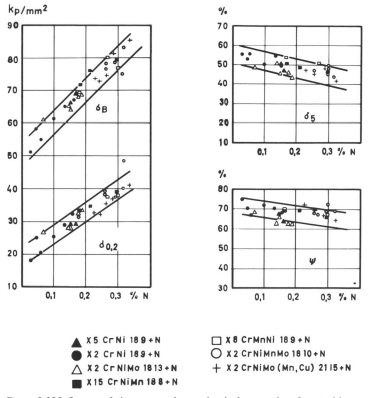

Figure 3.35 Influence of nitrogen on the mechanical properties of austenitic steels. Heat treatment: 30 min 1050°C/water quench (after Wessling, Kraatz and Bock[12])

low-impact region shows a brittle fracture. It is this type of fracture that was found in connexion with the catastrophic failures occurring in ships and bridges (and were then in the 1940s regarded as absolute mysteries).

The simplest way to reduce the dangerous influence of N is to kill the steel with Al which combines with the N to form nitrides. At the same time the steel is rendered less susceptible to grain growth, which in turn leads to a displacement of the transition-temperature range to lower temperatures.

The influence of nitrogen on ageing is discussed in some detail in Sections 5.6 and 5.7

3.2.3   Oxygen

Rate of diffusion

Oxygen has a very small diffusion coefficient, in fact it is the smallest one of the gases at present under consideration. For γ-iron the following value is given:

$$D_\gamma = 1 \cdot 2 \exp\left(-\frac{50\,000}{R\,T}\right) cm^2/s$$

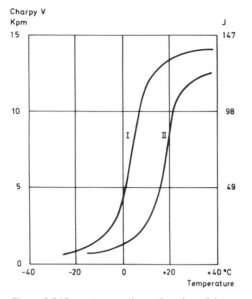

Figure 3.36 Impact strength as a function of the
testing temperature before (I) and after (II)
ageing (schematic diagram)

In unalloyed and low-alloy good-quality constructional steel the oxygen
concentration is about 60 ppm. By means of refining processes
(deoxidation) or vacuum degassing the oxygen can be reduced to about
30 ppm and by vacuum remelting to below 10 ppm.

Influence of oxygen on the finished steel

Investigations into the influence of oxygen on the mechanical properties of
pure iron have shown that in amounts up to saturation level it has no
noticeable effect on the ultimate tensile strength, yield point or hardness at
temperatures down to −73 °C. However, values of elongation and
reduction of area diminish rapidly as the oxygen increases and the
temperature falls. Further, impact strength is reduced and the
transition-temperature range is raised as oxygen increases.

Some harmful effects, e.g. poor surface finish and impaired
machinability, may be produced by otherwise acceptable oxygen contents
since some alloying elements form oxides that may remain in the steel as
non-metallic inclusions. The amount of micro-inclusions, which may be
estimated on ground and polished samples in the metallurgical microscope,
is an approximate measure of the oxygen content. On the other hand,
macro-inclusions derive generally from some local oxidation in connexion
with tapping and are not directly related to the mean oxygen content of the
steel. The amount of macro-inclusions may be estimated with reference to
various scales, one of which is shown in *Figure 3.37*. The samples, in the

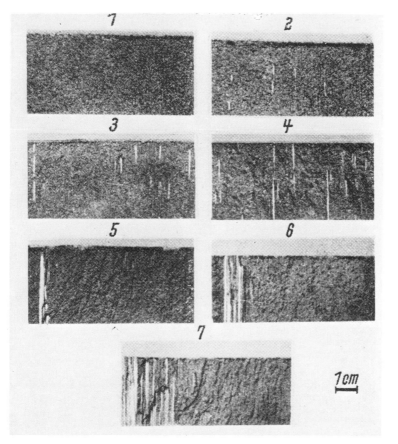

Figure 3.37 Example of a slag scale for estimating the macro-inclusions from
the inspection of a blue-tempered fractured surface (after Hoff)

shape of discs, are hardened, tempered and then fractured and
blue-tempered which reveals the inclusions very clearly. Alternatively the
discs may simply be heated in the blue-brittle range, 300–350 °C, and
broken in two at this temperature since less energy is then required. Should
the colour not develop, a blue-tempering treatment may be required after
fracturing. The amount of macroslag may also be estimated by inspecting
bars subjected to the step-down test method, i.e. by machining them down
to consecutively smaller diameters.

Macro methods for assessing slag inclusions in steel are described in SS
11 11 10, which is based on proposition ISO/TC 17N 1348 as a tentative
international standard. The amount and types of macro-inclusions may be
assessed from Jernkontoret's Slag Inclusion Chart, the 'JK' chart
(*Figure 3.38*), which has gained international recognition. Compare also
ASTM Designation E 45: *Determining Inclusion Content of Steel*, and
ISO 4967–1979 *Steel Determination of Non-metallic inclusions
—Micrographic Method Using Standard Diagrams.*

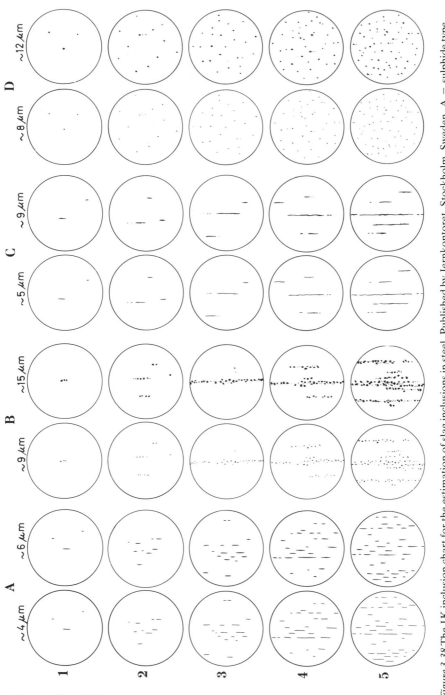

Figure 3.38 The J K inclusion chart for the estimation of slag inclusions in steel. Published by Jernkontoret, Stockholm, Sweden. A = sulphide type. B = alumina type, C = silicate type, D = globular type oxides. The thickness (μm) indicated are only arbitrary examples

3.3 New steelmaking processes

For the production of high-quality steel based on scrap the basic electric-arc furnace has been the predominant unit. The furnace time is divided into a melt-down period, an oxidizing and dephosphorizing period and finally a deoxidizing and desulphurizing period. Since the introduction of oxygen lancing the oxidizing and dephosphorizing periods have been shortened. On the other hand it has not been found possible to obtain a more economic or efficient oxygen and sulphur removal in the electric-arc furnace than the ordinary deoxidizing and desulphurizing practice. As a consequence there has been intensive research during the last decade into steelmaking practices in order to develop steel production methods that will yield steel of high cleanness. The primary aim is to reduce the oxygen and sulphur contents, since oxides and sulphides have a deleterious influence on the properties of steel, including surface finish, impact strength and fatigue strength. During the 1960s and 1970s several such methods were developed. These newer steelmaking practices have had a significant influence on the properties of steel and therefore a brief account of these methods is given in the following.

Concurrently a number of ore-based processes in which the furnaces are charged with hot metal have gained increased importance, notably the LD process, due not least to its adaptability to sequence casting during the continuous-casting operation. Several of the new steel-cleansing methods are found to fit in well with the LD practice and as a result the production of high-grade quality steel can be based on this process.

3.3.1   Vacuum remelting

Vacuum-melted steel is produced in a specially designed high-frequency induction furnace under a reduced pressure of about 10^{-3} torr (1 torr = 1 mm Hg) or by melting conventional air-melted steel in a special electric-arc furnace at a reduced pressure of about 10^{-3} torr. *Figure 3.39* illustrates schematically a vacuum electric-arc remelting furnace (also called consumable-electrode furnace). A heavy-section air-melted steel electrode is progressively melted down in an electric arc, while the molten steel is allowed to solidify in a water-cooled copper mould. The reduced pressure promotes certain metallurgical reactions which, for the remelted steel, result in a large reduction in the contents of hydrogen and oxygen, and also nitrogen to some extent. Vacuum-remelting is applied to steels and alloys intended for elevated-temperature service as well as to constructional and tool steels for which a high degree of cleanness is stipulated. The quality of the steel is enhanced in several respects, in particular the mechanical properties in the transverse direction, i.e. at right angles to the direction of rolling or forging. The increased cleanness of this steel is conducive to its being stipulated for parts requiring surfaces free from non-metallic inclusions, for example parts to be highly polished or electroplated. Its high fatigue properties make it attractive as material for ball-bearings, springs, rolls for cold rolling, etc. The remelting and re-solidification of steel in the vacuum electric-arc furnace result in a reduced amount of segregation. Hence, subsequent operations involving

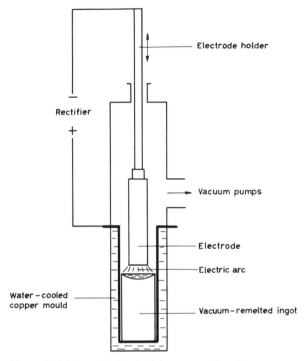

Figure 3.39 Schematic sketch of the vacuum electric-arc
remelting furnace at the Bofors company

rolling or forging are facilitated which is of especial advantage for steels
and alloys intended for elevated-temperature service. For such
vacuum-remelted steels there are additional advantages, e.g. enhanced
creep properties and reduced notch sensitivity. *Figures 3.40a–f* show
examples of improvements obtained as a result of changing from
air-melted to vacuum arc remelted steel.

3.3.2   Electroslag refining

This process is called the ESR process (Electroslag Refining or Remelting)
in English-speaking countries and also in Sweden. The German
designation is ESU (Elektroschlacke-Umschmelzung). The method was
developed during the 1960s and towards the end of this period some
400000 tonnes per annum were produced in the USSR where the process
had reached its greatest development.

The principle of ESR is illustrated in *Figure 3.41*. A consumable steel
electrode dips into a hot slag bath and melts progressively. An ingot is
thereby built up in a water-cooled mould. The method has certain
similarities to vacuum arc remelting (VAR) but a distinctive difference is
that in the ESR process the metallurgical reaction involved in purifying the
melting electrode takes place in the slag bath, first between this and the
thin liquid-steel film on the electrode tip and then as the steel droplets drip
through the slag. The purifying reaction in the VAR process takes place as

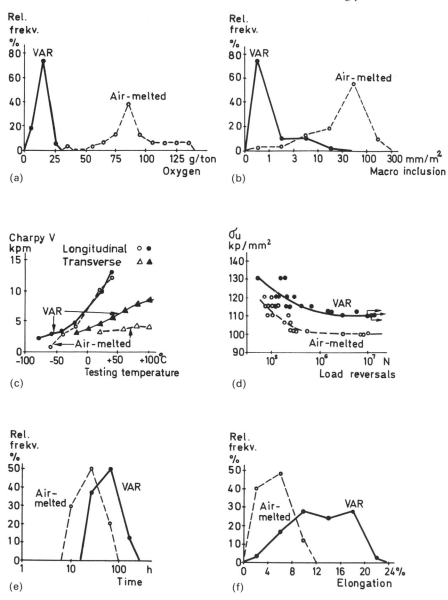

Figure 3.40 Comparison between the properties of air-melted steel and vacuum electric-arc remelted (VAR) steel. (a) Oxygen content of high-temperature 13% Cr steel; (b) Macro-inclusion content; (c) Impact strength of Cr–Ni–Mo steel BS 817M40 (En 24), Dimension 100 mm ϕ. UTS 930 N/mm^2; (d) Wöhler diagram for a Cr–Ni–Mo steel. Yield point 520 N/mm^2. UTS 1600 N/mm^2; (e) Time to creep rupture. High-temperature steel A 286; (f) Creep rupture elongation of high-temperature steel A 286

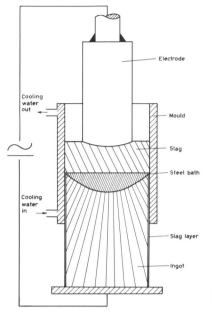

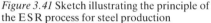

Figure 3.41 Sketch illustrating the principle of the ESR process for steel production

a result of the vacuum. In the ESR process the heat necessary to melt the electrode is created in the liquid slag by its acting as an electric resistance to the current passing through it. In the case of the VAR process free electric arcs are struck in the vacuum between the electrode and the ingot. Direct current is necessary for this process whereas the ESR can use either dc or ac, the latter being preferable.

The slag is usually made up of the three ingredients: quicklime (CaO), fluorspar (CaF_2) and alumina (Al_2O_3). The lime is present mainly to remove sulphur and oxygen from the steel; the fluorspar lowers the melting point of the slag and increases its fluidity; the alumina increases the electric resistance of the slag which in turn causes its temperature to rise and thereby increases the melting rate of the steel electrode. The relative proportions of the constituent ingredients are generally a compromise between cleaning effect, melting rate and ingot surface. A standard composition is 60% CaF_2, 20% CaO and 20% Al_2O_3 but the optimum composition varies widely for different grades of steel.

By optimizing conditions so that an acceptable melting rate, ingot structure and surface are obtained it is possible to effect, for most grades of steel, a removal of 50–70% of the sulphur and about 50% of the oxygen. Hereby the amounts of sulphide and oxide inclusions respectively are reduced in direct proportion in the remelted steel. However, the hydrogen and nitrogen contents are not changed by the ESR process. The evenly fine-grained structure of ESR ingots ensures that a smaller forging reduction can be accepted than would be the case for conventionally teemed ingots. Thanks to the absence of pipe a considerably greater yield is obtained.

Tests carried out with ESR steel have shown that, compared with air-melted steel, noteworthy improvements in physical properties are obtained, in particular fatigue properties, impact strength and ductility. The improvements have been most striking in transverse samples taken from central areas. In comparison with air-melted steel the degree of scatter in the values obtained has in some instances diminished appreciably. *See Figure 6.47*

3.3.3   Vacuum degassing

Vacuum degassing can be carried out by several methods. In principle the molten steel is subjected to a reduced pressure of about 1 torr (1 mm Hg), either while it is being tapped from the furnace or after completed tapping. When ladle degassing, for example, the ladle is placed in a vacuum chamber or alternatively, the ladle is fitted with a vacuum-tight cover and the degassing is started. The degassing unit used at Bofors (ASEA–SKF method) consists of a ladle-furnace which, during the actual degassing operation, is placed inside an induction coil stand in order to set up a stirring action in the melt. This in turn facilitates gas evolution and also gives an even distribution of the alloying elements in the melt. When the chemical composition has been adjusted the steel can be reheated by means of the electric-arc electrodes (*see Figure 3.42*). The degassing process removes about one half of the amount of dissolved gases, which

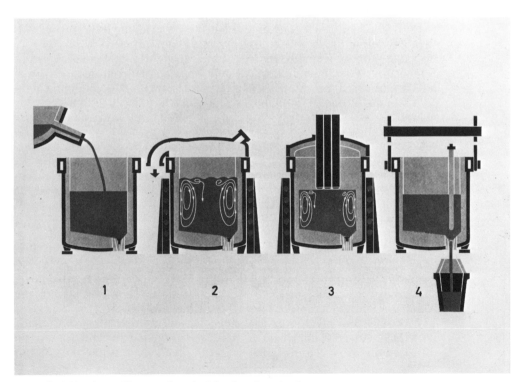

Figure 3.42 Sketches to illustrate the principle of steel production sequence according to the ASEA–SKF method. (1) Tapping the molten steel into the ladle-furnace. (2) Degassing. (3) Reheating. (4) Teeming

implies that vacuum degassing produces roughly the same cleanness as deoxidized and hydrogen-annealed steel, i.e. steel containing about 30 ppm of oxygen and 2·5 ppm of hydrogen.

3.3.4   The Bofors method of sulphur removal

When processing steel by the ASEA–SKF method, no sulphur is removed in the ladle-furnace. When it is stipulated that the steel shall have a low sulphur content it must be desulphurized before it is transferred to the ladle-furnace. In order to develop a method whereby desulphurization could be carried out in the degassing unit, Bofors instituted a joint development project with ASEA, the Royal Institute of Technology in Stockholm (KTH) and the Metallurgical Research Station at Luleå. Based on the results obtained from this investigation the first trial melts were carried out in the Company's 50-ton capacity ladle-furnace[15]. The outcome was the Bofors method of sulphur removal.

The desulphurization process was primarily developed to operate in conjunction with the ASEA–SKF process but it is adaptable to other processes which incorporate a strongly basic fluid slag and good stirring facilities. In principle the method consists of obtaining an initial S removal in the carefully deoxidized steel by means of a strongly basic slag, using an ASEA–SKF ladle-furnace. Then, if required, the S content may be reduced still further by the addition of some agent that has a strong affinity for S, e.g. Ca or Ce. It is essential that the oxygen activity in the steel be kept low during the whole course of the desulphurizing process and right up to the point of teeming. By means of a strongly basic slag alone, a 50% S reduction is obtainable, which is quite satisfactory in most cases. By an addition of misch metal (an alloy of the rare-earth metals containing about 53% Ce) the S content of unalloyed steel may be reduced to extremely low values, e.g. below 0·001%. Alloy constructional steels are available with S below 0·003%. The mechanical properties are enhanced by a reduction of the S content, in particular ductility, impact strength and fatigue strength in the transverse direction of the steel. This resulting diminished anisotropy is due to the very large increase in the cleanness of the steel and also to the modification of the sulphide inclusions as a result of the treatment with Ca or Ce.

3.3.5   Injection metallurgy

Some of the previously mentioned desulphurizing processes are rather costly but they may still be worth their price when applied to highly alloyed structural steels and tool steels since, in addition to the desulphurizing action, they produce a more homogenous material.

With regard to plain structural steels as well as high-strength weldable ones (HSLA steels) improved properties are in rising demand, e.g. for the off-shore industry and for gas and oil pipelines for low-temperature service[16].

In order to be able to treat the large tonnages involved the comparatively simple yet effectual process of injection metallurgy was developed towards the end of the 1960s and the beginning of the 1970s.

Deoxidation is brought about by injecting calcium compounds (CaSi, CaC_2) by means of a carrier gas (usually argon) into the molten bath which is covered with a basic reducing slag to which Al has been added. The powder injection is maintained for some 5 to 10 minutes, using from 1 to 5 kg of powder per tonne of steel. The bath being vigorously stirred, the chemical reactions take place very rapidly. The oxygen concentration is reduced to about the Al–O equilibrium concentration. The sulphur content is reduced by at least 50%.

When very low sulphur contents ($<0.003\%$) are stipulated the starting heat must have a relatively low sulphur content. For ore-based steel this is most simply attained by injection-treating the hot metal.

By treating the steel with calcium it is also possible to prevent the formation of Type II manganese sulphides (*see* Section 3.3.6), thereby enhancing the mechanical properties of the steel in the transverse direction. In addition, the hard abrasive oxides of type Al_2O_3 are converted to $CaO \cdot Al_2O_3 \cdot SiO_2$ inclusions which will give reduced wear when machining with high-speed steel-cutting tools. A further reduction in the abrasive wear properties is obtained if the inclusions are enveloped in a softer mantle of (Ca, Mn)S. The calcium treatment also gives rise to inclusions which will facilitate the formation of an oxide film, when cutting speeds are employed, between the sinter-carbide cutting tool and the workpiece. This oxide film will restrict diffusion and thus reduce tool wear.

The above-mentioned inclusions are mainly globular in shape and are decidedly harder than ordinary manganese sulphide. They are not deformed during hot working and as a result the transverse properties of the steel are not impaired, which they would be if there were manganese sulphides present since they are elongated in the rolling direction during the hot working.

Al_2O_3 inclusions form clusters that easily get lodged in the pouring nozzle and thereby hinder the pouring or may even render pouring impossible through small nozzles. The calcium aluminates, on the other hand, pass through the nozzles without blocking them. Thereby the maintenance costs in the steel mill are reduced and in addition Al-killed steel can be cast continuously in dimensions about 150 mm square and even smaller. The modification process of MnS by means of the calcium treatment is very complicated since besides other factors the oxygen concentration in the steel plays some part in the process. In order to obtain all the above-mentioned improvements, further research is necessary to determine suitable processing methods; also further investigation into the shape and composition of the inclusions is required.

Injection treatment to electric-furnace steel cuts out the time-consuming reduction process in the arc furnace since the refining step made possible by injection will take place in the ladle. By doing so, higher productivity in the arc furnace is established, as well as an improved recovery on ferro alloys. Injection treatments are also very favourable for open hearth steels, and in fact, it opens a new improved quality product programme for these types of furnaces.

During the 1970s the technique of injection treatment attracted a great deal of attention and a considerable amount of research work has been carried out at Jernkontorets (the Swedish Ironmasters Association) research station at Luleå. Three international conferences have been held there, viz. Scaninject I[17], Scaninject II[18] and Scaninject III[19].

As an example of the methods of injection metallurgy may be mentioned the TN-process (Thyssen-Niederrhein, W. Germany), *Figure 3.43*. Another is the Scandinavian Lancers, Sweden, *Figure 3.44*. These processes operate more or less along the same lines and differ principally in the location of the dispensers. The latest units are fitted with several dispensers, thereby enabling the steel to be alloyed with several elements at the same time as the CaSi injection is taking place.

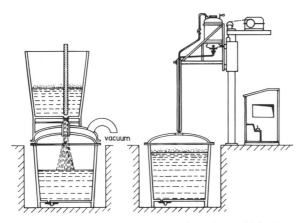

Figure 3.43 Arrangement for vacuum degassing and injection of powdered materials of Thyssen-Niederrhein AG

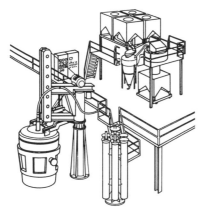

Figure 3.44 Scandinavian Lancers multi-container injection system

3.3.6   Effect of sulphur on the properties of steel

The above heading really should come under the main heading 'Solids' but since the subject matter is based on some of the results of the desulphurizing methods just described it has been included along with the descriptions of these methods instead.

Steel always contains large numbers of non-metallic inclusions which, depending on their shape, size and composition, are able to influence its properties in various degrees, Manganese sulphides of varying composition are common inclusions in steel. The amount of sulphides is predetermined by the sulphur content of the steel whereas their shape is influenced very largely by the deoxidation practice and by the composition of the steel.

Sims[20,21] has put forward the following sulphide classification:

Type I   Randomly dispersed globular sulphides; individual inclusions sometimes associated with an oxygen-rich second phase. *Figure 3.45a*

Type II   Fan or chain-like distribution of fine inclusions usually described as eutectoid and confined to the interdentritic regions. *Figure 3.45b*

Type III Angular inclusions, usually randomly dispersed. *Figure 3.45c*

Baker and Charles[22] have carried out detailed studies of the morphology of the manganese sulphides.

By means of a special technique Fredriksson and Hillert[23] have studied the separation of sulphides, which led to their discovering a fourth type. Also, after deep-etching the specimens they studied the sulphides in the scanning electron microscope (SEM). *Figure 3.45b* shows an example of Type II.

It is generally assumed that the non-metallic inclusions will impair the mechanical properties of steel, in particular fatigue strength, impact strength and ductility[24]. On the other hand, an increase in the sulphur content has a favourable effect on machinability[25]. See *Figure 3.46*. Free-cutting steels contain about 0·2% sulphur.

The tensile strength of steel is scarcely affected by the sulphur content but, as seen in *Table 3.2,* the values of elongation and reduction of area are improved as the sulphur is reduced. This is particularly noticeable in transverse specimens.

Figure 3.47 shows how the sulphur content in a conventional steel of type 24 CrMo 4 affects the fatigue strength (alternating stress: tension–compression) in the transverse direction. Even though the degree of scatter is considerable the improvement registered for the lowest sulphur contents is unmistakable[26].

Table 3.2 Mechanical properties of normalized steel SS 1650 (C 50). Specimen 150 mm square, L = longitudinal, T = transverse
(Mean values of test results obtained from samples taken from top, middle and bottom of ingot)

Sulphur content %	Position of specimen	Yield point kp/mm²	Ultimate tensile strength kp/mm²	Elongation L = 5xd = 6.64√4 %	Reduction of area %
<0·001	L	38	67	24	52
	T	38	67	23	47
0·016	L	39	68	23	52
	T	39	69	20	36
0·032	L	37	68	22	44
	T	36	64	10	11

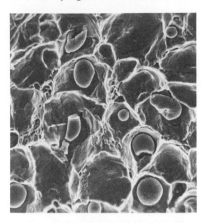

(a) Type I sulphides.
Fractured surface of casting. 200 ×.
(After Baker and Charles[22])

(b) Type II sulphides.
Normal monotectic colony. Globules
formed at ends of Mn S rods. 300 ×.
(After Fredriksson and Hillert[23])

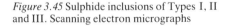

(c) Type III sulphides.
Fractured surface from casting. 300 ×.
(After Baker and Charles[22])

Figure 3.45 Sulphide inclusions of Types I, II
and III. Scanning electron micrographs

It is fairly well known that the impact strength increases with decreasing
sulphur content. This is clearly shown in *Figure 3.48* and *Figure 3.49*,
which apply to a hardened and tempered steel of composition

C	Si	Mn	P	Ni	Cr	Mo	%
0·30	0·25	0·80	0·020	2·5	0·75	0·45	

S varying between 0·005 and 0·179%[26]

For very low sulphur contents even small differences produce changes in impact strength; see *Figure 3.50*[28].

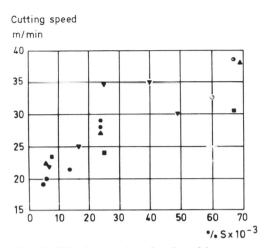

Figure 3.46 Cutting speed as a function of the sulphur content. Tools of high-speed steel used for machining steel 41Cr4, heat treated to 85–90 kp/mm². Different symbols represent different heats (from Randak, Knorr and Vöge[25])

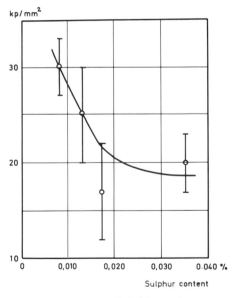

Figure 3.47 Endurance limit (alternating stress: tension–compression) in transverse direction v. varying sulphur contents of steel. SS 2225 heat treated to 250–290 HB. Nominal analysis 0·28% C, 1% Cr, 0·25% Mo (after Arwidson[26])

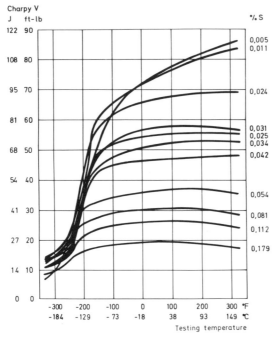

Figure 3.48 Transition curves for longitudinal specimens from ½ in plate, no cross-rolling and heat treated to 30 HRC (after Hodge, Frazier and Boulger[27])

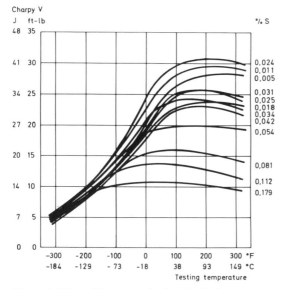

Figure 3.49 Transition curves for longitudinal specimens from ½ in plate with no cross-rolling and heat treated to 40 HRC (after Hodge, Frazier and Boulger[27])

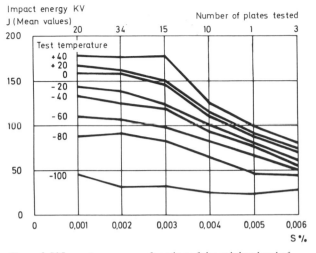

Figure 3.50 Impact energy as a function of the sulphur level of steel, grade A PI X 70 CC. $R_e > 500$ N/mm². Plate thickness 16 mm (after Haastert, Maas and Richter[28])

Figure 3.51 shows typical impact-strength curves for a martensitic stainless 12% Cr steel. A reduction in the sulphur content has resulted in a marked improvement in the longitudinal impact strength which would imply a displacement of the transition-temperature range by about 40 °C towards lower temperatures. At the same time there is an increase in the absolute values of the impact strength in the upper range. The impact strength of steel SS 2225, hardened and tempered, is shown in *Figure 3.52*. A reduction of sulphur from 0·012 to 0·005% has not affected the longitudinal impact strength but the transverse impact value has been practically doubled. *Figure 3.53* shows how the transverse impact strength of a degassed 0·35% C Cr–Ni–Mo steel, hardened and tempered increases as the S is reduced.

As is apparent from *Figure 3.54* the fracture toughness is also affected by the sulphur content[29].

Sulphur is used as an alloying element to improve the machinability of a large number of structural steels. Therefore it might be expected that a reduction in the content should impair machinability. The results of some machining tests are reproduced in *Table 3.3* in which the

Table 3.3 Data obtained from machining tests of normalized steel. SS 1650 (C 50) and SS 2225 (25 CrMo 4)

Sulphur content %	SS steel	Hardness HB	Drilling High-speed steel B_{BS} m/min	Turning High-speed steel B_{SS} m/min	Cemented carbides B_{SH} m/min
<0·001	1650	195	40	32	140
0·016	1650	197	48	32	135
0·032	1650	198	48	33	140
0·005	2225	229	18	33	160
0·012	2225	241	20	32	155

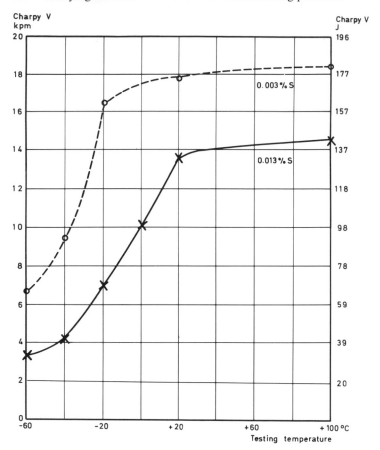

Figure 3.51 Impact strength of steel SS 2303, BS 420S37 (En 56C) with different sulphur contents. Longitudinal specimen, 20 mm diameter heat treated to 235 HB (After Arwidson[26])

machinability-index figures refer to the standard test method used by Bofors. B_{BS} and B_{SM} denote the cutting speeds for a tool life of 30 minutes and B_{SS} for a tool life of 45 minutes. For the steel concerned the results indicate that whereas the machinability by drilling is somewhat impaired as the sulphur is reduced the machinability by lathe turning is not affected appreciably.

When the sulphur content is brought down by means of calcium and cerium the abrasive oxidic inclusions are partly reduced in number and partly modified by the process on page 130. This result compensates the unfavourable effect of the reduced sulphur content on machinability.

3.3.7   Continuous casting

It is estimated that for 1983 about 35% of the total world output of steel will be produced by continuous casting and that this upward production trend is not likely to diminish for the rest of the 1980s. About 80% of the

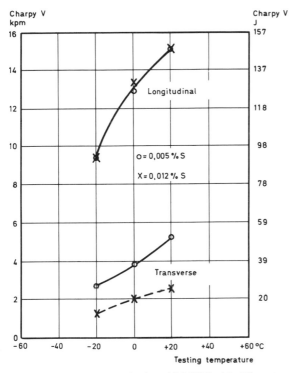

Figure 3.52 Impact strength of steel S S 2225 with different sulphur contents. Specimen 100 mm diameter. Heat treated to 222 H B. Longitudinal and transverse specimens. Nominal analysis 0·28% C, 1% Cr, 0·25% Mo (after Arwidson[26])

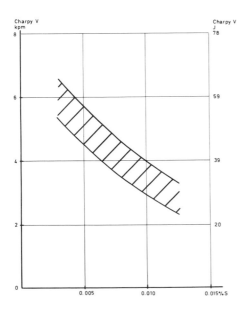

Figure 3.53 Impact strength of transverse specimens, hardened and tempered to 350 H B, of steel containing 0·35% C, 1·4% Cr, 3% Ni, 0·4% Mo and varying sulphur contents. Testing temperature −40°C (after Arwidson[26])

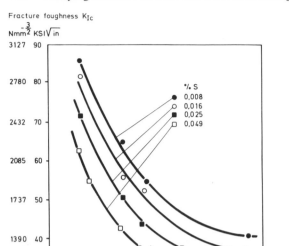

Figure 3.54 Influence of sulphur level on plane strain fracture toughness of A I S I steel (after Birkle, Wei and Pellessier[29])

Japanese steel production in 1982 was by continuous casting. Up till 1980 mainly simple straightforward steel grades were made by this process but progress has been rapid throughout the following years. Since the beginning of the 1980s high-grade structural steels and spring steels are being produced by continuous casting and trials involving tool steels are under way. Some doubt as regards continous-cast steel has prevailed, and perhaps rightly so, and is still widespread in many quarters. This section aims at informing steel users and heat treaters about continuous casting in general and about the latest improvement in this field.

The process of continuous casting begins by first transferring the steel from the ladle to a tundish whence it is teemed into water-cooled copper moulds in which it solidifies on the surface. The steel is continuously drawn out into a strand. In the case of vertical casting the strand is ordinarily bent into a large-radius arc to emerge horizontally onto a cooling bed. *Figure 3.55* shows in outline the principle of continuous casting for two somewhat different procedures. Trials with horizontal continuous casting have taken place since the beginning of the 1980s.

The advantages of continuous casting compared with conventional ingot casting are very great. The reduction of ingots to billets is eliminated since it is possible to cast direct to required billet size. Besides, the yield is greater.

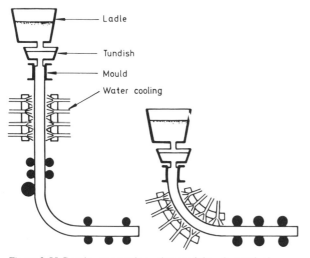

Figure 3.55 Continuous casting using straight, alternatively
bent, moulded

Although there have been many mechanical and metallurgical problems
with continuous casting, they have been resolved at a steady pace. One of
the difficulties encountered during teeming and casting was oxygen
absorption by the steel streams, which created large macroslag inclusions
both at the surface and in the body of the steel. By providing protection for

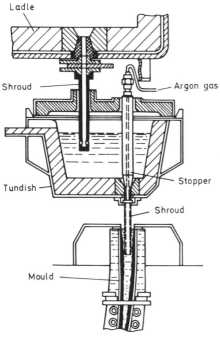

Figure 3.56 Continuous casting with
arrangement for protection of pouring stream

the teeming and casting streams it is now possible to make continuous-cast steel without objectionable macroslag inclusions. Besides, the feasibility of producing steel containing strong oxygen-binding agents such as Al and B is greatly enhanced.

The principle of protected pouring is shown in *Figure 3.56*. An immersion tube is situated between the ladle and the tundish and there is another one between the tundish and the mould.

When pouring into moulds about 150 mm square and smaller the Al-content in the ladle must not be more than about 0·010%, otherwise the steel will 'freeze' in the ladle and tundish nozzles owing to the accumulation there of sticky Al-compounds. By injecting calcium silicide, as described in previous sections, freely flowing calcium aluminates are formed which do not stick in the nozzles. This enables steels alloyed with Al to be continuous-cast into moulds about 150 mm square.

The production of alloy steels for high-duty service by continuous casting involves, besides stringent requirements of structural stability for the equipment, also exacting demands of temperature control and cooling during the whole process. In addition, a rather close inspection of billets and finished steel is called for with regard to macroslag and mechanical properties. Intensive research work is currently being carried out in these fields. Reliable methods of inspection have been introduced and confirmed as normal routine operations.

References

1. HOUDREMONT, E., *Handbuch der Sonderstähle*, Springer-Verlag (1956) (in German)
2. RAPATZ, F., *Die Edelstähle*, Springer-Verlag (1962) (in German)
3. *De Ferri Metallographia*, Part I, HABRAKEN L. and DE BROUWER, J., *Fundamentals of Metallography*, Presses Academiques Européenes S. C. Bruxelles (1966). Part II, SCHRADER A., and ROSE, A., *Structure of Steels*, Verlag Stahleisen m.bH., Düsseldorf (1966). Distributed by W. B. Saunders Company, Philadelphia and London
4. BAIN, E. C., and PAXTON, H. P., *Alloying Elements in Steel*, ASM (1961)
5. IRVINE, K. J., PICKERING, F. B., and GLADMAN, T., 'Grain-refined C-Mn-Steels', *JISI*, 161–182 (February 1967)
6. STEVENS, W., and HAYNES, A. G., 'The Temperature of Formation of Martensite and Bainite in Low-alloy Steels', *J. Iron & Steel Inst.*, **183**, 349–359 (1956)
7. STUHLMANN, W., 'What the TTT-diagrams Tell Us', *Härterei-Techn. Mitt.*, **6**, 31–48 (1954) (in German)
8. JAFFE, L. D., and GORDON, E., 'Temperability of Steels', *Trans ASTM*, **49** (1957)
9. GRANGE, R. A., HRIBAL, C. R., and PORTER, L. F., 'Hardness of Tempered Martensite in Carbon and Low-alloy Steels, *Met. Trans.*, **8A**, 1775–1785 (1977)
10. BUNGARDT, K., MÜLDERS, O., and LENNARTZ, G., 'Influence of Chromium, Molybdenum, Tungsten and Vanadium on the Properties of Steels Containing ⁻0·3% C', *Arch. Eisenhüttenwes.* **32**, 823–891 (1961) (in German)
11. HOBSON, I. D., 'The Removal of Hydrogen by Diffusion from Large Masses of Steel', *J. Iron & Steel Inst.*, **191**, 342–352 (1959)
12. WESSLING, W., KRAATZ, D., and BOCK, H. E., 'Nitrogen-alloyed High-strength Chrome–Nickel Steels for the Chemical Industry', *Molybdän–Dienst.* **69** (November 1970) (in German)
13. IRVINE, K. J., and PICKERING, F. B., 'High-strength 12%-Chromium Steels' *High Alloy Steels*, ISI Spec. Rep. **86**, 34–48 (1964)
14. LANNER, C., 'The Binding Action of Micro Additions for Nitrogen', *Jernkont. Ann.*, **153**, 277–285 (1969) (in Swedish)

15. CARLSSON, L. E., GREVILLIUS, N. F., and HELLNER, L., 'The Bofors Method for the Desulphurizing of Steel with the ASEA–SKF Process', *ASEA J.*, **43**, No. 5. 98–100 (1970)
16. GRAY, J. M., and WILSON, W. G., 'Molycorps Develops X–80 Arctic Pipeline Steel', *Pipeline and Gas J.*, 50 (December 1972)
17. *Scaninject,* International Conference on Injection Metallurgy, Luleå, Sweden (1977), Jernkontoret, Stockholm
18. *Scaninject II,* 2nd International Conference on Injection Metallurgy, Luleå, Sweden (1980), Jernkontoret, Stockholm
19. *Scaninject III,* 3rd International Conference on Refining of Iron and Steel by Powder Injection, Luleå, Sweden (1983), Jernkontoret, Stockholm
20. SIMS, C. E., and DALE, F. B., 'Effect of Aluminium on the Properties of Medium Carbon Cast Steel', *Trans. Am. Foundrymen's Ass.,* **46**, 65–132 (1938)
21. SIMS, C. E., 'The Non-Metallic Constituents of Steel', *Trans. AIME,* **215** 367–393 (1959)
22. BAKER, T. J., and CHARLES, J. A., 'Morphology of Manganese Sulphide in Steel', *JISI,* 702–706 (September 1972)
23. FREDRIKSSON, H., and HILLERT, M., 'On the Formation of Manganese Sulphide Inclusions in Steel', *Scand. J. Metallurgy,* **2**, 125–145 (1973)
24. DAHL, W., HENGSTENBERG, H., and DÜREN, C., 'Behaviour of the Different Types of Sulphides During Shaping and their Effect on the Mechanical Properties', *Stahl. u. Eisen,* **86**, 796–817 (1966) (in German)
25. RANDAK, A., KNORR, W., and VÖGE, H., 'Some Metallurgical Possibilities of Influencing the Machinability of High-grade and Special Constructional Steels', International Symposium on Influence of Metallurgy on Machinability of Steel, Tokyo, Japan, The Iron and Steel Institute of Japan, American Society for Metals (September 26–28, 1977)
26. ARWIDSON, S., 'The Importance of Extra-low Sulphur in Steel in Relation to Mechanical Properties', *Scandinavian J. Met.,* **1**, 167–170 (1972)
27. HODGE, J. M., FRAZIER, R. H., and BOULGER, F. W., The Effects of Sulphur on the Notch Toughness of Heat-treated Steels', *Trans. AIME,* **215**, 745–753 (1959)
28. HAASTERT, H. P., MAAS, H., and RICHTER, H., 'Application of the TN-Process for Steels with Special Toughness Properties', *Scaninject II* (1980) 26:1–26:13 and *Stahl. u. Eisen,* **100**, No. 22, 1298–1303 (1980)
29. BIRKLE, A. J., WEI, R. P., and PELLESSIER, G. E., 'Analysis of Plane-strain Fracture in a Series of 0·45 C–Ni–Cr–Mo Steels with different Sulphur Contents', *Trans. ASM,* **59**, 981–990 (1966)
30. KIESSLING, R., *Non-metallic Inclusions in Steel,* The Metals Society, London (1978)
31. 'Swedish Symposium on Non-Metallic Inclusions in Steel'. Edited by NORDBERG, H., Uddeholm Tooling, and SANDSTRÖM, R., Swedish Institute of Metal Research (1981)

4

Hardenability

4.1 General remarks

Provided the rate of cooling is greater than the critical cooling rate, i.e. the rate at which the formation of pearlite or bainite is just suppressed, the hardness obtained on quenching depends principally on the C content of the steel. If the cooling rate is lower than the critical rate the amount of martensite is reduced, thus lowering the overall hardness of the steel. Carbon content in this context means the amount of C dissolved in the austenite. Carbon which remains combined as carbide after the austenitizing treatment does not take part in the martensite reaction and has therefore no influence on the hardness of the martensite. The relationship between the hardness, carbon content and amount of martensite is shown in *Figure 4.1*. Hardness after quench hardening must not be confused with the concept of hardenability.

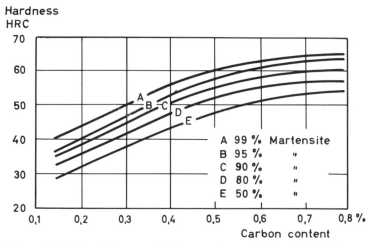

Figure 4.1 Relationship between hardness, carbon content and amount of martensite (after Hodge and Orehoski)

144

Hardenability is the ability of the steel to harden by the formation of martensite on quenching. The hardenability determines the depth of hardening obtained on quenching, which is usually specified as the distance below the surface where the amount of martensite has been reduced to 50%, or more precisely to 50% martensite and bainite. The characteristic property of a steel possessing high hardenability is that it shows a large depth of hardening or that it hardens through entirely in heavy sections. Without electron microscopy it may be difficult to distinguish martensite from lower bainite.

Since the depth of hardening is of great importance to tools and constructional parts it is customary to indiciate this property by means of diagrams of the type shown in *Figure 4.2.* When the depth of hardening is measured in this way the cooling medium must also be stated. In this instance the plain carbon steel, W 1, was quenched in water and the other steels in oil. The heavier the section to be hardened the smaller is the depth of hardening and the lower is the core hardness, which is illustrated in *Figure 4.3.* The reason why a steel is harder at the surface than at the centre is explained by referring to a continuous-cooling-transformation (CCT) diagram. By studying the schematic CCT diagram in *Figure 4.4* it is obvious that since the surface cools at a considerably faster rate than the centre, the cooling curve representing the surface will pass in front of the ferrite and bainite noses and as a result only martensite is formed. At the centre which cools more slowly some bainite will be formed, as may be inferred from the figure, and this will result in a lower hardness in the core.

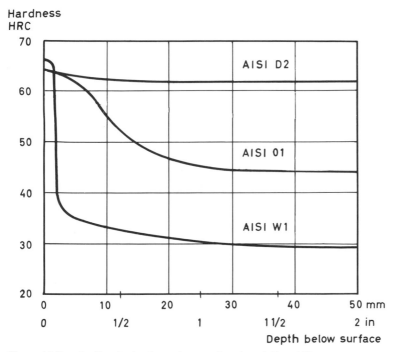

Figure 4.2 Depth of hardening for various grades of steel. Bars 100 mm diameter. Steel W1 water quenched, the rest oil quenched

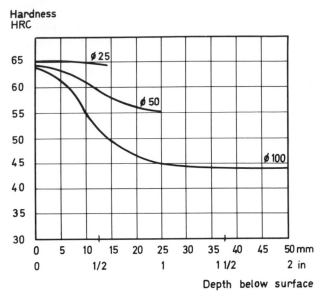

Figure 4.3 Depth of hardening in different dimensions after oil
quenching (AISI 01). Test-piece 25 mm diameter hardened from
800°C in oil. Test-piece 50 mm diameter hardened from 820°C in oil.
Test-piece 100 mm diameter hardened from 840°C in oil

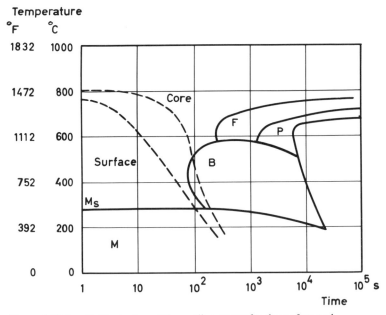

Figure 4.4 Schemtic illustration of the cooling curves for the surface and
core, respectively, of an oil-quenched bar, 95 mm diameter. The surface is
wholly martensitic; the core contains some upper bainite

As the dimensions of the steel increase, the rate of cooling decreases and the core hardness will be still further reduced owing to the formation of ferrite and pearlite. The hardness will also decrease when the cooling curve is so displaced as to be to the right of the critical cooling curve.

Fundamental hardenability data are of considerable use to the steel consumer and heat treater and therefore a number of simple methods have been developed whereby hardenability can be determined. Some of the best known methods are described below.

4.2 The Grossmann hardenability test

To determine hardenability according to Grossmann's method[1], a number of cylindrical steel bars of different diameters are hardened in a given cooling medium. By means of metallographic examination the bar that has 50% martensite at its centre is singled out and the diameter of this bar is designated as the critical diameter (D_0), the unit generally being inches.

The cooling intensites of the different cooling media have been determined and are called the H-factors. The values of H are given in *Table 4.1*. Using the approriate value of the H-coefficient of the cooling medium under consideration, the D_0-value can be converted to the ideal diameter D_i which is defined as the bar diameter which, when the surface is cooled at an infinitely rapid rate $(H = \infty)$, will yield a structure, at the centre, containing 50% martensite. The diagrams correlating D_0 and D_i are shown in *Figure 4.5*.

Table 4.1

		Coefficient of severity of quench H	
Agitation		*Cooling medium*	
	Oil	*Water*	*Brine*
None	0·25–0·30	0·9–1·0	2·0
Mild	0·30–0·35	1·0–1·1	2·0–2·2
Moderate	0·35–0·40	1·2–1·3	
Good	0·4 –0·5	1·4–1·5	
Strong	0·5 –0·8	1·6–2·0	
Violent	0·8 –1·1	4·0	5·0

The value of D_i obtained is hence a measure of the hardenability of the steel and is independent of the cooling medium. In practice the D_i-values are used to determine the values of D_0 for bars quenched in various cooling media, using the diagrams in *Figure 4.5*.

Example

By subjecting a steel whose D_i-value is 2·0 in to an oil quench, the H-coefficient of which is 0·4, it would yield a D_0-value of 0·8 in. Familiarity with concept of hardenability enables a good indication of the hardenability of a steel to be obtained from its D_i-value since this is a useful figure for comparison purposes.

148

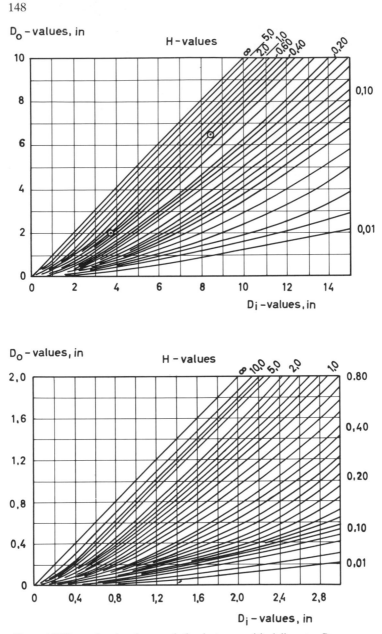

Figure 4.5 Charts showing the correlation between critical diameter D_0, ideal diameter D_i and H-value. The lower diagram is an enlargement of the lower, left-hand portion of the upper diagram (after Grossmann, Asimow and Urban[1])

4.2.1   Calculation of D_i-values from the chemical composition

The hardenability may be calculated from the composition of low-alloy and medium-alloy steels, taking into account only the amount of each element in solution at the austenitizing temperature. The austenite grain size must also be taken into consideration. The smaller the grain size the lower is the hardenability. This is due to the fact that the total surface area of the grain boundaries increases as the grain size decreases, thereby in turn giving rise to an increasing number of nuclei which serve as inititating points for pearlite formation.

The computation starts from the C content and the grain size. By means of *Figure 4.6* a 'base' hardenability characteristic for D_i is obtained. For the other alloying elements the curves in *Figure 4.7* indicate the multiplying factor that corresponds to each alloy content. The factors given in this diagram have been selected from a number of test results obtained by different research workers and approved by the American Iron and Steel Institute.

Figure 4.6 is applicable to C contents above 0·8%, but only on the assumption that all carbides are in solution at the austenitizing temperature. However, this is generally not the case since an unnecessarily high temperature would then have to be employed. Further, complete dissolution of the carbides might result in deleterious grain growth effects and high retained austenite content in the steel. Consequently if conventional hardening temperatures are used for low-alloy steels lying in the higher C ranges of the equilibrium diagram, a falling-off in the hardenability must be expected when the C exceeds 0·8%. This is because the 'excess carbon' combines with carbide and hardenability-inducing elements such as Cr and Mo. In spite of this reduction in hardenability, steels are still alloyed with about 1% C and more, but under these circumstances the carbides are beneficial in increasing the wear resistance of the steel.

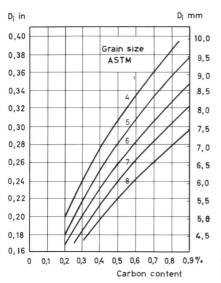

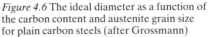

Figure 4.6 The ideal diameter as a function of the carbon content and austenite grain size for plain carbon steels (after Grossmann)

Multiplying factor   Multiplying factor

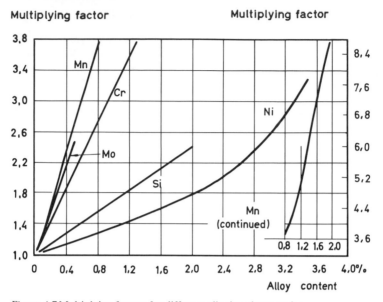

Alloy   content

Figure 4.7 Multiplying factors for different alloying elements for
hardenability calculations (after A I S I)

Examples of hardenability calculations

1. Steel S S 2225, A S T M grain size 7, has the following composition:

C	Si	Mn	Cr	Mo	
0·25	0·3	0·7	1·1	0·2	%

From *Figure 4.6* the base value of D_i is 0·17 in. On multiplying this value
with the appropriate factors we obtain:

$$D_i = 0·17 \times 1·2 \times 3·3 \times 3·4 \times 1·6 = 3·7 \text{ in}$$

2. Steel S S 2541 (B S 816M40), A S T M grain size 6, the following
composition:

C	Si	Mn	Cr	Ni	Mo	
0·35	0·3	0·7	1·4	1·4	0·2	%

$$D_i = 0·22 \times 1·2 \times 3·3 \times 4·0 \times 1·5 \times 1·6 = 8·4 \text{ in}$$

The D_i-values obtained may be converted to D_0 by means of diagrams in
Figure 4.5 as described above. For example, by quenching in oil with
moderate agitation ($H = 0·40$) the critical diameter of steel S S 2225 is
$D_0 = 2$ in and of steel S S 2541, $D_0 = 6·4$ in. These values are marked in
the diagram, *Figure 4.5*.

In order to calculate D_i slide rules are obtainable which have scales
graduated according to the different multiplying factors. Values of D_i,
calculated as above, are only approximate but they are useful as a means of
comparing the hardenability of different grades or heats.

The hardenability factors proposed by Grossman have been re-examined by Moser and Legat[2] and a more precise relationship between grain size and C content has been obtained and this is presented in *Figure 4.8*. The revised multiplying factors by Moser and Legat are shown in *Figure 4.9*.

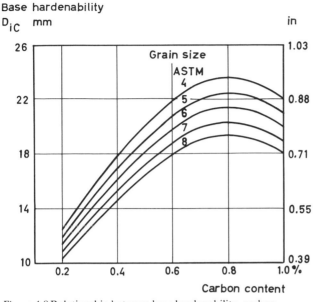

Figure 4.8 Relationship between base hardenability, carbon content and grain size as obtaining in actual practice (after Moser and Legat[2])

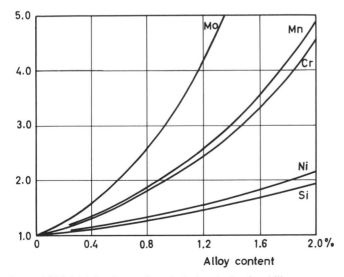

Figure 4.9 Multiplying factors for calculating the hardenability as influenced by Mo, Mn, Cr, Si and Ni (after Moser and Legat[2])

Using the results obtained by Moser and Legat the hardenability may be calculated from the following expression:

$$D_i = D_{iC} \times 2\cdot21 \begin{array}{c} \%\ Mn \\ \times\ 1\cdot40 \end{array} \begin{array}{c} \%\ Si \\ \times\ 2\cdot13 \end{array} \begin{array}{c} \%\ Cr \\ \times\ 3\cdot275 \end{array} \begin{array}{c} \%\ Mo \\ \times\ 1\cdot47 \end{array} \begin{array}{c} \%Ni \end{array}$$

Recalculated values of D_i for the steels in the examples are as follows. (Grossmann's values are given in brackets.)

SS 2225 $D_i = 2\cdot5$ in ($3\cdot7$ in)
SS 2541 $D_i = 8\cdot0$ in ($8\cdot4$ in)

The agreement is not very good for the first steel but is acceptable for the second.

Kramer, Siegel and Brooks[3] have also examined Grossmann's equations and have published a diagram that is practically identical with the one in *Figure 4.8*. Jatczak[4] has extended this diagram by adding multiplying factors for other alloying elements. Fairly close agreement with the values of Moser and Legat were obtained in the case of low- and medium-carbon steels. Jatczak has also published multiplying factors for steels containing C between 0·90% and 1·10% and which take account of the hardening temperatures. For these cases calculations can proceed straight from the composition of the steel without the necessity of applying any corrections for such alloying elements as are not in solution in the austenite. A comparison beteen the diagrams in *Figures 4.10* and *4.11* shows the effect of raising the hardening temperature.

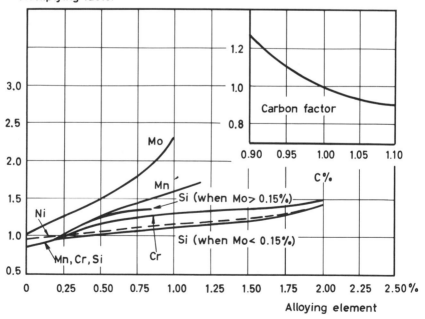

Figure 4.10 This chart is used for determining the multiplying factors for high-carbon steels austenitized at 830°C (1525°F) (base D_i is 1·13) (after Jatczak[4])

Multiplying factor

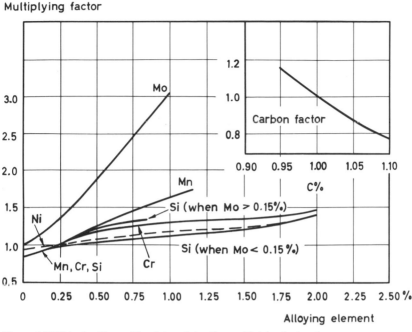

Figure 4.11 This chart is used for determining the multiplying factors for high-carbon steels austenitized at 860°C (1575 °F) (base D_i is 1·35) (after Jatczak[4])

As a practical example illustrating the effect produced by the C content in the range mentioned, it was found that the depth of hardening in grindingmill rods, made from a shallow-hardening steel, increased from 11 to 15 mm as the C content was reduced from 0·90% to 0·85%.

Jatczak's factors for Si, however, are not in agreement with the results obtained at Bofors where it has been shown beyond question that the hardenability of steels containing about 1% C first falls as the Si content increases to about 1% and then rises as the Si increases to about 2%. The changing influence of Si in this range of composition is illustrated in *Figure 4.12* which shows the hardenability, denoted by the Jominy distance to 500 HV, as a function of the Si content. As the Mn content increases, the troughs of the curves are displaced towards higher Si contents.

4.3 The Jominy end-quench hardenability test

On account of the high cost of Grossmann's method it is used nowadays only to a very small extent, albeit it must be regarded as the most exact test. The most commonly used method at present has been developed by Jominy[5]. For this test a round bar specimen is used, 25 mm in diameter and 100 mm in length. The specimen is heated to the hardening temperature of the steel with a holding time of usually 20 min. One end-face of the specimen is quenched by spraying it with a jet of water, as illustrated in

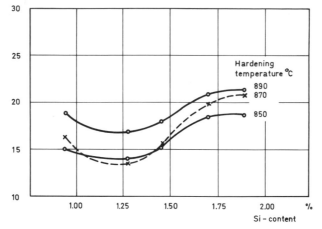

Figure 4.12 Variations of hardenability with Si content in a steel with nominal composition 0·90% C, 0·75% Mn and 1·0% Cr

Figure 4.13. Hereby the rate of cooling decreases progressively from the quenched end along the length of the bar. When it is cool, two diametrically opposite flats, 0·4 mm deep and parallel to the axis of the bar are ground and the hardness is measured along the flats. The hardness values are plotted in a diagram against their distances from the quenched end. Jominy curves for some steel grades are shown in *Figures 4.14* to *4.23*. The upper curve represents the maximum hardness values corresponding to the upper composition limit of the steel and the lower curve the minimum hardness values corresponding to the lower limit of the composition range. Together the curves form what is called a Jominy or hardenability band.

4.3.1   Calculation of Jominy curves from the chemical composition

Using regression analysis and proceeding from the chemical composition, Just[6] has derived expressions that enable the hardness at different Jominy distances to be calculated direct. It is found that all alloying elements have

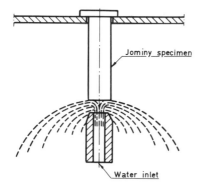

Figure 4.13 Hardening of Jominy end-quench specimen

an increasing influence up to a Jominy distance of about 10 mm, but beyond this their influence remains practically constant. Carbon starts off at a Jominy distance zero with a factor of 50 but all other alloying elements have factor 0 at distance zero. This implies that the hardness at Jominy distance zero is solely governed by the carbon content. Using the derived factors it is possible to present a hardenability expression for each Jominy distance. Since the factors vary by only negligible amounts for Jominy distances exceeding 6 mm, a single expression can be formulated for the whole test specimen except for distances shorter that 6 mm.

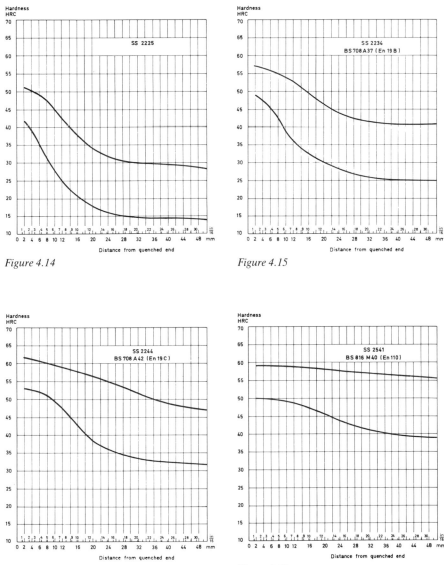

Figure 4.14

Figure 4.15

Figure 4.16

Figure 4.17

The equation given below is taken from a report of a paper by Just[7].

$$J_{6-80} = 95\sqrt{C} - 0.0028\,s^2\sqrt{C} + 20\,Cr + 38\,Mo + 14\,Mn \\ + 6\,Ni + 6\,Si + 39\,V + 96\,P - 0.8\,K - 12\sqrt{s} \\ + 0.9\,s - 13\,HRC$$

where J = Jominy hardness, HRC
        s = Jominy distance, mm
        K = ASTM grain size

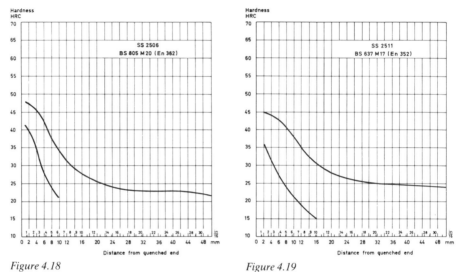

Figure 4.18

Figure 4.19

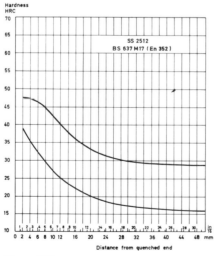

Figure 4.20

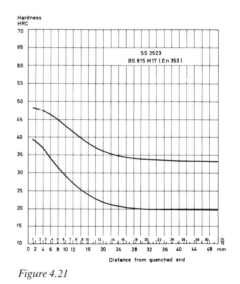

Figure 4.21

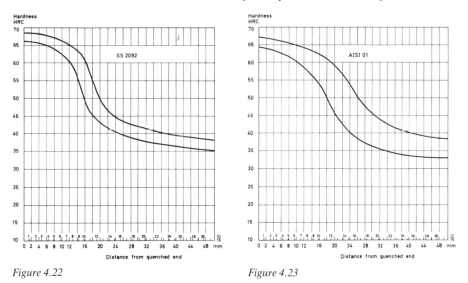

Figure 4.22   Figure 4.23

The equation is valid for Jominy distances from 6 to 80 mm. The following limits are specified for the alloying elements:

C < 0·6%, Cr < 2%, Mn < 2%, Ni < 4%, Mo < 0·5%, V < 0·2%.

For Jominy distances shorter than 6 mm the influence of alloying elements other than C on the hardness is negligible. Hence the following equation holds for the end-face of the Jominy specimen:

$$J_0 = 60 \times \sqrt{C} + 20 \, HRC \quad (C < 0 \cdot 6\%)$$

It is found that the C content on the one hand and the rest of the alloying elements on the other interact mutually on the hardness. For instance, Cr acts more powerfully in medium-carbon steels than in low-carbon, i.e. case-hardening steels.

The article[6] already referred to, contains a number of expressions based on available Jominy curves. These expressions are to a great extent dependent on the origins of these curves. Thus, for steels specified by S A E, U S S–Atlas and M P I–Atlas, the expressions have different forms. Medium-carbon (steels for hardening and tempering) and low-carbon (case-hardening) steels have been given different expressions for hardenability, as seen below:

J_{6-40} (case-hardening steel)

$$= 74\sqrt{C} + 14 \, Cr + 5 \cdot 4 \, Ni + 29 \, Mo + 16 \, Mn - 16 \cdot 8\sqrt{s}$$
$$+ 1 \cdot 386s + 7 \, HRC$$

J_{6-40} (steel for hardening and tempering)

$$= 102\sqrt{C} + 22 \, Cr + 21 \, Mn + 7 \, Ni + 33 \, Mo - 15 \cdot 47\sqrt{s}$$
$$+ 1 \cdot 102s - 16 \, HRC$$

In order to assess the accuracy of the first-mentioned expression a calculation has been carried out for two case-hardening steels and two steels for hardening and tempering, for which Jominy curves have been determined according to SS. As may be seen from *Figures 4.24* to *4.27*, in some instances the conformity is reasonably satisfactory.

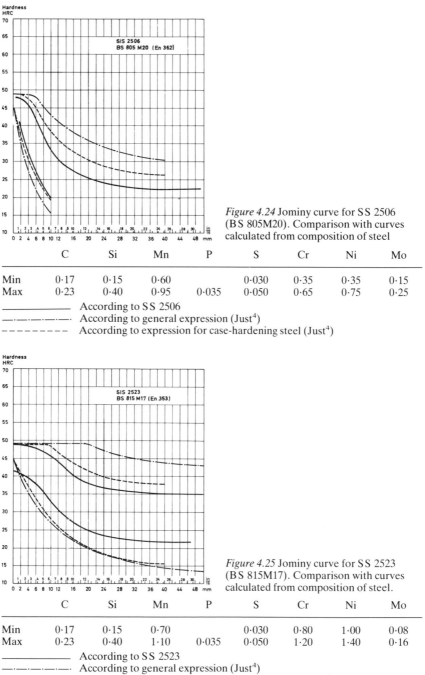

Figure 4.24 Jominy curve for SS 2506 (BS 805M20). Comparison with curves calculated from composition of steel

	C	Si	Mn	P	S	Cr	Ni	Mo
Min	0·17	0·15	0·60		0·030	0·35	0·35	0·15
Max	0·23	0·40	0·95	0·035	0·050	0·65	0·75	0·25

———————— According to SS 2506
—·—·—·—·—· According to general expression (Just[4])
— — — — — — According to expression for case-hardening steel (Just[4])

Figure 4.25 Jominy curve for SS 2523 (BS 815M17). Comparison with curves calculated from composition of steel.

	C	Si	Mn	P	S	Cr	Ni	Mo
Min	0·17	0·15	0·70		0·030	0·80	1·00	0·08
Max	0·23	0·40	1·10	0·035	0·050	1·20	1·40	0·16

———————— According to SS 2523
—·—·—·—·—· According to general expression (Just[4])
— — — — — — According to expression for case-hardening steel (Just[4])

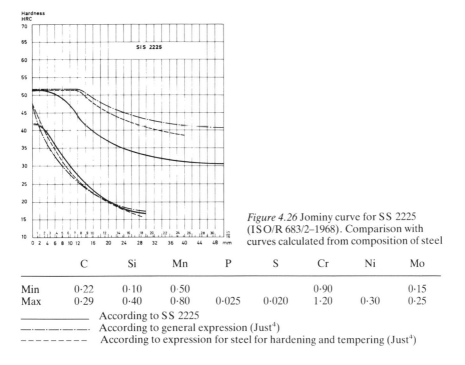

Figure 4.26 Jominy curve for SS 2225 (ISO/R 683/2–1968). Comparison with curves calculated from composition of steel

	C	Si	Mn	P	S	Cr	Ni	Mo
Min	0·22	0·10	0·50			0·90		0·15
Max	0·29	0·40	0·80	0·025	0·020	1·20	0·30	0·25

———————— According to SS 2225
—·—·—·—·—. According to general expression (Just[4])
– – – – – – – According to expression for steel for hardening and tempering (Just[4])

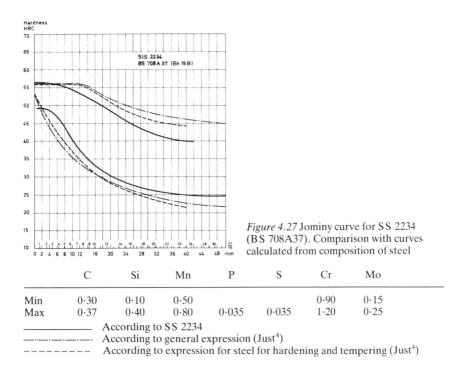

Figure 4.27 Jominy curve for SS 2234 (BS 708A37). Comparison with curves calculated from composition of steel

	C	Si	Mn	P	S	Cr	Mo
Min	0·30	0·10	0·50			0·90	0·15
Max	0·37	0·40	0·80	0·035	0·035	1·20	0·25

———————— According to SS 2234
—·—·—·—·—. According to general expression (Just[4])
– – – – – – – According to expression for steel for hardening and tempering (Just[4])

Kirkaldy[8,9] has developed a method for calculating Jominy diagrams by means of computers. He starts from the assumption that the nucleation of ferrite, pearlite and bainite takes place instantaneously; the problem is then to calculate the rate of growth $v(T)$. He has derived, analytically, an expression for the cooling rate at various Jominy distances. He assumes that the cooling rate producing 0·1% of transformed phase at the temperature corresponding to the maximum $v(T)$ also corresponds to the Jominy inflection point at which the martensite content is 50%. *See Figure 4.28.*

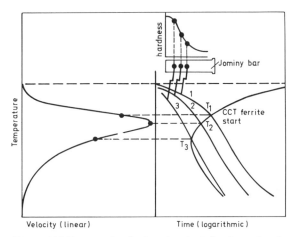

Figure 4.28 Construction for locating the inflection point of a Jominy curve (after Kirkaldy[8, 9])

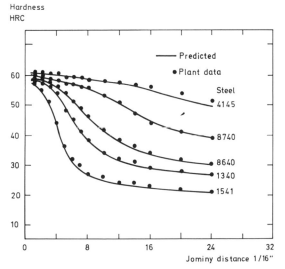
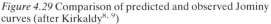

Figure 4.29 Comparison of predicted and observed Jominy curves (after Kirkaldy[8, 9])

By proceeding from the carbon content that gives the maximum hardness and Grossmann's D_i-value, which is an expression for the hardenability, and finally from the location of the point of inflection, it is possible to compute the whole Jominy curve. According to Kirkaldy the result obtained from such an exercise is shown in *Figure 4.29*.

Ericsson[10], in a report published by Jernkontoret (The Swedish Ironmasters' Association), has tested Kirkaldy's formulae by applying them to, *inter alia*, 25 steel grades from three different steel mills. One result that aims at giving the hardnesses at 1/16th inch Jominy distances is shown in *Figure 4.30*. It may be regarded as satisfactory but can be improved further by making use of correction factors that have been adapted to current steel-making processes. From practical experience it was found that although they may be of the same type, each individual melting furnace is able to influence in some way the hardenability of the steel even if the chemical composition—as determined by ordinary standard methods—is the same. The difference in the results between the various mills, as may be seen in *Figure 4.30*, is, in fact, quite appreciable.

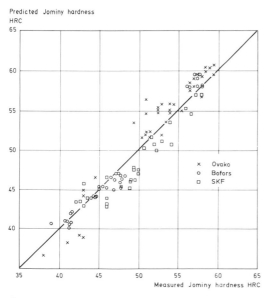

Figure 4.30 Hardness at Jominy distances of 1/16 in. As observed and predicted (after Kirkaldy[8, 9])

4.3.2 Practical applications of Jominy curves

Jominy curves may be used in the first instance to compare the hardenabilities of different heats of the same grades of steel, hereby serving as a valuable method of quality control. For more than 30 years there have been available in the USA special so-called *H*-steels for which there are specifications defining their hardenability bands. (BS 970:Part 2:1970 and BS 970:Part 3:1971 also contain *H*-steels.)

The curves may also be used to predict the expected hardness distribution obtained on hardening steel bars of different dimensions by cooling in various quenching media. The rates of cooling prevailing at different distances in the Jominy specimen may be compared with the cooling rates prevailing in bars of different diameters cooled in various quenching media. Such a comparison may be made by referring to the curves drawn in *Figure 4.31*. From this it may be seen that the rate of cooling at a Jominy distance of, say, 14 mm is the same as that prevalent at a point 2 mm below the surface of a 75 mm diameter bar, or at 10 mm below the surface of a 50 mm diameter bar, or at the centre of a 38 mm diameter bar; all the bars being quenched in moderately agitated oil. Using the Jominy curves and *Figure 4.31* it is therefore possible to construct a diagram showing the hardness distribution, after hardening, in the manner shown in *Figure 4.32*.

If a tempering curve for the Jominy specimen is available, a diagram showing the transverse hardness of a hardened and tempered bar may be constructed in a similar way. The form such Jominy tempering diagrams take is shown in *Figure 4.33*.

Figure 4.31 applies to *one* rate of cooling only. The diagram may be supplemented to include various rates of cooling, the values of which correspond to those given in *Table 4.1*. Examples of such diagrams are

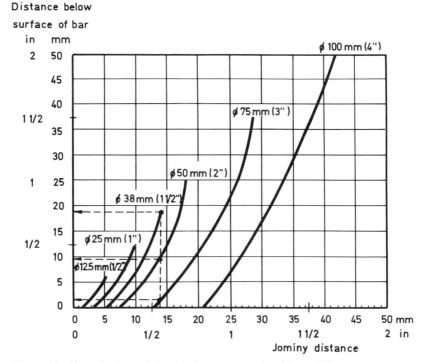

Figure 4.31 Chart showing relationship between rate of cooling at different Jominy distances and rate of cooling in moderately agitated oil. Curves apply to round bars from 12·5 to 100 mm (½–4 in) diameter. (Diagram based on data taken from ASTM Standards)

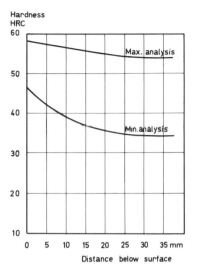

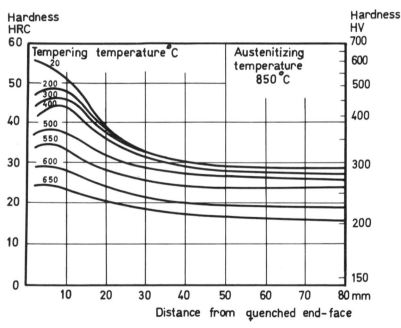

Figure 4.32 Transverse hardness in a 75 mm diameter test-piece, steel BS 708A42. Construction from Jominy curves and *Figure 4.31*

Figure 4.33 Jominy curves for steel DIN 42CrMo4, corresponding to BS 708A42 after tempering at different temperatures

given in *Figures 4.34* to *4.39*[11]. This type of diagram is constructed in a different way and its mode of applications will therefore be explained by means of an example. Suppose we wish to know for a certain grade of steel the transverse hardness in a bar, say 100 mm in diameter, that has been quenched in well-agitated oil ($H = 0·5$). The first step is to consult *Figure 4.34*. From this we obtain a Jominy distance of 12 mm, which refers

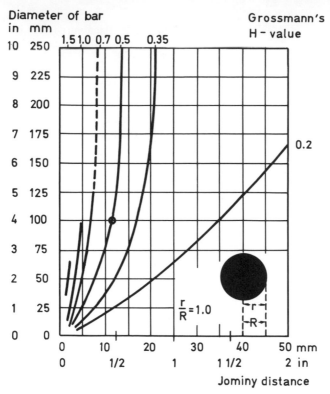

Figures 4.34 and Figure 4.35 Curves showing correlation between rates of cooling in the Jominy specimen and rates of bars cooled in various quenching media. The diagrams apply to the surface of the bar and to points situated $0{\cdot}9R$ from the centre of the bar respectively (after Lamont[11])

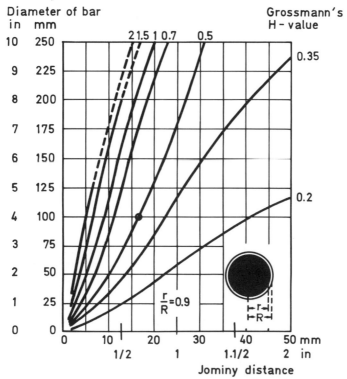

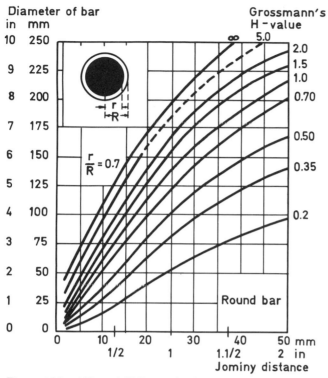

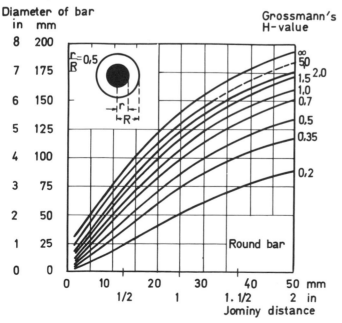

Figures 4.36 and Figure 4.37 Curves showing correlation between rates of cooling in the Jominy specimens and rates of cooling in bars cooled in various quenching media. The diagrams apply to points situated at distances of 0·7R and 0·5R from the centre of the bar respectively (after Lamont[11])

166

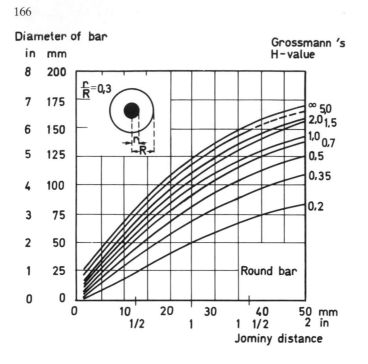

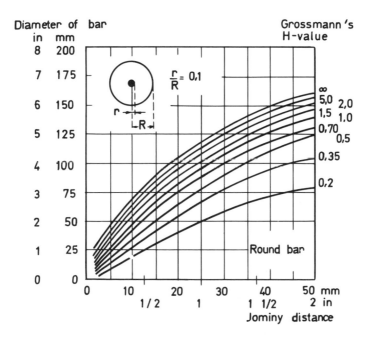

Figures 4.38 and Figure 4.39 Curves showing correlation between rates of cooling in the Jominy specimen and rates of cooling in various quenching media. The diagrams apply to points situated at distances of $0.3R$ and $0.1R$ from the centre of the bar respectively (after Lamont[11])

to the hardness at the surface of the bar. From the curves in *Figure 4.85* we obtain a Jominy distance of 17 mm which refers to a zone situated $0.9 \times R$, i.e. $0.9 \times 50 = 45$ mm from the centre, i.e. 5 mm below the surface. In a similar manner we consult the other diagrams to obtain the correlated Jominy distances and hardness zones. The actual hardness values are then obtained from the Jominy curve of the steel grade concerned. A diagram may now be drawn showing the hardness as a function of the distance below the surface. More examples of the application of the diagrams will be given in greater detail later in this chapter.

Figure 4.31 as well as *Figures 4.34* to *4.39* are valid for Jominy distances up to 50 mm only. Therefore, the diagrams can be used to assess the core hardness of bars only to about 150 mm in diameter. Of course, the diagrams may be extended but their general limit is about 80 mm which is usually the maximum Jominy distance. To evaluate bars of larger cross-sections other methods must be used and these will be described in the next section.

It is possible to combine the Jominy diagrams direct with the information contained in *Figure 4.31* and *Figures 4.34* to *4.39*, thereby enabling a rapid assessment to be made as to whether it is advisable to use a given type of steel or a steel with a given Jominy diagram in a certain dimension. Such combined diagrams may look like the one shown in *Figure 4.40*. From this diagram, which also holds for various cooling conditions, it may be deduced, for example, that a bar, 50 mm in diameter, after being quenched in strongly agitated oil, will probably obtain a hardness of 40–49 HRC at the centre and 48–52 HRC at the surface.

If the quenching were to take place in moderately agitated oil an unsatisfactory hardness would most likely to be the result. A 75 mm diameter bar of this steel is not suited for oil quenching whereas water quenching will give acceptable hardness results.

If a Jominy diagram for the actual heat itself is available the assessment will obviously be more reliable.

The information given in *Figures 4.34* to *4.39* applies to round bars only. When bars of square or rectangular cross-section are hardened these sections must be converted by estimation to equivalent round cross-sections for which diagrams are available.

In connexion with the compilation of an international steel grades standard a conversion diagram has been constructed, by means of which square and rectangular cross-sections may be converted to equivalent round ones, *Figure 4.41*. For example, a 38 mm square and a 25×100 mm rectangular cross-section are each equivalent to a 40 mm diameter round section; a 60×100 mm cross-section is equivalent to an 80 mm diameter round one.

4.4 Practical application of the TTT and the CCT diagrams

4.4.1 How should TTT diagrams be used?

The familiar TTT diagrams, which show the progress of the transformation at constant temperature (isothermal transformation) are useful when isothermal processes are concerned, such as isothermal

annealing, austempering or martempering. These processes are described
in the chapter on thermal treatments. However, TTT diagrams of this type
are suitable only for an approximate assessment of the hardenability of the
various steels.

4.4.2   Various types of CCT diagram

Continuous-cooling-transformation diagrams (also called the CCT
diagram) are so constructed as to enable the personnel dealing with heat

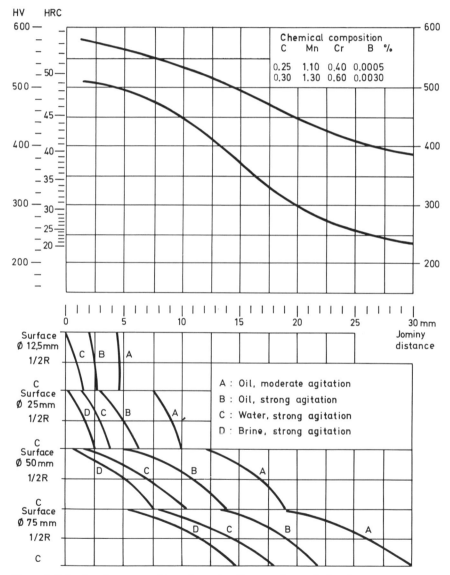

Figure 4.40 Jominy diagrams and transformation diagrams for predicting the
hardness of bar steel after hardening by means of various media

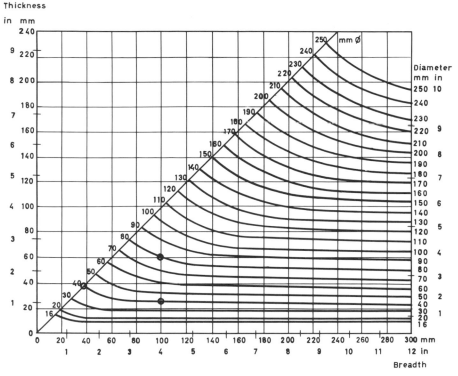

Figure 4.41 Correlation between rectangular cross-sections and their equivalent round sections (according to ISO)

treatment to predict which constituents will be produced as the rate of cooling varies. Provided that each individual point on the Jominy specimen corresponds to a known rate of cooling, it has been possible, by using data from the Jominy test, to constuct CCT diagrams on which are superimposed cooling curves corresponding to the rates of cooling prevalent at various Jominy distances[12, 13]. Such a diagram is shown in *Figure 4.42*. The rate of cooling at each point on the Jominy specimen is equivalent to some definite rate of cooling in a steel bar—*see Figure 4.31*—and hence it is possible to predict from this type of transformation diagram which structural constituents will form in the steel bar, as well as the transverse hardness, the latter being computed from the Jominy diagram. However, it is not quite correct to assume that the same cooling curves apply to all grades of steel, since different steels have different thermal conductivities.

During the 1960s a series of CCT diagrams was published in *Metal Progress*. One of these diagrams is reproduced in *Figure 4.43*. Let us follow the constitutional transformations in the steel grade A I S I 3140 at a point 19 mm from the quenched end-face of the Jominy specimen. In 25 s ferrite begins to form, in 30 s pearlite begins and in 45 s, bainite. 50% of the austenite has been transformed in 90 s. In about 140 s from the start of the cooling of the test-piece the temperature has fallen to 315°C and martensite formation begins.

Based on data obtained from dilatometric investigations, however, Wever, Rose and co-workers[14], have constructed diagrams with continuous-cooling curves. The hardness obtained for each cooling rate is indicated on the appropriate cooling curve. In addition, cylindrical test-pieces have been cooled in various quenching media and the rate of cooling has been measured at points 0·5 mm below the surface and at the centre. The values so obtained were plotted in diagram form on a sheet of transparent paper which is to be laid on top of the CCT diagram. In order to assess the hardness from these diagrams select the hardness figure marked at the end of that cooling curve (drawn on the CCT diagram) that comes closest, immediately below the M_s line, to the cooling curve drawn on the transparent sheet that represents the bar dimension concerned.

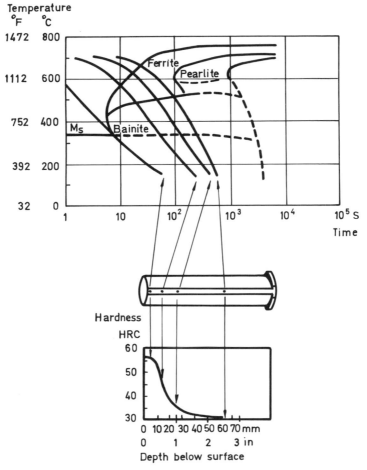

Figure 4.42 CCT diagram with some cooling curves for corresponding Jominy distances. The lower diagram shows the hardness values along a Jominy test specimen

Example

From *Figure 4.44*, on which the cooling curves for a bar, 95 mm in diameter, have been superimposed, it can be seen that on quenching this bar in water, the surface will contain 2·5% bainite and have a hardness of 52·5 HRC. At the centre, 5% of the austenite will have transformed to ferrite and 75% to bainite before the start of the martensite formation. The final hardness will be 34 HRC. In the case of oil quenching it is more difficult to asses the surface hardness, which will be somewhere between 34 and 52 HRC. On air cooling the hardness will lie between 220 and 230 HB at both surface and centre. The method just described must be regarded as a rather rough one, but all the same it is a step in the right direction towards making practical use of CCT diagrams.

In the book *Phase Transformation Kinetics and Hardenability of*

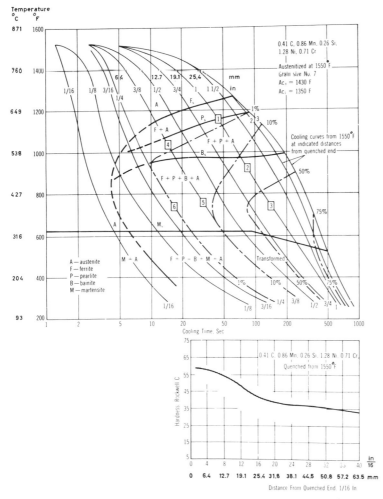

Figure 4.43 CCT diagram and Jominy hardenability band for AISI 3140 (*Metal Progress*, Oct. 1963, p. 134)

Medium-Carbon Alloy Steels issued by Climax Molybdenum, there are a number of CCT diagrams, four of which are reproduced in *Figures 4.45–4.48*. The diagrams show the influence of the molybdenum content on the transformations of a steel having a base analysis of 0·38% C, 0·34% Si, 0·84% Mn, 1·46% Ni and 0.74% Cr. These diagrams, like many others, are based mainly on linear cooling and hence they do not always represent the actual cooling conditions existing in steel parts which may be cooled in various ways.

The German Stahl-Eisen Werkstoffblat Nos. 200–69 and 250–70 contain CCT diagrams which indicate: (a) numerical values of the time, in seconds × 10^{-2}, taken to cool between 800 and 500°C—e.g. 0·22, 2·0 and 10 (the figures are to be multiplied by 100 in order to give the time in seconds) and (b) the cooling rate in degrees C per minute between 800 and 500°C. One of these diagrams is reproduced in *Figure 4.49*. It has superimposed on it the cooling curves (oil quench) for the centres of bars 50 mm, 100 mm and 200 mm in diameter, respectively.

In order to calculate the depth of hardening in large cross-sections Bandel and Haumer[15] have carried out exhaustive theoretical and

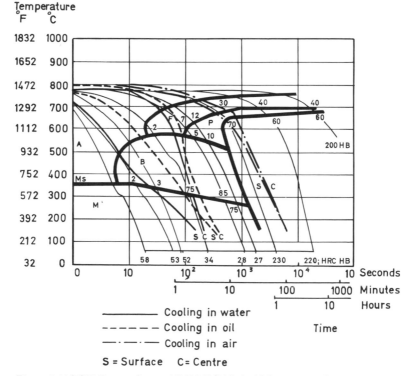

Figure 4.44 CCT diagram for steel DIN 42CrMo4 which corresponds to BS 708A42. Cooling curves for 95 mm diameter bar are superimposed and represent cooling curves on the transparent overlay sheet in *Atlas zur Wärmebehandlung der Stähle*. Figures attached to the lines in the diagram indicate percentage amount of structural constituent. For example fourth thin line from the left indicates that 2% ferrite has formed when bainite starts and that 75% bainite has formed when martensite starts to develop

empirical calculations. Russel's [16] theoretically deduced cooling curves have been compared with the results from a number of experiments with thermocouples placed inside steel bars, the diameters of which ranged from 250 mm to 950 mm. From the data obtained they have drawn amended curves pertaining to cooling in water, oil and air, *Figure 4.50*. The shape of the curves does not alter with austenitizing temperature

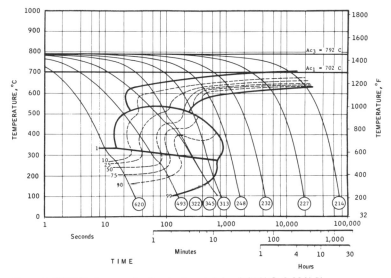

Figure 4.45 CCT diagram for the steel containing 0·38% C, 0·33% Si, 0·85% Mn, 1·46% Ni, 0·74% Cr and 0·01% Mo, austenitized at 820°C (1510°F) for 20 min (by courtesy of the Climax Molybdenum Company)

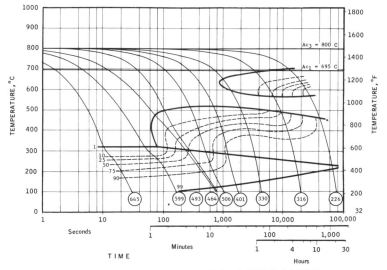

Figure 4.46 CCT diagram for the steel containing 0·38% C, 0·35% Si, 0·85% Mn, 1·45% Ni, 0·73% Cr and 0·24% Mo, austenitized at 830°C (1525°F) for 20 min (by courtesy of the Climax Molybdenum Company)

provided the appropriate temperature scale is chosen. Obviously the curves must be regarded as representing mean values since it is not possible to take into consideration those aberrations and irregularities that occur during cooling. By their investigations Bandel and Haumer have been mainly concerned in finding, with the aid of cooling curves and CCT diagrams, some connexion between Grossmann's D_i-values and the

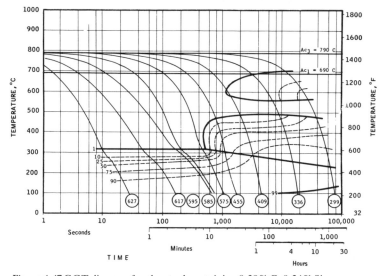

Figure 4.47 CCT diagram for the steel containing 0·38% C, 0·34% Si, 0·84% Mn, 1·46% Ni, 0·73% Cr and 0·48% Mo, austenitized at 820°C (1510°F) for 20 min (by courtesy of the Climax Molybdenum Company)

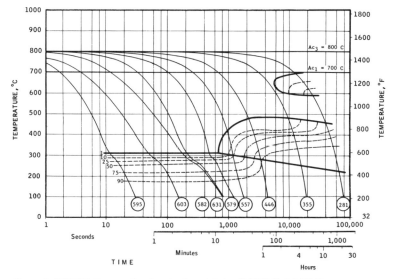

Figure 4.48 CCT diagram for the steel containing 0·38% C, 0·34% Si, 0·82% Mn, 1·46% Ni, 0·75% Cr and 0·74% Mo, austenitized at 830°C (1525°F) for 20 min (by courtesy of the Climax Molybdenum Company)

diameter of the bar that, on oil cooling, becomes ferrite-free at the centre. The calculations apply to heat-treatable steels and have resulted in the curves shown in *Figure 4.51* in which the decisive influence of grain size can be seen. A certain heat-treatable steel, the D_i and grain size of which are 200 mm and 7–8 ASTM respectively, becomes ferrite-free at the centre on hardening in oil when its diameter is 350 mm. When the grain size is 3–4 ASTM the corresponding dimension is about 600 mm. Owing to the long heating times used for large forgings it is not unusual for the grains to grow to a size corresponding to 3–4 ASTM, unless the steel has been effectually fine-grain treated.

By making use of the diagrams in *Figure 4.50* it is a fairly simple matter to estimate the hardness and structural constitution from the CCT diagrams provided that the cooling curves can be made to conform to the curves in the diagrams. Below is given an example how this is carried out.

Steel grade X 40 CrMoV 51 (*Figure 4.49*). Austenitizing temperature 1020°C.

If we choose the cooling curve marked 8·5 s $\times$ 10^{-2} we see that it intersects the 350°C horizontal in 2 $\times$ 10^3 s. From *Figure 4.50* we see that the dimensions of a bar giving this time-temperature relationship at the surface and centre on oil quenching are roughly 500 mm in diameter and 300 mm in diameter respectively. If the same structure is required on air cooling the

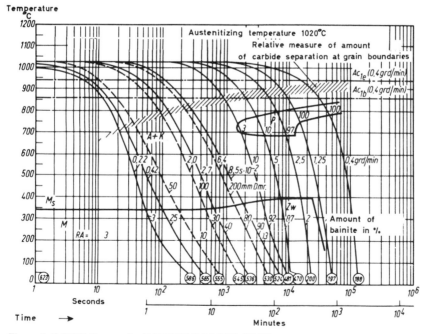

Figure 4.49 CCT diagram for DIN X 40 CrMoV 51 (H 13)
(Stahl-Eisen-Werkstoffblatt 250–70)

................Cooling curves for core of round bars, 50,100 and 200 mm diameter, cooled in oil

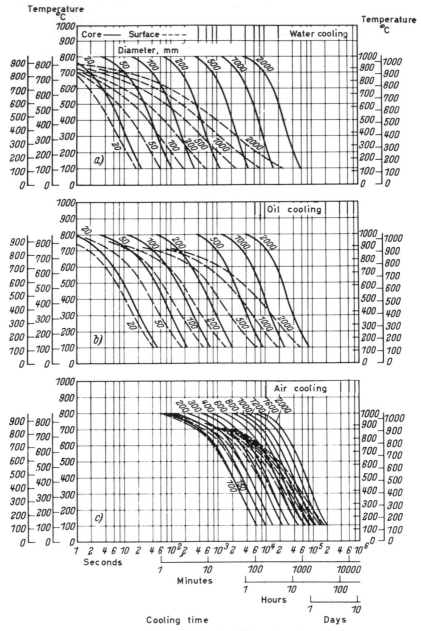

Figure 4.50 Cooling time for surface and core for different bar dimensions on cooling in water, oil and air respectively (after Bandel and Haumer[15])

bar diameter must be less than 100 mm. The maximum bar diameter that will result in a non-pearlite structure on air cooling is about 125 mm.

Atkins and Andrews[17] have carried out a very large number of trials with steel bars of various diameters. Specimens were prepared by inserting thermocouples at the centre, at positions 0·5 radius and 0·8 radius, reckoned from the centre. The specimens were hardened in water, oil and air. The cooling curves thus obtained were then simulated in dilatometric tests in order to identify the transformation temperatures, microstructures and hardness. Tests showed that the cooling rate through the temperature region immediately above the zone of transformation to ferrite, pearlite and bainite exerts a decisive influence on the hardness and microstructure obtained at room temperature. The data obtained was then brought together in diagram form of the type shown in *Figure 4.52*. This diagram applies only to the central region of round bars but with the aid of tables it is also adaptable to other regions. A compromise was arrived at in that the diagram, which is to read off vertically, is valid for bars cooled in air, oil or water. For example, from it may be read off that a specimen, 2 mm in diameter, on being cooled in air corresponds to a bar somewhat less than 40 mm in diameter being cooled in oil or to a bar 50 mm in diameter quenched in water. In the upper part is shown the amount and type of microstructures and in the lower part the hardness after hardening and tempering at various temperatures. Diagrams representing 172 steel grades have been collected and published as a handbook[18] which is most useful when a choice of material and heat treatment is being discussed.

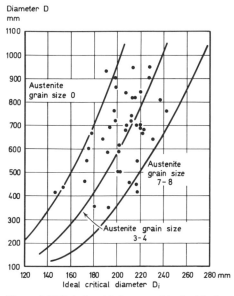

Figure 4.51 Relationship between ideal critical diameter D_i, austenite grain size and bar diameter D (after Bandel and Haumer[15])

4.4.3   A new generation of CCT diagrams

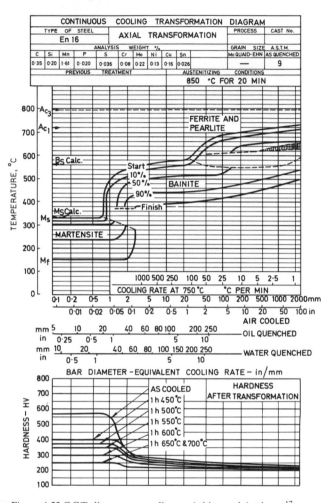

Figure 4.52 CCT-diagram according to Atkins and Andrews[17].
The vertical lines in the upper diagram give the cooling rate for
the centre of bars with different diameters when quenching in
different media (by courtesy of the British Steel Corporation)

Earlier in this chapter it was stated that the cooling rate at various Jominy
distances corresponds to the actual cooling rate at certain definite depths
below the surface of bars of various dimensions. This holds good only with
the proviso that the cooling rate at a certain temperature, usually 700°C, is
given. However, this cooling rate cannot affect the transformations taking
place at other temperatures. The CCT diagrams shown in this chapter are
based on data either from linear cooling or from some sort of 'natural' or
free cooling according to Newton's Law of Cooling. Furthermore, it has
been mainfested that these CCT diagrams have their limitations when they
are referred to for information regarding hardening results attainable for

the actual steel in question. However, Atkins's diagrams constitute an important exception even if they contain a number of compromises.

Within the Research Organization of Jernkontoret a committee has been working with these questions. As an introduction to this project three steels of different types were examined by constructing their Jominy diagrams, CCT diagrams from linear cooling and by hardening specimens of various dimensions. The examinations were aimed at finding out whether there exists a temperature range through which the steels could be cooled by the three test methods either at the same rate or for the same cooling time and then result in the same hardness for each of the steels subjected to the three tests. However, no such temperature range could be identified.

In a later project undertaken by the same committee[19] dilatometric tests involving 'natural' cooling according to Atkins's curves were carried out. Comparison with those of Hengerer, Strässle and Bremi [20] showed good agreement. Subsequently, CCT diagrams of quite a new type were constructed which are in better agreement with actual conditions. Examples may be taken from *Figure 4.53*. From this diagram it is easy to read off which structural constituents and hardness exist at $0.8 R$ and centre of bars of various dimensions after being quenched in both water or oil. When the results of the actual hardening trials with bars of various dimensions from the same heat were studied some of the results differed somewhat from the values shown in *Figure 4.53*. This is due in part to the fact that the cooling curves in the diagram are not always applicable to the many variants of a cooling medium since the cooling capacity of both oil and water are difficult to specify exactly. Futhermore, as a consequence of the hardening operation the internal residual stresses created in the steel exert some influence on the result, as shall be discussed later; likewise lesser variations in the shape of the cooling curve play their part.

4.4.4 Mechanisms that control hardenability

Influence of nucleation and growth

As a basis for the understanding of TTT and CCT diagrams there is in Section 1.3.7 a schematic figure that serves to explain simply what happens during isothermal transformation. This figure is reproduced here (*Figure 4.54*) in a somewhat new shape. It presupposes that the rate of nucleation is due mainly to the degree of supercooling, whereas growth is due mainly to diffusion.

If cooling is rapid down to the temperature range of the nose and the specimen then held there (cooling mode 1) nucleation and growth both take place quite fast and the transformation proceeds as shown in the diagram. Should the cooling be taken to a lower temperature, nucleation will start after a shorter interval. If the temperature is now raised (mode 2) growth will be activated and the time taken for complete transformation will be shorter than if the transformation had followed mode 1. When the TTT and CCT diagrams for the same steel are compared it is quite obvious that when the steady rate of cooling is interrupted in the temperature range in which transformation by diffusion takes place the progress of the transformation is stimulated by the interruption. On the

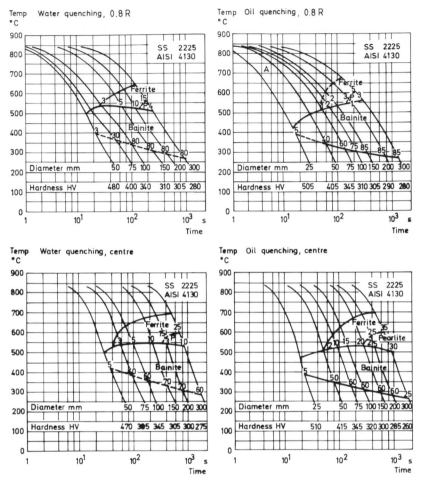

Figure 4.53 CCT diagram for SS 2225, AISI 4130 and DIN 25 CrMo 4.
Chemical composition: 0·26% C, 0·33% Si, 0·86% Mn, 0·012% P, 0·013% S,
1·17% Cr, 0·30% Ni, 0·23% Mo, 0·05% Al, 0·010% N

other hand, an uninterrupted steady cooling is unfavourable to the
transformation. This observation is illustrated by a test carried out with the
spring steeel En 45 (55 Si 7). Two continuous rates of cooling were
simulated and superimposed on the TTT diagram for this steel; *see Figure
4.55*. The first cooling rate gave 98% of martensite and the other one,
75%. When the cooling was interrupted at 600°C for about 30 seconds and
then continued, only 6% of martensite resulted and the hardness was
341 HV, i.e. about the same is indicated on the TTT diagram at 600°C.
Figure 4.56 also shows how the hardness is affected when at first the
cooling rate is constant down to the bainitic range and then when the
cooling rate is varied. The highest cooling rate resulted in a hardness of
495 HV, the lowest, 340 HV.

If the cooling through the upper metastable austenite range is slow nucleation proceeds slowly, which enhances hardenability. Should the slow cooling continue till just before the nucleation range and then change over to rapid cooling, the steel passes into a temperature range where any nuclei previously formed will grow slowly. Such a cooling mode, mode 3 in *Figure 4.54*, should give a high hardenability.

In order to verify this hypothesis *Figure 4.57* was constructed in which various rates of cooling (dashed lines) down to 700°C were simulated. From this temperature the cooling progressed at a constant rate. These curves were superimposed on a CCT diagram, constructed from linear cooling data (solid lines). It was found that the slower the cooling proceeded in the first stage the higher became the hardness of the steel; an observation that may appear surprising.

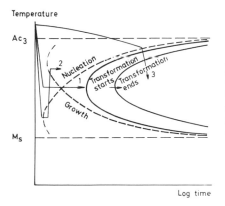

Figure 4.54 Schematic representation of nucleation and growth during isothermal transformation in an eutectoid carbon steel. The following modes of cooling which give rise to different rates of transformation are indicated in the diagram.
(1) Conventional mode of cooling for TTT diagrams;
(2) Mode of cooling that increases rate of transformation (reduces hardenability);
(3) Mode of cooling that reduces rate of transformation (enhances hardenability)

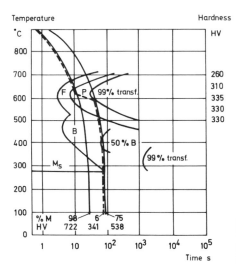

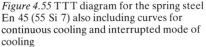

Figure 4.55 TTT diagram for the spring steel En 45 (55 Si 7) also including curves for continuous cooling and interrupted mode of cooling

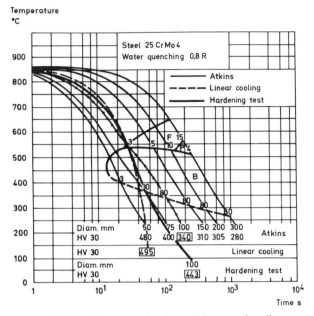

Figure 4.56 The influence on hardness of the rate of cooling through the bainite region. Simulated curves for water quenching, according to Atkins: also included are curves showing linear cooling and hardening tests with steel S S 2225 (25 CrMo 4)

Similar observations were made by Loria[21]. The left-hand part (a) of the diagram in *Figure 4.58* contains the conventional CCT diagram while the right-hand part (b) contains a composite CCT diagram. This diagram holds for air cooling to approximately A_{c1} and then for water quenching from there. A part of the large displacement of the CCT diagram was occasioned during the two-stage cooling procedure by the increased time taken for air cooling to reach A_3 as compared with water quenching and also by the fact that no transformations take place above A_3, i.e. in the range where austenite is stable. It had been found in practice that steel plates, half inch to one inch in thickness did not take on an even hardness after only water quenching. When samples were subjected to the duplex air-and-water cooling operation full and even hardness was imparted to sections from one inch to 1⅛ inch.

Hardening trials with the tool steel S S 2092, using specimens 70 mm in diameter × 95 mm, showed that the depth of hardening as obtained by the duplex air–oil hardening operation increased when the holding time in air was increased from 5 to 20 seconds. *See Figure 4.59.* Investigations of a similar nature have been referred to in the literature[22, 23] but so far they have not been sufficiently observed. Section 5.3.6, Quenching media, will revert to this point and will present further data to support the need for more information on this important matter.

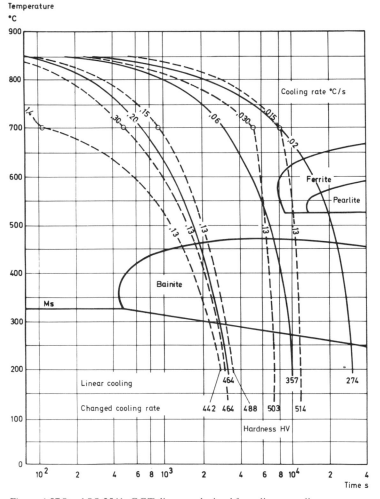

Figure 4.57 Steel SS 2541. CCT diagram derived from linear cooling.
Curves representing prolonged cooling above 700°C are drawn in the
diagram. Austenitized at 850°C for 10 min

Effect of internal stresses

It has been known for a long time that the formation of martensite is
favoured by tensile stresses and retarded by compressive stresses. That this
state of affairs also holds for other transformations has been demonstrated
very clearly by Schmidtmann, Grave and Klauke[24]. *See Figure 4.60.*

In a dilatometric test a specimen was subjected to a controlled rate of
cooling simulating that taking place at the centre of a 250 mm diameter
bar, grade SS 2541, during hardening. It was found that the hardness at the
centre of the bar was appreciably lower than the hardness of the
dilatometric test specimen, both samples having been taken from the same
heat. It was assumed that this difference is due to the influence of the stress
situation existing during the cooling.

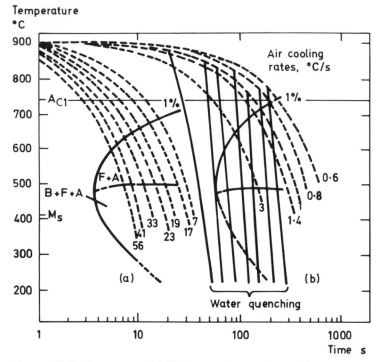

Figure 4.58 Cooling curves and CCT diagrams for a steel containing
0·20% C, 0·78% Mn, 0·60% Cr, 0·52% Ni and 0·16% Mo. Austenitizing
temperature 900°C (after Loria[21]). (a) Conventional CCT diagram; (b) Air
cooling followed by water quenching

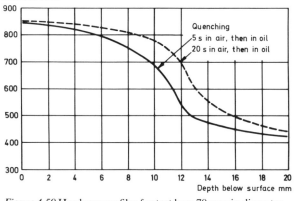

Figure 4.59 Hardness profiles for test bars 70 mm in diameter
× 95 mm of steel SS 2090 after delayed quenching in oil

An acceptable explanation of the cause of this lower hardness was
offered with the aid of a simple model for assessing the stress situation
prevailing during the hardening operation[19]. According to *Figure 4.61*,
which is applicable to the steel and specimen size under discussion, there
are tensile stresses acting at the centre while the bainite is being formed.
This concurrence serves to facilitate the progress of bainite formation and
as a conseqence a relatively low hardness is the result.

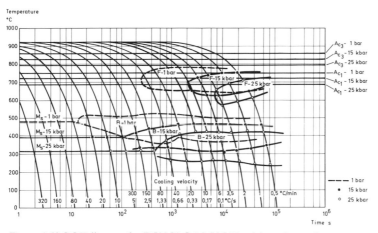

Figure 4.60 CCT diagram for DIN 21 CrMoV 5 11 subjected to various pressures. Chemical composition: 0·19% C, 0·43% Si, 0·009% P, 0·014% S, 1·42% Cr, 0·96% Mo, 0·49% Ni, 0·30% V. Austenitization 925°C, 25 min (after Schmidtmann, Grave and Klauke[24])

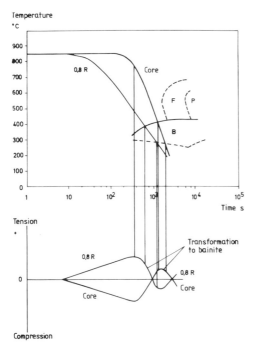

Figure 4.61 Hypothetical stress patterns existing in a 300 mm diameter sample of steel SS 2541, on quenching in oil. Schematic diagram[21]

Influence of the heat of transformation

When diffusion-dependent transformation products are being formed during the process of cooling, some heat is liberated which may influence the subsequent cooling process in various ways. In *Figure 4.62* it is shown how the rate of cooling in oil, as actually measured on a 50 mm diameter test-piece, differs from simulated cooling in oil for this dimension; also that the hardness of the test-piece takes on a lower value. This difference, which is due to the liberated heat of transformation, sets in even before the predicted start of the bainite formation, as deduced from the diagram; the reason being that the heat of transformation liberated from those parts of the specimen situated away from the area under consideration affects those parts already in the transformation zone.

If a steel has its pearlite nose situated rather high and if the cooling rate of the surface layer of the specimen is impeded by the heat liberated during the pearlite transformation it may happen that at the centre the rate of cooling through the zone of metastable austenite is slowed down, which enhances the hardenability of the central region. *Figure 4.63* shows both a cooling rate that gives a lower hardness at the surface than at the centre and the cooling rate for normal hardening. Schimizu and Tamura[25] have reported this phenomenon and have given it the name Inverse Hardening. It has also been reported by Tensi and Schwalm[26].

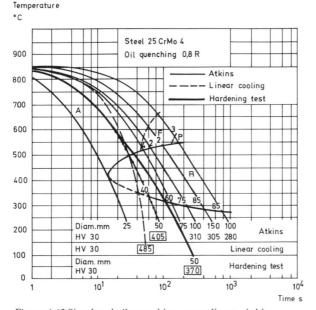

Figure 4.62 Simulated oil quenching, according to Atkins, including curves for linear cooling and of hardening tests with steel S S 2225 (25 CrMo 4)

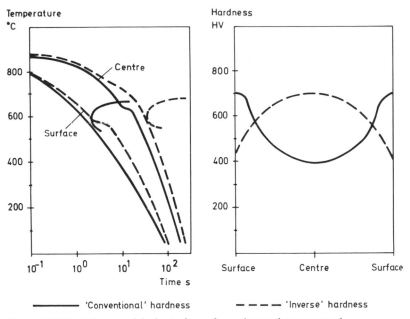

Figure 4.63 The influence of the heat of transformation on the progress of cooling and transformation and on the hardness

4.4.5 Computing CCT diagrams from TTT diagrams

TTT diagrams are quite easy to construct. Hence several attempts have been made to compute CCT diagrams by mathematical methods from TTT diagrams. Very successful work along these lines has been accomplished by Hildenwall[27] who, by applying his computer program has succeeded in computing the temperature trend, phase transformations and hardness. *Figure 4.64* shows the flow sequence for the program and how it functions during the course of a straightforward computation.

The computer program consists in the main of three substages at which the following is worked out:

the temperature distribution in the specimen as a function of time and depth below the surface;
the transformation of austenite to martensite, ferrite, pearlite and bainite or cementite as a function of time and depth below the surface;
the build-up of internal stresses as a function of time and depth below the surface

The cooling time is divided up into intervals and the specimen into small segments. For each time interval and specimen segment the temperature, the amount of newly formed phase and the internal stress are calculated. A number of numerical checks are added to the program. For each time interval and for each combination of temperature and phase content all pieces of information that have any bearing on subsequent computations

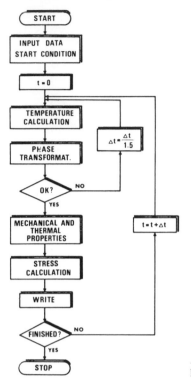

Figure 4.64 Schematic flowchart of the computer program (after Hildenwall[27])

are updated. When the ultimate state has been computed it is possible to work out the hardness profile for the specimen.

Temperature distribution is computed by solving the general equation for heat conduction adapted to those geometrical shapes that lend themselves to such treatment. The input data consist of such physical data as are functions of the temperature, i.e. the rate of heat flow between steel and the cooling medium, the specific heat capacity and heat conductivity along with the heat of transformation of austenite.

The rate of heat flow must be determined by cooling tests. When this has been found the temperature at all points of the specimen may be computed as a function of the time.

Computation of phase transformation is carried out according to the principle of dividing the cooling curve into a number of steps. By starting off from the mean temperature of each step the transformation taking place within the length of time set by the time interval is computed, using Avrami's exponential equation as a basis for the computation[28]. The transformation increment is treated as if the transformation of austenite to ferrite, pearlite and bainite took place isothermally.

The transformation to martensite is computed according to an equation developed by Koistinen–Marburger[29]. The input data required, besides the cooling curve, is a TTT diagram for the steel being investigated. It is possible to compute the amount of each phase for each segment of the specimen as a function of the time. By knowing which phases are present at every point in the body of the steel specimen it is possible to estimate the hardness of each phase in turn, given the alloy content of the steel and the rate of cooling. The total hardness may then be calculated from the expression below

$$H_{tot} = V_m \cdot H_m + V_{p+f} \cdot H_{p+f} + V_b \cdot H_b$$

where H = hardness (HV)
        V = volume fraction
        m = martensite
        b = bainite
        p + f = pearlite + ferrite

Internal stresses are computed by resolving the elastic–plastic problem. The effective elastic strain in each phase contributes to the total work hardening. It is assumed that the condition of flow, as stated by von Mises, holds as well as an associated flow law, propounded by Prandtl-Reuss. For each structural component the mechanical properties, i.e. modulus of elasticity, yield point and rate of strain hardening and their dependence on carbon content and temperature, are obvious components of the input data.

Figure 4.65 shows the result of a computation for the steel grade SS 2234 (34 CrMo 4). The agreement between the experimental and computed CCT diagram must be regarded as satisfactory.

The diagram, *Figure 4.66.* is based on investigations involving both dilatometric and conventional metallographic methods, the mean results of which are reported. The CCT diagram in *Figure 4.67* was computed by basing it on this diagram and assuming linear cooling. The points in the diagram represented the values obtained from the dilatometric tests with specimens of the steel under investigation.

It is possible to compute any mode of cooling. One such computation involving the different curves in *Figure 4.56* showed good conformity with the experimentally derived results. The reliability of the method may be enhanced by amending the input data.

Working on a research assignment within Jernkontoret, Hildenwall and Ericsson[30] have developed further the testing of computer programs for, *inter alia*, the computation of CCT diagrams.

4.5 Practical application of hardenability

There are two depth-of-hardening limits between which the hardenability should lie in order that the steel, after hardening, may be used to good technical and economic advantage. The upper limit is set by the hardenability that will result in an approximately 90% martensitic structure at the centre. A steel having a higher hardenability, i.e. a steel that in a heavier section would give the same hardness as in a smaller cross-section would be regarded as uneconomic when used in the smaller dimension, provided that only the price factor is considered and that price is equated with hardenability. For every tool and every constructional part there is a rather narrow range of hardenability that will give optimum results, but it is usually for highly stressed parts only that such results are required. From

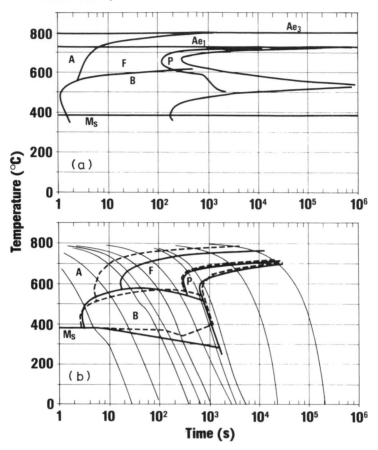

Figure 4.65 TTT diagram (a) and CCT diagram (b) for steel 34 CrMo 4.
The TTT diagram illustrates the input data representation. The CCT
diagram illustrated is computed (dashed lines) and experimentally
determined (solid lines) (after Hildenwall[27])

the point of view of stock-keeping and safety it often happens that steels
having a hardenability higher than necessary are used.

The lower limit, applicable mainly to unalloyed and low alloyed steels, is
set by that hardenability that will give a depth of hardening of about 3 mm,
this hardened zone having at least a 90% martensitic structure derived
from the heat treatment and the mass effect of the cross-section. (This does
not apply to case-hardening and other thermochemical surface treatment
processes.)

Which properties are concomitant with high hardenability and which
with low hardenability, respectively?

4.5.1   High hardenability

Where steels for hardening and tempering in large cross-sections are
concerned, the benefits of high hardenability are obvious. Such steels can
be used in heavy dimensions for parts that are later to be machined to

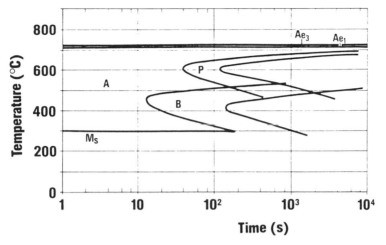

Figure 4.66 TTT diagram based on experimental isothermal test data for a steel with the following chemical composition: 0·39% C, 0·21% Si, 0·86% Mn, 0·72% Cr, 0·97% Ni, 0·04% Mo, 0·012% N, 0·026% Al (after Hildenwall[27])

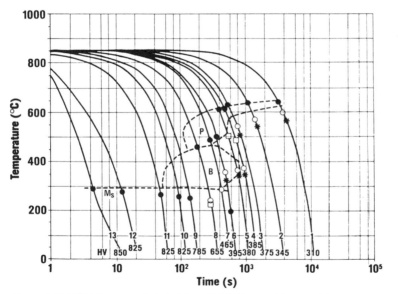

Figure 4.67 CCT diagram calculated from TTT diagram illustrated in *Figure 4.66* (after Hildenwall[27])

various cross-sectional areas and for which strength requirements are equally high in the smaller and in the heavier sections. Unfortunately the test methods as stipulated by both SS and ISO are concerned only with a zone extending to about 20 mm below the surface.

Even if the hardness is the same at the surface as at the centre in a hardened and tempered bar it must be accepted generally that the impact strength of the central regions of a bar are often lower than at the surface. This is because the surface zones, as a rule, consist of tempered martensite or lower bainite whereas at the centre the structure is composed mainly of upper bainite or pearlite. These two latter constituents have considerably lower impact strength and higher transition temperature ranges than tempered martensite or lower bainite. As an example of this, *see Figure 4.68* which shows the microstructure and impact values of two different bars, steel SS 2225 (25 CrMo 4), the composition and hardness of which were in close agreement. In spite of this, the steel (in *a*) which contained upper bainite, failed on welding. In contrast to this, the steel, the structure of which was mainly lower bainite (*b*), gave excellent service on being welded. In the first case the rate of cooling from the hardening temperature had not been rapid enough. When service requirements really do call for high impact values, not only should the composition of the steel be considered but also the microstructure which may be influenced by the heat treatment.

Another instance when high hardenability is required, concerns hardened and tempered forged blooms for drop-forging dies or pressure diecasting dies. These forgings are engraved subsequent to the heat treatment. Tools subjected to high pressures must also be made from steel that hardens through entirely, otherwise the softer core might be deformed. In such cases the surface layers will be exposed to excessive stresses which are liable to cause this layer to rupture.

4.5.2   Low hardenability

In many cases a low hardenability is to be preferred, both for machine parts and for tools. A shallow-hardening steel, such as W1, can give a

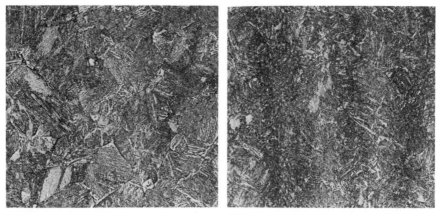

(a)   (b)

Figure 4.68 The influence of the microstructure on the impact strength of hardened and tempered steel SS 2225 with 0·25% C, 1% Cr and 0·2% Mo. (a) Structure composed mainly of upper bainite. Grain size 8·5 ASTM, hardness 285 HB, impact value Charpy V, 30 J, (b) Structure composed mainly of lower bainite. Grain size 9·5 ASTM hardness 285 HB, impact value Charpy V, 67 J

remarkably good performance. Long experience tells us that such steels can stand up to impact stresses extremely well and it is said they can do so because of the presence of a hard case and a tough core. This point will be dealt with further on in this chapter.

Tools with varying cross-sectional areas often need to have an even depth of hardness on all working faces. For such uses steel SS 2092 has shown itself to be eminently serviceable under working conditions. This has been subsequently verified by metallographic investigations and practical heat treatments. These investigations, which are of interest since they tend to verify the deductions drawn from data obtained from Jominy diagrams, are described below.

The steels used for cold pilger rolls are SS 2092 (~90 CrSi 5) and AISI O 1. In order to harden a pilger roll made from O 1, outer diameter 340 mm, cross-sectional area 145×82 mm^2, it was quenched in oil from 830°C. The hardness at the corners was 55 HRC but on the cylindrical surfaces and in the groove the hardness was only 45 HRC. The appearance of the fractured surfaces is shown in *Figure 4.69*. Since it was suspected that the hardening temperature used for this rather large roll was too low, a temperature of 860°C was tried next. A hardness of 62 HRC on most surfaces was the result. As may be seen in *Figure 4.70* the roll had hardened throughout the section, shown in the photograph on the right, whereas in the other section (*see* photograph on the left) the result of the hardening was unsatisfactory both in the central area of the roll and at one surface.

Yet another ring roll was hardened, a temperature of 890°C being used this time. Full hardness of 62 HRC was realized all over the roll but it cracked in service after a short time. Metallographic examination revealed

Figure 4.69 Fractured surfaces of ring roll made in steel O 1, oil hardened from 830°C. The dark areas on the top part of the photograph are rounded notches to initiate the fracturing of the roll

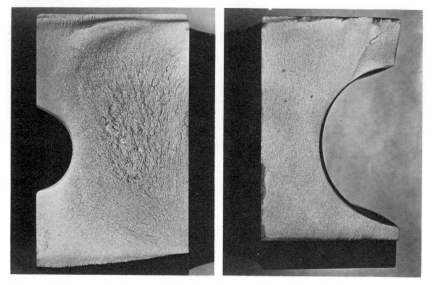

Figure 4.70 Fractured surfaces of ring roll made in steel O 1, oil hardened
from 860°C

that the roll had a very coarse fracture, viz. fracture No. 5 according to the
J K scale, and it had hardened throughout the whole section. It was
apparent that to find the ideal hardening temperature for this tool was
quite a problem.

Examination of a similar roll made from steel S S 2092 and hardened
from 860°C revealed the fracture as shown in *Figure 4.71*. The hardened

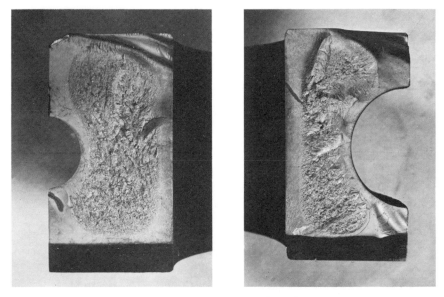

Figure 4.71 Fractured surfaces of ring roll made in steel S S 2092, oil
hardened from 860°C

zone has approximately the same depth all the way round. Rolls made in this steel give satisfactory service.

In order to obtain a proper insight into the behaviour of these various steels a systematic investigation was initiated which involved the hardening of Jominy test-pieces from different hardening temperatures. The results are shown in *Figure 4.72* and *4.73*. From these curves we see that steel SS 2092 has the lowest hardenability and that the depth of hardening and grain size are least influenced by the hardening temperature.

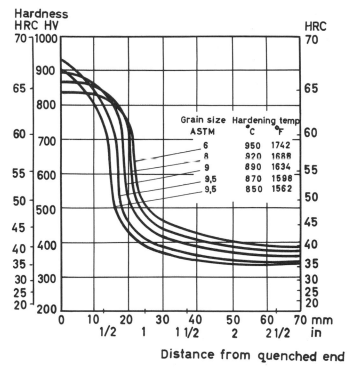

Figure 4.72 Steel S S 2092. Jominy curves and A S T M grain size obtained with different hardening temperatures

By referring back to the curves in *Figures 4.34* to *4.39* it is possible to construct depth-of-hardening diagrams for different rates of cooling. The different severities of quench were as follows:

$H = 0.5 = $ oil quench, good agitation,
$H = 0.7 = $ oil quench, strong agitation,
$H = 1.0 = $ oil quench, violent agitation.

The calculations were carried out for a bar diameter of 120 mm, which is roughly equivalent to the cross-sectional area of the ring rolls. *Figures 4.74* and *4.75* show the results for $H = 0.5$ and 1.0, respectively. Since the depths of hardening down to 60 HRC are of particular interest these values were read off from the curves in the above-mentioned figures (*see Table 4.2*). From this table it is clear that in the dimensions concerned, steel SS 2092 obtains the most uniform hardness penetration, 6–9 mm, from a hardening temperature of 890°C. We see that on quenching from this temperature the steel is independent, by and large, of cooling system rates lying between the limits $H = 0.5$ and $H = 1.0$. By increasing the

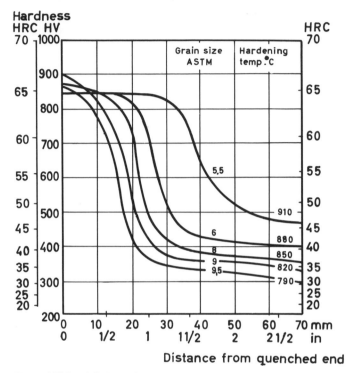

Figure 4.73 Steel O 1. Jominy curves and ASTM grain size obtained with different hardening temperatures

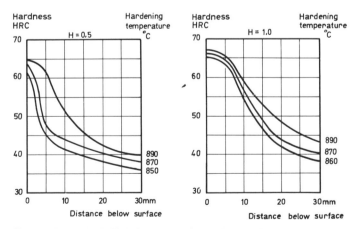

Figure 4.74 Steel SS 2092. Diameter of bar: 120 mm. Estimated transverse hardness curves after hardening from different temperatures and using various cooling rates

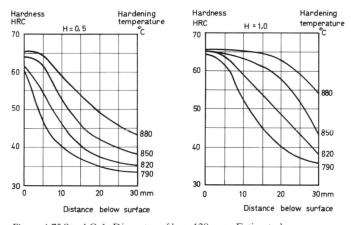

Figure 4.75 Steel O 1. Diameter of bar: 120 mm. Estimated
transverse hardness curves after hardening from different
temperatures and using various cooling rates

cooling rates to values lying in the range of, say, $H = 0.7$ to $H = 1.0$ a
hardening temperature between 850 and 890°C may be used. Should 890°C
not be high enough the temperature may be increased by some 20° without
causing undue grain growth in the steel, as may be seen in *Figure 4.72*.

As regards O 1, the intermediate hardening temperatures of 850°C or
860°C can be used but any change in the depth of hardening is largely
dependent on the rate of cooling. It is not advisable to try the higher
temperatures if the steels are to remain fine-grained. For sections
exceeding 120 mm in diameter and for which increasing depths of
hardening are required the hardenability of steel SS 2092 is not high
enough to render this steel suitable for oil quenching.

With regard to the steels under investigation only steel SS 2092 is
capable of being what is known as contour hardened. This concept will be
dealt with further in Chapter 6.

Table 4.2 Depth of hardening (mm) down to 60 HRC

Grade	SS 2092			O 1		
Hardening temperature°C	850	870	890	820	850	880
$H = 0.5$	1	2	6	2	7	9
$H = 0.7$	6.5	7	8	7	12	20
$H = 1.0$	7	8	9	8	18	24

4.6 The influence of the depth of hardening on the stress pattern

In the preceding section we have discussed the advantages attendant upon
a favourable depth of hardening. In addition, an instance has been quoted
illustrating the benefits accruing to a steel possessing a limited

hardenability. However, this is only one aspect of the matter. The residual stresses that are created as a consequence of the hardening operation are of such importance that they must be taken into consideration when a steel and its heat treatment are being selected.

As a result of the hardening operation, thermal and transformation stresses are set up in the steel. If we regard the thermal stresses only and reckon them as tangential stresses, we obtain a stress distribution as shown in *Figure 4.76,* which applies to a bar 100 mm in diameter and quenched in water from 850°C[31]. At the point of time *T* the temperature difference between the surface and the centre is greatest, i.e. some 500°C. This corresponds to a linear differential of about 0·6% and gives rise to a stress that exceeds 1000 N/mm². If the body of the material were elastic enough to accommodate this stress, the stress diagram for the surface layer would take the shape of curve *a.* After temperature equalization between the surface and the centre the stress should fall to zero. Since the yield point of steel is considerably lower at elevated temperatures than at room temperature the material will flow plastically, resulting in a surface stress diagram, such as curve *b.*

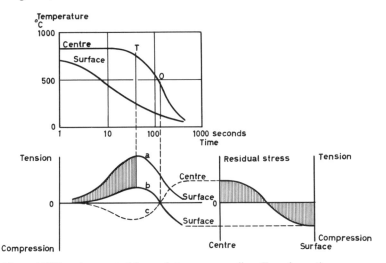

Figure 4.76 Development of thermal stresses on cooling. Transformation stresses not taken into account (after Rose[31])

At the centre the stresses will be such as indicated by curve *c.* When time *T* has elapsed, contraction in the core will exceed that at the surface. This implies that at time *O* there is complete stress neutralization and consequently at room temperature there will be compressive stresses in surface and tensile stresses at the centre.

When the bar has cooled to room temperature, the stress distribution will be as shown in the right-hand figure, i.e. tensile stresses at the centre and compressive stresses in the surface. The more effective the quenching medium and the larger dimensions of the bar, the greater will be the stresses.

The phenomena involved when steel *hardens* produce additional stresses due to the increase in volume that takes place as austenite transforms to other structures. The following increments of volume and linear expansion, *Table 4.3*, apply to a 1% plain carbon steel. If the main part of the transformation takes place in the centre before the surface starts to transform, which is the course of events in shallow-hardening steels, the same stress pattern as has been described in *Figure 4.76* is set up. If the transformation to, say, pearlite has proceeded to 100% completion at the centre before martensite has started to form in the surface, the compressive stresses increase considerably when martensite formation starts. Such an instance is shown schematically in *Figure 4.77*. Up till time t_1 tensile stresses are created in the surface and compressive stresses at the

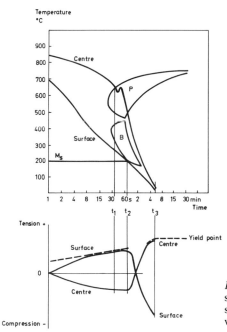

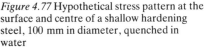

Figure 4.77 Hypothetical stress pattern at the surface and centre of a shallow hardening steel, 100 mm in diameter, quenched in water

centre. When pearlite formation starts after time t_1 there is an incremental volume increase in the core at the same time as the fall in temperature produces an incremental volume decrease. Which of these phenomena is the overriding one is quite immaterial since the yield strength of the surface layer determines the maximum stress there. After time t_2 in this hypothetical case, pearlite formation at the centre has been completed just

Table 4.3

Transformation	Change in volume %	Change in length %
Austenite to martensite	+4·2	+1·40
Austenite to lower bainite	+3·2	+1·07
Austenite to pearlite	+2·4	+0·80

as the martensite starts to form in the outer layers. The volume increase in the surface and the contraction of the centre both combine to produce a rapid stress reversal. The compressive stresses in the martensitic outer layers may assume very high values. The yield strength of the central pearlitic regions is comparatively low and as the thickness of the outer martensitic layers increases, the tensile stresses at the centre increase; in fact they may gradually reach such a critical value that fracture takes place, often explosively.

If the steel through-hardens the stress situation, as it is schematically illustrated in *Figure 4.78,* may quite well develop. As they cool down to M_s the outer layers contract, thereby building up tensile stresses and creating correspondingly compressive stresses at the centre. When, after time t_1 the

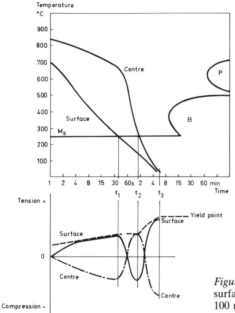

Figure 4.78 Hypothetical stress pattern at the surface and centre of a through-hardening steel, 100 mm in diameter, quenched in water

martensite transformation starts in the outer layers there is an increase in volume which reduces the tensile stresses somewhat. By the time t_2 is reached there are compressive stresses in the outer layers and tensile stresses at the centre. The fall in temperature at the centre, being greater there than in the surface layers, aggravates this stress situation. A critical state is thus created for the central region but since the latter has an austenitic structure its ductility is good and owing to plastic deformation some degree of stress relief takes place. For this reason crack formation in the central regions is a rare occurrence at this stage. When the temperature in this region has reached M_s a volume expansion sets in and as the temperature continues to fall the outer layers are subjected to tensile stresses which may become so severe that the steel cracks, often longitudinally. The risk that such cracking will occur increases with increasing section size and hardenability. Therefore the greater the hardenability of the steel the milder should be the quenching medium in order to avoid hardening cracks in through-hardening steels.

From the above discussion it follows that there are two size groups that are most critical from the hardening point of view. One such size group among which most cracking actually occurs is the maximum section size in which the steel through-hardens. In this instance tensile stresses exceeding the tensile strength of the steel may build up in the outer layers. Cracking due to such causes is favoured by stress raisers on the surface and also by a high cooling rate. The other size group is the one where cracks start at the centre. This takes place when sufficiently high compressive stresses have been built up in the outer layers and correspondingly high tensile stress at the centre. In addition, this type of cracking is favoured by flaws, specially in the central regions of the steel and also by a high rate of cooling. *Figure 4.79* shows in diagrammatic form the two critical section size groups discussed, the type of crack formation that may occur and the prevailing stress patterns.

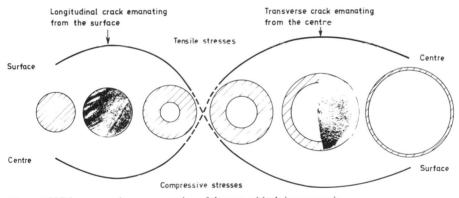

Figure 4.79 Diagrammatic representation of the two critical size groups in which cracking may occur when the steel is being fully heated and quenched. Hatched areas represent martensite.

Rose and Bühler have carried out measurements on different specimens. One such series of measurements, shown in *Figure 4.80,* deals with bars of various diameters, quenched in water. We can see how the dimension is instrumental in shaping the stress pattern.

The stress distribution in the 100 mm bar corresponds roughly to that due solely to thermal stresses. If the nose of the pearlite curve had been located farther to the left an additional increment of stress would have resulted.

With regard to the 30 mm diameter bar the start of the bainite formation at the centre coincides approximately with that of the martensite in the surface. Compressive stresses were measured both in the surface and at the centre and tensile stresses in the intervening zones.

The 10 mm diameter bar is hardened through completely. The temperature difference between the surface and the centre is smaller than that obtaining in larger-diameter bars. Furthermore, the main part of the martensite formation in the centre takes place when the surface-zone

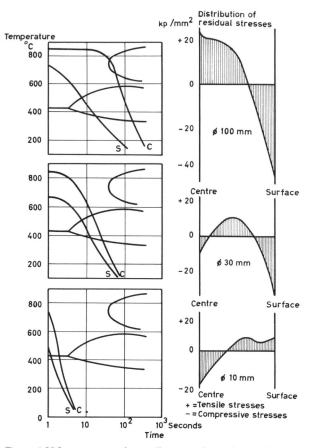

Figure 4.80 Stress pattern for cooling, transformation and residual stresses in round test bars of different dimensions in steel DIN 22 CrMo 44 (after Rose[31])

the creation of tensile stresses in the surface and compressive stresses at the centre.

The development of similar stress patterns may be envisaged if the alloy content of the steel is increased or if the hardening temperature is raised, the dimensions of the bars being kept constant (*see Figure 4.81*). There are endless variations possible along these lines. Let us limit our studies to two cases only.

1. Hardening of a plain-carbon steel so that the result will be compressive stresses in the surface and tensile stresses at the centre.
2. Hardening of an alloy steel so that the result will be tensile stresses in the surface and compressive stresses at the centre.

In the first case the steel hardens at the surface and in the other case it hardens through entirely. The stress distribution is shown in *Figure 4.82*.

We shall discuss these cases with regard to the incidence of crack initiation on bending, assuming that the strength of the surface layer is the

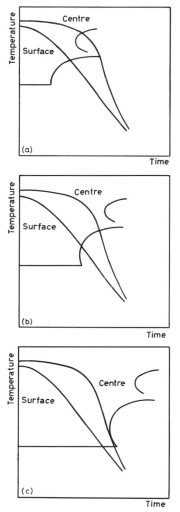

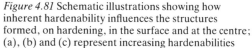

Figure 4.81 Schematic illustrations showing how inherent hardenability influences the structures formed, on hardening, in the surface and at the centre; (a), (b) and (c) represent increasing hardenabilities

same for both bars and the hardening depth is at least 10% of the bar diameter for the surface-hardened bar. On subjecting the bars to bending forces, as illustrated in *Figure 4.82* tensile stresses are induced in the lower layers and compressive stresses in the upper ones. A crack is more easily initiated in the layer subjected to tensile stresses. In order to balance the compressive stresses in the underside of the surface-hardened bar a definite bending force must be applied. This force, if applied to the through-hardened bar, will simply cause the tensile stresses on the underside to increase. On increasing this bending force additional tensile stresses are induced in the underside of this bar which will therefore be the first one to fracture, always assuming that there is an ample depth of hardening in the surface-hardened bar. What has been said above holds in a higher degree for fatigue and impact stresses.

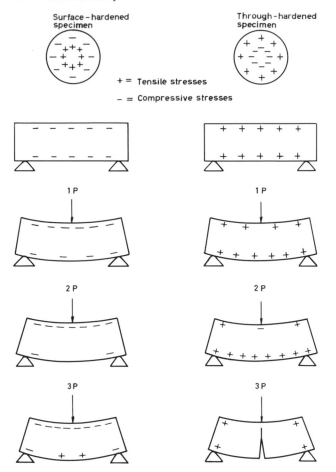

Figure 4.82 Schematic sketch showing stress distribution, produced on bending, in surface-hardened and through-hardened test bars

In practice it has often been observed that surface hardening is an excellent means of increasing the endurance limit, whether by making use of shallow-hardening steel or by special methods of surface hardening, such as induction hardening or case hardening. Similarly, a long service life has been given by surface-hardened steels when used as tools subjected to impact stresses. A necessary condition is, however, that all working faces have been hardened.

To sum up, it may be said that whenever practicable from the point of view of the heat-treatment process, surface-hardened or alternatively not completely through-hardened tools and machine parts are to be preferred, due consideration of course being given to the service that the parts are subsequently to be subjected to.

References

1. GROSSMANN, M. A., ASIMOW, M., and URBAN, S. F., 'Hardenability of Alloys Steels', Cleveland, Ohio, USA, *ASM* (1939)
2. MOSER, A., and LEGAT, A., 'Determining Hardenability from Composition', *Härterei-Techn. Mitt.*, **24**, No. 2, 100–105 (1969) (in German)
3. KRAMER, J. R., SIEGEL, S., and BROOKS, J. G., 'Factors for the Calculation of Hardenability', *AIME Trans.*, **167**, 670–697 (1946)
4. JATCZAK, C. F., 'Determining Hardenability from Composition', *Met. Prog.* 60–65, (September 1971)
5. JOMINY, W. E., and BOGEHOLD, A. L., 'A Hardenability Test for Carburizing Steel', *Trans. ASM*, **26**, 574–606 (1938)
6. JUST, E., 'Hardenability Formulae', *Härterei-Techn. Mitt.*, **23**, No. 2, 85–100 (1968) (in German)
7. JUST, E., 'Determining Hardenability from Composition', *Draht*, **19**, No. 3, 145–146 (1968) (in German)
8. KIRKALDY, J. S., 'Prediction of Alloy Hardenability from Thermodynamic and Kinetic Data', *Met. Trans.* **4**, 2327 (1973)
9. KIRKALDY, J. S., PAZIONIS, G., and FELDMAN S. E., 'A Universal Predictor for the Hardenability of Hypo-eutectoid Steels', *Heat Treatment*, **76**, Metals Society, London (1976)
10. ERICSSON, T., *Calculation of Hardenability Indices*, Jernkontoret's Research Report No. D 259 (1978) (in Swedish)
11. LAMONT, J, L., 'How to Estimate Hardening Depth in Bars', *Iron Age*, **152**, 64–70 (October 1943)
12. LIEDHOLM, C. A., 'Continuous Cooling Transformation Diagram from Modified Endquench Method', *Met. Prog.*, **45**, No. 1 (1944)
13. BLICKWEDE, D. J., and HESS, D. C., 'On the Cooling Transformation in Some 0·40% Carbon Constructional Alloy Steels', *Trans. ASM*, **49**, 427–444 (1957)
14. WEVER, F., ROSE, A., PETER, W., STRASSBURG, W., and RADEMACHER, L., 'Atlas zur Wärmebehandlung der Stähle', *Teil I und II* (1954/56/58)
15. BANDEL, G., and HAUMER, H. C., 'Determining by Calculation the Full-hardening Characteristics of Long-section Forgings', *Stahl u. Eisen*, **84**, 15, 932–946 (1946) (in German)
16. RUSSEL, T. F., *First Report on the Alloy Steel Research Committee*, Iron Street Inst. Spec. Rep. No. 14, 149–187, London (1936)
17. ATKINS, M., and ANDREWS, K. W., 'Continuous Cooling Transformation Diagrams I. Cooling Curve Analysis for Transformation Diagrams and Heat Treatment', *BSC Report* SP/PTM/6063/–/7/C/
18. ATKINS, M., *Atlas of Continous Transformation Diagrams for Engineering Steels*, British Steel Corporation, BSC Billet, Bar and Rod Product, Sheffield (1977)
19. THELNING, K–E., 'CCT-diagrams with Natural Cooling', *Scand. J. Metall*, **7**, 252–263 (1978)
20. HENGERER, F., STRÄSSLE, B., and BREMI, P., 'Computing the Course of Cooling in Oil and in Air of Cylindrical and Flat Specimens of Alloyed Heat-treatable Steels by Means of an Electronic Computer', *Stahl u. Eisen* **89**, 641–654 (1969) (in German)
21. LORIA, E. A., 'Transformation Behaviour on Air Cooling Steel in A_3–A_1 Temperature Range', *Metals Technology*, 490–492 (October 1977)
22. GRANGE, R. A., KILHEFNER JR, P. T., and BITTNER, T. P., 'Austenite Transformation and Incubation in an Alloy Steel of Eutectoid Carbon Content', *Tans. ASM*, **51**, 495–519 (1959)
23. KLIER, E. P., and YEH, T. H., 'The Decomposition of Austenite in 4340 Steel during Cooling', *Trans. ASM*, **53**, 75–93 (1961)
24. SCHMIDTMANN, E., GRAVE, H., KLAUKE, H., 'Influence of Compressive Stresses on the Course of the Transformation when Heavy Steel Forgings, Grade 21CrMoV511, are being Quenched', *Stahl u. Eisen*, **96**, No. 23, 1168–1175 (1976) (in German)
25. SHIMIZU, N., and TAMURA, J., 'An Examination of the Relation between Quench-hardening Behaviour of Steel and Cooling Curve in Oil', *Trans. ISIJ*, **18**, 445–450 (1978)
26. TENSI, H. M., and SCHWALM, M., 'Resulting Effect of Liquid Quenching Media, with Reference to Specific Aqueous Solutions of Synthetic Plastics (Polyethylene Oxide)', *Härterei-Techn. Mitt.*, **35**, 122–131 (1980) (in German)

27. HILDENWALL, B., 'Prediction of the Residual Stresses Created During Quenching', *Linköpings Universitet*, Dissertation No. 39 (1979)
28. AVRAMI, M., 'Kinetics of Phase Change', *J. Chem. Phys.*, **7**, 1103–1112 (1939), 212–224 (1940), 177–184 (1941)
29. KOISTINEN, D. P., and MARBURGER, R. E.., 'A General Equation Prescribing the Extent of the Austenite-Martensite Transformation in Pure Iron Carbon Alloys and Plain Carbon Steels', *Act. Metall.*, **7**, 59–60 (1959)
30. HILDENWALL, B., and ERICSSON, T., *Verifying Computer Programmes for Determining Cooling Rate, Residual Stress and Distortion in Steel*, Jernkontoret's Research Report No. D 357 (1981) (in Swedish)
31. ROSE, A., 'International Stresses Resulting from Heat Treatment and Transformation Processes', *Härterei-Techn. Mitt.*, **21**, No. 1, 1–16 (1966) (in German)
32. THELNING, K-E., 'Why Does Steel Crack on Hardening?, *Härterei-Techn*. Mitt., **25**, 271–281 (1970) (in German)
33. DOANE, D. V. and KIRKALDY, J. S., 'Hardenability Concepts with Applications to Steel', *AIME* (1978)

5

Heat treatment—general

The preceding chapters have dealt with the theoretical background of the heat treatment of steel. The following chapters are based mainly on practical experience and applications. This general section describes the standard heat-treatment on which are based the special methods described in the next chapter.

5.1 Annealing

The general purpose of an annealing treatment is to reduce the hardness of steel or to produce a structure that facilitates the progress of subsequent manufacturing operations. The term *annealing* by itself usually implies *full annealing*, which in most cases involves solution treatment in the austenite phase followed by slow cooling. For other types of annealing prefixes that indicate the nature of the process are generally used.

5.1.1 Spheroidizing anneal or annealing for maximum softness

The theory of this apparently simple heat-treatment operation is still rather obscure but on the basis of practical experience and on data derived from isothermal (TTT) and continuous cooling (CCT) transformation diagrams, almost ideal annealing practices have been developed. The classic spheroidizing annealing temperatures for carbon steel are show in *Figure 5.1*, where it can be seen that for hypo-eutectoid steels annealing below A_1 is suggested. However, annealing above A_1 followed by transformation just below A_1 is to be recommended in many instances as outlined below.

When a steel with a pearlite structure is annealed the cementite lamellae after a short time assume a 'knucklebone' shape. On continuing the annealing the lamellae form globules at their ends and split up into spheroids; hence the name spheroidizing *see Figure 5.2. Figure 5.3* is typical of the fully spheroidized structure of a tool steel. If the annealing treatment is carried out below A_1 the rate of cooling does not influence the final hardness at room temperature. If annealed above A_1 and followed by

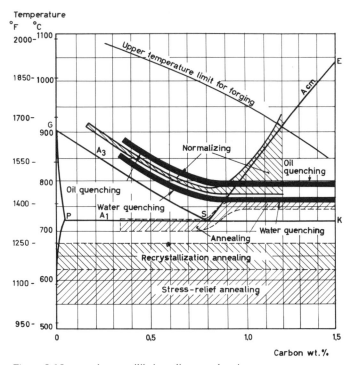

Figure 5.1 Iron-carbon equilibrium diagram showing temperature regions for various heat-treatment operations

slow cooling, the spheroidized structure is still largely maintained provided that the temperature was only a few tens of degrees above A_1. As the steel cools, the carbon dissolved in the austenite will separate out on the carbide spheroids. If the annealing temperature is higher, a greater amount of carbide will dissolve and the cementite will separate out as lamellae. *Figure 5.4* shows how the amount of lamellar structure diminishes as the annealing temperature is reduced. For the steel in question an annealing treatment at 750°C results in 100% spheroidization and at the same time gives the lowest hardness. However, such a microstructure may be rather unsuited to certain machining operations, such as milling, drilling or reaming. A higher annealing temperature would probably result in a

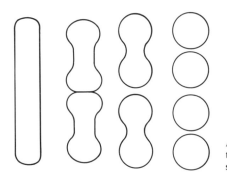

Figure 5.2 Schematic representation of the transformation of a cementite lamella to spheroids during full annealing

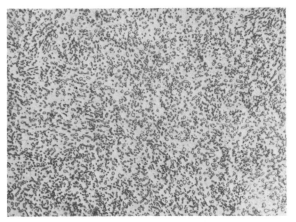

Figure 5.3 Typical annealed structure showing completely spheroidized carbides

microstructure better suited to such operations. The actual temperature below A_1 at which the transformation takes place has an important bearing on the appearance of the structure obtained. The closer the temperature of transformation lies to A_1 the coarser and softer will be the spheroidized structure. The farther away below A_1 the transformation takes place the

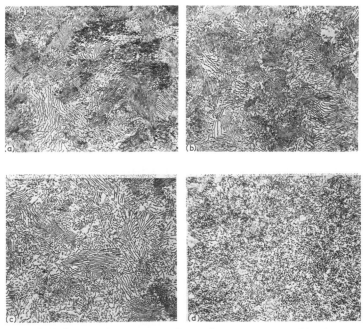

Figure 5.4 Influence of temperature of annealing on structure and hardness of a steel 0·60% C, 0·60% Cr. Holding time: 2 h. Rate of cooling: 10–15°C/h to 650°C. (a) Annealing temperature 820°C, hardness 205 HB; (b) Annealing temperature 800°C, hardness 195 HB; (c) Annealing temperature 775°C, hardness 185 HB; (d) Annealing temperature 750°C, hardness 174 HB. All 250 ×

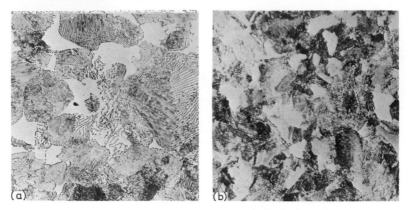

Figure 5.5 Influence of transformation temperature during isothermal annealing on structure and hardness of steel BS 708A42. On account of the high annealing temperature, the pearlite is practically wholly lamellar. (a) Annealing temperature 900°C, transformation temperature 700°C, hardness 194 HB; (b) Annealing temperature 900°C, transformation temperature 625°C, hardness 225 HB. Both 250 ×

finer, more lamellar and harder will be the structure, which is apparent from *Figure 5.5*. In this instance the annealing temperature was unusually high and as a result practically no spheroidization took place.

The major rule to apply when undertaking a spheroidizing anneal of plain-carbon and low-alloy steels may be summarized as follows:

Austenitize at a temperature not more than 50°C above A_1 and transform at a temperature not more than 50°C below A_1.

Such an annealing with partial austenitizing may be carried out according to either of the two following operations:

1. *Annealing followed by slow cooling through A_1*. This annealing operation, which is the most usual one, is very time-consuming but is simple to carry out in a conventional furnace. An annealing cycle according to this operation, designated 1, is drawn in the TTT diagram, *Figure 5.6*, using the following values:

Partial austenitizing at 750°C (4 h)
Cooling to 700°C at the rate of 10°C/h
Free cooling to room temperature (approx. 2 h)

2. *Annealing followed by rapid cooling to a temperature below A_1*. The holding time required for the steel at the temperature below A_1 may be read off from the appropriate TTT diagram. This isothermal annealing operation requires less time than the above-mentioned conventional operation but it requires furnace facilities for rapid temperature control. A pusher-type furnace with different temperature zones is ideal for this sort of annealing process.

Two isothermal annealing treatments, designated (2) and (3) are drawn in *Figure 5.6*. The upper curve (2) which gives the lower hardness is based on the following data:

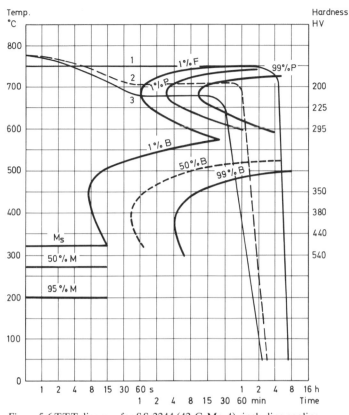

Figure 5.6 TTT diagram for SS 2244 (42 CrMo 4), including cooling curves for a conventional full-annealing treatment (1) and isothermal treatment (2) and (3)

Austenitizing temperature 775°C, holding time 1 h
Rapid cooling to 710°C, holding time 1 h
Free cooling to room temperature, approx. 2 h

After this treatment the hardness was lower than 200 HV.
For the lower curve (3) the following applies:

Austenitizing temperature 775°C, holding time 1 h
Rapid cooling to 675°C, holding time approx 15 minutes
Free cooling to room temperature, approx 2 h

After this treatment the hardness was about 210 HV.

For annealing treatments it is not quite correct to refer to conventional TTT diagrams that are intended for hardening treatments, i.e. for treatments at temperatures considerably higher than those used for annealing. As a consequence of the incomplete carbide dissolution usually resulting from the annealing treatment, the isothermal transformations take place more rapidly than they would following a complete carbide dissolution. However, owing to the fairly slow cooling through the upper metastable temperature range the transformations will be displaced towards longer times and since the rather fast cooling rates implied in curves (2) and (3) are normally not attainable, the conventional TTT diagrams may therefore be employed for practical purposes.

Hülsbruch and Theiss[2] have carried out extensive annealing trials with a ball-bearing steel of composition:

1·0% C, 0·3% Si, 0·6% Mn, 1% Cr

They found that the required hardness, approximately 200 HB, and a well-spheroidized cementite are obtainable from an isothermal annealing operation such as, for example:

Austenitizing treatment at 800°C for 1 h
Transformation at 700°C for 2 h

Subsequent cooling took place in water in order to show that the speed of cooling after transformation had no influence on either the hardness or microstructure. The conventional full-annealing process used for this ball-bearing steel took considerably longer owing to more time-consuming operations, viz.

Austenitizing treatment at 780°C for 4 h
Cooling to 720°C—8 h
Holding time at 720°C—4 h
Furnace cooling to 580°C
Free cooling to room temperature

Hyper-eutectoid steels must be annealed above A_1 in order to spheroidize the grain-boundary cementite within a reasonable time. Even so, to obtain acceptable annealing results in a steel with a pronounced network of grain-boundary cementite may be quite a problem. A prior normalizing treatment may help to dissolve the network. After the steel has cooled from the normalizing temperature the carbides are present in a more finely dispersed form.

Full annealing of cold-worked steel

As mentioned above the conventional full-annealing treatment is very time consuming if a 100% spheroidized structure is aimed at. By performing a cold-work operation before annealing the annealing time may be shortened considerably, the time reduction increasing as the degree of cold work increases. The reason for this is that the derangement of the crystal lattice caused by the cold work promotes recrystallization during the subsequent annealing (*see* Section 5.1.2) which thereby facilitates the spheroidization of the cementite[3, 4, 5]. The cold-work deformation increases the solubility of ferrite for carbon and consequently increases its rate of diffusion.

Figure 5.7 shows how the effect of various amounts of cold work on carbon steels (0·14–0·64%) reduces the required annealing time at 697°C[3].

For certain low-carbon steels a cold-work reduction of less than 20% occasions no shortening of the annealing time, in fact, annealing produces a higher hardness in the steel than it had before the cold work[6].

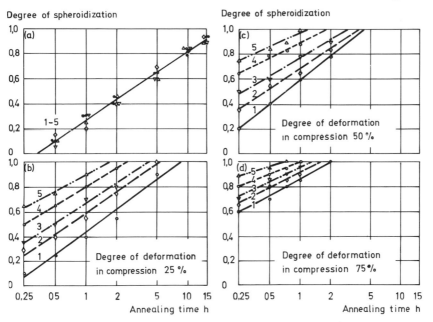

Figure 5.7 Influence of degree of deformation in compression and annealing time at 697°C on the degree of spheroidization for steels of various carbon content (after Köster and Fröhlke[3])

Steel	%C	Si	Mn	P	S
1	0·14	0·22	0·49	0·031	0·029
2	0·24	0·19	0·48	0·022	0·024
3	0·37	0·27	0·70	0·027	0·024
4	0·47	0·24	0·64	0·023	0·021
5	0·64	0·24	0·69	0·028	0·022

Full annealing of high-speed steel

With regard to high-speed and other high-alloy steels, where A_1 temperatures are around 800°C, the annealing temperature must be in the region of 875°C in order that the steels may obtain their lowest hardness. The grain size resulting from the hardening of high-speed steel is dependent on the temperature of the preceding annealing process. A low annealing temperature gives rise to a fine-grained structure and consequently a greater hardness, as is shown in *Figure 5.8*. The annealing temperature has a very definite influence on the grain size, which is also affected by other factors such as the degree of reduction during hot working. In general, the smaller the dimensions of the finished steel, the finer is the resulting grain size.

In principle, most steels can be rehardened without any intervening annealing treatment. Each successive hardening operation may be regarded as a normalizing treatment that has a grain-refining effect. However, this does not apply to high-speed steels nor to steels with high tungsten contents, e.g. hot-work steels[7]. If such steels are rehardened without an intervening and thorough full annealing treatment they are liable to become coarse-grained. This abnormal grain growth depends on the hardening temperature. As is shown in *Figure 5.9*, for each hardening temperature above approximately 1040°C there corresponds, during the second hardening operation, a critical temperature above which abnormal grain growth will start. If the first hardening treatment is carried out at a low temperature, say 1150°C, then during the second hardening treatment, grain growth will already have started as the temperature approaches 1100°C. If, however, the first hardening takes place from 1290°C there is no grain growth during the second solution treatment for hardening until the temperature exceeds about 1230°C.

The reason for this anomaly has not been definitely accounted for. One possible explanation is that the dissolution of carbides (mainly $M_{23}C_6$), which started during the first hardening treatment, is continuing during the

Intercept grain size

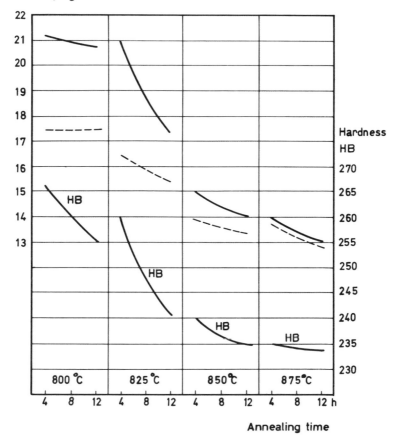

Figure 5.8 High-speed steel M 2. Influence of annealing temperature and time on hardness after annealing and on grain size after hardening. Daimeter of test-piece 18 mm

———————   1200°C hardening temperature
– – – – – –   1230°C hardening temperature

Intercept grain size

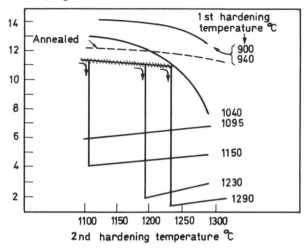

Figure 5.9 Dependence of grain size on hardening temperature when double-hardening high-speed steel, type T 1 (after Kula and Cohen[7])

second heating for hardening but that now the amount of carbide remaining is too small to prevent grain growth. A form of carbide precipitation occurs around 550°C and the higher the preceding hardening temperature has been, the more pronounced is this precipitation. Assuming that the first hardening temperature has been relatively high carbides will begin to be precipitated during the heating-up for the second hardening as a temperature of 550°C is approached. This precipitation might constitute an impediment to grain growth and as a consequence the critical temperature should be displaced upwards. If an annealing treatment is introduced at this stage it should cause the carbides to be precipitated more completely, thus impeding grain growth during subsequent hardening treatments. The phenomenon described above, however, does not manifest itself in other medium- or high-alloy steels such as H 13 or D 6.

Annealing for optimum machinability

The above heading appears in certain specifications. Apparently the implication is that the steelmaker is to carry out some annealing treatment that, along with giving the steel its best machinability is also the most economic one. The specification may imply, too, that there exists an understanding between the steelmaker and the consumer to the effect that a certain structure or a certain annealing treatment has been agreed upon.

An example of the above is the case-hardening steel B S 637 M12 (En 35), for which the maximum hardness has been set at 217 HB. Should this hardness be exceeded after rolling, an annealing treatment at 650°C for 1 h is normally carried out. If the annealing temperature is higher the hardness may become so low that the steel will be 'sticky' in certain machining operations. Similar steels with somewhat higher carbon contents possess the best machinability, e.g. in gear and spline shaping when they have been isothermally annealed (*see* later sections).

The phrase 'annealing for improved machinability' is also used. This treatment is applied to low-alloy steels that are to be subjected to forging operations after rolling, and usually consists of an anneal at 650–730 °C for 1 h. In such cases the maximum hardness must be stipulated, often 250 HB, in order that the steel may be cold sheared or sawn. Several grades of low-alloy structural steels possess optimum machinability when the pearlitic constituent consists of roughly equal parts of lamellar and spheroidized forms. For alloy steels it is a maxim that the lower their hardness, the better is their machinability. The structure should consist of completely spheroidized carbides in a ferritic matrix. The presence of grain-boundary bands of carbide globules or traces of lamellar pearlite drastically impairs machinability.

5.1.2   Recrystallization annealing

When steel is subjected to cold work its hardness increases and its ability to endure continued cold work diminishes. By annealing the steel above 600 °C recrystallization occurs, i.e. new, stress-free grains are formed which grow at the expense of the deformed original grains.

A prerequisite to recrystallization on annealing is that the degree of deformation has been large enough to produce the required number of defects in the crystals to initiate nucleation which is then followed by grain growth.

In general, for recrystallization to take place, the deformation should be equivalent to a reduction exceeding about 20%, but it is obviously dependent on the chemical composition of the steel. *Figure 5.10* shows how the amount of deformation in compression and the holding time on annealing affect the hardness of En 14A, a low-alloy steel[6].

Usually the holding time for a recrystallization anneal is ½–1 h at 650 °C. The time required for continuous annealing may be considerably shorter, but in that case a higher temperature must be used and more stringent stipulations with regard to the minimum degree of reduction are necessary.

Figure 5.11 shows the microstructure of a mild low-carbon steel that has been cold-work reduced by 20%. This reduction was necessary in order to initiate recrystallization during the subsequent annealing treatment, the latter being brought about by induction heating at 750 °C for 10 seconds. The microstructure after this treatment is shown in *Figure 5.12*.

Low-carbon steels, on being subjected to cold-work reduction of about 4%, may develop excessive grain growth when annealed slightly below A_{c3}.

Annealing for recrystallization is most commonly applied to cold-rolled, low-carbon sheet or strip steel. Austenitic steels, such as the 18/8 stainless and the 13% manganese (Hadfield) steels, also recrystallize on being annealed after cold work (*see* Section 5.1.5, Quench annealing). For these steels the temperature of recrystallization is higher than for carbon and low-alloy steels.

During the hot working of steel, grain deformation takes place which results in recrystallization. This occurs during the rolling operation and takes place between the passes or, when forging, between the forging blows.

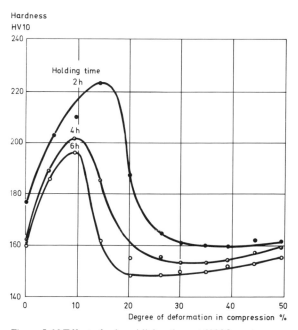

Figure 5.10 Effect of spheroidizing time at 680 °C on the hardness of En 14A deformed in compression (after Cooksey[6])

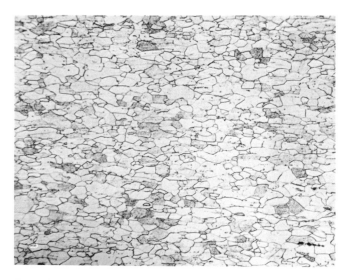

Figure 5.11 Low-carbon steel, 0·05% C, 20% reduction by cold work. Hardness 135 H V 200 ×

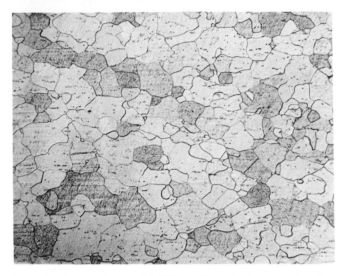

Figure 5.12 Low-carbon steel, 0·05% C, 20% reduction by cold work, subsequently annealed for 10 s at 750°C. Hardness 75 H V 200 ×

5.1.3   Stress-relief annealing

When steel is machined or deformed plastically, stresses are induced in to the cold-worked surfaces. These stresses may give rise to increased hardness which is apt to render continued working of the steel increasingly difficult. In addition they may cause the steel to distort during the subsequent heat-treating operation and should therefore be reduced or eliminated, which can be done by a 1–2 h stress-relief anneal. For plain-carbon and low-alloy steels a temperature of 550–650°C is required, for hot-work and high-speed steels, 600–750°C. This treatment will not cause any phase changes, but recrystallization may take place. In order that thermal stresses are not induced during cooling, it is good practice to allow the parts to cool slowly in the furnace to approximately 500°C after which they may be taken out and allowed to cool freely in air.

When very large tools or machine components requiring the greatest possible freedom from residual stresses are being stress-relieved, the cooling rate at the start must be very slow, i.e. a few degrees C only per hour. As the temperature gets lower the cooling rate may be increased but not until a temperature of about 300°C is reached is it advisable to cool freely in air. The reason for recommending an initial slow cooling when the steel is at its highest temperature is that the yield point is low and there is a danger that it might be exceeded by the stresses induced if too great a temperature difference exists between the surface and centre. If the thermal stresses cause permanent deformation, new stresses will remain after cooling to room temperature.

Hardened and tempered steel may be stress relieved at a temperature about 25°C below that used for tempering. Tools and machine components that are to be heat treated should be left with a machining allowance sufficient to compensate for any warping resulting from stress relieving.

Forgings should also be stress relieved if internal stresses are induced in the steel on account of rapid or uneven cooling, or if the forging has been subjected to severe straightening. In some instances a stress-relief treatment after normalizing is stipulated, for example for castings and weldments, especially those of alloy steel.

If the greatest possible freedom from internal stresses is required the annealing treatment should be carried out near the upper limit of the appropriate temperature range indicated above. Unfortunately this might not always be practicable since a high temperature may give rise to an unacceptably large amount of superficial oxidation or to too much softening in a hardened and tempered steel. In such cases it is necessary to choose a lower stress-relief temperature but which is as high as is compatible with a scale-free surface. However, a certain amount of residual stress must then be accepted.

Rosenstein[8] has suggested a method of assessing the effectiveness of the stress-relief treatment. The method has been tested on 13 steels having yield points ranging from 25 to 125 kp/mm^2, at temperatures between 482 and 593°C and for times up to 24 h. The specimens were inserted in a creep-testing machine and after being heated to the testing temperature they were loaded, to their approximate yield points at the temperature concerned. As stress relaxation proceeded progressively at the testing temperature the machine automatically reduced the applied stress so that the gauge length remained constant. The type of diagrams obtained is shown in *Figure 5.13*. Steel HY–100 has the following composition:

C	Cr	Ni	Mo	
0·18	1·65	2·91	0·42	%

The specimens were hardened and tempered at 620°C to give a yield strength of 74 kp/mm^2. During the first hour, according to *Figure 5.13*, stress relaxation occurs very rapidly but after that it slows down considerably.

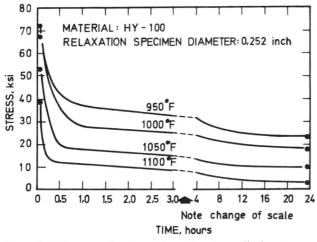

Figure 5.13 Stress as a function of time after stress relieving at various temperatures (after Rosenstein[8])

In Chapter 1, p. 23 it was shown how the relation between hardness and temperature and time of tempering may be represented by means of an expression derived by Hollomon and Jaffe. Rosenstein has suggested that this expression (which is also known as the Larsson–Miller parameter) can be applied equally well to stress relaxation when annealing for stress relief. Rosenstein's original diagram which contains values for all the steels tested, is shown in *Figure 5.14*. In *Figure 5.15* the diagram has been reconstructed and has been incorporated with a new type of tempering diagram showing the interdependence of time and temperature, since this type of diagram is easier to interpret. For further information *see Figure 5.96*.

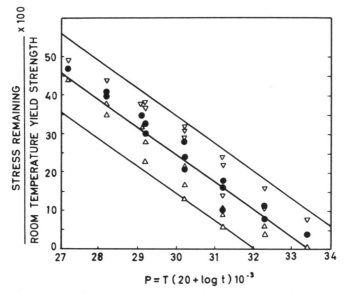

Figure 5.14 The Larson-Miller parameter for stress relieving. Diagram based on data obtained on heating to temperature between 900°F and 1100°F for 124 h. *T* in degrees Rankine; *t* in hours (after Rosenstein[8])

According to the mean curve in the diagram it can be seen that an anneal of 1 h at 450°C results in about 50% stress relief only. For complete stress relief it would be necessary to anneal at, say, 650°C for 1 h or at 600°C for 15 h. Rosenstein's results agree reasonably well with stress measurements carried out at Bofors on steel before and after annealing for stress relief.

When carrying out stress relief or tempering subsequent to a hardening treatment, a considerable stress reduction is obtained by heating to some lower temperature, provided that there is a decomposition of martensite. When a hardened and tempered steel is to be stress relieved the stress-relief diagram may be combined with the tempering diagram for the steel concerned, since both diagrams are functions of time and temperature. In this way it is possible to optimize the time and temperature for stress-relief annealing without jeopardizing the hardness.

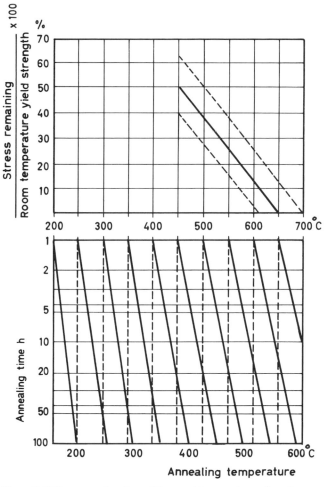

Figure 5.15 Stress as a function of time and temperature when stress relieving

5.1.4 Isothermal annealing

This treatment involves heating the steel at two different temperatures: first an austenitization treatment at a suitable temperature above A_1 and then a rapid lowering to the transformation temperature below A_1. When the transformation is complete the steel may cool freely to room temperature. Isothermal annealing curves are drawn in *Figure 5.6*. For practical reasons it is not possible to carry out such a heat treatment in a conventional muffle furnace since its large heat capacity does not permit the required rapid fall in temperature. Instead, use is made of two muffle furnaces or of continuous furnaces with different temperature zones in each of which comparatively small portions only of the charge are present at the one time. By resorting to isothermal annealing, the time of treatment can be shortened considerably in comparison with the time taken for conventional full annealing. This has already been discussed in Section 5.1.1.

Alloy case-hardening steels such as BS 637M17 (En 352) are often subjected to an isothermal anneal. The austenitization is carried out at 930–940°C, i.e. at a temperature somewhat higher than that subsequently used for the carburization. This treatment is intended to reduce any distortion that may occur during the actual carburization treatment. The transformation is made to start at 610–680°C and takes 2–4 h to completion. The structure obtained consists of ferrite and pearlite which is entirely suitable for most machining operations.

Isothermal annealing is also resorted to during the various stages of steel processing. If an ingot or rolled billet of alloy (air-hardening) steel is allowed to cool freely to room temperature, cracks are very likely to appear if the core transforms to martensite. Should this happen, tensile stresses are set up in the surface layers which by that time have already been transformed. The warm ingots or billets are therefore placed in isothermal annealing furnaces at about 700°C, this being the temperature at which transformation to pearlite takes place. When this is complete the steel may cool freely to room temperature.

5.1.5   Quench annealing

This treatment is used for austenitic steels only and consists in fact of a homogenizing anneal or, in certain cases a recrystallization anneal. Slow cooling or heating will cause carbides to precipitate in austenitic steels.

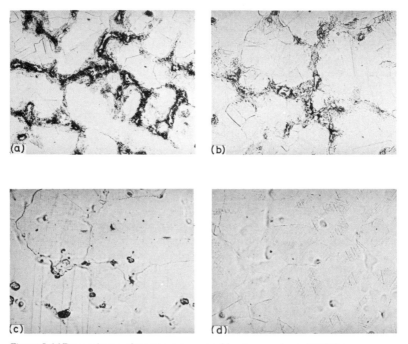

Figure 5.16 Dependence of structure on quenching temperatures (950°C to 1100°C) for a Hadfield steel (18% Mn): (a) 950; (b) 1000; (c) 1050; (d) 1100°C. All 100 ×

This precipitation generally takes place at the grain boundaries. For stainless steels (type 18/8) this implies a reduction in the chromium content in the proximity of these grain boundaries. Hence the corrosion resistance there will diminish; thereby facilitating the incidence of what is called intergranular corrosion. This carbide precipitation can also take place by heating the steel in the temperature range 500–800°C. The austenitic manganese steel, which is generally used for castings, is subject to intense carbide precipitation on being allowed to cool in the mould from the casting temperature. This gives the steel low impact strength.

The above-mentioned ill effects are eliminated by solution heat treatment and quenching, i.e. quench annealing. This consists of heating the steel to a temperature at which the austenite becomes homogeneous, usually around 1000–1100°C, followed by rapid cooling (quenching), in water for instance. For light sections, air cooling will suffice to give the steel a satisfactory structure.

The stages of carbide dissolution and homogenization of the structure for a manganese steel containing 18% Mn are shown in *Figures 5.16a–d*. The dependence of the hardness of this steel on the quenching temperature is shown in *Figure 5.17*.

Neither on heating nor on cooling does the matrix of austenitic steels undergo any phase transformation that would result in grain refinement. Solution treatment may instead give rise to some grain coarsening, but this has generally no detrimental effect on the toughness of the steel. Solution treatment of austenitic manganese steel causes its toughness to increase considerably. This improvement is more noticeable in heavy dimensions in which the carbide precipitation is more pronounced on account of the slow cooling. Even in light sections the improvement can be quite apparent, as is shown in *Figure 5.18*.

On quench annealing cold-worked austenitic steels it is found that they recrystallize, which gives them their lowest hardness. If they have been subjected to a critical strain level, an abnormally large grain growth results.

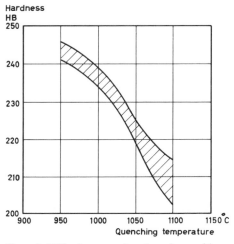

Figure 5.17 Hardness as a function of quenching temperature for a Hadfield steel

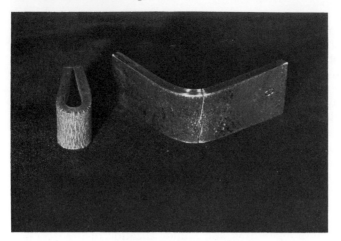

Figure 5.18 Bend test on a manganese sheet strip 6 mm in thickness.
Right: as rolled, followed by slow cooling. Left: as quench annealed

5.1.6   Homogenizing annealing

When the steel, after teeming, has solidified, its structure is inhomogeneous. The subsequent heating and hot working equalize the structure to a large extent. This process of homogenization can be speeded up by a so-called diffusion or homogenizing anneal which takes place at a high temperature, approximately 1100°C, and for a rather long time in order to produce the desired result. Section 1.5 discusses an instance of diffusion annealing, the purpose of which was to even out concentration differences of manganese. The high temperatures and long treatment times entail very high costs and this makes the treatment applicable only to very special cases.

5.1.7   Hydrogen annealing

The hydrogen dissolved in the molten steel may give rise to internal cracks, the so-called 'flakes', in the steel after it has cooled from the hot-working operation. By subjecting the steel to an annealing treatment at 600–650°C for several days, followed by slow cooling, the hydrogen content can be reduced and the danger of flakes formation removed. Hydrogen annealing was discussed in more detail in Section 3.2.1.

5.1.8   Hydrogen expulsion

While being pickled in non-oxidizing acids, or being cathodically cleaned or plated, steel is liable to pick up hydrogen. This hydrogen may cause the steel to fail under a stress considerably lower than that normal for the steel in question. This phenomenon is called hydrogen embrittlement. After ordinary pickling a large amount of the hydrogen diffuses out of the steel at room temperature but after cathodic electroplating with Cr, Zn, Cd, Ni, Sn, Pb, and Ag the hydrogen leaves much less readily. The current practice, which is to heat to 170°C for 15 h or to 200°C for 5 h, causes the hydrogen to be driven off. In some instances a higher temperature, e.g. up to 300°C, is required, but then the treatment must take place in a controlled atmosphere to protect the plating from oxidation. At a steel hardness of around 350 HB the danger of fracture from hydrogen embrittlement is already apparent; at a hardness of more than 400 HB hydrogen removal is recommended; at hardness around 60 HRC hydrogen must be expelled without delay, directly following the surface treatment, otherwise the steel may crack in a few hours' time.

5.2 Normalizing

This treatment is also called normalizing anneal. It consists of heating the steel to a temperature corresponding roughly to its hardening temperature, holding it there for some 10–20 min and then allowing it to cool freely in air. The normalizing treatment refines the grain of a steel that has become coarse-grained as a result of its having been heated to a high temperature, for example for forging or welding. In *Figure 5.19* is shown the grain-refining effect produced by normalizing a steel containing 0·50% C.

When hypo-eutectoid steels are being normalized the first stage of grain refinement takes place as the transformation to austenite is proceeding. During the subsequent cooling, ferrite grains first separate out from the austenite grains and then pearlite is formed whereby the grain refinement is further enhanced.

Normalizing is applied mainly to carbon and low-alloy steels. However, the normalizing of hyper-eutectoid steels may be mentioned as a special case. These steels in the as-rolled condition and especially in heavy sections have an elongated grain boundary structure of carbides. This structure is not broken down during a conventional full-annealing treatment nor during the austenitizing operation prior to hardening. By applying a combination of normalizing and annealing the elongated carbide structure is broken down and a normal full-annealed structure is obtained. *See Figure 6.40*.

The hardness resulting from this treatment depends on the dimensions and composition of the steel. The difference in the rate of cooling between the surface and centre during normalizing is but slight. *Figure 5.20* shows cooling curves for various dimensions at different depths below the surface. *See also Figure 4.50*. In order to calculate the tensile strength of normalized plain-carbon and low-alloy steels, use can be made of several current expressions for this purpose but these do not take any account of the dimensions of the steel. The expression used at Bofors is as follows:

$$Cp = C\,[1 + 0\cdot5\,(C - 0\cdot20)] + Si \times 0\cdot15 + Mn\,[0\cdot125 + 0\cdot25\,(C - 0\cdot20)]$$
$$+ P\,[1\cdot25 - 0\cdot5\,(C - 0\cdot20)] + Cr \times 0\cdot2 + Ni \times 0\cdot1$$

Cp is the sum of the carbon potentials.

The ultimate tensile strength in kp/mm^2 after normalizing is approximately:

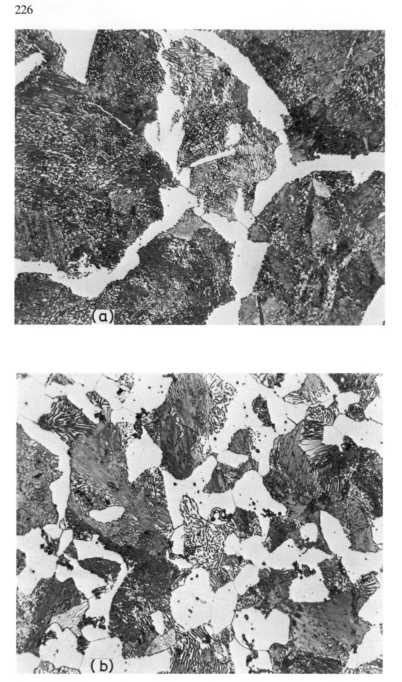

Figure 5.19 Influence of normalizing anneal on grain size. Carbon steel
0·50% C. (a) As rolled or forged. Grain size 3 ASTM; (b) Normalized.
Grain size 6 ASTM. Both 500 ×

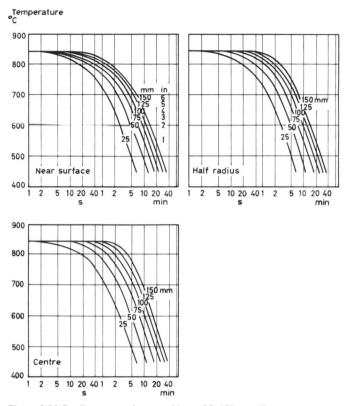

Figure 5.20 Cooling curves for round bars, 25–150 mm diameter,
normalizing treatment (courtesy The Mond Nickel Co. Ltd)

$27 + 56 \times Cp$ for hot-rolled steel
$27 + 50 \times Cp$ for forged steel
$27 + 48 \times Cp$ for cast steel

In some instances a double normalizing treatment is stipulated. This
treatment consists of first heating the steel to a temperature some
50–100°C above the normal temperature, whereby a more complete
dissolution of the constituents is obtained. The second normalizing anneal
is carried out near the lower limit of the temperature range and is solely
aimed at giving the steel a fine-grained structure. This double treatment is
sometimes stipulated for steels to be used under subzero conditions.
Generally speaking, a fine-grained steel has a lower transition temperature
range than a coarse-grained one. Plain-carbon or low-alloy steel castings
should always be normalized.

5.3 Hardening

The theoretical background of the process of hardening has been discussed
in detail in Section 1.3. The present section will deal mainly with the

practical aspects of hardening. The structure of a steel as a result of hardening may be regarded as a starting structure which is capable of modification by subsequent tempering. The result of the hardening depends on several factors, each of which may be of decisive importance. Hence it is of the utmost importance that each of the various steps of the hardening process is given the same care and consideration.

5.3.1   Heating media

During the heating-up stage it is necessary to provide against any unintentional carburization or decarburization. Should this happen the superficial hardness measured after hardening will be misleading, which may result in the choice of an incorrect tempering temperature. Variations in the carbon content of the surface layers may cause the steel to crack during hardening. If the surface oxidation is too severe it is possible that the tolerances set for the tool might not be kept after descaling and grinding. To protect the surface of the tool the necessary precautions to be observed depend on the type of furnace used.

Salt-bath furnaces

These offer good protection against variations in the superficial carbon content, owing to the short heating time required and the neutral character of the bath. The bath must of course be maintained in good condition. As a rule, manufacturers of heat-treatment salts do not specify the composition of the salts. On the other hand they give detailed information on their properties and uses. For purposes of information, however, some typical compositions and working temperatures for neutral salt baths are given below:

$45\% \text{ NaCl} + 55\% \text{ KCl}$ 675–900°C
$20\% \text{ NaCl} + 80\% \text{ BaCl}_2$ 675–1060°C
$100\% \text{ BaCl}_2$ 1025–1325°C

During use the bath takes up iron from the material being heat treated. This iron is oxidized since the bath is in contact with the atmosphere. The iron oxide thus formed has a decarburizing action on the steel charge and hence the bath must be regenerated. Follow the recommendations given by the salt manufacturers! Barium chloride baths with certain additions are used as heating media for high-speed steel and certain other high-alloy steels. Such baths may be regenerated by means of a few pieces of silica brick, which combines with the iron dissolved in the bath. The sludge thus formed must be removed regularly. By means of a graphite rod immersed into the salt bath it is easy to check whether iron is present. Any iron oxide in the bath will be reduced by the graphitic carbon and small bright beads of metallic iron will form on the rod. The carburizing or decarburizing action of a salt bath may be tested by means of steel foil which is dipped into the bath for a few minutes and then quenched. If the foil is soft after quenching, the bath is decarburizing; if more brittle than it normally should be the bath is carburizing. Foil of varying carbon contents is available.

Gas- or oil-fired muffle furnaces

These allow the amounts of air of combustion to be controlled. Hence the atmosphere inside the muffle can be adjusted so that for the treatment temperature and grade of steel concerned the superficial carbon content of the charge will not be appreciably affected.

Electrically-heated muffle furnaces

These, if not operated with controlled atmospheres, should be used with annealing boxes in which the steel charge is packed with some protective material, which should be as neutral as possible. Unfortunately there is no general-purpose substance so versatile that it is suitable as a packing material for all grades of steel since these materials are only neutral towards steels of a definite carbon content and at a definite temperature. One purpose of the packing material is to prevent the ingress of air to the steel. For this purpose cast-iron chips are very suitable since the air that is sucked into the box on account of temperature variations preferentially oxidizes the chips. Practical tests and scientific investigations have shown that cast-iron chips are practically inert towards plain-carbon steel and low-alloy steel containing carbon between 0·6 and 1%. They are also inert towards high-alloy chrome steels of the type D 2 and D 6, provided the customary hardening temperatures are employed. With regard to hot-work tool steel containing carbon around 0·3% cast-iron chips have a slightly carburizing action. At about 1000°C there is already a danger that cast-iron chips may begin to sinter and they should not be used at temperatures above 1050°C. If the tool is wrapped in a thin layer of newsprint before being packed in cast-iron chips the surface is also protected against mechanical damage.

Wood-charcoal or coke fines (but not carburizing compounds) act as decarburizers on plain-carbon and low-alloy steels containing medium to high carbon, but act as carburizers on high-alloy chrome steels and hot-work steels. However, it is the temperature and not the composition of the steels that is the factor that determines whether carburization or decarburization will take place. The result of an investigation into the influence of packing media on the hardness, carbon content and amount of retained austenite in some tool steels is shown in *Figures 5.21* to *5.24*. The normal hardening temperature for each steel was used; heating-up time was about 2 h and holding time, 15 min[9].

Under certain conditions borax may be used as a protective agent against decarburization. The borax powder is sprinkled over the surface to be protected and on melting, it forms a covering film. Before hardening, this film should be brushed off, otherwise it retards the rate of cooling of the steel. Borax is often used to protect the engraved surface of coining dies when they are being hardened. On water-hardening steels the borax film spalls off when the steel is quenched and leaves a bright clean surface.

There are also some commercially available protective pastes on the market. These pastes are applied to the protected surface prior to heat treating. The protective efficiency of these pastes is best assessed by testing them on a piece of steel of the grade for which they are recommended. One

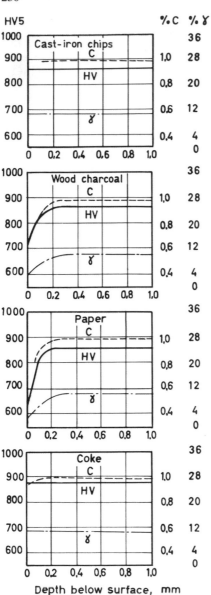

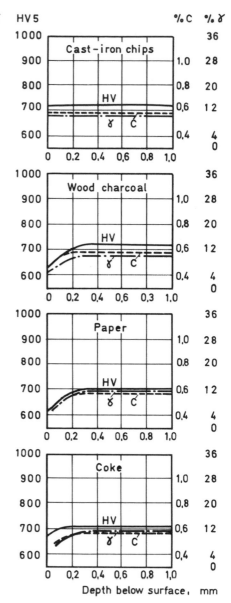

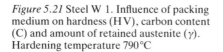

Figure 5.21 Steel W 1. Influence of packing medium on hardness (HV), carbon content (C) and amount of retained austenite (γ). Hardening temperature 790°C

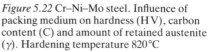

Figure 5.22 Cr–Ni–Mo steel. Influence of packing medium on hardness (HV), carbon content (C) and amount of retained austenite (γ). Hardening temperature 820°C

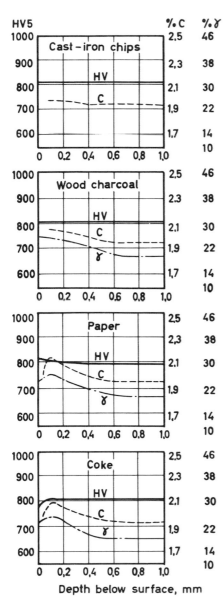

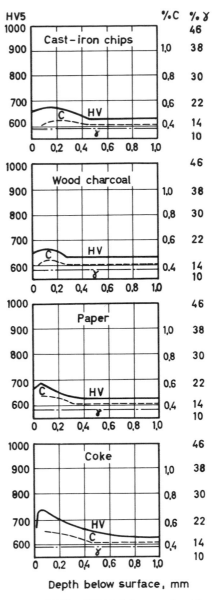

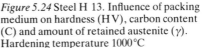

Depth below surface, mm

Figure 5.23 Steel D 6. Influence of packing medium on hardness (HV), carbon content (C) and amount of retained austenite (γ). Hardening temperature 980°C

Figure 5.24 Steel H 13. Influence of packing medium on hardness (HV), carbon content (C) and amount of retained austenite (γ). Hardening temperature 1000°C

half of the test-piece is coated with the paste, placed in the hot furnace and held at the normal hardening temperature for, say, 1 h. After hardening, the Vickers hardness in the surface, both treated and untreated, is measured, using varying loads, or using, say, a 5 kp load on a taper section. Examples of such tests are shown in *Figures 5.25* and *5.26*.

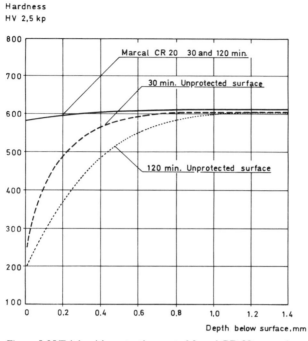

Figure 5.25 Trials with protective paste Marcal CR 20 on steel H 13. Hardening temperature: 1010°C. Holding times: 30 min and 120 min. Cooling: in air

Heat-resisting steel foil has also been put to use as a protective measure against decarburization. The foil is supplied as sheet or bags in which the steel parts are wrapped.

Controlled atmospheres[10]

These specially produced protective gases have shown themselves, both technically and economically, to be highly suitable when used for the heat treatment of costly tools and component parts. This is particularly so for large-scale production hardening. The gas is made in a separate plant, whence it is led into the hardening furnace which must be gas-tight to prevent any leakage. Protective gas for hardening is made according to the following three main principles:

1. *Exothermic gas* is produced by the exothermic combustion (i.e. without the addition of heat) of gas and air; the combustion not going to completion. As the gas contains large volumes of CO_2 and H_2O the greater part of these must be removed by chemical absorption and drying,

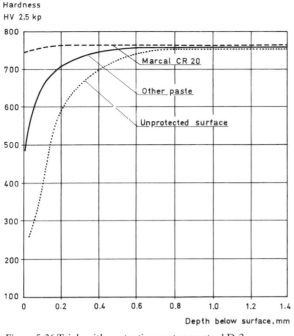

Figure 5.26 Trials with protective pastes on steel D 2.
Hardening temperature: 1050°C. Holding time: 2 h.
Quenching: in oil. Tempering: 180°C, 2 h

respectively. If both H_2O and CO_2 are removed, the gas contains CO, H_2 and N_2. This is called monogas.

2. *Endothermic gas* is produced by endothermic combusion, i.e. energy must be supplied. The gas, which is usually propane or propylene, is mixed in carefully balanced proportions with air and passed through a gas generator in which a so-called *carrier* gas is produced. The production of endothermic gas will be discussed in Section 6.5, Gas carburizing on p. 462. The carrier gas usually has a carbon potential of about 0·40%. This means that a great number of steels can be heated in direct contact with this gas without any risk of changes taking place in their surface carbon contents. When treating higher-carbon steels the carbon potential of the gas may be increased by introducing undiluted propane or propylene direct into the furnace. A requirement for a successful result is that a close check is kept on the carbon potential. In many cases a satisfactory check may be obtained by measuring the dew point. To obtain a more exact check and hence a better control of the carbon potential it is recommended that an instrument be used capable of measuring the CO_2 content by means of infrared radiation and converting direct to a figure indicating the carbon potential. (The definition of carbon potential is given on p. 445.) The required carbon potential of the gas should be about the same as that of the carbon dissolved in the steel at the hardening temperature. *Table 5.1* gives a selection of carbon potentials that have been used with various steels with good results.

3. *Inert gas* is a protective gas that, as regards its carbon, oxygen and nitrogen contents, remains unreactive to the steel. From a chemical point of view it is only the Inert Gases, such as argon, that truly satisfy this criterion. The most commonly used unreactive gas consists mainly of nitrogen. On a large scale it is produced by the exothermic combustion of crude gas and air and then cleaned according to various methods. For further information, readers are referred to the literature, for example an article by Grassel and Wünning[11] in which different methods of producing inert gas are described.

A fairly simple method applicable to air-hardening steels is to place the tools in a gastight box fitted with an inlet and an outlet tube. Nitrogen gas from a nitrogen cylinder is led through the box during the whole treatment cycle. When the box is taken out of the furnace the rate of cooling may be increased by placing the box in a current of air from a fan. A large number of such hardening treatments have been applied to tools made from steel H13.

Table 5.1 Selection of carbon potential for various steels

Grade of Steel AISI, BS	SS	Carbon content %	Carbon potential %
W1	1880	1·0	0·80
—	2092	1·0	0·80
O1	2140	0·90	0·75
—	2550	0·55	0·45
A2	2260	1·0	0·70
D2	2310	1·50	0·70
D6	2312	2·00	0·80
H13	2242	0·40	0·30

This method is a simplified form of 'hardening in a protective atmosphere' and the necessary equipment can be installed without large capital expenditure.

Analysis and control of the furnace atmosphere

The following four methods are employed for this purpose. They measure and control the carbon potential either directly or indirectly.

Direct: by measuring the specific resistance of a thin iron wire;
Indirect: by measuring the dew point;
Indirect: by means of an infra-red analyser;
Indirect: by means of an oxygen probe.

For a more detailed study of the various measuring methods, please consult the literature, references[12, 13, 14, 15].

Vacuum hardening

This has become increasingly popular during the last two decades. The charge is placed in a cold furnace which is then evacuated by means of vacuum pumps. For most cases an operating pressure of 10^{-2} torr is acceptable. Should a higher vacuum be required, diffusion pumps are necessary which will increase the cost of the furnaces. After the charge has been heated and austenitized, the furnace and contents are cooled to room temperature, usually in a nitrogen atmosphere but in special cases argon is used. New furnace types have been developed in which the charge can be quenched straight into oil while under vacuum. For this purpose special oils have been developed. These oils can be heated to about 200 °C, which enables an austempering treatment to be carried out.

Figure 5.27 shows one type of vacuum furnace in which the charge can be cooled with gas or quenched in oil in the underlying tank. *Figure 5.28* shows the profile of the heating and cooling curves of a cube-shaped tool, the cube length of which was 380 mm (15 in) and weighing 350 kg (775 lb).

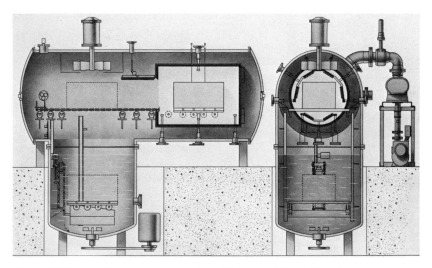

Figure 5.27 Vacuum furnace. Charge cooled with gas or oil (after Hayes)

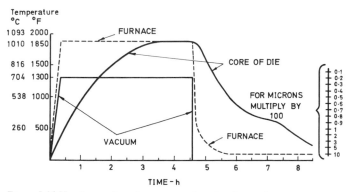

Figure 5.28 Heating and cooling curves of cooling by gas in a vacuum furnace. The piece tested was a block of H 11 steel. Cube length 380 mm (15 in) and weighing 350 kg (775 lb). The thermocouple was located at the centre of the block (after Brennan[16])

The greatest benefit resulting from vacuum hardening is that the surfaces of the parts comprising the charge are not altered by the hardening treatment, i.e. the parts are completely free from surface decarburization or oxidation. There are now available advanced furnace designs that can compete successfully with protective gas furnaces from the points of view of both economy and speed of operation. Extensive literature on vacuum hardening is now available[16, 17, 18].

5.3.2   Rate of heating

If the rate of heating to the hardening temperature is very fast the steel may crack or warp, since sections having different dimensions heat up at different speeds. These differences may be diminished by preheating the tool to a temperature usually lying close below the transformation temperature A_{c1} of the steel. When heating for hardening in a muffle furnace, preheating is not so necessary as when a salt bath is used. Even in this latter case preheating may be omitted if the parts to be treated are small and symmetrical. However, if a salt bath is used, a preheat to about 100°C is always advisable in order to remove any moisture adhering to the steel.

During the initial heating-up stage the surface of the steel is at a higher temperature than the centre. The closer the temperature of the steel approaches the preset temperature, the smaller is this temperature difference, which is shown in *Figures 5.29* and *5.30*. In actual practice it can therefore be assumed that when the surface has reached the temperature of the furnace, the steel is heated right through.

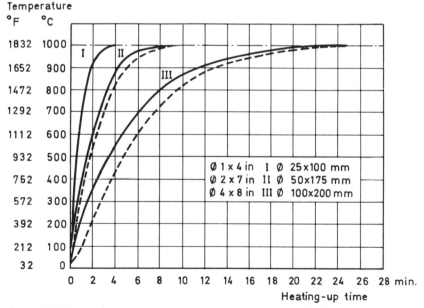

Figure 5.29 Time-temperature curves for steel bars heated in a salt bath at 1000°C. Full line: temperature at surface. Dashed line: temperature at centre

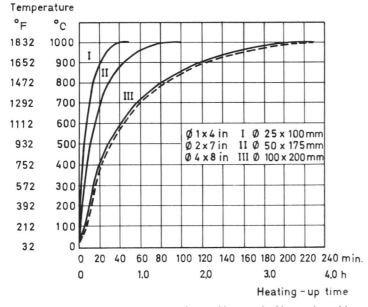

Figure 5.30 Time-temperature curves for steel bars packed in cast-iron chips,
heated in a muffle furnace at 1000°C. Diameter of boxes: 50, 75 and
140 mm. Full line: temperature at surface. Dashed line: temperature at
centre

This assumption runs contrary to the generally widespread belief that the
surface of the steel reaches the temperature of the furnace considerably
faster than the centre. This belief is based on the observation that the
corners of steel objects become warm more quickly than other sections.
The diagrams in *Figures 5.29* and *5.30* were obtained by inserting
thermocouples at the centre of test specimens, thereby ensuring that the
heating effect from the end surfaces would not influence the readings. That
the 'corner effect' really does exist has been shown in a heating trial carried
out on a large die block. One thermocouple was located at a corner, 50 mm
below the surface; another at the middle of the die, also 50 mm below the
surface and a third thermocouple right at the centre. As is seen from
Figure 5.31 the corner heated up much more quickly than did the other
parts of the block. As regards the other surfaces it is seen that even in this
large section the temperature as registered in the middle followed closely
that at the surface, once the latter had reached the temperature of the
furnace.

The heating-up time may be calculated from standard expressions[19, 20, 21].
Based on trials and with the help of these expressions the heating-up time
necessary in an oil-fired muffle furnace has been calculated for various steel
dimensions and plotted in a diagram as a function of the cross-sectional
area. *Figure 5.32.* This diagram is valid for plain-carbon and low-alloy
steels. A presupposition is that the length of the part is at least twice the
size of the diameter or diagonal. The dimension 950 × 500 mm
(*Figure 5.31*) is equivalent to a round bar 700 mm in diameter. According
to *Figure 5.32* the heating-up time for a round bar of this diameter is 20 h,

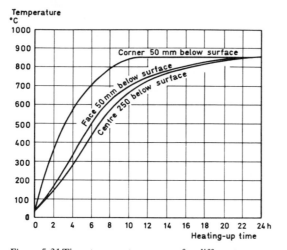

Figure 5.31 Time-temperature curves for different positions in a die block, 2300 × 950 × 500 mm, on being heated in an oil-fired muffle furnace

which is in quite good agreement with the 22 h indicated in *Figure 5.31*.

The time of heating-up depends not only on the dimensions of the steel but also on the heating capacity of the furnace and the degree of packing. Hence it is obviously necessary to gain experience with specific furnaces and to compile tables or diagrams for bars of various dimensions. An example illustrating this is shown in *Figure 5.33* which traces the heating curve of the centre of a 75 mm diameter bar made from steel D2. Preheating was carried out in an 80 kW convection furnace measuring 800 mm diameter by 2000 mm. The final heating took place in a 160 kW

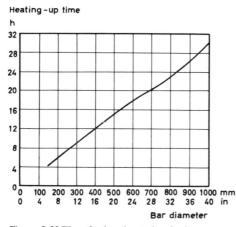

Figure 5.32 Time for heating to hardening temperature, 850°C, for bars of varying diameters up to 1000 mm. Length 2 × diameter. Starting temperature 20°C. Oil-fired muffle furnace, calculated values

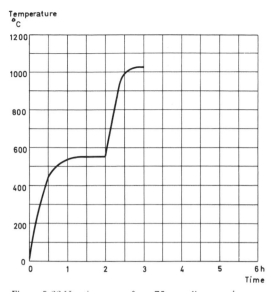

Figure 5.33 Heating curve for a 75 mm diameter bar, preheated in a furnace 800 mm diameter × 2000 mm, rated at 80 kW. Final heating in a furnace of the same dimensions, rated at 160 kW

furnace of the same dimensions. Note the rather long time to reach the preheating temperature (1·5 h) compared with the time taken to reach the hardening temperature (about 0·75 h). This is due to the increase in the rate of heat transfer as the temperature increases. Another five trials were made with the same steel but of different dimensions. The results containing the heating times from 550 °C to 1000 °C are reproduced in the curve in *Figure 5.34*. With regard to times for heating to the preheating temperature, *see Figure 5.93*.

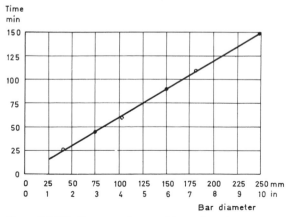

Figure 5.34 Heating curve for round bars of varying diameters from 550 to 1000 °C, in a furnace 800 mm diameter × 2000 mm, rated at 160 kW

5.3.3   Hardening temperature

For each grade of steel, on the basis of a series of practical trials, a range of temperatures has been established to which the steel is to be heated for hardening. This temperature range, also called the quenching range, is chosen so as to give maximum hardness and at the same time maintain a fine-grained structure in the steel. *Figure 5.35* illustrates how to determine the hardening temperature for steel SS 2092, the specimen dimension in this case being 30 mm diameter × 100 mm and a holding time of 20 min at the hardening temperature. The micrographs were taken 2 mm below the surface of the specimen and show that 850°C is the lowest possible temperature that can be used along with the rate of cooling actually employed. A lower hardening temperature would give rise to the formation of pearlite and bainite and the hardness at the surface would be inadequate.

As the temperature increases, so does the grain size and also the amount of retained austenite. The trial run shown in *Figure 5.36* illustrates this fact. The specimens were tempered at 200°C in order to bring out more clearly the retained austenite. At 920°C and 970°C the retained austenite may be discerned as light angular areas. An untrained observer may confuse the retained austenite with carbides which in *Figure 5.36a* are clearly visible as round or oval grains. The acicular areas in *b* and *c* are martensite needles. When these are distinctly visible in the microscope, it usually implies that the fractures may be assessed as No. 7 or coarser, on the JK scale.

Based on actual trials the quenching range for this steel has been fixed at 850–880°C. The hardening temperatures for the most commonly occurring steels are given in Chapter 6 in connection with each type of steel.

As has been previously mentioned, by raising the hardening temperature the hardenability of steel can be increased. This increase is due to the greater amount of carbide going into solution and to increased grain size. When high-speed steel is being hardened, a temperature of some tens of degrees below the melting range is used. Only a slight increase above the ordinary temperature will cause incipient fusion round the carbides. This molten phase will 'permeate' through the grain boundaries and hence reduce the toughness. A further increase in temperature may lead to pronounced melting. The steel is then said to be burnt, *Figure 5.37*. The danger of melting increases with the degree of segregation. The larger the dimension the greater the segregation. For this reason a lower hardening temperature should be chosen for high-speed steel tools made from heavy sections. Even for dimensions corresponding to 100 mm diameter a hardening temperature near the lower limit of the customary temperature range should be chosen.

5.3.4   Holding time at temperature

When the steel has reached the hardening temperature it is austenitic provided that the temperature has been correctly chosen. The time of holding at the hardening temperature depends on the desired degree of carbide dissolution and acceptable grain size. Since the amount of carbide

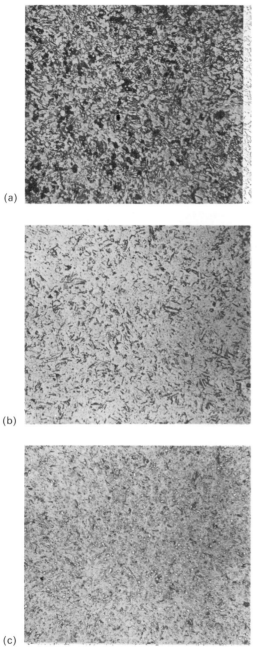

(a)

(b)

(c)

Figure 5.35 Microstructure of test-pieces of steel
SS 2092 after hardening from 800 to 850 °C.
Dimensions of test-pieces: 30 × 100 mm. All 400 ×.
Chemical composition: 1% Cr, 1·5% Si, 0·8% Mn,
1% Cr.

	(a)	(b)	(c)
Hardening temperatures	800	825	850 °C
J K Fracture number	10	10	10
Hardness (HRC)	55	61·5	66

242

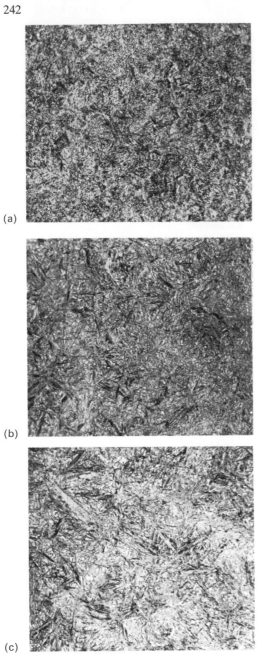

(a)

(b)

(c)

Figure 5.36 Microstructure of test-pieces of steel
SS 2092 after hardening from 870 to 970°C, followed
by tempering at 200°C. Dimensions of test-pieces:
30 mm diameter × 100 mm. All 400 ×

	(a)	(b)	(c)
Hardening temperatures	870	920	970°C
J K Fracture number	9	7	5
Hardness (H R C)	62·5	62	61
Retained austenite %	12	20	28

Figure 5.37 High-speed steel (M 42). Owing to partial
melting, carbides having 'permeated' through the
grain boundaries. Hardening temperature has been
too high, viz. 1260°C. Tempering twice at 640°C.
Etched in 1% nital, 700 ×

is different for different types of steel the time of holding is also dependent
on the grade of steel.

However, the holding time is not only dependent on the hardening
temperature but also on the rate of heating. With very slow heating,
hypo-eutectoid steels are completely austenitic immediately above A_{c3} and
hence no holding time should be necessary. With more rapid heating some
holding time should be required to ensure temperature equalization and
carbide dissolution. Alternatively, a higher hardening temperature might
be used.

The influence of the rate of heating and the holding time on carbide
dissolution, grain growth and hardness after hardening have been treated
in detail in *Atlas zur Wärmebehandlung der Stähle,* Volumes 3 and 4, from
which the following examples are taken, by permission of Verlag
Stahleisen mbH, Düsseldorf[22, 23].

The heating was carried out both isothermally and continuously. In the
first case the steel was heated at the rate of 130°C/s to the temperature in
question, held there for certain predetermined times and then cooled
rapidly. This is the same procedure as that shown schematically in
Figure 1.7. The continuous heating was carried out at various constant
rates, ranging from 0·05 to 2400°C/s. *Figure 5.38* shows such a diagram.
With a continuous heating rate of, for example, 0·22°C/s homogeneous
austenite was obtained when 860°C was reached after about 4000 seconds
(i.e. about 1 h). If the heating rate had been 1°C/s the temperature and
time required would have been 880°C and 14 minutes, respectively. With a
rate of 10°C/s, 920°C and 1·5 minutes would have been required, etc.

However, the important feature with the austenitizing treatment is
whether maximum hardness is attained on quenching to room
temperature. This information may be gleaned from *Figure 5.39* which
shows that the maximum hardness, 770 HV, would be obtained with a
heating rate of 1°C/s up to 875°C (heating-up time about 14 minutes). This

244

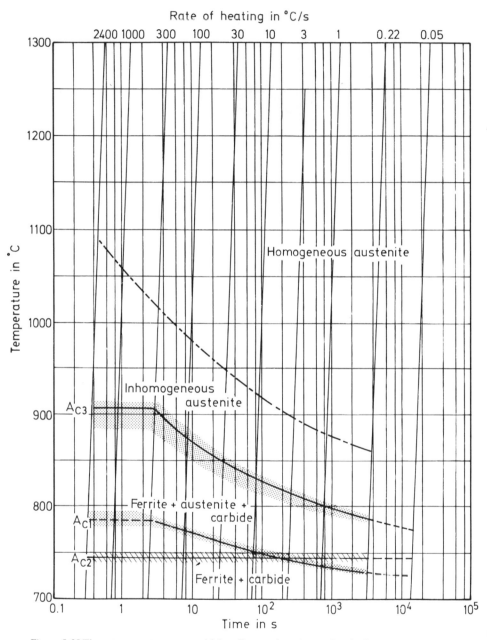

Figure 5.38 Time–temperature–austenitizing diagram (continuous heating) for steel Cf 53 (0·51% C, 0·38% Si, 0·64% Mn, 0·022% Al, 0·005% N). Initially fully annealed for 5 h at 700°C, furnace cooled (Reproduced by permission of Stahleisen mbH, Düsseldorf)

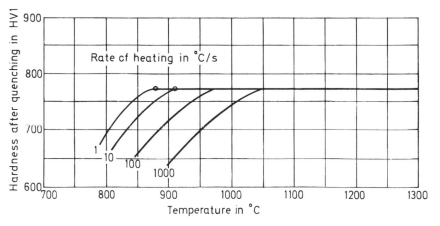

Figure 5.39 Hardness after quenching as a function of the rate of heating and austenitizing temperature for fully annealed steel, type Cf 53 (Reproduced by permission of Stahleisen mbH, Düsseldorf)

indicates that in this instance a homogeneous austenitic structure would be required to give maximum hardness. If the heating rate were 10°C/s a temperature of slightly more than 900°C would be required for maximum hardness. Heating-up time in this instance would be 5 minutes. Thus the dissolution of carbide in this steel is so rapid that no holding time should be required for the tests reported above.

The amount of dissolution is also dependent on the initial structure of the steel. According to *Figure 5.39* the structure of the steel investigated consisted of spheroidized cementite. If, on the other hand, a start were made on a hardened and tempered structure such as that shown in *Figure 5.40*, a heating rate of 1000°C/s i.e. shorter than one second's heating-up time, could be used and on quenching the steel to room temperature a hardness of 770 HV would still be attained. Most low-alloy

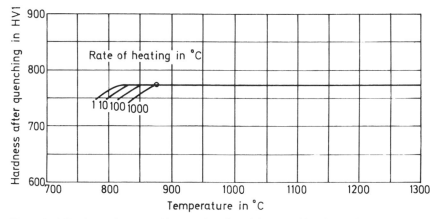

Figure 5.40 Hardness after quenching as a function of the rate of heating and austenitizing temperature for hardened and tempered steel, Cf 53 (Reproduced by permission of Stahleisen mbH, Düsseldorf)

hypoeutectoid steels require only similarly short times for carbide dissolution, as do straight carbon steels containing about the same carbon content.

The situation is somewhat more complicated for eutectoid and hyper-eutectoid steels. Consider, for example the ball-bearing steel 100 Cr 6 (1·00% C, 0·22% Si, 0·24% Mn, 1·52% Cr). *Figure 5.41* which applies to continuous heating, shows that maximum hardness, i.e. 900 HV after quenching, is attained on heating to a very narrow temperature range and furthermore that the range is displaced towards higher temperatures as the heating rate increases. If the steel is quenched from temperatures higher than this range the resulting hardness is reduced which is due to the presence of an increasing amount of retained austenite in the structure.

This state of affairs is more clearly illustrated in *Figure 5.42* which reproduces details from the isothermal diagram. If the steel is heated rapidly to 950°C a holding time of 2–4 seconds is sufficient for the attainment of maximum hardness on quenching. At a temperature of 900°C the holding time would be 30–60 seconds and at 850°C, between 800 and 1200 seconds—13 and 20 minutes—must be allowed.

Should a temperature of 800°C be used, a holding time of several hours must be reckoned with in order that the steel may attain maximum hardness on quenching. Hence it is rather obvious that in practice a hardening temperature of around 850°C should be used for this steel. However, this inference is not immediately obvious on referring to the heating diagrams in *Figures 5.43* and *5.44* from which much higher temperatures ought to be chosen in order to obtain either homogeneous or inhomogeneous austenite. For this steel, however, such temperatures would result in too much retained austenite after quenching.

Similar characteristics are displayed by high-chromium steels, i.e. the hardenable stainless steels as well as tool steels of type D 2. *Figure 5.45* illustrates a practical case in point from which it is seen that 1020–1040°C should be a suitable hardening temperature range for D 2 and a holding time of about 20 minutes. For the example submitted, the heating-up time was about 10 minutes.

The diagram in *Figure 5.46* applies to the hot-work steel X 38 CrMoV 51 which is approximately equivalent to H 13. From this diagram it is seen that maximum hardness on quenching, 750 HV, is attained after holding for 10^3 seconds (17 minutes) at 1000°C. If the temperature is raised to 1100°C, 60 seconds' holding time is sufficient. At 1300°C less that one second is required and the holding time may be increased to 17 minutes without causing any variation in the hardness. In order to see what might occur in such a case the diagram in *Figure 5.47* should be studied. In it the grain size is set off against the temperature and holding time. Earlier it has been pointed out that the grain size should preferably not be larger than 7 ASTM and consequently a holding time of 17 minutes at 1000°C is quite on the safe side in this respect.

The tests as reported in *Atlas zur Wärmebehandlung der Stähle* were carried out with tubular test pieces, $D_o/D_i = 8/7$ mm and were produced from bar stock 10–20 mm diameter. In such dimensions the carbides in the steel are very small and evenly distributed, which is apparent from the accompanying micrographs. Especially for the high-alloy steels the times

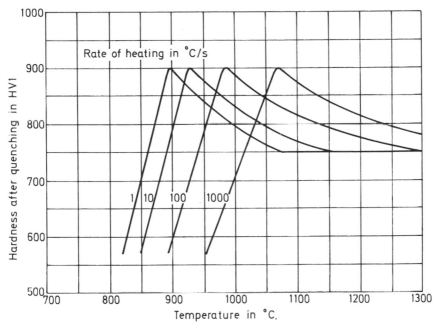

Figure 5.41 Hardness after quenching as a function of the rate of heating and austenitizing temperature for steel 100 Cr 6: initially fully annealed (Reproduced by permission of Stahleisen mbH, Düsseldorf)

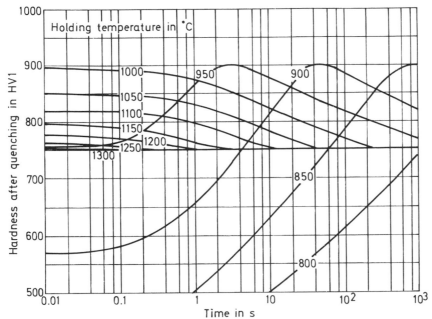

Figure 5.42 Hardness after quenching as a function of holding time and austenitizing temperature after a heating rate of 130°C/s for steel 100 Cr 6; initially fully annealed (Reproduced by permission of Stahleisen mbH, Düsseldorf)

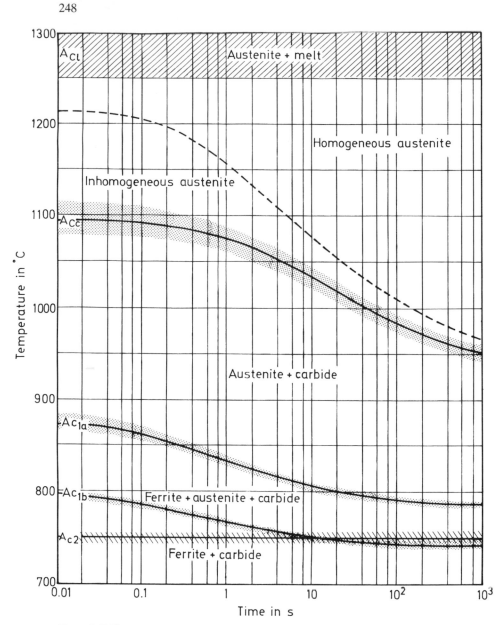

Figure 5.43 Time–temperature–austenitizing diagram (isothermal) for steel 100 Cr 6. Rate of heating to holding temperature: 130 °C/s (Reproduced by permission of Stahleisen mbH, Düsseldorf)

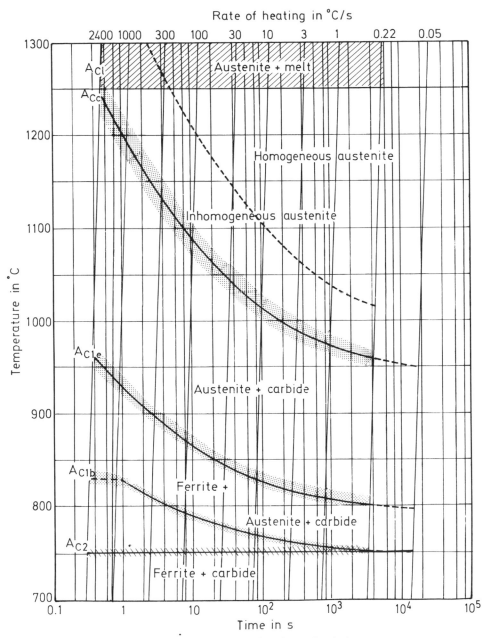

Figure 5.44 Time–temperature–austenitizing diagram (continuous heating) for steel 100 Cr 6 (Reproduced by permission of Stahleisen mbH, Düsseldorf)

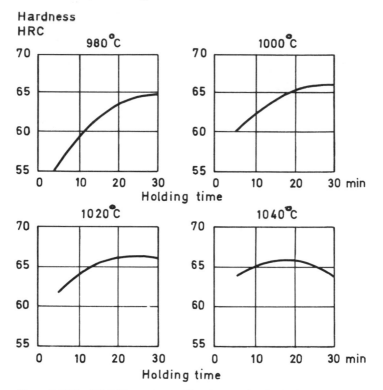

Figure 5.45 Steel D 2. Dependence of hardness on hardening temperature
and holding time. Test-piece 30 mm diameter

for dissolution will thus be considerably shorter than for the heavy
dimensions since the carbide particle size increases as the dimensions of the
steel stock increases. On the other hand the heating-up times for heavy
sections are generally much longer than for smaller dimensions. However,
it should be brought to general attention that the maximum hardness, as
given in the diagrams, may be up to 100 H V higher than what is attainable
in practice.

The information obtainable from *Atlas zur Wärmebehandlung der Stähle*
is in good agreement with the recommendations for holding times given in
the preceding edition of this book. Hence these recommendations are
repeated with practically the same numerical values in this edition.

Plain-carbon and low-alloy structural steels which contain easily
dissolved carbides require only a few minutes' holding time after they have
reached the hardening temperature. In order to make certain that there
has been sufficient carbide dissolution, a holding time of 5–15 min is quite
sufficient.

For *medium-alloy structural steels* a holding time of 15–25 min is
recommended, irrespective of the dimension.

For *flame and induction hardening* which enable the setting of accurately
determined heating times, considerably higher hardening temperatures are
used than for conventional hardening. Besides, the time of holding can be

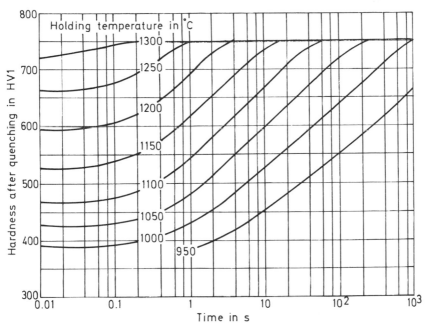

Figure 5.46 Hardness after quenching as a function of the austenitizing temperature and holding time for steel X 38 CrMoV 5 1 (H 13) after heating at rate of 130°C/s: initially fully annealed (Reproduced by permission of Stahleisen mbH, Düsseldorf)

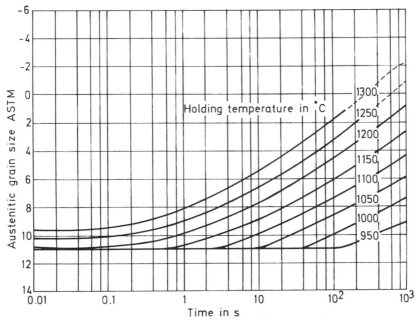

Figure 5.47 Austenitic grain size as a function of the holding time and austenitizing temperature for steel X 38 CrMoV 5 1 (H 13) after heating at a rate of 130°C/s: initially fully annealed (Reproduced by permission of Stahleisen mbH, Düsseldorf)

reduced to a few minutes and in some instances even to a second or so.

For *low-alloy tool steels* a definite holding time is necessary to ensure that the required hardness will be attained. A suggested figure for the holding time is 0·5 min per millimetre of section thickness; the minimum and maximum time, however, should be 10 and 30 min respectively. Why the holding time should increase with section thickness is due to the fact that the size of the carbide grains and hence their reluctance to go into solution increases as the section dimension increases. Besides, most low-alloy steels generally have insufficient hardenability which, however, may be improved, particularly in heavy sections, by promoting the increased dissolution of carbides. This may be brought about by increasing the time of holding at temperature.

High-alloy chrome steels require the longest time of all the tool steels. To a very great extent, however, the holding time for those steels is dependent on the hardening temperatures, *see Figure 5.45*, which applies to steel D 2. Each hardening temperature requires a definite holding time in order that maximum hardness should be obtained on hardening. Too short a time results in a lower hardness, caused by an insufficient amount of carbide going into solution; too long a time also results in a lower hardness due to the presence of retained austenite. A suggested figure for the holding time is 0·5 min per millimetre of section thickness, with minimum and maximum times of 10 min and 1 h respectively.

Hot-work steels contain carbides that do not go into solution until about 1000 °C. At these high temperatures grain growth is fairly rapid and for this reason the time of holding must be limited. If the heating is carried out in a salt bath, in which it is easy to see when the steel has reached the desired temperature, *15–30 min is a normal holding time* which is by and large independent of the dimensions of the part being treated. If the tool is packed in a box it is more difficult for the heat treater to judge when it is at the right temperature. Under such circumstances the best temperature to use is the one at the lower limit of the hardening range since a longer holding time may then be used, provided it does not exceed 1 h.

The *high-speed steels* are the most highly alloyed of all tool steel. Their hardening temperatures lie between 1200 and 1300 °C. In order to avoid excessive grain growth, which may easily occur at such high temperatures, the current practice is to use holding times of a few minutes only. In the case of high-speed steel, the normal heat treatment of which is preheating to 800–1000 °C followed by immersion in the salt bath, the time of holding

Table 5.2 Hardening of high-speed steel.
Influence of section thickness on immersion time

Section thickness mm	Immersion time min
Less than 10	3
10–20	4
20–30	5
30–40	6
40–50	7

at the hardening temperature is taken as the time of immersion in the salt bath. The times given in *Table 5.2* apply to tools heated to conventional hardening temperatures after having been preheated to 850°C.

The times given in *Table 5.2* can be exceeded by 50% without adverse consequences. When high-speed steel is hardened from lower temperatures, e.g. 1000–1100°C, the times of immersion should be trebled.

Holding time vs heating-up time over A_{c3}

For the sake of simplicity the times given in *Table 5.3* refer to holding times, after preheating, in a salt bath—short heating-up time—and in a muffle furnace—very long heating-up time, specially if pack heating. The heating-up time necessary, in the case of a muffle furnace, to raise the temperature of the steel through the final ten or so degrees increases sharply with increasing section thickness. Carbide dissolution, however, begins some 50–100°C below the ordinary hardening temperature and this fact should be taken into consideration when using long heating times. For the large die block referred to in *Figure 5.31* the heating-up time between 825 and 850°C is about 4 h. Hence, when heating in a muffle furnace it should be possible, on economic grounds, to reduce the time of holding to half of that previously reported; always remembering to keep to the stipulated minimum number of minutes.

Table 5.3 Heating up times between 950 and 1000°C from figures 5.29 and 5.30

Diameter of section mm	Salt bath min	Muffle furnace min
25	1	15
50	4	30
100	8	60

When heating-up a piece of steel in a muffle furnace the heat treater might well make use of the very rough and ready approximation of the holding time for all steels, except high-speed steels, viz. 20 min. This time is quite independent of the section thickness and the hardening temperature should be the one in the middle of the recommended temperature range. Therefore it is important for reasons of economy to study the characteristics of the furnaces at the disposal of the heat treater and, as suggested previously, to draw up recommended times for the heating-up of different dimensions and batch weights.

5.3.5   Methods of cooling

The cooling methods used for hardening depend on the grade of steel, the shape of the part and the properties to be imparted to the steel. The three following methods are described below: direct quenching to martensite, martempering and austempering (*Figure 5.48*).

254

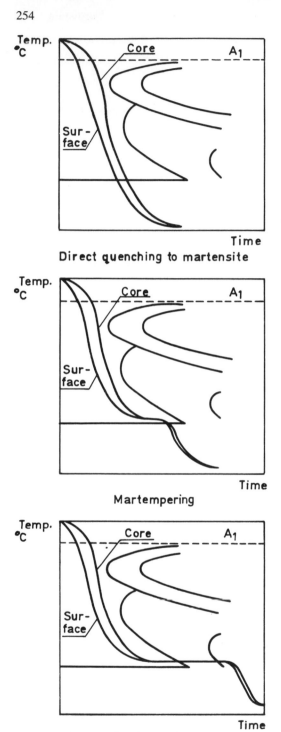

Direct quenching to martensite

Martempering

Austempering

Figure 5.48 Diagrams showing principles of different methods of cooling for hardening

Direct quenching to martensite

Direct quenching was doubtless the original hardening method and it is still the commonest one. According to this method the steel is cooled or quenched straight from the hardening temperature to room temperature or to the slightly higher temperature of the cooling medium being used. This medium may be water, oil or air.

Martempering

The principle underlying martempering is most clearly understood by referring to a TTT diagram (*see Figure 5.48*). In this hardening treatment the cooling takes place in two stages. First the tool is quenched in a molten salt or metal bath kept at a temperature somewhat above that at which the austenitic structure of the steel starts to transform to martensite (i.e. the M_s-temperature). The steel is held there until temperature equalization is complete. Next, the tool is allowed to cool freely in air, which results in the formation of martensite. Due to this method the temperature difference between the surface and the centre during the martensite formation is much smaller than it is in conventional hardening. Consequently the austenite transforms to martensite practically simultaneously throughout the various parts of the tool even if the latter has widely different section thicknesses. As a result there is less residual stress and a minimum of distortion. The final structure and properties obtained by martempering the steel are, in principle, the same as those obtained from conventional hardening but the amount of retained austenite is generally larger after martempering.

In order to carry out martempering successfully, the composition of the steel must be such that the steel obtains a martensitic structure—if this is what is required—when it has been quenched in the martempering bath and then air cooled. Therefore this heat-treatment process is confined mainly to the oil-hardening and air-hardening grades. *Table 5.4* lists the temperatures of quenching and martempering as applicable to some popular tool steels. Included in the table are also the largest dimensions suitable for martempering, i.e. such as will develop full superficial hardness on being quenched in a *clean,* well-stirred martempering bath. 300 mm is set as the maximum diameter. In some instances larger dimensions can be martempered but in that case large-capacity baths and effective stirring equipment are required.

The heating to the temperature indicated in *Table 5.4* is carried out in the same way as for conventional hardening. When sufficiently austenitized, the tool is transferred to the martempering bath, the capacity of which should be large enough to take up the heat from the tool without appreciably raising the temperature of the bath. The tool is kept moving and submerged in the bath long enough for the core to reach the temperature of the bath. However, the time of holding should not be so long that the austenite starts to transform to bainite, for if this should happen the tool will not attain full hardness. A suitable holding time is usually 1–3 min per each 10 mm of section thickness. The shorter time is applicable to low hardening temperatures, say around 850°C and the longer time for higher temperatures, *ca* 1000°C.

Table 5.4 Martempering. Austenitizing and martempering temperatures

Designation of steel AISI		Austenitizing temperature	Martempering temperature	Maximum diameter	
BS	SS	°C	°C	mm	in
O1	2140	810–840	225–250	50	2
A2	2260	950–980	225–250	150	6
D2	2310	1000–1020	225–500	200	8
D6	2312	960–1000	225–500	200	8
—	2550	820–840	225–300	125	5
S1	2710	880–920	350–400	25	1
H13	2242	1000–1050	300–500	300	12
H10A		1999–1050	300–500	100	4
M2	2722	1220–1240	450–550	100	4

Since the steel is still in the austenitic state when it is removed from the bath it can be subjected to any necessary straightening operation during the subsequent air cooling. This capacity is particularly valuable when hardening long tools such as reamers, mandrels, etc. The straightening is commenced as soon as the tool has been taken out of the bath but generally not above about 350°C and should be completed before the temperature has fallen to about 150°C. After the tool has been removed from the bath and straightened, if necessary, it is allowed to cool in air. It is during this cooling that the hard martensite is formed. Then, as in conventional hardening, the tool is tempered when the temperature has fallen to 75°C or 50°C. The curve in *Figure 5.49* shows the profile of the complete heat treatment programme for a roll made from steel D2, diameter 103 mm, length 1000 mm. The thermocouple that indicated the temperature was

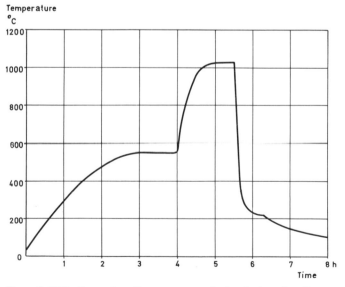

Figure 5.49 Heating and cooling programme for hardening of a roll, diameter 103 mm, made from steel D 2

located at the centre of the roll. After being preheated to 550°C the roll was heated in a protective atmosphere of nitrogen gas in a pit furnace rated at 160 kW. In this instance the hardening temperature was high, viz. 1030°C, and the time of holding at this rather high temperature was 30 min. The roll was quenched in a salt bath at 220°C, held there for 45 min and then allowed to cool freely in air. The total heat-treatment time was some 8 h. To this should be added the time required for tempering plus the time taken for straightening, if the latter operation should have been deemed necessary.

Figure 5.50 shows how a pressure die-casting die, made from steel H 13, is being transferred to the martempering bath. This tool is a rather special case. Its predecessor had suffered a large amount of distortion and after but a few shots it had begun to crack. The die shown in the photograph had been transported some 1500 km to be hardened by heat-treatment experts, who had it martempered. The result was negligible distortion and an above-average service life for the die.

Figure 5.50 Martempering of a pressure die-casting die, made from H 13

Austempering

By applying a heat treatment termed austempering it is possible, in certain cases, to obtain greater toughness than that obtained after conventional hardening and tempering to the same hardness. There are no excessively large or harmful stresses induced in the tools and, consequently, distortion is small. As in the case of martempering, the parts to be hardened are quenched in a salt bath, the temperature of which is kept above that of the martensite formation. The holding time in the bath, however, is so long that the austenite is practically transformed to bainite (*see Figure 5.48*). The resulting hardness depends on the temperature of the salt bath; a high temperature gives a low hardness. Diagrams of isothermal transformations (TTT diagrams) often indicate the hardness obtained at different temperatures after the formation of 98–99% bainite. Conventional punching tools made, for example from steel O1 are seldom able to utilize fully the increased toughness resulting from austempering. Very thin-walled tools, such as hollow punches austempered to 58–60 HRC, have given remarkably good service. This also applies to the type of punching tools used for making tin cans. When worn, the tool is renewed by upsetting and then ground to size. *Figure 5.51* exemplifies the austempering temperature–time programme for steel O1.

All steels and dimensions that are capable of being martempered as described in the preceding section can also be austempered. In addition, it is common practice to austemper certain low-alloy structural steels, e.g. spring steel, up to about 10 mm section and plain-carbon steel up to a maximum diameter of 5 mm. The length of time necessary for complete transformation restricts the application of the austempering treatment of high-alloy steels, since the time needed will often be uneconomically long. Cooling from the austempering temperature takes place freely in air. No tempering is required.

Austempering usually improves the toughness, particularly in that hardness range in which conventionally hardened and tempered steels are susceptible to a reduction in their impact strength. In ordinary low-alloy steels this reduction generally occurs when they are tempered around 300°C.

Owing to the rather high treatment costs, austempering is used mainly for expensive tools for which there are stringent requirements of dimensional accuracy and where the risk of cracking must be kept to a minimum. The austempering of machine components and structural parts is resorted to only when carefully conducted tests have shown that the generally unavoidable distortion during austempering is so small that the cost of this process is justified.

Patenting

The actual patenting treatment may be regarded as a variant of austempering. It is applied mainly to unalloyed and low-alloy steel wire with a carbon content of 0·6–1·1%. The treatment takes place at about 500°C in a molten lead bath through which the wire passes continuously. At this temperature very fine lamellar pearlite is formed. In spite of its high hardness this structure has shown itself to be well suited for wire drawing since it can accommodate large reductions of area without fracture.

5.3.6 Quenching media

When hardening tools and machine components the heat treater generally aims at obtaining a martensitic structure, at least in the surface layers of the steel. Hence the rate of cooling must be controlled so that the formation of ferrite, pearlite or bainite is avoided.

Since the dimensions of the part have generally been decided beforehand, the depth of hardness must be determined by a judicious choice of steel and cooling medium. The choice of the latter for hardening is often just as important as the choice of the former. With regard to the steel, a knowledge of its hardenability is a good aid when choosing a suitable grade. For the cooling medium, however, there are not available the same good criteria concerning the cooling capacity of the medium. Hence the simple rule as usually applied: unalloyed steel is quenched in water; alloy steels in oil and high-alloy steels cooled in air.

When it becomes desirable to introduce less sophisticated steel grades

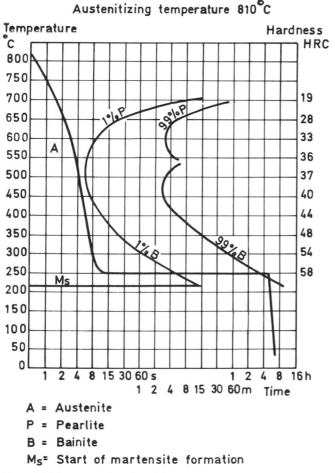

A = Austenite
P = Pearlite
B = Bainite
Ms = Start of martensite formation

Figure 5.51 Austempering of steel O 1

and more automation of the heat treating process, specialized knowledge is required both of the hardenability of steel and the cooling capacity of cooling media. A successful final result is also due in large measure to the mutual interplay between these two factors;cf. Section 4.4.4.

The three stages of quenching

During the quenching of steel in liquid media the whole operation may be split up into three stages, viz.,

　　The vapour-blanket stage
　　The boiling stage
　　The convection stage

Figure 5.52 shows a typical cooling curve for an oil and at the same time shows what happens at the surface of a steel that is being quenched from the hardening temperature.

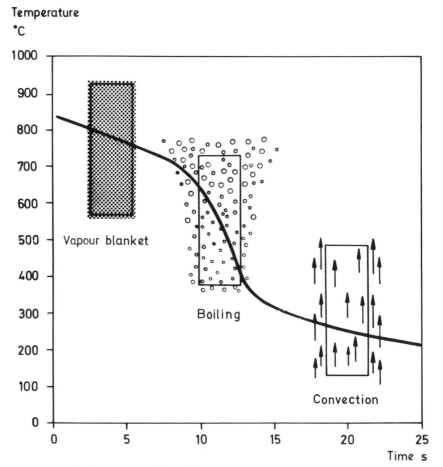

Figure 5.52 The three stages of the cooling curve

During the vapour-blanket stage a thin vapour film forms which is heat insulating and thus prevents heat abstraction. Therefore the curve has a rather flat appearance in this range.

During the boiling stage there is a high rate of heat abstraction which in part is due to the energy consumed by the vapour formation and in part to the vigorous agitation.

During the convection stage the heat abstraction takes place more slowly and the curve is flat once again.

In practically all literature on quenching media it is stressed that the shorter the vapour-blanket stage is kept the better will be the cooling effect. This is going to be examined in more detail in the following section.

Methods for evaluating quenching media

There are a great number of different methods for evaluating quenching media. An example of the type of apparatus used for such tests is shown in *Figure 5.53*.

Originally a silver ball, 20 mm in diameter, was employed, but for several reasons a cylindrical body of either steel or silver is used nowadays. A thermocouple is located at the centre of the testing body and is connected to a recording instrument. This instrument traces a record of the

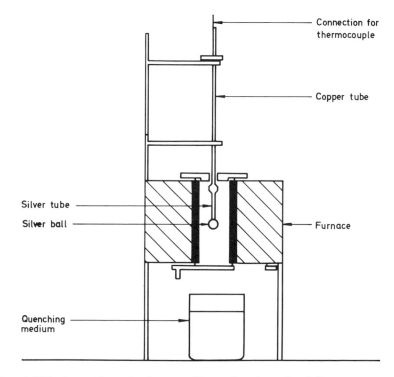

Figure 5.53 Equipment for evaluating quenching media using a silver ball

progress of the temperature change as the testing body (after being heated to the required temperature) is immersed in the cooling medium. Cooling curves obtained in this way are shown in *Figure 5.54*[24]. However, for a number of reasons they are rather difficult to interpret and put to practical use. The interpretation is facilitated if the rate of cooling during the whole course of the cooling process is registered, in which case the curves take the shape shown in *Figure 5.55*. From looking at *Figure 5.54* it is clearly seen that brine (3% common salt solution) produces the greatest cooling effect, also that the cooling capacity is reduced as the curves are displaced towards longer times. To a certain extent this observation is verified by *Figure 5.56* which shows the hardness profile from surface to centre of testing specimens, 50 mm in diameter, of a steel equivalent to 50M7. The specimens were hardened in the various quenching media mentioned, except brine. Of the various quenching media tested, water gave the

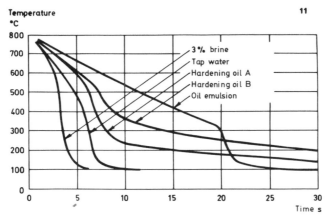

Figure 5.54 Cooling curves for various quenching media, derived by means of a silver ball. Testing temperature 40°C (after Rogen and Sidan[24])

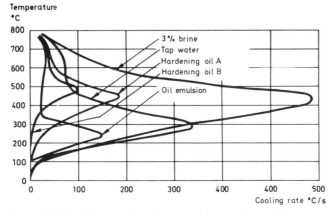

Figure 5.55 Cooling rate of various quenching media tested by means of a silver ball (after Rogen and Sidan[24])

highest hardness; of the oils, quench oil A yielded better results than oil B, but the oil emulsion gave a higher hardness than oil A. However, this last point is not directly obvious from *Figures 5.54* or *5.55*.

Liščić[25] has designed a piece of equipment that can evaluate quenching media systematically—*see Figure 5.57*. In this apparatus testing bodies of various diameters may be used to evaluate different quenching media, the temperature and rate of flow of which may be varied at will. In addition, the course of the cooling of the testing bodies may be recorded by means of thermocouples that are located at various distances below the surface of the testing body. Evaluation of the results obtained may be carried out in different ways.

An example of the result of testing a steel equivalent to DIN 34 CrMo 4 is shown in *Figure 5.58*. By increasing the flow rate an increased hardness was obtained, which was to be expected. What was not expected was, however, that a quenching-oil temperature of 70°C gave somewhat lower hardness values than an oil temperature of 20°C. This is contrary to the usual evaluation of the results obtained by the silver ball method.

Comparison between cooling curves obtained from silver and steel testing bars, respectively

The conditions controlling the course of the temperature changes during the cooling of silver and steel bars, respectively, are fundamentally different. The heat conductivity of silver is considerably higher than that of

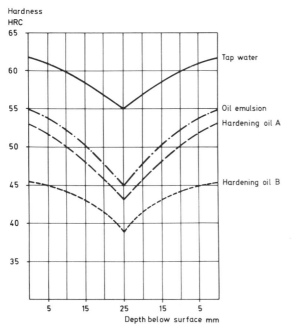

Figure 5.56 Hardness profiles across test bars, 50 mm diameter, in steel equivalent to 50 M 7, after hardening in various quenching media (after Rogen and Sidan[24])

stccl. In the case of silver there are no phase transformations that might influence the rate of cooling, such as there are in steel, viz. the transformation from austenite to ferrite, pearlite or bainite; transformations that all liberate heat and thus may have an effect on the progress of cooling and hence on the hardening result.

Beck[26] and co-workers have investigated the cooling capacity of various hardening oils by constructing cooling curves for testing bars—16 mm diameter × 48 mm—of both silver and steel. Hardness profiles of the steel bars were drawn from the surface to the centre. The steel used was grade XC 48, the nominal composition being: 0·48% C, 0·25% Si, 0·65% Mn. The cooling curves from this test are shown in *Figures 5.59* and *5.60*.

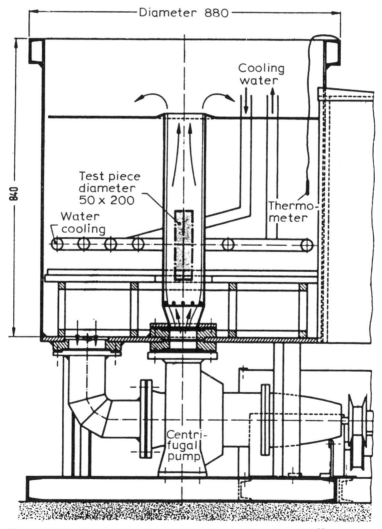

Figure 5.57 Equipment for evaluating quenching media (after Liščić[25])

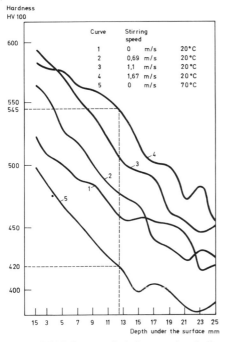

Figure 5.58 Influence of stirring speed and oil temperature on hardness profiles across test bars, 50 mm diameter, of steel equivalent to 34 CrMo 4. Quenching medium: mineral oil without additives, (after Liščić[25])

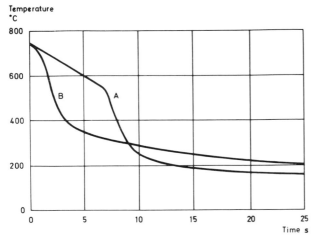

Figure 5.59 Cooling curves for the centre of silver bars, 16 mm diameter × 48 mm on being cooled in mineral oils A and B respectively (after Beck *et al.*[26])

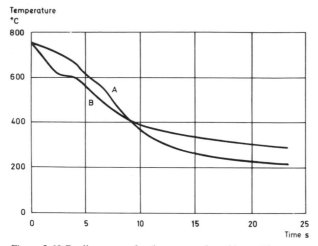

Figure 5.60 Cooling curves for the centre of steel bars, 16 mm diameter × 48 mm on being cooled in mineral oils A and B respectively, Steel XC 48 (after Beck *et al.*[26])

During the course of the tests with silver bars the boiling stage set-in considerably later in hardening oil A than in oil B. In the temperature range around 500 °C the rate of cooling was higher in oil A than in oil B. As regards the steel bars, the progress of the cooling was in principle the same as for the silver bars, but at 600 °C there was a break in the cooling of about 2 seconds in oil B.

The practical results of the tests are shown in *Figure 5.61*. Here it is seen that oil A yielded a higher hardness than oil B. The authors had expected that oil B would yield better results than oil A.

Rogen and Sidan[24] have cooled silver balls in various media and from the data obtained they drew the curves shown in *Figure 5.54*. From these it is seen that the emulsified oil gave slow cooling down to about 300 °C. On the other hand, steel testing bars (diameter 50 mm) were cooled in three of the media, and *Figure 5.56* shows the cooling curves and hardness values obtained. These cooling curves were superimposed on the CCT-diagram for the steel being tested—*see Figure 5.62*. The emulsified oil gave the highest hardness that could not be expected by referring to the cooling curves in *Figure 5.54*.

Evaluation of the cooling curves with the aid of the transformation diagrams

Judging from the examples given above it is quite clear that it may be difficult to evaluate cooling curves obtained from silver testing bars, and also that it is not possible to evaluate the curves for, say, certain emulsified oils according to the rules applied hitherto. (This applies also to polymer cooling media.) It is also open to doubt whether the evaluation made concerning steel testing bars is correct in all respects. What is missing from the examples quoted, except from the last one, is an evaluation of the cooling curves taken in conjunction with the respective CCT diagrams for the steels being investigated. It is important to emphasize that it is quite

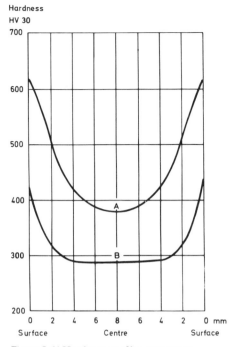

Figure 5.61 Hardness profiles across test
bars, 16 mm diameter × 48 mm of steel
XC 48 after being quenched in mineral oils
A and B respectively (after Beck *et al.*[26])

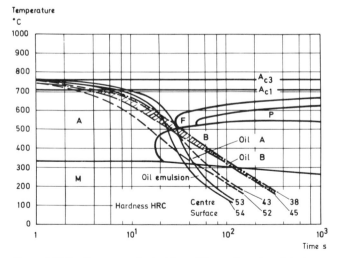

Figure 5.62 Cooling curves for test bars, 50 mm diameter of steel
50 M7, after being cooled in various media from hardening
temperature 850°C. The curves are superimposed on a CCT diagram
for 50 M7 (construction after Rogen and Sidan[24])

irrelevant to superimpose arbitrary cooling curves on such existing CCT diagrams as have been constructed from linear cooling data.

In Chapter 4—Hardenability—Section 4.4.4, the mechanisms that control hardenability have been discussed rather fully. The examples quoted above supplement these discussions and confirm that certain long-standing opinions must be considered. The most usual assertion that a short vapour-blanket stage enhances the hardening result must be regarded as incorrect. That time-honoured opinion was probably based on the fact that water has a short-blanket stage along with the observation that water quenching gives rise to a high hardness.

Of decisive importance, however, is the rate of cooling through the temperature ranges in which the diffusion-dependent transformations take place. When the effectiveness of a quenching medium is being appraised it is consequently necessary also to consider the characteristics of the steel being quenched. This conclusion has been confirmed by, *inter alia,* Schimizu and Tamura[27] who carried out hardening tests with four steels, the composition and CCT diagram of each being shown in *Figure 5.63.*

The quenching was carried out in three oils characterized by the cooling curves shown in *Figure 5.64.* In *Figure 5.65* are shown the quenching results for some critical bar dimensions. From the diagrams it is seen that for the unalloyed steels SK3 and S43C, the CCT diagrams of which

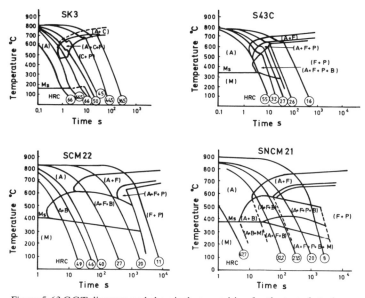

Figure 5.63 CCT diagram and chemical composition for the tested steels (after Schimizu and Tamura[27])

Chemical composition %:

Steel grade	C	Si	Mn	Ni	Cr	Mo
SK3	1·05	0·24	0·68	0·04	0·08	—
S43C	0·44	0·27	0·32	0·05	0·07	—
SCM 22	0·22	0·24	0·72	0·06	1·02	0·16
SNCM 21	0·21	0.26	0·85	0·47	0·57	0·22

Temperature
°C

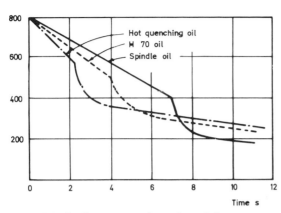

Figure 5.64 Cooling curves at the surface of silver
cylindrical specimens of 10 mm diameter × 30 mm
quenched from 800°C into several coolants (measured by
the method of JIS–K2526) (after Schimizu and Tamura[27])

showed incipient ferrite or pearlite formation respectively after about one
second at 600°C, the best results were obtained with those cooling media
the boiling stage of which started at a high temperature. In the two alloy
steels, SCM 22 and SNCM 21, ferrite formation starts in about 10
seconds and their bainite noses lie around 500°C. When samples were
quenched in the spindle oil, the boiling stage of which starts at 400°C as
determined by the silver ball method, it was found that this oil gave the
highest hardness values to the steels.

Our knowledge of hardenability and quenching media is still not
sufficiently advanced to enable us to master all problems in this field. The
relatively new concepts presented above should enable us to make a more

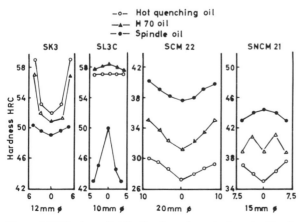

Figure 5.65 Hardness distribution in various steels austenitized
at 880°C for 40 min and quenched into several coolants (after
Schimizu and Tamura[27])

critical review and obtain a better understanding of the available literature concerning this subject. Within the International Federation of Heat Treatment (IFHT) there is a committee on quenching media engaged in studying these questions. Similar work is being undertaken within the Heat-Treatment Committee of Sveriges Mekanförbund (the Swedish Association for Metal Transforming, Mechanical and Electromechanical Engineering Industries).

The following description dealing with various cooling media is mainly of a conventional nature. The points discussed above are not treated in detail since they require further study.

Water

Water is probably the oldest cooling medium used for hardening and it has remained the major coolant throughout the ages. Pure water, however, is rather unsuitable as a cooling medium since its greatest cooling efficiency

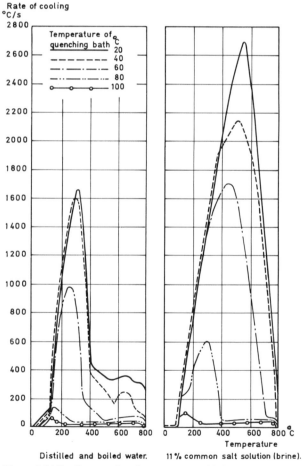

Distilled and boiled water.   11 % common salt solution (brine).

Figure 5.66 Cooling capcity of pure water and brine. Test carried out with 20 mm ball of silver (after W. Peter[28])

occurs at 300°C, i.e. the temperature at which martensite formation starts in many steels. By adding 5–10% of common salt or soda to the water, its cooling capacity is increased very considerably and at the same time its greatest heat-extracting capacity occurs at 500°C. These points are illustrated in *Figure 5.66* which also shows the rapid fall in cooling capacity as the temperature of the water rises above 60°C. The over-all best result is obtained when the water temperature is between 20°C and 40°C. The great drawback with water, as mentioned above, is that the rate of cooling is high in the temperature range of martensite formation. This exposes the steel to the simultaneous influence of transformation stresses and thermal stresses, the combined effects of which will increase the risk of crack formation.

The danger of cracking occurring during water quenching may be reduced if the steel is removed from the water when it has cooled to some 200–400°C and is rapidly transferred to an oil bath. Incidentally, this is a very neat way of increasing the depth of hardening of oil-hardening, low-alloy steels. *Figure 5.67* shows the cooling curves for different dimensions on being water quenched.

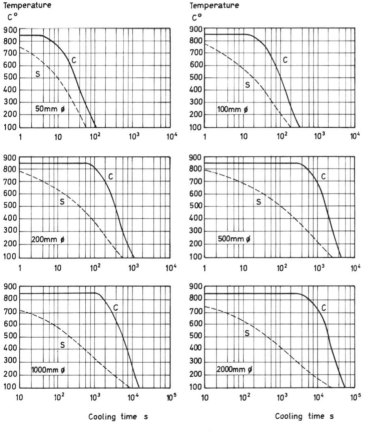

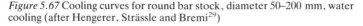

Figure 5.67 Cooling curves for round bar stock, diameter 50–200 mm, water cooling (after Hengerer, Strässle and Bremi[29])

Oil

Oil cooling is much slower than water cooling (*see Figure 5.68*). Oil is not a well-defined cooling medium as regards cooling capacity. The simplest and cheapest hardening oil is a standard mineral oil with the lowest possible viscosity (spindle oil). By introducing certain additives to such oils it has been possible to produce oils having considerably increased cooling capacity, as was shown in the preceding section. Furthermore, there are oils that may be used at elevated temperatures, e.g. at 200 °C. Proprietary oils specify, *inter alia,* viscosity and flash point. The rate of cooling is generally greatest at about 600 °C and is relatively slow in the range of martensite formation. Since oil has a rather low capacity for heat extraction relative to water, its use as a coolant for medium to low-alloy steels is restricted to light sections. For example, on quenching steel O 1 or En 19C (42 CrMo 4) in still oil, it is not certain that full superficial hardness is obtained in dimensions around 100 mm. A reliable and tested way of increasing the cooling capacity of oil is by agitating vigorously the bath or

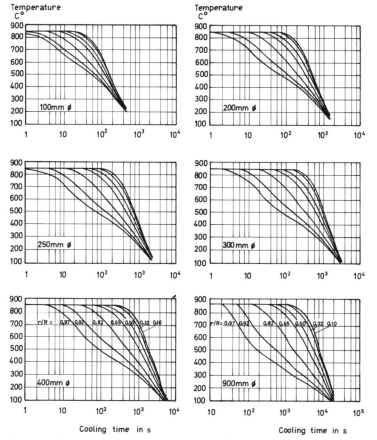

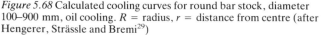

Figure 5.68 Calculated cooling curves for round bar stock, diameter 100–900 mm, oil cooling. R = radius, r = distance from centre (after Hengerer, Strässle and Bremi[29])

the charge. This simple technique should not be underrated. *Figure 5.69* shows the effect the agitation of the oil has on the hardness increase in a 0·50% C steel.

In an attempt to increase the agitation in the oil, trials with blowing in air were conducted. However, this resulted in the entrapment of air bubbles in pockets in the charge and on account of the uneven cooling caused by these air pockets hardening cracks occurred.

In another trial the oil was agitated mechanically in such a manner that very small air bubbles were intimately intermixed with the oil. During the following quenching the air bubbles formed a heat-insulating film round the components of the charge and the hardening result was unsatisfactory. The air bubbles remained in supension in the oil for several hours, thereby allowing hardening trials to be carried out in tanks containing both aerated and unaerated oil. Definite proof was thus obtained of the inhibiting effect of the air bubbles.

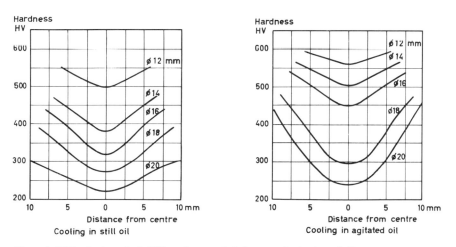

Figure 5.69 Hardening of a 0·50% carbon steel. Influence of agitation of oil on depth of hardening. Temperature of oil 60 °C

Another way of increasing the cooling capacity is to raise the temperature of the oil to 50–80 °C. This increase in temperature makes the oil more fluid and hence increases its cooling capacity. *Figure 5.70* shows how the cooling capacities of a conventional mineral oil and a fast-quenching oil are influenced by their respective temperatures.

Observe, however, that for the standard mineral oil the cooling rate around 500 °C is lower when its temperature is 80 °C than when it is 32 °C. This fact may explain the inferior result obtained in the elevated-temperature test as recorded in *Figure 5.58* in which also standard mineral oil was used. The fast-quenching oil gives rise to a considerably higher cooling rate at 500 °C than the standard mineral oil. The temperature increase would also enable a fast-quenching oil to maintain a higher cooling rate in this range.

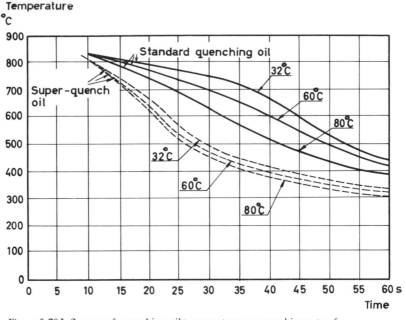

Figure 5.70 Influence of quenching-oil temperature on quenching rate of plain carbon steel (0·45% C) in standard quenching oil and super-quench oil. Test piece 25 mm diameter × 75 mm (from a Gulf publication)

Mixtures of water and oil (emulsions)

By emulsifying water and 'water-soluble' oil in various proportions it is possible to obtain cooling media of various cooling capacities. In this respect, however, such media might be inferior to the oil itself. *Figure 5.71* shows that an emulsion of 90% oil and 10% water has cooling properties inferior to those of unadulterated oil. An emulsion containing 90% water and 10% oil has a much higher cooling rate between 400 and 500°C. This may produce a satisfactory hardening result for a steel having its bainite nose within this temperature range (cf. also the cooling curve for the oil emulsion in *Figure 5.54* and oil A in *Figure 5.59*).

If water is inadvertently added to ordinary quenching oil (*Figure 5.72*), this may lead to cracking on hardening particularly if the steel is a deep-hardening one because then the martensite formation will start at the surface considerably in advance of the centre which, when it transforms, will increase the stresses in the surface. If the water and oil are not properly emulsified, the former having collected at the bottom of the tank instead, a rapid heating-up of the water followed by steam generation may give rise to an explosion.

The presence of water may be ascertained by dipping a narrow-bore tube, preferably of glass and temporarily closed at its upper end, into the tank at its lowest point. On opening the tube the operator lets the quenching fluid rise in the tube. The opening is then again closed, the tube withdrawn and its contents examined.

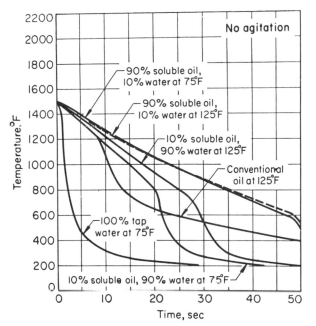

Figure 5.71 Comparison of water, oil and emulsions of water
and soluble oil. Centre cooling curves for still-quenched 18/8
stainless steel specimens ½ in diameter × 2½ in long,
indicating comparative cooling characteristics of plain water,
conventional oil and two emulsions of water and soluble oil, at
temperatures of 75°F (24°C) and 125°F (50°C). (Reproduced by
permission from *Metals Handbook,* **2,** American Society for
Metals, 1964)

Polymer quenching media

By mixing water with about 10% of water-soluble polymers (e.g.
polyalkylene glycol or polyvinyl alcohol) a quenching medium is obtained
that has a cooling capacity between that of water and oil. Intensive
research and development work in this field is being pursued at several
centres. Besides other advantages arising from the use of this quenching
medium, a reduced dependence on oil and an eliminated risk of fire are
gained. In *Figure 5.73* are shown cooling curves for a polymer quenching
medium called Aquaquench in various concentrations. Included are also
cooling curves for water and for oil. The curves were derived from trials
with 0·5% C steel bars, 70 mm diameter × 150 mm.

With just a moderate addition (about 2%) of this substance to the water
it has been shown that during the quenching of unalloyed steel the risk of
cracking is reduced without any loss of hardness being incurred. An
addition of 15% produces a quenching medium having almost the same
cooling properties as a quenching oil, with the added advantage that the
danger of fire is completely eliminated. During quenching, the
temperature must not rise above 30°C in still baths or above 45°C in
agitated baths. When in use it is found that the pH of the bath decreases

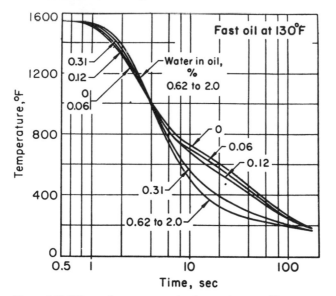

Figure 5.72 Effects of water contamination on the quenching power of fast oil at 130°F (55°C) in quenching type 304 stainless steel specimens ½ in diameter × 4 in long. Oil was not circulated and specimens were not agitated. Thermocouples were located at the geometric centre of specimens. (Reproduced by permission from *Metals Handbook*, **2,** American Society for Metals, 1964)

and hence it must be adjusted by means of neutralizing additions to keep the pH from falling below 7·5, otherwise there is the corrosion danger to contend with.

The polymer concentration in these quenching media must be kept under close control since it tends to fall as the baths are used. When this happens the cooling capacity increases, thereby increasing the risk of hardening cracks. Furthermore, precipitated and extraneous matter must be removed from the bath since such unwanted material affects the cooling conditions[30]

Air

Low-alloy steels in light sections and high-alloy steels may be successfully hardened by means of compressed air or still air. The advantages of using air are that distortion is negligible and that the steel can easily be straightened during the cooling process. So that it should cool evenly the part is rotated in a steady current of compressed air. One drawback here is that the surface may be oxidized during the cooling. Nowadays it is not so usual for structural parts to be made in air-hardening steel on account of the high cost of the necessary alloying elements and also for the fact that such steels must inevitably be annealed after forging or rolling before subsequent machining.

A typical composition of a conventional air-hardening structural steel is: 0·30% C, 1.2% Cr, 4% Ni, 0·3% Mo. In dimensions up to 100 mm

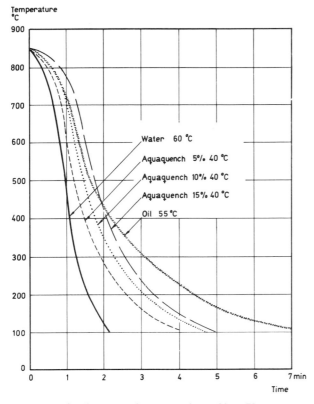

Figure 5.73 Cooling curves for centre of round bar, 70 mm
diameter × 150 mm, quenched in various cooling media

diameter such a steel would be wholly martensitic after air hardening.

In the group of high-alloy tool steels, A2, D2 and D6 have hardenability
sufficiently high to enable them to be air hardened in quite heavy sections.
In *Figure 5.74* a comparison is made between D6 and D3 with different
manganese contents. The most popular of the air-hardening steels is H13.
This steel can be hardened in still air in sections up to 200 mm in diameter.
Figure 5.75 shows the air-cooling curves for various dimensions (cf.
Figure 4.50).

Salt baths

The most popular salt baths used for cooling purposes consist of
approximately equal parts of sodium nitrite and potassium nitrate. They
are used in the temperature range 160–500°C. There are also salts for
500–600°C. A salt bath is the ideal quenching medium for a steel having a
reasonably good hardenability and not too heavy a section. The cooling
capacity down to about 500°C is high and then decreases as the
temperature of the steel continues to fall. The lower the temperature of the
bath and the greater the agitation, the better is its cooling capacity (*see*

278

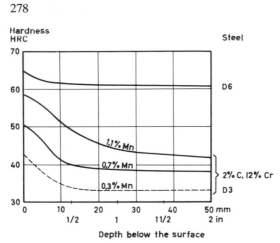

Hardness
HRC

Steel

D6

1,1% Mn

0,7% Mn

0,3% Mn

2% C, 12% Cr

D3

Depth below the surface

Figure 5.74 Hardenability of steel D 3 containing various
Mn contents and steel D 6. Steel specimen 100 mm (4 in)
diameter hardened from 1000 °C in still air

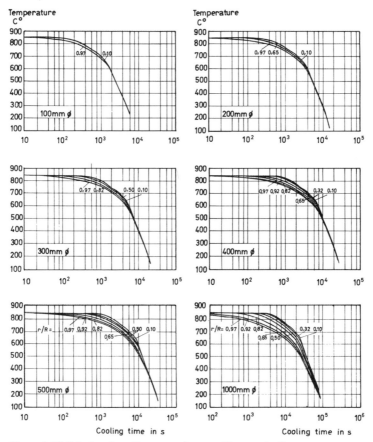

Cooling time in s

Cooling time in s

Figure 5.75 Calculated cooling curves for round bar stock, diameter
100–1000 mm air, cooling. R = radius, r = distance from centre (after
Hengerer, Strässle and Bremi[29])

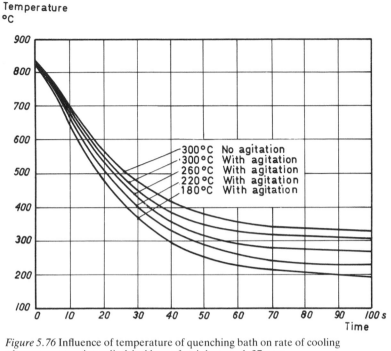

Figure 5.76 Influence of temperature of quenching bath on rate of cooling
when martempering cylindrical bars of stainless steel, 37 mm
diameter × 100 mm (courtesy I C I)

Figure 5.76). In *Figure 5.77* is illustrated the effect of agitation on the rate
of cooling and on the hardness of a plain-carbon steel containing 0·5% °C.

If the salt bath becomes contaminated its cooling efficiency is very much
reduced and when the bath is stirred the foreign particles are kept in

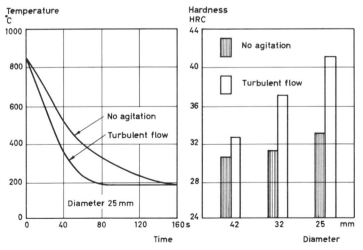

Figure 5.77 Influence of turbulent flow on rate of cooling and on hardness
after quenching in salt bath at 190°C. Specimen of 0·50% plain carbon steel

suspension in the molten salt. They are apt to attach themselves to the parts being treated and hence impede heat transfer. In such cases it is better to let the impurities sink to the bottom and instead move the parts smartly about in the clear supernatant molten salt. A recommended holding time in the salt bath is 1–3 min per centimetre of section thickness. The shorter time applies to low hardening temperatures and light sections; the longer time to high temperatures and heavy sections.

The cooling capacity may be increased by adding water to the salt bath. The amount of water must be kept within close limits and should amount to 0·3–0·5%. The cooling capacity will then be approximately doubled. The water, as steam, leaves the surface of the bath continuously and should therefore be replenished, either all the time or as periodic additions. A convenient way of adding water is to introduce it into the vortex created by a circulation impeller. It will then immediately dissolve in the salt without any danger of explosion. The water content may be checked by measuring the density of the bath by means of a hydrometer which has been calibrated in a subsequently solidified salt, the water content of which is determined by weighing. The presence of water may be detected by the characteristic sizzling sound that is caused by the evolution of steam when the charge is lowered into the salt. This steam also sets up a stirring action at the surface of the bath.

Parts that have been heated in a cyanide bath containing more than 10% cyanide must not be quenched in a nitrite–nitrate bath on account of the danger of explosion.

To complete this section, *Figure 5.78* shows a graphical compilation of the cooling capacities of various cooling media.

5.3.7   Quenching equipment

The result of the hardening operation by quenching in the above mentioned liquid media is dependent on their temperature and degree of agitation.

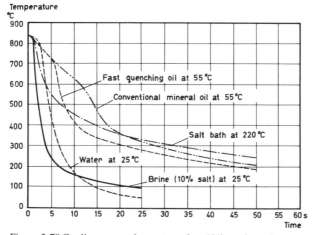

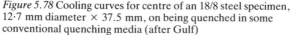

Figure 5.78 Cooling curves for centre of an 18/8 steel specimen, 12·7 mm diameter × 37.5 mm, on being quenched in some conventional quenching media (after Gulf)

Temperature

The best temperature range for water is 20–40 °C and for oil, 50–80 °C. The temperature should be checked continually. For small-scale operations it is common practice to raise the temperature of the oil, before the first quenching proper, by cooling some hot pieces of steel in the bath. Another possibility is to have an electric heating element immersed in the bath when it is not in use. There are also ready-made quenching tanks commercially available, complete with temperature control. This control is effected by letting hot or cold water flow through heat exchangers fitted in the bath. Such equipment is quite a matter of course in a modern heat-treatment shop.

Agitation

When quenching large pieces, the heat treater is generally not able to create the required stirring action by moving the piece. Modern quenching tanks have a built-in impeller or pump which provides the agitation of the liquid medium. Illustrations of such equipment are shown in *Figures 5.79* and *5.80.*

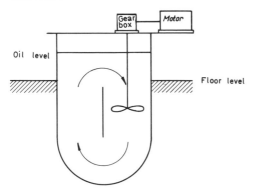

Figure 5.79 Quenching tank with propeller agitator

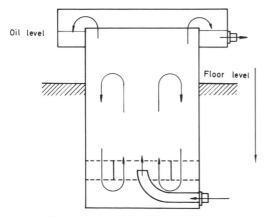

Figure 5.80 Quenching tank with pump circulation

Cooling in quenching jig

This method is used for several reasons. Two common quenching practices may be mentioned. The first one, called press quenching, is used for rings, gears and similar cylindrical flat or elongated parts that are susceptible to warpage. After having been heated, the part is clamped in a jig and placed in a press, usually hydraulically operated. Simultaneously with the closing of the press, oil is made to flow over the part which thereby maintains its shape during the quenching. The jig should be preheated to about 50°C lest the rate of cooling at the surfaces in direct contact with the jig be too rapid and hence give rise to superficial cracks.

Parts made from sheet steel, especially steel containing less than about 0·3% carbon which on quenching is not particularly susceptible to cracking, may be hardened in a water-cooled jig in which water circulates through ducts in the jig. The parts are thereby subjected to a very rapid and effective cooling action.

The other type of jig quenching is applied to tools and other parts containing holes or cavities which for design purposes are to be hardened from the inside only. This hardening method is often used for upsetting dies made from plain-carbon steel and intended for the production of cold-headed bolts. The quenching principle is shown in *Figure 5.81*.

Other equipment

Tools made from plain-carbon or low-alloy steel may have such a form that their cooling is impeded. Such obstruction to the cooling may be caused, for instance, by undercut cavities or recessed holes in which the cooling medium cannot circulate adequately or in which air bubbles prevent rapid cooling. In such cases it is necessary to increase the rate of flow of the coolant if a martensitic structure is to be obtained, otherwise cracking may occur during service or even during the hardening operation.

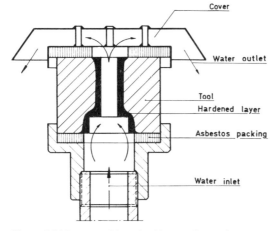

Figure 5.81 Jet quenching of cold-upsetting tool

Example 1

A coining die made from steel W 1 had cracked through the engraving after some time in use. The die was returned for examination but the engraving had been almost completely ground off, which somewhat hampered the investigation. Cracks were found at the bottom of the deepest engraving marks where very finely lamellar pearlite was observed in the otherwise martensitic structure (*see Figure 5.82*). The tensile stresses set up as the discs were being coined caused the die to crack since the pearlite has not the same tensile strength as martensite. Faults of this type may be avoided if, on quenching, the tool is immersed into the water bath and held above a tube which delivers a powerful current of water, as illustrated in *Figure 5.83*.

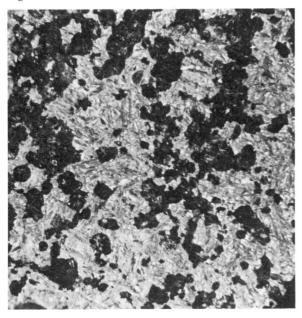

Figure 5.82 Duplex structure consisting of martensite and pearlite, 500 ×

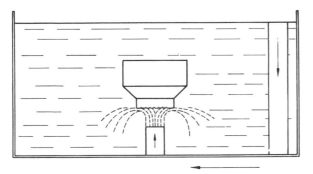

Figure 5.83 Submerged-jet quenching of engraved surface of coining die. Steel W 1

Example 2

A mandrel made from steel O1 had cracked while it was being heat treated, or more specifically during the tempering operation. The location of the cracks and the hardened surface zone are shown in longitudinal section in the sketch in *Figure 5.84*. Owing to an insufficient cooling effect in the hole, no martensite was formed there. The martensite in the outer zone of the mandrel had induced large tensile stresses at the bottom of the hole where the fillet radius was practically zero. The cracks had formed during the tempering, i.e. when the mandrel was being lowered into a vigorously agitated salt bath. Since the outer zone of the mandrel was the first part of it to expand, the tensile stresses on the inside were still further intensified. This failure could have been avoided if quenching oil had been circulated in the mandrel hole by means of some arrangement such as that shown in *Figure 5.83*. What further aggravated the situation was the insufficient fillet radius at the bottom of the hole as well as the sudden heating to the tempering temperature.

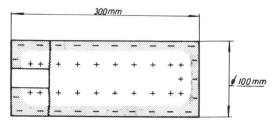

Figure 5.84 Depth of hardening and cracks in mandrel, steel O 1
+ = tensile stresses
− = compressive stresses

5.4 Tempering

When steel is hardened to a martensitic structure the toughness is rather low. On tempering, i.e. heating it to some temperature between 160°C and 650°C, the toughness increases considerably. In Section 1.4 the theoretical background has been described.

5.4.1 Heating to temperature

Tempering should be carried out in close proximity to the quenching operation and in certain cases as soon as the steel has cooled to 50–75°C. If it is allowed to cool to room temperature before being tempered the steel may crack. Heating for tempering can take place in a conventional muffle furnace, a convection-type furnace (*Figure 5.85*), i.e. a tightly closed furnace with a circulation fan, or in a salt-bath furnace of the same type as that used for martempering. The heating can also be done either by means of a gas flame or by electric induction. These last methods are used for local tempering.

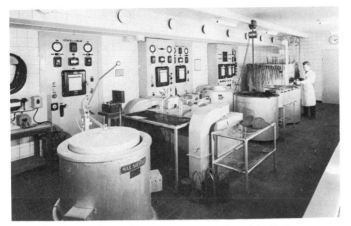

Figure 5.85 View of part of heat-treatment section of the Bofors metallurgical laboratory. In the foreground, convection-type tempering furnace

The heating of complicated tools must take place slowly. Rapid heating, as in an agitated salt bath, will cause the surface layers to increase in volume and this can induce unfavourable stress conditions that may result in cracking. Hardened parts that have been heated or quenched in a salt bath must be freed from any adhering salt before being tempered in a muffle or convection furnace. This salt is most conveniently removed by rinsing the charge in warm water. If the salt residues are not removed they may cause corrosion during the tempering but if this tempering is carried out in a salt bath there is no need to rinse the parts beforehand. Any adhering hardening salt may be very difficult to remove by rinsing but if the tempering is done in a salt bath of the nitrate–nitrite type, the hardening salt is dissolved. The tempering salt is very soluble in warm water.

5.4.2   Rate of heating

The superficial colour of unalloyed steel changes as the tempering temperature increases, according to the colour scale given in *Table 5.5*. When tempering is carried out in a salt bath at some temperature above 220 °C it is easy to decide from the clearly visible colour when the steel has taken on the temperature of the bath. When the tempering is carried out in a muffle furnace it is also possible to tell the temperature from the colour but in order to be able to distinguish the shade of colour the surface of the

Table 5.5

Temperature °C	Temper colour	Temperature °C	Temper colour
220	Straw yellow	310	Light blue
240	Light brown	325	Grey
270	Brown	350	Grey–purple
285	Purple	375	Grey–blue
295	Dark blue	400	Dull grey

steel must be ground clean, which, however, is not the usual practice with parts that are to be tempered. From the dull red colour produced by temperatures around 600°C it is possible to judge when the steel has reached the desired temperature.

There is an old rule-of-thumb that says that the tempering time should be 1–2 h for each inch of section thickness, but it does not say when to begin the timing, i.e. if it is the moment the charge is introduced into the furnace or when the charge is warm. To settle this question specific tests were performed. The test material consisted of round bars, the lengths of which were at least twice the diameter. The main part of the investigation was carried out in an 18 kW convection furnace, the internal dimensions of which were 360 mm diameter by 600 mm. Tests involving various steel grades, low-alloy as well as high-alloy, showed that the alloying elements had negligible effect on the time to reach the required temperature.

Furthermore it was observed that when tempering, the time taken to reach the tempering temperature is more or less independent of the temperature, whether it be high or low. Where high temperatures are concerned the rate of heating was somewhat more rapid on account of the effect of radiation. Another fact brought to light was that the centre of the test-piece reaches the temperature of the furnace at about the same time as

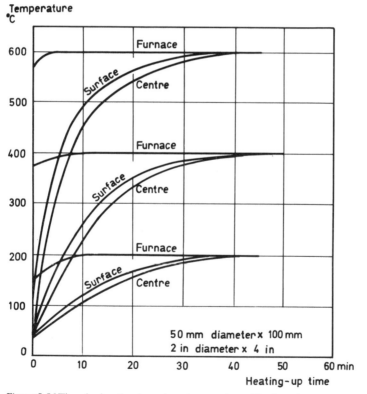

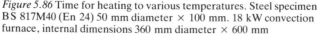

Figure 5.86 Time for heating to various temperatures. Steel specimen BS 817M40 (En 24) 50 mm diameter × 100 mm. 18 kW convection furnace, internal dimensions 360 mm diameter × 600 mm

the surface. In *Figures 5.86* and *5.87* are shown some of the results of the tests performed.

When the furnace is charged with several test-pieces at the same time the temperature obviously rises more slowly than if there is only one test-piece in it. Of course, the ratio of the furnace input to the weight of the charge is decisive in this case. In the tests reported in *Figures 5.88* and *5.89* the heating time to temperature was not prolonged by more than 15 min when, in the first instance the number of test-pieces increased from 1 to 15 (diameter 50 mm × 100 mm), and in the second, from 1 to 4 (diameter 150 mm × 300 mm). Similar test have been carried out in convection furnaces of various sizes, and reasonably good agreement with the above-mentioned tests has been obtained.

Often the parts are put into the furnace when its temperature is below the tempering temperature. The time taken to heat up will then be longer (*see Figure 5.90*) and in such cases the tempering time is taken from the moment the furnace has reached the preset temperature.

By making use of the data obtained from the tests performed, a diagram has been constructed showing the time for heating as a function of the diameter, up to 150 mm of the bar stock. The time may be taken either

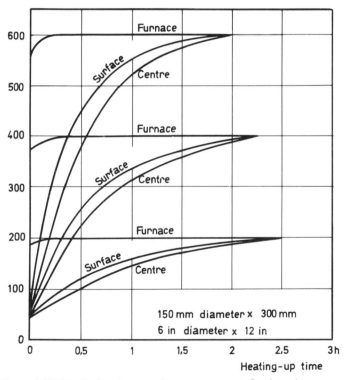

Figure 5.87 Time for heating to various temperatures. Steel specimen BS 817M40 (En 24) 150 mm diameter × 300 mm. 18 kW convection furnance, internal dimensions 360 mm diameter × 600 mm

from the moment when the charge is loaded into the preheated furnace or from the moment when the indicating instrument shows the preset temperature (*Figure 5.91*). This diagram cannot be expected to hold for all cases but it should not be out by more than some 30 min, which is generally of no practical significance.

The die block described in Section 5.3 has been used for time–temperature measurements and the values obtained have gone to the drawing up of *Figure 5.92*. The heating was carried out in an electric muffle furnace without forced circulation. *Figure 5.93* contains the results of theoretical calculations regarding the time taken to heat up bars of different dimensions. Round bars shorter than twice their diameter heat up more quickly, but the heating-up time is very little affected when the bar length is increased beyond three times its diameter.

5.4.3   Holding time

According to the description of the progress of the tempering process as given in Section 1.4, martensite and retained austenite are transformed to various products. This transformation is not only dependent on the temperature but also on the time factor which in some instances has a very great effect. Hollomon and Jaffe, amongst others have investigated this

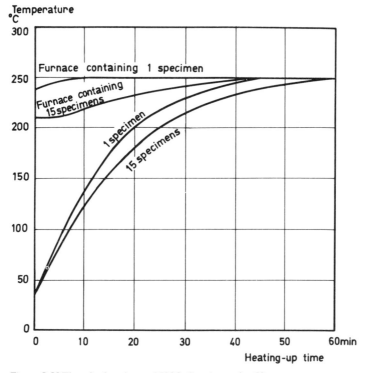

Figure 5.88 Time for heating to 250°C. Specimen size 50 mm diameter × 100 mm, 1 and 15 specimens respectively. 18 kW convection furnace

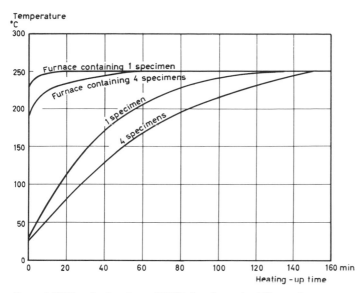

Figure 5.89 Time for heating to 250°C. Specimen size 150 mm diameter × 300 mm. 1 and 4 specimens respectively. 18 kW convection furnace

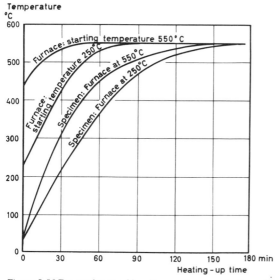

Figure 5.90 Dependence of heating-up time on starting-off temperature of furnace. Specimen size 150 mm diameter × 300 mm. 2 specimens. 18 kW convection furnace

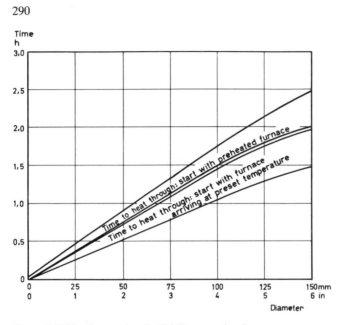

Figure 5.91 Heating-up time in 18 kW convection furnace as a function of bar stock diameter

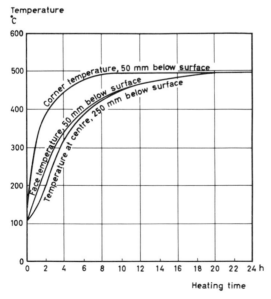

Figures 5.92 Heating-up time in electric muffle furnace for die block measuring 2300 × 950 × 500 mm. Dimensions of furnace: 6500 × 1400 × 1100 mm

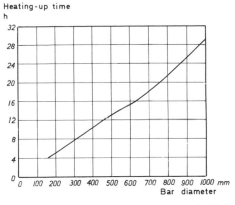

Figure 5.93 Time for heating from 100 to 500 °C
for bars of various diameters. Bar
length = 2 × diameter

time–temperature factor which has led to the development of the
expression for the tempering parameter *P* as described in Section 1.4, viz.

$$P = T(k + \log t)$$

Extensive tests performed at the Bofors Company have shown that this
expression can be applied to all the steels that have been investigated. A
figure of 20 for the constant *k* is valid for all the steels.

The time factor may produce a greater or smaller effect, depending on
the shape of the 'usual' tempering diagram, which is generally valid for a
tempering time of 1 to 2 h. This is brought out clearly in *Figure 5.94* which
applies to steel H 13 and which contains tempering curves for holding times
between 0·1 and 100 h.

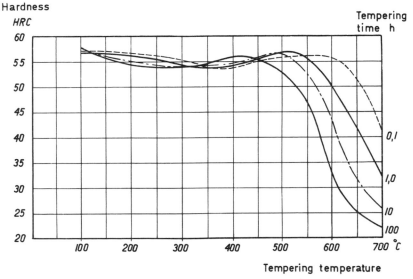

Figure 5.94 Tempering curves of steel H 13, held at temperature for various
times

For temperatures up to 450°C, time has very little influence on the hardness of this steel. At higher temperatures, however, the time factor is important. The values obtained from *Figure 5.94* can be condensed into a so-called 'master curve' which is shown in *Figure 5.95*. Since curves of this type are rather cumbersome for everyday use, a new type of diagram has been evolved by the Bofors Company and is shown in *Figure 5.96*. The upper part of the diagram is a conventional tempering curve and as such is valid for a 1 h tempering time. Should the effect of a longer tempering time be required, move vertically downwards to the lower diagram and follow the appropriate sloping temperature line until it intersects the time horizontal concerned. From this point of intersection move vertically upwards into the tempering diagram and read off the hardness in the usual way. The following example which applies to steel H13 and to the diagram in *Figure 5.96* gives an actual illustration of the procedure. After a tempering treatment of 1 h at 500°C the hardness is 56 HRC. If we wish to temper for 10 h at the same temperature, follow the 500°C line in the lower diagram until it intersects the 10-h horizontal. Proceed straight up to the top diagram and read off the hardness, viz. 55 HRC. In the same way, tempering for 100 h at 500°C gives a hardness of 53 HRC.

Tempering for 1 h at 600°C results in a hardness of 49 HRC. If the time is increased to 3 h the hardness falls to 46 HRC. This latter hardness is also obtained by tempering at 625°C for 1 h. By visualizing a temperature line at 625°C and following it down to the 10-h horizontal we obtain a hardness of 35 HRC, i.e. a hardness reduction of 11 units relative to the 1 h tempering. Notice the large hardness reduction obtained by tempering for 100 h at this temperature compared with the small reduction at 500°C after the same tempering time.

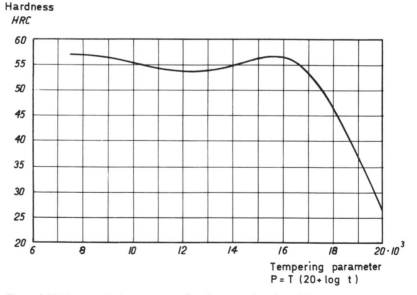

Figure 5.95 The so-called master curve (hardness as a function of the tempering parameter) for steel H 13

The high-alloy hot-work tool steels that show a sudden fall in hardness around a tempering temperature of 600 °C are obviously very sensitive to a few hours' difference in the holding time. There are untold instances of failures with such steels owing to the heat treaters' not being aware of the very strong influence of the time factor. Such failures are now avoidable if the new diagrams are used. Chapter 6 contains further examples of these new tempering charts. From the charts it is seen how the hardness is dependent on temperature and time. The question now crops up: how long

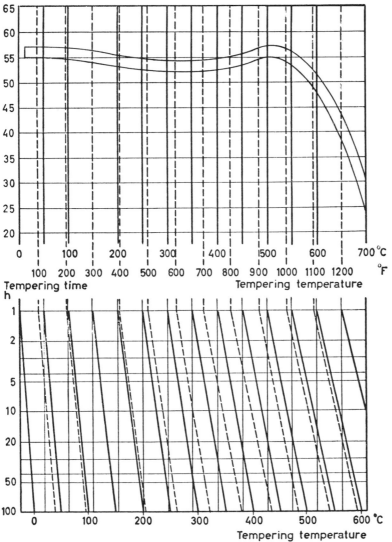

Figure 5.96 Time-dependent tempering diagram for steel H 13. Oil quenched or air cooled from 1025 °C

should the holding time be? There are many and varying opinions among heat-treatment experts on this question. For example, should it be a short holding time (½ h) at a high temperature or a long holding time (10 h) at a low temperature or is the more usual holding time of 2 h to be preferred? Results from investigations of a number of heat-treated steels have shown that figures for the yield point, ultimate tensile strength, elongation, reduction of area and impact strength have not been influenced in any special direction when, without going to extremes, the temperature and holding time have varied as permitted by the tempering parameter.

Figure 5.97 shows a diagram for steel En 29 B, the mechanical properties of which have been obtained by various combinations of treatment temperature—between 525°C and 700°C—and time—10 min to 8·5 h. Even longer holding times would yield the desired hardness but for some steels, impact strength may be considerably reduced (temper brittleness). In the absence of more detailed investigations *a holding time of 1–2 h* is recommended. For very large components the holding time may be prolonged by an hour or so in order to be certain that they have reached the desired temperature. Note that the holding time is regarded as beginning when the surface of the steel has reached the required temperature.

5.4.4   Double tempering

As was stated earlier, tempering should be carried out with a minimum of delay after the hardening operation or when the heat-treated part has cooled to between 75°C and 50°C. The martensite that forms between 75°C and room temperature may induce cracking. In some instances even at room temperature some martensite formation takes place isothermally. The temperature at which to interrupt the quenching process can be selected by referring to the position of the M_s-temperature in the TTT or CCT diagram, or better still, the M_f-temperature if it is included in the diagram. Compare *Figure 1.17*. The higher the alloy content of the steel, the lower are the M_s- and M_f-temperatures and hence the lower the temperature at which the cooling may be interrupted without risking the development of hardening cracks.

Steels of this type containing more than about 0·40% carbon and having M_s around 300°C are the ones most liable to crack, especially if they are through hardening types. Also, deleterious effects may result if the tempering is interrupted at too high a temperature.

If the cooling is interrupted at, say, 80°C and immediately followed by a tempering treatment at 170°C, the formation of martensite ceases and a fairly large amount of untempered martensite is produced as the steel cools to room temperature. For this reason a steel that has been subjected to interrupted cooling at some intermediate temperature should be tempered a second time. For example, a case was experienced when surface-hardened bars cracked during transportation. Metallographic examination showed that the surface layers contained about 30% untempered martensite—*see Figure 5.98a*. When tempered at 200°C the structure shown in *Figure 5.98b* was obtained.

Steels that form bainite on tempering or that already have a bainitic

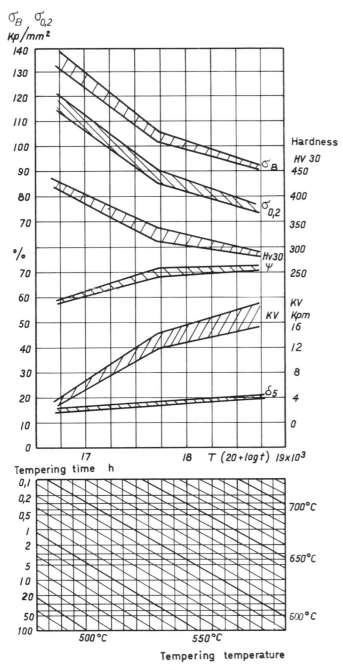

Figure 5.97 Master curves for steel En 29 B

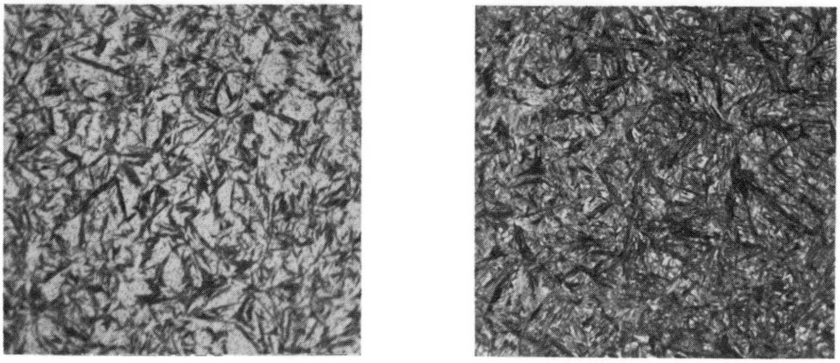

(a)   (b)

Figure 5.98 Microstructure resulting from insufficient hardening
immediately followed by tempering (a), and the same specimen after
another tempering treatment (b)
(a) 30% untempered martensite as a result of inadequate hardening
      practice;
(b) Same specimen after tempering at 200°C. Both 300 ×

structure after hardening, such as heat-treatable alloy steels in heavy
sections, need to be tempered once only. In high-alloy chrome steels and
high-speed steels the retained austenite is transformed to martensite on
cooling from a tempering temperature around 500°C. Therefore, such
steels should be retempered, which toughens the newly formed martensite.
If the required hardness has been reached after the first tempering
treatment the second tempering must take place at a lower temperature
lest the hardness of the steel should be reduced. In practice, this second
tempering is carried out some 10–30°C below the first temperature. A
suitable   retempering   temperature   may   be   deduced   from   the
treatment–time tempering diagrams. If the tempering treatment of a costly
tool cannot follow hard on the quenching, it should be kept between
50–100°C in a heating oven during the waiting period.

Tempering after grinding or electrical spark machining

When a hardened and tempered tool is subjected to grinding or spark
erosion machining operations, the heat developed in the surface of the
steel may lead to structural transformations and an unfavourable stress
pattern which may result in cracks and chipping after an exceptionally
short time. This ill effect can be avoided by an extra tempering treatment
introduced after the grinding or spark erosion machining operation.

5.4.5   Self-tempering (auto-tempering)

On quenching steels possessing a high M_s-temperature—around 400°C and
higher—which is the case for steels containing less than 0·30% carbon,
they produce a first generation of martensite which will be tempered during
the subsequent cooling. These steels are then no longer so susceptible to
cracking when cooled to room temperature. In particular, with those steels

having M_f around 100°C or higher, this tempering action will produce a favourable result. This process is called self-tempering (auto-tempering) and may have such a good effect that any subsequent tempering will not be required.

5.4.6   Temper brittleness

When steel is tempered it passes through a series of transformations that affect not only the hardness but also the toughness. The conspicuous reduction in toughness that occurs on tempering at certain temperatures is called temper brittleness. Due in part to the composition of the steel, this brittleness makes its appearance at different temperatures and by judging from the manifestations of embrittlement at various temperatures the four following regions may be distinguished.

1. 350°C
2. 500°C
3. 475°C
4. 500–570°C

1   350°C brittleness

This brittle range is found in most unalloyed and low-alloy steels, both tool steels and constructional steels, on being tempered between 250 and 400°C. This temperature range coincides with the range in which retained austenite is transformed to bainite and it is highly probable that some mechanism involved in this transformation is responsible for the loss of toughness.

It has long been known that phosphorus produces an unfavourable effect on the impact strength of steel after a hardening and tempering treatment, especially if the steel has been tempered in the 350°C range.

Materkowski and Krauss[31] have investigated the 350°C-brittleness phenomenon, *inter alia*, in steel AISI 4340, one sample of which contained low P (0·003%) and one high P (0·03%). In the low-P specimen the brittle fracture started at the cementite platelets produced by the decomposition of the retained austenite, and was propagated by means of cleavage cracks across the laths of the martensite aggregates.

In the high-P specimen the embrittlement occurred in conjunction with a large number of intercrystalline fractures. Even if the presence of grain-boundary carbides could not be detected it was thought that the mechanism of intercrystalline embrittlement was in some measure due to a combination of P-segregation and precipitation and grain-boundary carbides during the tempering in the critical range.

Investigations also showed that the P-content exerted an influence on the toughness of the steel even after hardening and that this influence persisted right through the whole of the temperature range being tested. (*See Figure 5.99*).

Banerji, McMahon and Feng[32] and Briant, Banerji and McMahon[33] suggested for this embrittling mechanism a model in which allowance was made for the segregation of certain alloying elements during the

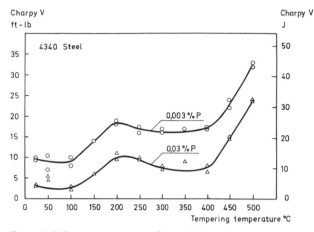

Figure 5.99 Room temperature Charpy V impact energy as a
function of tempering temperature for AISI 4340 steel with 0·003
and 0·03% P, austenitized at 870°C, oil quenched and tempered
1 h (after Materkowski and Krauss[31])

austenitizing stage, P in the first instance, and for the carbide separation
during the tempering. Sandberg and Westerhult[34] investigated steel
42CrMo4 and found that phosphorus affects the impact strength for the
whole of the tested range, viz. 200°C to 450°C. (*See Figure 5.100.*)
Molybdenum reduced temper brittleness but was not able to eliminate it
completely. The same applied to lanthanum (REM-treatment, i.e. Rare
Earth Metal), also shown in *Figure 5.100*. The mechanism is that P is
rendered innocuous by separating out as LaP.

Another 350°C brittleness is shown in *Figure 5.101* which gives the
impact values for steel SS 2550. There are many instances when an
incorrect tempering temperature has been the cause of costly failures.
Further examples are given in Chapter 6.

2   500°C brittleness

This brittleness is generally the one referred to when temper brittleness is
mentioned in connexion with hardened and tempered alloy constructional
steels. The development of this embrittlement has been found to depend
on both tempering temperature and time. Curves resembling conventional
TTT diagrams have been drawn up[35] and one such series of curves is
shown in *Figure 5.102*. The curves represent the impact transition
temperatures. A low transition temperature implies that the steel is tough
(*see* Section 3.2). According to *Figure 5.102*, if the tempering time is 10 h,
a tempering temperature of 450°C results in a transition temperature of
−60°C. If the temperature is raised to 485°C, the transition temperature
rises to −40°C, and if raised to 500°C the transition temperature is −30°C.
Improved impact strength (i.e. lower transition temperatures) is again
obtained as the tempering temperature is raised above 550°C.

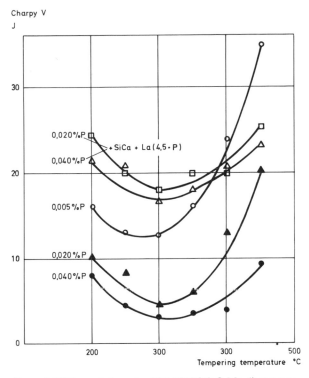

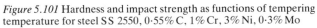

Figure 5.100 Impact strength of SS 2244 (42 CrMo 4) at room temperature with varying phosphorus contents as a function of the tempering temperature (after Sandberg and Westerhult[34])

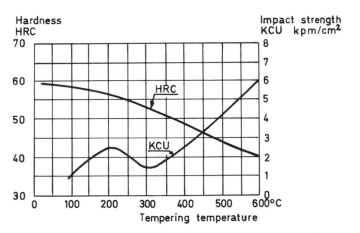

Figure 5.101 Hardness and impact strength as functions of tempering temperature for steel SS 2550, 0·55% C, 1% Cr, 3% Ni, 0·3% Mo

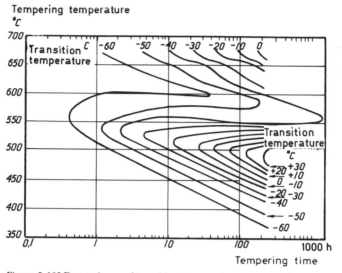

Figure 5.102 Dependence of transition temperature on tempering
temperature and time for steel SAE 3140 of the following
composition: 0·40% C, 0·80% Mn, 0·60% Cr, 1·25% Ni.
Water-quenched from 900°C. Double tempering: first tempering for
1 h at 675°C, water quenching. Second tempering—time and
temperature as in diagram (after Jaffe and Baffum[35])

This brittle range differs from the foregoing one in that the degree of
brittleness is dependent on the rate of cooling after tempering. Even if the
steel has been tempered at a higher temperature a slow cooling through
this critical range may result in a deterioration of the impact strength. For
this reason heat-treatable steels in heavy sections are particularly
susceptible to this type of embrittlement which, however, is fairly easy to
avoid simply by letting the steel cool in water or oil from the tempering
temperature. An alloy steel that has become embrittled may be restored to
its tough condition by being reheated above the brittle range. The
toughness may even be restored to a certain extent when the steel is heated
to the critical range but it must be quenched therefrom. This type
brittleness is called reversible brittleness.

Of the elements that tend to promote temper brittleness, the most
significant is P when present in amounts of more than about 0·015%. At
this P level, Cr and Mn each have an unfavourable influence too, which is
intensified if both elements occur together and in amounts of more than
about 1% each. If Ni also is present the susceptibility to embrittlement is
further increased although Ni by itself has no deleterious effect. The
presence of small amounts of As, Sb or Sn, approximately 0·05%, has in
some instances been found to be harmful. Contents in excess of 0·005% Sb
in 3·5% Ni steels may give rise to temper embrittlement. Susceptibility to
temper brittleness can be reduced by alloying the steel with Mo in amounts
from 0·15 to 0·5%[36].

Dumoulin and co-workers[37] found that in steel containing ≤0·02% P,
molybdenum in solid solution prevents the segregation of phosphorus

atoms to the grain boundaries by binding them as Mo-phosphides. In steel containing >0·03% P, the molybdenum located at the grain boundaries will reduce the embrittling effect of the phosphorus atoms.

3   475°C brittleness

The 475°C brittleness is the embrittlement that, accompanied by a hardness increase, appears in ferritic and semi-ferritic chrome steels, the Cr content of which lies over 13%. The brittleness is brought about by holding the steel in the temperature range 400–550°C. The brittleness is most noticeable at 475°C. In steels with Cr contents over 25% the embrittlement may occur after a few minutes heating at 475°C. There are several theories about the cause of this embrittlement which disappears on heating to a higher temperature than 600°C. The hardness increase is said to be due to the coherent precipitation of a complex Cr rich compound. This in turn causes a Cr-depletion of the matrix, resulting in reduced resistance to corrosion (*see* Sections 5.6 and 6.2.4).

4   500–570°C brittleness

In high-alloy steels, such as hot-work steels and high-speed steels, finely dispersed carbides are precipitated in this temperature range and during cooling martensite is formed. Both these factors increase the hardness and at the same time the impact strength is reduced. This brittleness is graphically illustrated in *Figure 5.103*.

5.5 Transformation of retained austenite

Austenite may remain in the structure after hardening in appreciable quantities and is called 'retained austenite'. Applied stresses may transform this austenite to martensite which in turn may cause the steel to

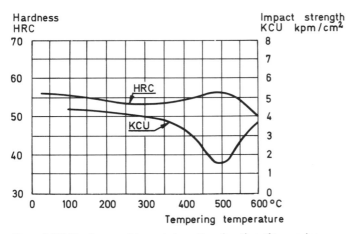

Figure 5.103 Hardness and impact strength as function of tempering temperature for steel H 13

crack. The retained austenite is transformed to martensite either by subzero treatment, i.e. cooling to below $-70\,^\circ\text{C}$, or by tempering. In the latter bainite or martensite is formed, depending on which temperature range is used for the treatment (*see Figure 1.22*). As mentioned previously, the amount of retained austenite increases with increasing hardening temperature. It also increases if an interruption in the cooling occurs in the vicinity of the M_s-temperature. Since the subject of retained austenite and its transformation is of paramount interest where tool steel is concerned it will be treated along with the individual tool steel grades in Chapter 6.

Mechanical transformation of retained austenite

In the introduction to this section it was stated that retained austenite can be transformed to martensite by the application of high stresses[38]. For this reason, tools and machine components that are liable to be subjected to high stresses should be tempered in order to transform the retained austenite. Recommendations for suitable tempering temperatures for each individual steel grade are given in the following chapter.

Hadfield steel, i.e. manganese steel containing 12–14% Mn and 1·2–1·4% C is completely austenitic after a quench annealing treatment. By subjecting the surface to cold work the austenite is transformed to martensite, resulting in a very considerable increase in surface hardness. According to Schumann[39] the cold work gives rise to dislocations which create stacking faults which then act as nuclei for ε-martensite. This constituent later changes to α-martensite.

In order to increase the hardness of manganese steel intentionally by cold working, a charge of explosive paste may be detonated on it. Shot-blasting by steel sand will also produce a considerable increase in surface hardness (*see Figure 5.104*). The formation of martensite also in this case results in a volume increase.

Stabilization of retained austenite

If after hardening, the steel is kept at room temperature for some time or is heated to the temperature range corresponding to the first tempering stage (*see* Section 1.4) the austenite is stabilized which implies that it has become more difficult to transform when subjected to a subzero treatment. The stabilizing effect increases as the tempering temperature and time increase. The diagrams in *Figures 5.105a, b* and *c* show how much of the retained austenite is transformed by a subzero treatment at $-180\,^\circ\text{C}$ after various pretreatments[40].

After quenching from, say, $840\,^\circ\text{C}$ there is 18% retained austenite. If the subzero treatment is carried out within 5 min after the temperature of the steel has reached $20\,^\circ\text{C}$, i.e. almost immediately after quenching, about 70% of the retained austenite will be transformed. If a lapse of 40 min is allowed before the treatment, 60% is transformed and after 50 h only 30% of the retained austenite will respond to the subzero treatment. If the steel is heated to $120\,^\circ\text{C}$ for 10 min after quenching and then subzero treated, only 30% of the retained austenite will be transformed.

The initial amount of retained austenite is dependent to a very large

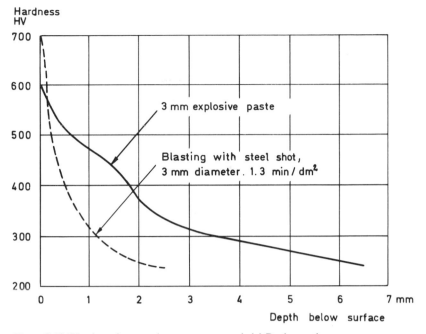

Figure 5.104 Hardness increase in manganese steel: (a) By detonating explosive paste; (b) By steel-shot blasting

extent on the hardening temperature. As this temperature is raised there is an increase in the amount of retained austenite that will be transformed after it has been given a specific stabilizing treatment. This is clearly shown in the diagram, *Figure 5.106*, which shows the stabilizing effect at 20°C. From the diagram it can be seen that, for example, after holding it for 1000 min at 20°C the retained austenite, on subzero cooling to −180°C, will undergo about 10% transformation if the hardening temperature had been 780°C. For temperatures 840°C and 900°C the corresponding figures are 40% and 62% respectively. By raising the hardening temperature it is thus possible to transform increasing amounts of retained austenite by a subzero treatment. By estimating from *Figures 5.105* and *5.106* the amounts of retained austenite remaining after the subzero treatment it is found that for the hardening temperatures investigated, they are practically the same, viz. at 6–10% for stabilizing times between 10 and 1000 min.

If the hardening quench is interrupted somewhere around the M_s-temperature a similar stabilization of retained austenite is obtained. If the cooling to room temperature is then continued the same effect, in principle, results as that obtained by the subzero treatment in the example given above, viz. a transformation of retained austenite to martensite. The amount of martensite will however in this case be less than after a direct quench to room temperature. The numerical values quoted in the above description are valid only for the steel examined.

304

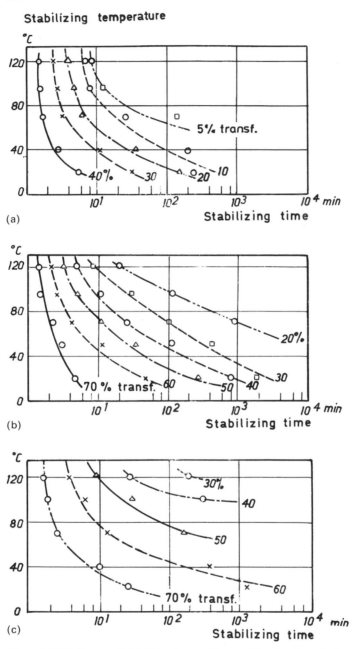

Figure 5.105 Influence of stabilizing temperature and time on amount of retained austenite that transforms on being subzero treated at $-180°C$. Ball-bearing steel AISI 52 100 (after Liebmann[40]).
(a) Temperature of quench: 780°C. Retained austenite after quenching: 9·4%;
(b) Temperature of quench: 840°C. Retained austenite after quenching: 18%;
(c) Temperature of quench: 900°C. Retained austenite after quenching: 27%

Amount of
transformed austenite

%

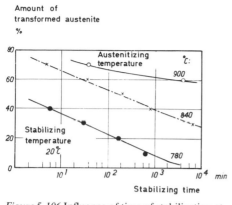

Stabilizing time

Figure 5.106 Influence of time of stabilization at
20 °C on amount of retained austenite, obtained
by quenching specimens from various
temperatures, which transforms on being
subzero treatment at −180 °C. Ball-bearing steel
AISI 52 100 (after Liebmann[40])

5.6 Precipitation hardening

This type of hardening normally involves two treatments, viz. solution
treatment and ageing. It can be applied to low-carbon unalloyed steels,
albeit for such it is of hardly any practical use. To introduce the subject,
however, the example below is taken from this group of steels. The
solution treatment consists of heating the steel to a temperature just below
A_1 where the solubility of C in ferrite is greatest, 0·02% (*see Figure 5.107*).
Any N present is also dissolved, a similar diagram being applicable to N.

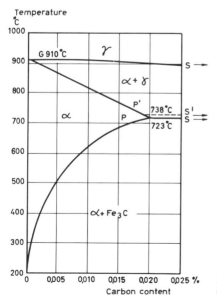

Carbon content

Figure 5.107 Solubility field for carbon in α-iron
(after Horstmann)

When the steel is quenched to room temperature, C and N are still in solution. After the lapse of some time (ageing) at this or at some slightly higher temperature a finely dispersed precipitation of ε-carbide and nitride or carbo-nitride—if N is present—takes place. This precipitation induces an appreciable hardness increase, as shown in *Figure 5.108* and at the same time the impact strength of the steel decreases. By tempering the steel at some moderate temperature a hardness reduction and an increased impact strength are obtained. This relatively high increase in hardness is due to the occurrence of the precipitate at the dislocations. Hereby their mobility is impeded which manifests itself as an increase in hardness. As the temperature is raised the precipitated particles increase in size at the expense of their numbers. The dislocations become increasingly mobile and as a consequence the hardness fails. This type of ageing, which is also called quench-ageing, occurs after the part has been quenched from a temperature at which the alloying elements under consideration are in solid solution. Carbon and nitrogen produce an adverse effect on the impact strength of low-carbon steels subjected to quench-ageing. Their effect may be reduced if the steel is alloyed with small amounts of, for example, V, Nb or Ti since these elements have a greater affinity than iron for carbon and nitrogen. Aluminium has the same effect but it binds nitrogen only.

The precipitation-hardening effect is utilized with advantage in alloy steels. Some examples of this are given below and, at the same time, they serve to illustrate the mechanism still further. Useful properties resulting

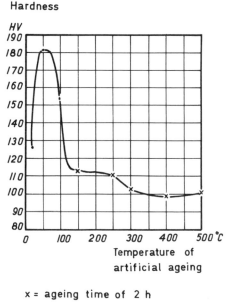

Hardness

x = ageing time of 2 h

• = ageing time of 50 h

Figure 5.108 Hardness variations obtained by artificially ageing a steel containing 0·05% C. Prior treatment: heating to 700°C and quenching

from precipitation hardening can be conferred on nickel-base alloys by additions of Al, Ta, Ti or Mo. Subsequent ageing gives considerable hardness increases. Since this ageing takes place at rather high temperatures these alloys are well adapted to high-temperature service.

With the aid of the electron microscope it is possible to follow the progressive growth, with increasing time and temperature, of the precipitated particles. *Figure 5.109* illustrates this for an austenitic alloy designated Pyromet 860[41] which has the following composition:

C	Mn	Si	Cr	Ni	Mo	Co	Ti	Al	B	Fe
max.	max.	max.	12.0	40	5·0	3·5	2·5	0·75	0·008	rem.
0·10	1·0	1·0	16·0	45	7·0	4·5	3·5	1·5	0·012	

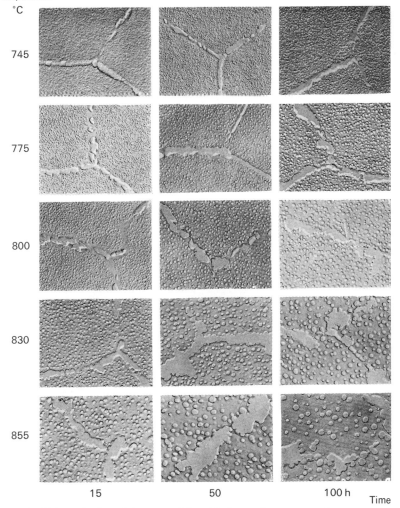

Figure 5.109 Growth of precipitated particles with time and temperature in a solution-treated. precipitation-hardening alloy. 10 000 × (after Maniar *et al.*[41])

A solution treatment at 1095°C for 2 h followed by water quenching gave the alloy a hardness of 160 HB. Ageing it for various times and temperatures gave the hardness indicated in *Table 5.6*. Other examples of austenitic precipitation-hardening alloys are A 286 and the nickel-base alloys designated Nimonic. These alloys are solution treated and quenched from about 1000°C, followed by ageing at about 700°C for 16 h. The hardness increases from 200 HB to 300 HB. Fe–Ni alloys with less than 25% Ni, on cooling from the solution temperature, produce a martensite called nickel-martensite[42] (*see* micrographs *Figure 1.15*).

The effect of nickel on the transformations is shown in *Figure 5.110*. This diagram holds for normal rates of heating and air cooling. If a ferritic alloy containing 18% Ni, for example, is heated, it is found that at 600°C, 10% austenite is formed and at 630°C there is 90% austenite. On cooling, the transformation to α-iron—in this instance to martensite—is suppressed to a lower temperature so that when 315°C is reached, 10% α-iron has formed. At 260°C, 90% of the austenite has been transformed to martensite.

Table 5.6 Hardness HRC after ageing

Ageing time h	Ageing temperature 6°C				
	745	775	800	830	855
5	35–36	36–37	36	36	33–34
15	38–39	37–38	35–36	35–36	31–33
50	38–39	37–38	35–36	33–34	30–31
100	38–39	37–38	34–35	33–34	30–31

If the steel is alloyed with, say, Ti for precipitation-hardening purposes, it can be aged at temperatures up to about 600°C without any appreciable transformation of martensite to austenite taking place.

Nickel-martensite is fairly soft and is probably not subject to the type of tempering effect associated with the more familiar type of martensite. When Ni-martensite is aged, Ni_3Ti is precipitated. Since the dislocation density in this martensite is very high the precipitate is highly dispersed, which explains the large hardness increase that is produced on ageing these so-called maraging steels. Yet another contribution to the hardness increase comes from the coherency strains that arise owing to the similarities in atomic structure between the martensite and the Ni_3Ti precipitate, *Figure 5.111*[43].

The maraging steel, grade 250, is supplied solution treated, hardness 280 HB. (The treatment consists of annealing at 820°C for 1 h followed by air cooling.) The precipitation hardening temperature is 480°C, holding time 3 h. The mechanical properties given below are obtained:

$R_{p0.2}$ N/mm^2	R_m N/mm^2	A_5 %	Z %	HB
1720	1770	10	50	470

The maraging steels are also discussed in Section 6.2.6.

Precipitation hardening or dispersed-phase hardening has been discussed in Section 1.8 where it is also stated that the hardening

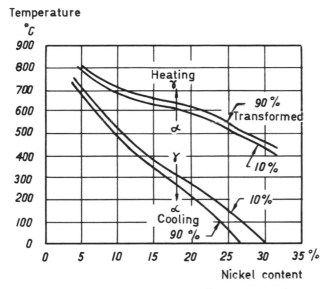

Temperature
°C

Figure 5.110 Temperature-transformation diagram for Fe–Ni alloys. N.B. not an equilibrium diagram (after Jones and Pumphrey)

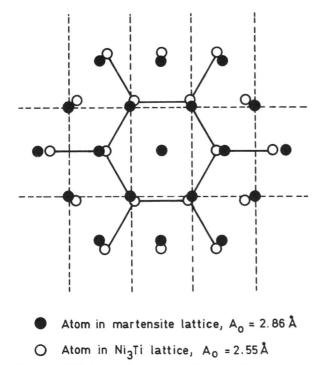

● Atom in martensite lattice, A_0 = 2.86 Å

○ Atom in Ni$_3$Ti lattice, A_0 = 2.55 Å

Figure 5.111 Coherency between the (110) crystallographic plane of martensite and the (0001) plane of Ni$_3$Ti in a maraging steel (after Pitler and Ansell in[43])

mechanism may take place without subsequent ageing. This statement applies to the high-strength low-alloy (HSLA) steels which are discussed in Section 6.4.

5.7 Strain-ageing

This type of ageing is rather similar to quench-ageing, which was discussed in the foregoing section and, like it, there are both negative and positive features of this phenomenon. The negative feature is well known and occurs mainly in commercial straight-carbon steels and low-alloy manganese steels. If the steel is plastically cold deformed the impact curve decreases, resulting in a displacement of the impact-strength curve towards higher temperatures. This tendency is further emphasized during a subsequent ageing treatment. The embrittlement is caused by the diffusion during the ageing process of the carbon and nitrogen atoms to the dislocations which were produced in great numbers by the deformation and which thereby are locked in situ. Besides impaired impact strength there is, on the other hand, an increase in the yield and tensile strengths; the former developing the greater proportional increase. The nitrogen-ageing proceeds rapidly at temperatures between 60 and 100°C whereas the carbon-ageing does not yield maximum properties until the steel is heated to about 250°C.

In order to find out whether a steel is susceptible to strain-ageing the sample is cold deformed by about 10%, aged for ½ h at 250°C and then impact tested as V-notch impact test pieces. Independent of the steel composition, the deformation itself gives rise to an embrittlement which manifests itself as an increase in the transition temperature of the steel, cf. *Figure 1.46*. The actual ageing embrittlement does not manifest itself until after the heat treatment. This is very convincingly shown by a classical diagram[44] in which an Al-killed open-hearth steel is seen to be practically

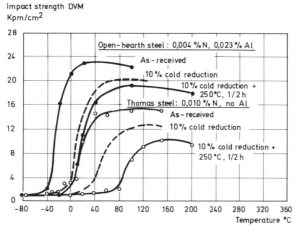

Figure 5.112 Impact strength of two types of steel, as received, after cold reduction and after cold reduction plus ageing (after Heller and Stolte[44])

immune to strain-ageing whereas a rimming (untreated) Thomas steel responds very strongly (*see Figure 5.112*).

As a general statement it may be asserted that the susceptibility of a steel to strain-ageing may be assessed by means of the impact test in its as-rolled or as-normalized condition. This point is made in a paper by Hussmann and Krisch[45] who examined seven steels of composition in accordance with *Table 5.7*

Table 5.7 Chemical composition, section area and grain size of seven steels in as-normalized condition (after Hussmann and Krisch[45])

Steel No.	Steel grade	% C	% Si	% Mn	% P	% S
1	TUSt 37-1	0·094	≤0·01	0·33	0·041	0·021
2	MUSt 37-1	0·133	≤0·01	0·37	0·013	0·028
3	RTSt 37-1	0·117	0·02	0·62	0·049	0·018
4	RSt 37-1	0·160	0·02	0·63	0·023	0·016
5	RSt 37-2	0·115	0·17	0·29	0·009	0·009
6	St 37-3	0·135	0·24	0·41	0·009	0·021
7	St 52-3	0·194	0·39	1·37	0·020	0·020

Steel grade	% N total	%N soluble	%Al total	% Al soluble	Section area mm × mm	Grain size ASTM
1	0·0100	0·0089	0·007	0·004	60 × 15	5
2	0·0039	0·0034	0·002	0·001	60 × 15	5–6
3	0·0106	0·0102	0·004	0·003	60 × 15	5–6
4	0·0039	0·0015	0·012	0·004	60 × 15	5–6
5	0·0032	0·0023	0·005	0·001	120 × 15	5–6
6	0·0051	<0·0001	0·022	0·021	120 × 15	7–8
7	0·0052	<0·0001	0·039	0·038	120 × 15	8–9

The impact values from longitudinal as-normalized samples are shown in *Figure 5.113*. Steels Nos. 6 and 7, which showed good impact values, were both from Al-killed steel and had grain size 7 ASTM. The other steels

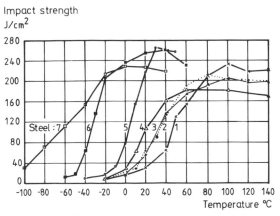

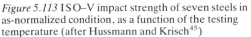

Figure 5.113 ISO–V impact strength of seven steels in as-normalized condition, as a function of the testing temperature (after Hussmann and Krisch[45])

The most efficient way of carrying out the straightening operation is to take it in combination with the hardening process and while the steel is still in the wholly austenitic state. The most convenient hardening process from the point of view of subsequent straightening is martempering, i.e. when the steel has assumed the temperature of the bath. The straightening is started immediately the steel is taken out of the bath and should be completed when the temperature has fallen to about 150°C. Straightening during air hardening is also convenient.

Most of the high-alloy steels are easy to straighten as quenched, since in their untempered state they have a high content of retained austenite and their yield points are relatively low. Straightening may also be performed during the actual tempering process. Warped tools can be clamped in a jig and become straightened as a result of the stress relaxation that sets in during the tempering. If there are several warped tools it may be practicable to clamp them together in pairs, placing washers between them if necessary. Straightening by means of local heating with a welding torch requires much experience and a high degree of discrimination on the part of the operator. Such straightening should be avoided if the steel is harder than 50 HRC. It should preferably have been tempered to at least 400°C. The bent part is heated on the convex side, the torch being moved in one or several straight passes at right angles to the direction of curvature of the bent part. The heated length expands thermally, thereby becoming upset. When it cools, tensile stresses are induced and these tend to straighten the part but there is the risk that the stresses may cause cracking.

5.9 Machining allowances

Tool steel

Even if tools do not require clean surfaces from a functional point of view it is a general rule that if the tool is to be hardened all its surfaces must be machined clean. The reason for machining before hardening is to remove scale, surface cracks and decarburization. Surface cracks may initiate further crack growth. A decarburized surface may easily give rise to cracks during hardening since martensite formation starts in the surface layers containing the lower carbon content. Subsequently, as more and more martensite continues to be formed in the core the latter increases in volume and may eventually cause the initially formed martensite case to rupture. For high-speed steel the stipulated maximum acceptable depth of decarburization is about 1% of the bar diameter + 0·1 mm per surface. According to the AISI the maximum allowable decarburization per surface for hot-rolled tool steel is about 2% of the bar diameter + 0·2 mm. For further information see ASTM A600–79 and A681–76.

Constructional steel

Structural components that are required to meet exacting demands of high strength, in particular fatigue strength, must be machined clean before hardening. Sometimes this machining is confined to the most highly stressed surfaces only. Structural steels generally have lower carbon contents than tool steels and are therefore not so susceptible to quenching cracks as the latter.

Table 5.8 Minimum machining allowance per side for round, hexagonal and octagonal bars

Specified dimension mm	Allowance per side mm			
	Hot rolled	Forged	Rough machined	Cold drawn
≤ 13	0·4	—	–	0·4
> 13–25	0·8	—	—	0·8
> 25–50	1·2	1·8	—	1·2
> 50–75	1·6	2·4	0·5	1·6
> 75–100	2·2	3·0	0·6	2·2
> 100–125	2·8	3·7	0·8	—
> 125–150	3·8	4·3	1·0	—
> 150–200	5·1	5·1	1·2	—
> 200–250	—	5·1	1·8	—

Table 5.9 Minimum machining allowance (A and B) per side for hot-rolled square or flat bars

Specified width mm		Specified thickness mm				
		< 13	13–25	25–50	50–75	75–100
≤ 13	A	0·6	—	—	—	—
	B	0·6	—	—	—	—
> 13–25	A	0·6	1·1	—	—	—
	B	0·9	1·1	—	—	—
> 25–50	A	0·8	1·1	1·7	—	—
	B	1·0	1·3	1·7	—	—
> 50–75	A	0·9	1·3	1·7	2·2	—
	B	1·3	1·5	1·8	2·2	—
> 75–100	A	1·0	1·4	1·8	2·2	2·9
	B	1·7	1·9	2·2	2·5	2·9
> 100–125	A	1·1	1·5	1·8	2·2	2·9
	B	2·0	2·4	2·7	3·0	3·2
> 125–150	A	1·3	1·7	1·9	2·2	2·9
	B	2·4	2·9	3·2	3·4	3·6
> 150–175	A	1·4	1·8	1·9	2·3	2·9
	B	2·7	3·3	3·7	3·9	4·3
> 175–200	A	1·5	1·9	2·0	2·5	3·2
	B	3·0	3·8	4·2	4·3	4·8
> 220–225	A	1·5	1·9	2·4	2·5	3·2
	B	3·3	3·9	4·3	4·8	4·8
> 225–250	A	1·5	1·9	2·5	2·5	3·2
	B	3·6	3·9	4·3	4·8	4·8

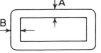

For the conversion of inches to millimetres the factor 25·0 has been used.

Nitriding steel

Nitriding steels, alloyed with aluminium, require a larger machining allowance than other steels since if there is any surface decarburization it invariably gives rise to a rough surface and spalling when nitriding. The normal machining allowance should be increased by 25% for these steels.

General

The machining allowance that must be compensated for is dependent chiefly on the depth of carburization. Out-of-roundness and deformation that can occur in rolled and forged steel should also be reckoned with. For tools and machine components having elongated shapes more machining allowance is required than for compact ones. Nominal values of minimum allowance per side as published by AISI[51] are given in *Tables 5.8* and *5.9*.

References

1. PAYSON, P., 'The Annealing of Steel', *Iron Age*, June 24, July 8, 15 and 22 (1943)
2. HÜLSBRUCH, W. and THEISS, E., 'Full Annealing of High-carbon Steel by Means of Isothermal Annealing', *Stahl u. Eisen*, **72**, 123–133 (1952) (in German)
3. KÖSTLER, H. J., and FRÖHLKE, M., 'Influence of Various Carbon Contents on the Spheroidization of Lamellar Cementite and its Dependence on the Degree of Cold Work and Time of Annealing', *Arch. Eisenhüttenw.*, **46**, 655–659 (1975) (in German)
4. NIJHOF, G. H., Influence of Cold Working on the Spheroidization of Cementite in a Steel with Lamellar Pearlite, *Härterei-Techn. Mitt.*, **35**, No. 2, 59–68 (1980) (in German)
5. NIJHOF, G. H., 'Spheroidization of Cementite in Cold-worked Steel Containing Lamellar Pearlite, Mechanism and Kinetics, *Härterei-Techn. Mitt.*, **36**, No. 5, 242–247 (1981) (in German)
6. COOKSEY, R. J., 'The Cold Extrusion Pressures. Effect of Composition, Cold Work and Heat Treatment', *Metal Forming*, **35**, 98–106, 111 (1968)
7. KULA, E., and COHEN, M., 'Grain Growth in High Speed Steel', *Trans. ASM*, **46**, 727 (1954)
8. ROSENSTEIN, A. H., 'Interpretation of Stress Relaxation Data for Stress Relief Application', *J. Mat.*, **6**, No. 2, 265–281 (1971)
9. LUNDGREN, Å., The Effect Produced by Packing Material on the Tool Surface during Hardening', *Härterei-Technik und Wärmebehandlung*, No. 5, 205–214 (1961) (in German)
10. FAIRBANKS, L. H., and PALETHORPE, L. G. W., *Controlled Atmospheres for the Heat Treatment of Metals*, Sp. Rep. 95, The Iron and Steel Institute (1966)
11. GRASSEL, D., and WÜNNING, J., 'Increased Use of Inert Gas in Heat Treatment', *Härterei-Tech. Mitt.*, **25**, No. 1, 23–24 (1970) (in German)
12. 'Furnace Atmospheres and Carbon Control', *Met. Handbook*, **2**
13. WÜNNING, J., 'Rapid Method of Checking the C and N Potentials in Furnace Atmospheres', *Härterei-Techn. Mitt.*, **24**, No. 1, 35–38 (1970) (in German)
14. HUGHES, R. L., 'Atmosphere Control by Infrared CO_2', *Met. Eng. Quart.*, **11**, No. 2, 1–3 (1971)
15. RECORD, R. G. H., 'Oxygen Potential Monitoring of Treatment Atmospheres, *Metals and Materials*, **5**, 1, 25–30 (1971)
16. BRENNAN, H. F., 'Development of Vacuum Oil Quench Furnaces for Modern Heat Treatment', *Metallurgia*, April, 139–144, May, 193–200 (1969)
17. NEUMÜLLER, E., 'Recently developed Vacuum Units for Heat Treating and Soldering', *Härterei-Techn. Mitt.*, **35**, No. 5, 245–250 (1980) (in German)
18. LAUDENBERG, H. J., 'Recently developed Units for Heat Treatment in Vacuum and in Nitrogen as Protective Atmospheres up to Temperatures around 1250°C', *Härterei-Techn. Mitt.*, **36**, No. 1, 23–28 (1981) (in German)
19. UHLITZSCH, H., 'Mathematical Treatment of the Rate of Change of Temperature on

Heating and Cooling a Finite Solid Cylinder and its Application to the Heat Treatment of Large Forgings', *Neue Hütte*, No. 5, 277–287 (May 1959) (in German)

20. JENSFELT, P. N., 'Method of Calculating the Rate of Change of Temperature on Cooling of Flat Bars', *Com. J. K. Techn. Council*, **21**, No. 268, 229–248 (1962) (in Swedish)

21. BANDEL, G., and HAUMER, H. C., 'A Note on the Estimation by Calculation of the Depth of Hardening by Heat Treatment of Heavy-section Forgings, *Stahl u. Eisen*, No. 15, 932–946 (1964) (in German)

22. ORLICH, J., ROSE, A., and WIEST, P., *Atlas zur Wärmebehandlung der Stähle*, Band 3, Verlag Stahleisen mbH, Düsseldorf (1973)

23. ORLICH, J., and PIETRZENIUK, H-J., *Atlas zur Wärmebehandlung der Stähle*, Band 4, Verlag Stahleisen mbH, Düsseldorf (1976)

24. ROGEN, G., and SIDAN, H., 'Testing the Quenching Capacity of Liquid Quenching Media, Particularly Oils, *Berg- und Hüttenmännische Monatshefte*, **117**, No. 7, 250–258 (1972) (in German)

25. LIŠČIĆ, B., 'Assessing the Absolute Quenching Capacity During Hardening from the Surface Temperature Gradient', *Härterei-Techn. Mitt.*, **33**, No. 4, 179–191 (1978) (in German)

26. BECK, G., DUMONT, C., MOREAUX, F., and SIMON, A., 'Guiding Principles in Choosing and Selecting a Hardening Oil', *Härterei-Techn. Mitt.*, **30**, No. 6, 346–358 (1975) (in German)

27. SCHIMIZU, N., and TAMURA I., 'An Examination of the Relation between Quench-hardening Behaviour of Steel and Cooling Curve in Oil,' *Transactions ISIJ*, **18**, 445–450 (1978)

28. PETER, W., 'The Cooling Capacity of Liquid Quenching Media', *Arch. Eisenhüttenw.*, **20**, No. 7/8, 263–274 (1949) (in German)

29. HENGERER, F., STRÄSSLE, B., and BREMI, P., 'Calculation of the Progress of Cooling during Oil and Air Hardening of Cylindrical and Flat Tool-steel Tools, using an Electrical Computer, *Stahl u. Eisen*, **89**, No. 12, 641–654 (1969) (in German)

30. BURGDORF, E. H., 'Effect of Contaminations in Polymer Solutions on the Quenching Properties', *Härterei-Techn. Mitt.*, **35**, No. 6, 294–297 (1980) (in German)

31. MATERKOWSKI, J. P., and KRAUSS, G., 'Tempered Martensite Embrittlement in SAE 4340 Steel', *Met. Trans.*, **10A**, 1643–1651 (1979)

32. BANERJI, S. K., MCMAHON, C. J., and FENG, H. C., 'Intergranular Fracture in 4340-Type Steels: Effect of Impurities and Hydrogen', *Met. Trans.*, **9A**, 237–247 (1978)

33. BRIANT, C. L., BANERJI, S. K., and MCMAHON, C. J., AIME Fall Meeting, Chicago (1977)

34. SANDBERG, O., and WESTERHULT, P., *Influence of Phosphorus on the Properties of Cr-Mo and Cr-Ni-Mo Heat-treatable Steels*, Swedish Institute for Metal Research (1983) (in Swedish)

35. JAFFE, L. D., and BUFFUM, D. C., 'Upper Temper Embrittlement of a Ni-Cr-Steel, *Trans. AIME*, **209**, 8–19 (1957)

36. BARON, G. H., and TURNER, S., 'Influence of Residual Elements on Brittleness in Hardened and Tempered 3% Ni-Cr-Mo-Steel', *J. Iron Steel Inst.*, **203**, 1229–1236 (1965)

37. DUMOULIN, P. H., GUTTMANN, M., FOUCAULT, M., PAHNIER, M., WAYMAN, M., and BISONOI, M., 'The Role of Molybdenum in Phosphorus Induced Temper Embrittlement', *Met. Sci. J.*, **14**, 1–15 (1980)

38. BELL, T., and BRYANS, R. G., 'The Effects of Prior Transformation and Prestrain on the Habit Planes of Acicular Iron–Nickel Martensite', *Met. Sci. J.*, **5**, 135–138 (1971)

39. SCHUMANN, H., 'Investigation into the Cause of the High Rate of Hardening of High-manganese Steel', *Neue Hütte*, **12**, No. 4, 220–226 (1967) (in German)

40. LIEBMANN, G., 'What Happens to the Retained Austenite?', *Zeit. für wirtschaftliche Fertigung*, **61**, No. 5, 235–238 (1966) (in German)

41. MANIAR, G. N., WHITNEY G. R., and JAMES, H. M., 'Microstructure of Pyromet 860, a Cobalt-containing Iron–Nickel-base Heat-resistant Alloy', *Cobalt*, No. 31, 87–92 (June 1966)

42. BRYANS, R. G., BELL, T., and THOMAS, V. M., 'The Morphology and Crystallography of Massive Martensite in Iron–Nickel Alloys', *The Mechanism of Phase Transformations in Crystalline Solids*, Institute of Metals, Monograph 33, London (1969)

43. DETERT, K., 'Investigation of the Precipitation Reaction in High-strength Martensitic Precipitation-hardening Steels', *Arch. Eisenhüttenwes.*, No. 7, 579–589 (1966) (in German)

44. HELLER, W., and STOLTE, E., 'Present State of Knowledge of the Ageing of Steel', *Stahl u. Eisen*, **90**, 909–916 (1970) (in German)

45. HUSSMANN, W. and KRISCH, A., 'Influence of a Small Amount of Cold-work Deformation and Various Ageing Treatments on the Notch Toughness of Commercial Structural Steels', *Arch. Eisenhüttenw.*, **8**, 613–618 (1972) (in German)
46. DAHL, W., Contribution to Discussion on Paper by Heller and Stolte. See Ref. No. 44.
47. LANNER, C., 'The Binding Action of Micro Additions for Nitrogen', *Jernkont. Ann.*, **153**, 277–285 (1969) (in Swedish)
48. KÖSTER, W., and KAMPFSCHULTE, G., 'Quality Criteria for Low-carbon Steel from Measurements of Damping Capacity', *Arch. Eisenhüttenw.*, **32**, 809–822 (1961) (in German)
49. ERASMUS, L. A., 'Effect of Aluminium Additions on Forgeability, Austenitic Grain Coarsening Temperature and Impact Properties of Steel', *J. Iron Steel Inst.*, **202**, 32–41 (1964)
50. ERASMUS, L. A., 'Effect of Small Additions of Vanadium on the Austenitic Grain Size, Forgeability and Impact Properties of Steel', *J. Iron Steel Inst.*, **202**, 128–134 (1964)
51. *Steel Products Manual Tool Steels*, American Iron and Steel Institute (April 1976)

6

Heat treatment—special

This chapter is specifically written for designers and heat treaters actively engaged in practical work. It contains recommendations for suitable heat treatments for various types of steel as well as suggested applications for a number of well-known steels. The chapter recapitulates some of the material of the preceding one and also extends its contents in the context mentioned above. Each section contains an introductory survey of the steels discussed in the chapter.

This review contains the steel designations as laid down by ISO, AISI, BS, DIN and SS. The letters SIS previously used to designate Swedish standardized steels have now been replaced by SS. Designations of the type 42 CrMo 4 are used in several places in the text and in various connexions. Such designations or very similar ones are often employed in several standards. According to DIN the first digits specify the mean carbon content multiplied by 100. In order to assess the content of the other alloying elements there is a set of different multiplying factors, one for each element. This type of designatory system is applied by ISO in certain cases, e.g. ISO 4957–1980.

6.1 Hardening and tempering of tool steels

Reckoned on a tonnage basis, tool steel represents only a few percent of the total quantity of steel produced but its importance to the industry as a whole is immense. Regrettably this fact is seldom sufficiently appreciated. Perhaps in greatest measure this applies to the heat treatment of tool steel. The cost of the steel and its heat treatment amount generally to less than a quarter of the total cost of the whole tool. A wrong choice of steel or faulty heat treatment may give rise to serious disruption of production and higher costs. It is the author's hope that the contents of this section will increase his readers' knowledge of tool steels and their heat treatment.

In the text, tool steels are designated mainly by the type letter and number as used in the USA and the UK for standardized tool steels, e,g, H13, O1. These designations are so well known by steel consumers all over the world that no special institutional designations are necessary.

Steels for which there are no AISI or BS specifications are designated according to DIN or SS standards. Tool steels are standardized according to ISO 4557–1980, which lists 48 tool steel types. Fifteen of these will be discussed in this book together with other standardized tool steels.

6.1.1   Carbon steels and vanadium-alloyed steels

The hardening of these steels, which are made with carbon contents between 0·80% and 1·20%, is quite straightforward. Since the rate of carbide dissolution proceeds rapidly, the holding time is consequently short and therefore small tools may generally be heated without any extra precautions against atmospheric oxidation. The hardening temperature is about 780°C. Quenching is carried out direct into brine with tempering following immediately. The quenching operation is the most critical part of the heat treatment since too slow a rate of cooling might give rise to either soft spots or quenching cracks. If the tool is designed to contain hardened areas around holes or re-entrant angles the cooling effect must be very intensive at these areas. Manual stirring will often suffice but in many cases the coolant must be sprayed on to the tool (*see* Sections 5.3.6 and 5.3.7). For sections heavier than 20 mm the depth of hardening, i.e. the distance from the surface to the 550 HV level, is about 4 mm. Sections lighter than about 8 mm in thickness will harden through.

Figure 6.1 The successful heat treatment of an expensive tool is often a team-work job

Table 6.1 Tool steel. Steel designations and nominal composition

Type of steel	AISI	BS	DIN	SS	ISO 4957	Uddeholm	C	Cr	Mo	W	V	Others
			Steel designations				Nominal composition %					
Plain carbon steel	W 1	BW 1	C 100 W 1	1880	—	20	1·0	—	—	—	—	
V-alloy steel	—	—	—	—	TcV 105	16 Va	0·85	—	—	—	0·1	
Low-alloy cold-work steel	O1	BO 1	105 WCrV	2140	95 MnCrW 1	ARNE	0·95	0·5	—	0·5	—	1·2 Mn
	—	—	~ 90 CrSi 5	2092	—	SR 1855	1·0	1·0	—	—	—	1·5 Si
Low-alloy cold-work and hot-work steel	—	—	~ 60 NiCrMo 4	2550	—	GRANE 1	0·55	1·0	0·3	—	—	3·0 Ni
	S 1	BS 1	45 WCrV 7	2730	—	REGIN 3	0·50	1·2	0·3	2·4	0·2	1·0 Si
High-alloy cold-work steel	A 2	BA 2	X 100 CrMoV 5 1	2260	100 CrMoV 5	RIGOR	1·0	5·5	1·1	—	0·2	
	D 2	B 2	X 165 CrMoV 12	2310	160 CrMoV 12	SVERKER 21	1·5	12	0·8	—	0·9	
	D 3	BD 3	X 210 Cr 12	—	210 Cr 12	—	2·0	12	—	—	—	
	D 6	—	X 210 CrW 12	2312	210 CrW 12	SVERKER 3	2·0	13	—	1·0	—	
Hot-work steel	—	BH 13	X 40 CrMoV 5 1	—	—	PREGA	0·40	3·0	0·5	—	1·0	
	H 13	BH 10 A	—	2242	40 CrMoV 5	ORVAR	0·40	5·3	1·4	—	0·5	
	—	—	—	—	—	QRO 45	0·30	2·8	2·8	—	1·2	
	—	—	—	(2730)	—	QRO 80	0·40	2·6	2·0	—	0·4	
	H 21	—	—		30 WCrV 9	—	0·30	3·0	—	9·0	—	2·8 Co
High-speed steel	T 1	BT 1	S 18–0–1 (B 18)	—	HS 18–0–1	Speed steel KT 1	0·75	4·0	—	18	1·0	
	M 7	—	S 2–9–2 (BMo 9 V)	(2782)	HS 2–9–2	KM 7	1·0	4·0	8·8	2·0	2·0	
	M 2	BM 2	S 6–5–2 (DMo 5)	2722	HS 6–5–2	KM 2	0·85	4·0	5·0	6·0	2·0	
	M 3:2	—	S 6–5–3 (EMo5V3)	—	HS 6–5–3	ASP 23	1·2	4·0	5·0	6·0	3·0	
	M 35	—	S 6–5–2–5 (EMo5Co5)	2723	HS 6–5–2–5	KM 35	0·9	4·0	5·0	6·2	2·0	5·0 Co
	—	—	S 10–4–3–10 (EW9Co10)	2736	HS 10–4–3–10	WKE 4	1·3	4·0	3·5	9·5	3·3	10·0 Co
	M 42	BM 42	S 2–10–1–8	—	HS 2–9–1–8	KM 42	1·1	4·0	3·5	1·5	1·2	8·0 Co

The designations are tentative and will be subject to alteration when the relevant International Standards have been established

Table 6.2 Tool steel. Recommendations for heat treatment

Hardness Rockwell C after hardening and tempering (experimental results)

Hardening treatment Temperature range °C	Cooling medium	Dimension mm	Hardening temperature °C	Tempering temperature °C 1h 100	150	200	250	300	400	500	600	AISI (Uddeholm)
770–800	Water	25	770	67	65	63	59	56	47	38	—	W 1
770–800	Water	25	790	66	64	63	60	57	48	38	—	(16 Va)
800–840	Oil	25	820	64	63	62	60	58	53	47	40	O 1
850–890	Oil	25	860	66	64	63	62	61	58	50	42	(SR 1855)
810–840	Oil	100	820	58	58	57	55	53	48	44	40	(GRANE)
830–850	Air	100	840	50	50	49	48	47	44	41	38	
860–900	Water	20	880	62	60	58	57	56	53	49	45	S 1
880–920	Oil	20	900	59	58	57	57	56	53	49	45	
950–980	Oil, salt bath or air	25	960	64	63	61	60	58	58	57	52	A 2
1000–1025		25	1000	63	63	62	60	59	59	58	50	D 2
950–980		25	970	64	63	62	61	60	57	54	50	D 3
960–1000		25	980	64	63	62	61	60	59	58	50	D 6

Hardening treatment Temperature range °C	Cooling medium	Dimension mm	Hardening temperature °C	Tempering temperature °C 1h 200	500	550	575	600	625	650	675	AISI (Uddeholm)
880–914	Oil	100	900	52	52	52	52	52	51	47	39	(PREGA)
	Air	100	900	45	45	45	45	45	45	43	38	
1000–1050	Oil,	50	1025	54	55	53	51	48	44	36	32	H 13
1000–1050	salt bath	50	1040	51	52	52	51	49	46	48	38	(QRO 45)
1050–1100	or air	50	1100	52	53	54	53	52	50	47	44	(QRO 80 M)
1260–1280	Oil, salt bath or air	25	1270	64	64	65	65	64	63	58	53	T 1
1200–1220		25	1210	65	64	65	66	63	62	58	54	M 7
1200–1220		25	1210	65	64	65	66	65	64	61	57	M 2
1100–1180		25	1180	63	63	67	65	63	61	58	56	M 3:2
1180–1220		25	1210	66	65	66	65	64	61	57	52	M 35
1160–1200		25	1180	62	63	66	67	67	66	65	63	M 42

For awkward tools, hardenability may be a crucial factor and under such circumstances the composition of the steel must be adjusted in accordance herewith, in particular as regards the alloying elements Mn and Cr which have a powerful influence on hardenability. The diagram in *Figure 6.2* shows how the hardening temperature affects the depth of hardening and fracture number on a W 1-type steel of conventional composition. The V-content is only 0·04% which implies that the grain coarsening is initiated when the hardening temperature exceeds 815°C.

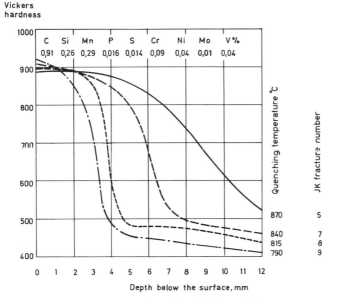

Figure 6.2 Depth of hardening for a C steel, 25 mm in diameter, corresponding to W 1. Quenched in water from various temperatures

In *Figure 6.3* are shown the results of corresponding trials with a steel containing somewhat larger amounts of alloying elements. The depth of hardening is considerably greater. Owing to the high content of V the steel remains fine-grained even when hardened from exceptionally high temperatures. The conspicuous increase in hardness after hardening from 840°C is due to the synergetic effect of Cr, Mo and V where V probably plays a dominant role. The very considerable toughness inherent in a plain-carbon steel, due to its shallow-hardening properties, is forfeited if the tool through-hardens locally at some sections because the cross-sectional area there is too small. For shearing tools or small tools generally, such as scissors, knives or letter die punches, which are not subjected to heavy impact blows, this drawback is of less importance. Tools operating under heavy blows, e.g. upsetting dies for cold-heading of bolts, must not be through-hardened. Coining and striking punches are other examples of carbon tool steels that require high wear resistance. Such tools may also be subjected to bending stresses and should therefore not be through-hardened. The tempering temperature normally used for

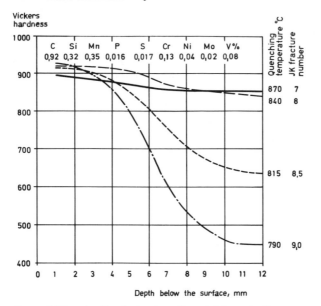

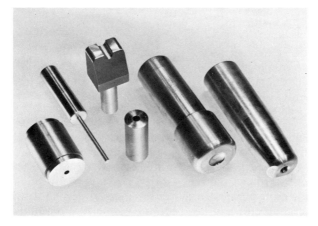

Figure 6.3 Depth of hardening for a C steel, 25 mm in diameter, corresponding to W 1. Quenched in water from various temperatures

Figure 6.4 Punches made from steel W 1

tools belonging to this group lies in the range 170–220°C, the hardness being generally about 60–64 HRC. Representative examples of tools made from grade W 1 are shown in *Figure 6.4.*

6.1.2 Low-alloy cold-work steels

Only two steels have been chosen from this group for discussion, viz. grade O 1 and SS 2092 (~DIN 90 CrSi 5).

When carbon steel is used for punching dies or cold hobbing tools the dimensions of the tool are bound by a ruling section that is determined by the load on the tool. A carbon–steel punch or die having a diameter of, say, 50 mm, will show rather poor resistance to sinking on account of the shallow depth of hardening. Should this resistance not suffice, another steel will have to be chosen, in this case grade O 1 or SS 2092. From the point of view of heat treatment, these two steels differ somewhat since their hardening temperatures are different. Steel SS 2092 requires 850–890°C whereas grade O 1 requires 800–840°C. Owing to its lower hardening temperature, O 1 has somewhat greater dimensional stability. This property makes it a first choice for blanking dies and other tools requiring a high degree of dimensional stability (*Figure 6.5*).

Figure 6.5 Blanking tool made from steel O 1

In both steels the depth of hardening decreases by roughly the same amount as the thickness of the section increases. In the diagram in *Figure 6.6* the hardening temperature was raised as the cross-sectional area increased in order to increase the hardenability of the steel. Tools having diameters greater than about 80 mm or equivalent sections in flat dimensions are difficult to harden to full hardness if there are re-entrant corners. For such designs it is advisable to choose SS 2092 since in spite of its showing, under certain circumstances, a shallower depth of hardening, it obtains full surface hardness more readily and gives a more regular depth of hardening in a tool with varying section thicknesses. This point is well illustrated in *Figure 6.7* which shows a section through a 'contour-hardened' Pilger roll.

As a rule both steels are oil quenched. For heavy sections, e.g. dimensions greater than 100 mm in diameter, it is best to use water quenching when dealing with SS 2092. When the surface temperature of the steel has fallen to between 400°C and 300°C the water quenching is interrupted by transferring the tool to an oil bath.

The temperature for both steel is generally in the range 170–200°C which gives a hardness of more than 60 HRC. As can be seen from *Figure 6.8*,

Hardness
HRC

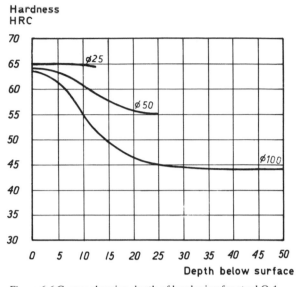

Depth below surface

Figure 6.6 Curves showing depth of hardening for steel O 1.
Specimen 25 mm diameter oil quenched from 800°C.
Specimen 50 mm diameter oil quenched from 820°C.
Specimen 100 mm diameter oil quenched from 840°C

S S 2092 has a greater resistance to tempering than grade O 1. On being tempered in the range 250–350°C the steel suffers a reduction in its impact strength which in turn increases the risk of chipping. For this reason tools that are subjected to impact stresses should not be tempered in this temperature range. The higher impact strength manifested after tempering at 170–200°C is due to the presence of retained austenite, viz. about 10%.

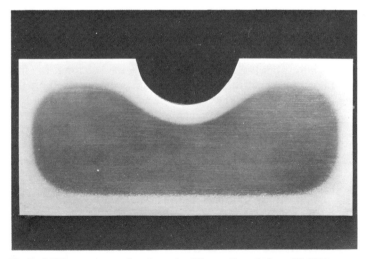

Figure 6.7 Transverse section through a Pilger roll made from S S 2092 (~ 90CrSi5) Size 50 × 120 mm

This soft retained austenite can accommodate impact stresses better than the harder constituents. Retained austenite is decomposed when it is tempered at about 300 °C. For the steel DIN 105 WCr 5 (1·05% C; 1% Mn; 1% Cr; 1·2% W), which closely approximates to grade O 1, the relation between tempering temperature and torsional fracture strength, torsional yield strength and energy of plastic deformation in torsion is shown in *Figure 6.9*. If a specimen is subzero treated before annealing to 200 °C it is found that the work required for plastic deformation (toughness) is less and the torsional yield strength is higher than if the specimen were not subzero treated.

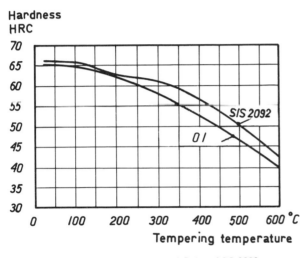

Figure 6.8 Tempering curves for steel O 1 and SS 2092

If, during service, some areas of the tool have to support excessive pressures, for example the shearing edge of circular slitting knives, retained austenite may be transformed to martensite, with spalling at the edge as a result. Should this occur, tempering at 300–400 °C is recommended.

In recent years SS 2092 has increasingly been used for so-called Pilger rolls which are in part made as rings and in part as dies. *Figure 6.10* shows one of the world's largest Pilger rolls, designed for cold-rolling 10 in tubes. The only steel suitable for this tool was SS 2092. Another field of application is for what are known as Yoder rolls. A sketch showing the principle of operation and tube manufacture is shown in *Figure 6.11*. For this mill unit, the wear resistance of rolls made from SS 2092 has shown itself to be on a par with that of grade D 2, in fact in some instances it has outlasted this grade. This observation is particularly striking when stainless steel tubes are being rolled, since there is no 'galling' when SS 2092 is being used.

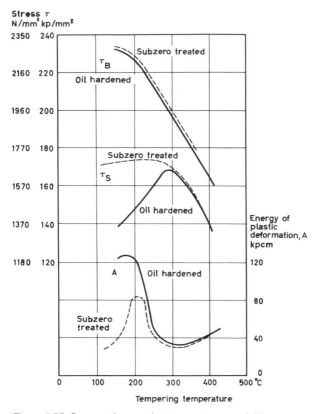

Figure 6.9 Influence of tempering temperature on yield strength, τ_s, fracture strength, τ_B, and energy of plastic deformation, A, in torsion of a steel approximately corresponding to steel O 1. Oil quenched from 840°C, subzero-treated at −195°C. Tempering time 1 h (Bungardt, Mülders and Schmidt)

6.1.3   Low-alloy cold-work and hot-work steels

The hardenability of SS 2550 (~DIN 60 NiCrMo 4) is considerably better than that of either of the two steels just discussed. SS 2550 is air hardening in fairly heavy sections, which is of advantage where dimensional stability is concerned. Due to the lower carbon content the toughness is greater than that of the previously mentioned steels. When used for cold-work tools, the steel is tempered at 170–250°C, the resulting hardness then being 55–58 HRC. With regard to impact strength this steel, too is susceptible to tempering treatments around 300°C (*see Figure 6.12*). In *Figure 6.13* is shown a tool that failed during use on account of its having been tempered at 300°C. A similar tool that was tempered at 200°C functioned satisfactorily.

SS 2550, after hardening and tempering at 200–250°C possesses very high tensile strenght and good impact strength. The values given below have been obtained on tensile test specimens that were oil quenched from 830°C and tempered at 250°C.

$R_{p0.02}$ = 140 kp/mm^2(1370 N/mm^2) A_5 = 8%
$R_{p0.1}$ = 155 kp/mm^2(1520 N/mm^2) Z = 33%
$R_{p0.2}$ = 166 kp/mm^2(1630 N/mm^2)   HRC = 54
R_m = 208 kp/mm^2(2000 N/mm^2)

Figure 6.10 Pilger roll made from S S 2092 for 10 in tube mill. Dimensions: 800 mm diameter × 400 mm, weight 800 kg (courtesy Messrs Mannesmann-Meer and Sandvik AB)

Figure 6.11 Sketch showing tube-mill operating principle for welded tubes (Yoder mill). Welding stage omitted

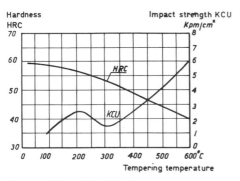

Figure 6.12 Steel S S 2550. Hardness and impact strength as functions of tempering temperature

Figure 6.13 Shear blade made from SS 2550. Failed in service owing to its having been tempered at 300°C

Such favourable mechanical properties make the steel suitable for tools subjected to large static and dynamic forces. Some typical applications are dies for tableware, shear blades for heavy plate, dies for plastic moulds for which is required a steel possessing a high degree of dimensional stability and excellent polishability.

SS 2550 is also used for hot-work tools working at moderate temperatures, e.g. drop-forging dies. Such tools are tempered between 400°C and 600°C, the exact temperature depending on the hardness required and the working temperature of the tool. For working temperatures above approximately 400°C the hardness of the steel falls relatively quickly; this is shown in *Figure 6.14*. If higher working temperatures are involved it is recommended to use the special hot-work steels that are discussed in Section 6.1.5.

Grade S 1 has both high wear resistance and high impact strength. The hardenability is inferior to that of the Cr–Ni–Mo steel SS 2550. This implies that for dimensions greater than 50 mm in diameter, this steel is contour-hardening which, in fact, further increases its toughness The normal hardening temperature is about 900°C but it may be raised to 950°C without any risk of grain growth being incurred. If a hardness higher than 50 HRC is required in dimensions up to about 50 mm in diameter the steel should be quenched in oil. For heavier dimensions a combined water-oil quenching procedure or only water quenching may be necessary. As regards the tempering temperature ranges the same principles as are applicable to the foregoing steels are valid in this case also.

Of the many cold-work applications for tool steel, special mention should be made of the cold punching of plate having a thickness greater than about 3 mm. If plate of increasing thickness is being punched and consequently the thickness measurement of the plate is approaching the diameter of the hole, the punches used show an increasing tendency to break if they are made from, for example, grades O 1, A 2 or D 2. For this type of punching work, grade S 1 has been shown to possess the best combination of toughness and wear resistance. A suitable hardness is 56–58 HRC. Wear resistance is further increased if, during the course of the hardening treatment, the tools are heated for some 20 mm in a cyanide bath. After this treatment no further finishing is required; at the most a very light finish grinding is permissable. Another example is the use of this steel as the impact hammer in nail guns, used for driving nails into concrete.

Owing to its high toughness in comparatively large dimensions, grade S 1 can successfully be used for tableware dies which, depending on their dimensions should either be quenched in oil or be heat treated according to the combined water–oil quenching procedure previously mentioned. Another field of application is shear blades for cold shearing of heavy plate. Owing to its rather good resistance to tempering, grade S 1 may also be used for hot shears, a suitable hardness for this latter use being about 45 HRC.

In the field of hot-work, grade S 1 has been superseded by other grades, e.g. H 13. However, mention should be made of an interesting and successful field of application for grade S 1, viz. as chisels used in connexion with the electrolytic reduction of aluminium from bauxite. The

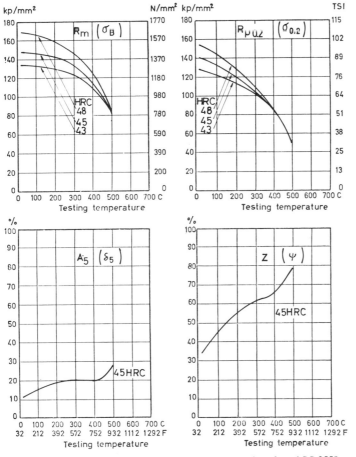

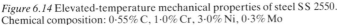

Figure 6.14 Elevated-temperature mechanical properties of steel SS 2550. Chemical composition: 0·55% C, 1·0% Cr, 3·0% Ni, 0·3% Mo

function of the chisels is to break up the hard alumina-containing crust which forms on the metal bath. During use the chisels also come into contact with the bath itself and are thus subjected to both high temperatures and impact stresses. A suitable chisel hardness is about 350 HB. *Figure 6.15* shows a worn-out chisel.

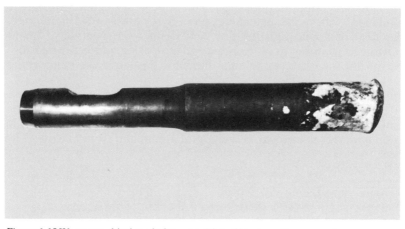

Figure 6.15 Worn-out chisel made from steel S 1. Chisel used in connection with aluminium smelting

6.1.4   High-alloy cold-work steels

This group comprises the following four steels:

Grade A 2	(ISO 100 CrMoV 5)
Grade D 2	(ISO 160 CrMoV 12)
Grade D 3	(ISO 210 Cr 12)
Grade D 6	(ISO 210 CrW 12)

D 6 is no longer an AISI standard grade but the steel is standardized in a number of countries other than the USA. These steels, except grade D 3, are air-hardening in dimensions up to ~100 mm in diameter. Grade D 3 air hardens up to about 50 mm in diameter. Grade D 6 has a hardenability superior to that of grade D 3, partly owing to its higher content of Mn and partly to the presence of W. The results of air-hardening trials with 12% Cr steels of various compositions are shown in *Figure 6.16*. The degree of hardenability can be controlled by means of the hardening temperature since these steels are very rich in carbides. Since progressively higher hardening temperatures bring about increasing grain growth and increasing amounts of retained austenite in the hardened steel, the upper temperature limit of the hardening range is determined by how much grain growth and retained austenite, respectively, are acceptable. Grade A 2 is more susceptible to grain growth than either of the grades D 2 or D 6 (*see Figure 6.17*).

Figure 6.18 shows the hardness obtained in specimens, 30 mm in diameter, on quenching them in oil from various temperatures after a

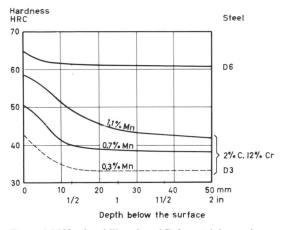

Figure 6.16 Hardenability of steel D 3 containing various Mn contents, and steel D 6. Steel specimen 100 mm (4 in) diameter hardened from 1000 °C in still air

holding time of 30 min. One series of specimens was subzero treated at −80 °C, another at −180 °C. The conclusions reached from the series of experiments described above were that maximum hardness was obtained after quenching from the following temperatures:

A 2   950 °C
D 2   1010 °C
D 6   980 °C

These temperatures are also the ordinary hardening temperatures, respectively, for the steel grades listed above. If the steels are treated at subzero temperatures after hardening, maximum hardness is obtained when the hardening temperature is somewhat higher. *Figure 6.19* shows how the amount of retained austenite in grade D 2 increases with increasing hardening temperature[1]. As has been pointed out in a previous section, an air-hardened specimen contains a larger amount of retained

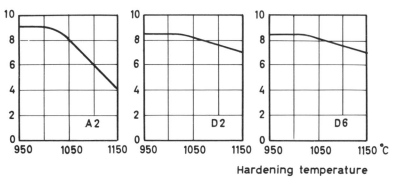

Figure 6.17 Influence of hardening temperature on fracture number for steels A 2, D 2 and D 6

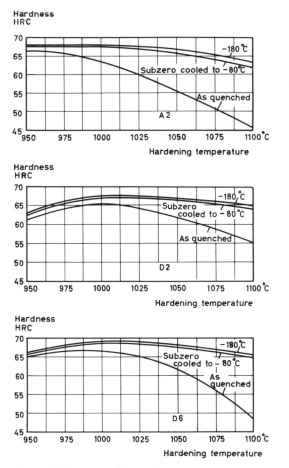

Figure 6.18 Influence of hardening temperature on hardness of the steels A 2, D 2 and D 6

austenite than other, more drastically hardened, specimens owing to the stablizing effect produced in the austenite as it slowly cools through the range of martensite formation.

By far the greater number of tools made from these steels are tempered at 180–120°C, which gives them a hardness of 61–63 HRC, provided that a normal hardening temperature has been used and that the section is not too heavy. The relatively high toughness shown by the steels is due to the presence of retained austenite, which usually amounts to some 20%. The hardness falls as the tempering temperature is raised but if the hardening temperature has been sufficiently high the tempering may produce a secondary hardening peak in the tempering curve. The temperature at which this peak occurs increases as the hardening temperature is raised. The results obtained on *tempering* the specimens that were hardened as described in *Figure 6.18* are shown in *Figures 6.20, 6.21* and *6.22*.

A similar investigation was carried out by Rapatz[2] who used a steel containing 1·75% C, 0·22% Mn and 13% Cr. The steel specimens were oil

Retained austenite
%

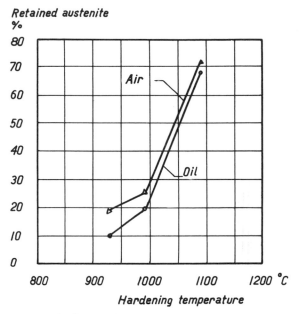

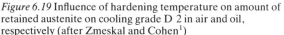

Hardening temperature

Figure 6.19 Influence of hardening temperature on amount of
retained austenite on cooling grade D 2 in air and oil,
respectively (after Zmeskal and Cohen[1])

quenched from temperatures ranging from 900 to 1300°C. One test series
was subzero treated in liquid air. After tempering at 450°C, 500°C, 550°C
and 600°C the hardness and amount of retained austenite were determined
(*see Figure 6.23*).

Tempering at 450°C. The amount of retained austenite is the same as after
quenching (*see Figure 6.19*). When a subzero treatment is introduced
before tempering, the amount of retained austenite is reduced, which
results in a hardness increase.

Tempering at 500°C. The amount of retained austenite in the specimen
quenched from 1000°C is reduced from 30% to 10%. There is also some
decomposition of the austenite that is retained in specimens quenched
from temperatures approaching 1200°C. Quenching from 1100°C produces
maximum hardness in the specimen. A subzero treatment preceding the
tempering does not influence the hardness until quenching temperatures of
1150°C and higher are used.

Tempering at 550°C. All the retained austenite resulting from the 1000°C
quench has been decomposed. The maximum feasible hardness, viz.
62 HRC, is obtained by stepping up the hardening temperature to 1175°C,
the retained austenite content then being about 50% after quenching. If a
subzero treatment precedes the tempering, the maximum hardness is only
57 HRC.

Tempering at 600°C. The retained austenite is completely decomposed in
all the specimens. A quenching temperature of 1150°C gives a hardness
maximum when the subzero treatment is omitted before the tempering.

336

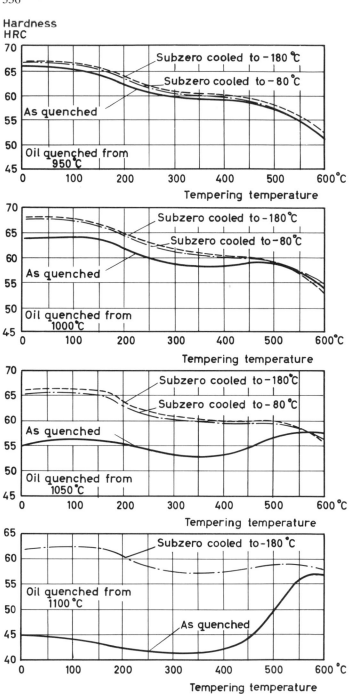

Figure 6.20 Steel A 2. Influence of tempering temperature on hardness of specimens (i) after oil quenching from various temperatures, and (ii) after oil quenching followed by subzero treatments at −80 and −180°C, respectively. Specimens 30 mm in diameter. Holding times: at hardening temperatures, 30 min; at tempering temperatures, 1 h

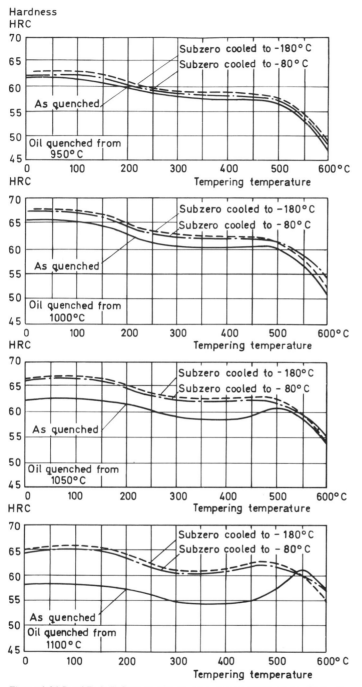

Figure 6.21 Steel D 2. Influence of tempering temperature on hardness of specimens (i) after oil quenching from various temperatures, and (ii) after oil quenching followed by subzero treatments at −80 and −180°C, respectively. Specimens 30 mm in diameter. Holding times: at hardening temperatures, 30 min; at tempering temperatures, 1 h

338

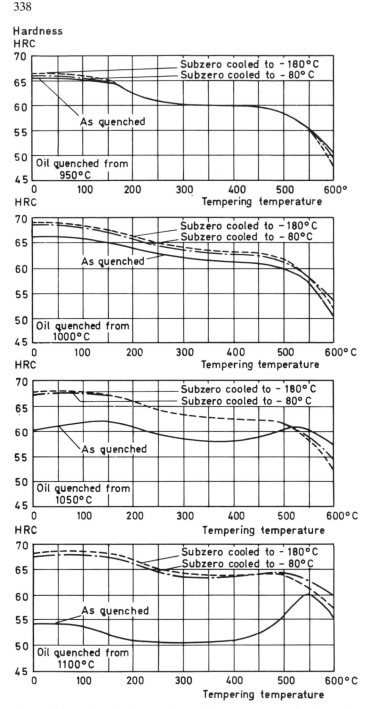

Figure 6.22 Steel D 6. Influence of tempering temperature on hardness of specimens (i) after oil quenching from various temperatures, and (ii) after oil quenching followed by subzero treatments at −80 and −180°C, respectively. Specimens 30 mm diameter. Holding time: at hardening temperature, 30 min; at tempering temperatures, 1 h

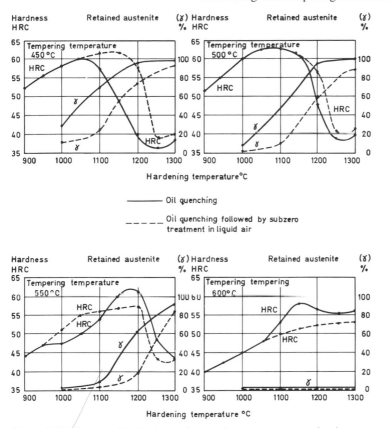

Figure 6.23 Influence of hardening and tempering temperatures on hardness and amount of retained austenite in a steel with 1·75% C, 0·22% Mn and 13% Cr. Tempering time 1 h (after Rapatz, Krainer and Swoboda[2])

The high hardness, around 64 HRC, found in light sections after quenching and tempering at 180°C is seldom obtainable in heavy sections, i.e. in sections more than 75 mm in diameter, when quenched in oil or in a martempering bath. This is because the amount of retained austenite increases as the cooling rate decreases. *Figure 6.24* shows diagrammatically the expected hardnesses in sections ranging from 10 to 150 mm in diameter when they have been quenched in an oil or martempering bath and then tempered. The upper line refers to the smallest dimensions, the lower line, the largest.

If grades D 2 and D 6 are quenched from a high hardening temperature, say about 1050°C, and then tempered around 525°C, the resulting hardness is more or less independent of the dimensions of the specimen (*see Figures 6.25* and *6.26*). The advantage of using such a treatment is that the greater part of the retained austenite is decomposed during tempering. Hence high working pressures may be applied to the steel without risking the manifestation of the adverse effects that result when untempered martensite is formed. Since the steel is heated to a high hardening temperature a large amount of carbide dissolution takes place and this increases the homogeneity of the steel as well as its toughness. Owing to the increased resistance to tempering imparted to the steels on their being quenched from high hardening temperatures they can also be used as hot-work steels, provided that the requirments of impact strength are not too severe. Their improved resistance to tempering, just mentioned, is an additional advantage if these steels are to be nitrided.

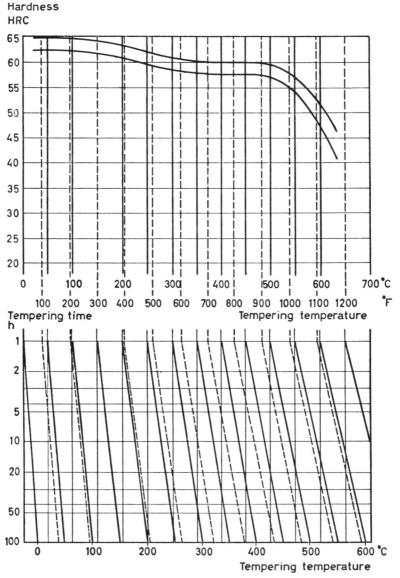

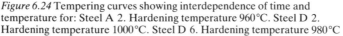

Figure 6.24 Tempering curves showing interdependence of time and temperature for: Steel A 2. Hardening temperature 960°C. Steel D 2. Hardening temperature 1000°C. Steel D 6. Hardening temperature 980°C

Figures for impact strength, as recorded from the Charpy U test, are very low for these high-alloy steels. Moreover, different heat treatments produce only insignificant differences in impact values. These facts are brought out in *Figure 6.27*, which applies to grade A 2. Consequently, tests have also been carried out with *unnotched* specimens, 8 mm in diameter × 55 mm, held in supports 30 mm apart. From *Figures 6.28* to *6.30* it may be seen that an impact strength maximum is obtained after a tempering treatment at 300–400°C and a minimum at around 500°C. If there is a

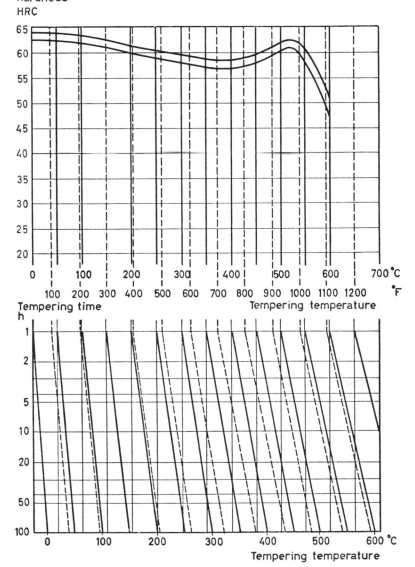

Figure 6.25 Tempering curves showing interdependence of time and temperature for steel D 2, quenched from 1050°C in still salt bath at 225°C. Diameter of specimen 50 mm

342

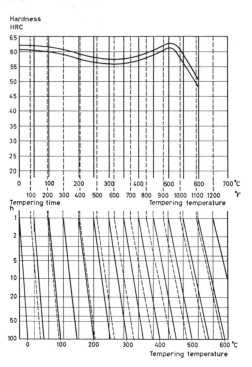

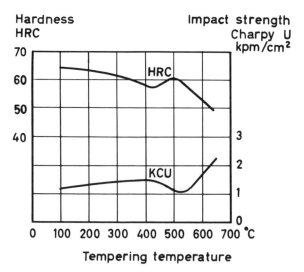

Figure 6.26 Tempering curves showing interdependence of time and temperature for steel D 2, quenched from 1050°C in still salt bath at 225°C. Diameter of specimen 150 mm

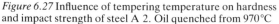

Figure 6.27 Influence of tempering temperature on hardness and impact strength of steel A 2. Oil quenched from 970°C

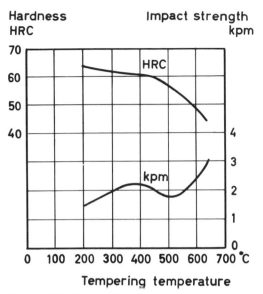

Figure 6.28 Influence of tempering temperature on hardness and impact strength of steel D 6 oil quenched from 950°C. Unnotched impact test specimen, 8 mm diameter × 55 mm

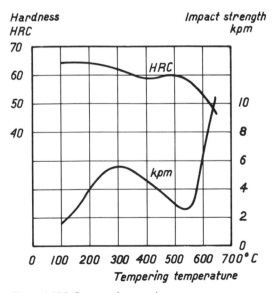

Figure 6.29 Influence of tempering temperature on hardness and impact strength of steel A 2, oil quenched from 970°C. Unnotched impact test specimen, 8 mm diameter × 55 mm

possibility that tool failure can occur on account of heavy impact blows, the risk may be anticipated by avoiding a tempering treatment for the tool concerned in the region around 500°C. However, actual trials with punching tools, tempered at 500°C, have demonstrated that this temperature range need not be regarded as particularly risky for grades D 2 and D 6.

The probable explanation of the relatively high impact values obtained in the 200–400°C range is that a favourable intermixture of retained austenite and tempered martensite is present in the steel. This is borne out by the fact that if specimens are subzero treated before being tempered at temperatures up to 450°C (*see Figure 6.30*) their impact strength is consistently lower than if they were not subzero treated.

Trials with punching and blanking tools made from various grades of steel have shown, according to Bühler, Pollmar and Rose[3] that grade D 3 (DIN X 210 Cr 12) suffers 20% more wear than grade D 6 (X 210 CrW 12). *Figure 6.31* contains results from this investigation. Steel 105 WCr 6 which corresponds to grade O 1 with an addition of 1% W, suffers about twice as much wear as the steels containing 12% Cr. The high-alloy Cr steels are used for blanking tools and drawing tools when large production series are concerned. A blanking tool for 1 mm mild steel sheet, if made from grade D 2 and D 6 can make about 100 000 stampings before being redressed. For grade O 1 the corresponding number of stampings is about 25 000. However, the hardness of the tool must be suited to the thickness of the sheet. The curves in the diagram (*Figure 6.32*) may be of help in choosing the best hardness.

To a large extent the steels are used for tools for shaping and forming. The so-called Sendzimir rolls, i.e. rolls forming a cluster-roll mill,

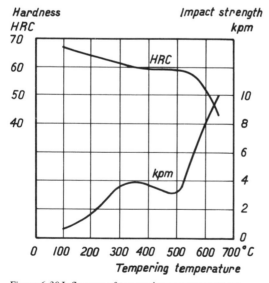

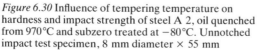

Figure 6.30 Influence of tempering temperature on hardness and impact strength of steel A 2, oil quenched from 970°C and subzero treated at −80°C. Unnotched impact test specimen, 8 mm diameter × 55 mm

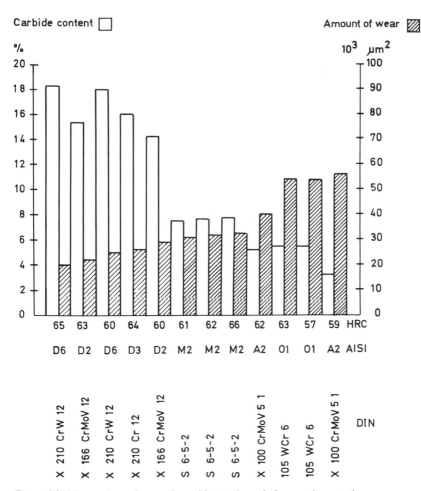

Figure 6.31 Mean values of wear of punching tools made from various steels containing different amounts of carbides and of different hardness. The bar chart is drawn from results published by Bühler, Pollmar and Rose[3]

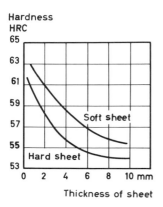

Figure 6.32 Recommended hardness figures for blanking and punching tools with reference to various sheet thicknesses. Steel A 2, D 2 and D 6

constitute an example of such use. The hardness of these rolls is in the range 58–64 H RC. Another example is mandrels for tube rolling by means of Pilger rolls. A suitable hardness for mandrels in grade D 2 is 52–54 H RC. *Figure 6.33* illustrates such mandrels. As mentioned above, the steels may also be used for hot work. This applies in the first instance to grade D 2 which finds an application as forging dies for stainless-steel knives. Valve seats for motorcar internal combustion engines are best made of grade D 6 steel which has been found to be superior to D 3 for this purpose.

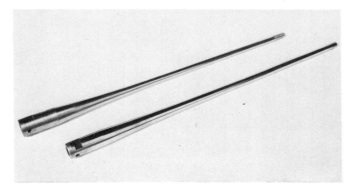

Figure 6.33 Mandrels made from steel D 2 for the production of copper tubing by cold pilgering

6.1.5   Hot-work steels

Among the hot-work tool steels may also be included, as mentioned previously, the Cr–Ni–Mo steel S S 2550. A similar steel, much used for drop-forging dies, is the slightly lower-alloy Cr–Ni–Mo–V steel, designated 55 NiCrMoV 2. These steels are often supplied heat treated to a hardness of 350 HB. The remaining hot-work tool steels included in *Table 6.1* on p. 321 have Cr and Mo as their principal alloying elements.

UHB PREGA

This steel has no direct equivalent among the standardized steels. It owes its existence primarily to its serving as a substitute for the more expensive grade H 13 which it has successfully replaced in a number of instances. The steel is normally supplied heat treated to a hardness of 330–360 HB. As may be seen from the curves in *Figure 6.34* it has excellent hardenability, i.e. on oil quenching, the steel is practically through-hardening in dimensions up to 200 mm in diameter. The mechanical properties of the steel, both as oil quenched and tempered and as air hardened and tempered, are shown in *Figure 6.35*. As may be seen from the curves in *Figure 6.36* the steel has high strength at elevated temperatures, i.e. up to about 500 °C.

One of the uses of the steel is for drop-forging dies (*see Figure 6.37*). In this capacity it often has a longer service life than grade H 13. This is

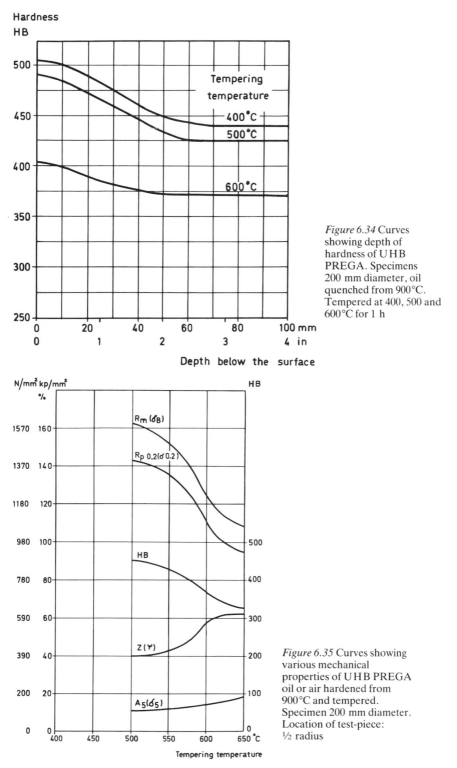

Figure 6.34 Curves showing depth of hardness of UHB PREGA. Specimens 200 mm diameter, oil quenched from 900°C. Tempered at 400, 500 and 600°C for 1 h

Figure 6.35 Curves showing various mechanical properties of UHB PREGA oil or air hardened from 900°C and tempered. Specimen 200 mm diameter. Location of test-piece: ½ radius

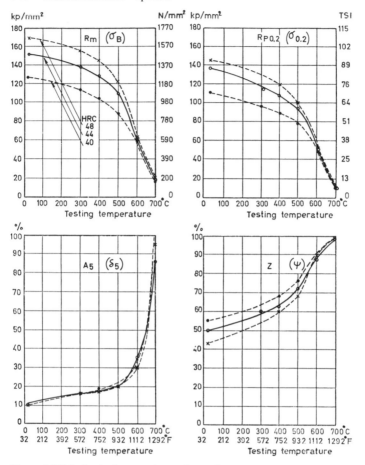

Figure 6.36 Mechanical properties at elevated temperatures of UHB PREGA. Air hardened from 900 °C and tempered to various hardnesses

probably due to the superior thermal conductivity of UHB PREGA. Another use to which the steel is put is as material for supporting dies in the extrusion of aluminium. The hardness should be 44–47 HRC. The steel is well suited to nitriding.

On account of its low hardness (about 170 HB) in its as-annealed condition the steel is suitable for cold hobbing.[4]

Steel H 13

This steel is one of the most popular of all hot-work tool steels and it is used for a multitude of tool types. Although the steel is well known, costly tool failures do happen which in many cases can be traced back to the heat-treater's insufficient knowledge of heat treatment. The hardening temperature lies between 1000 °C and 1050 °C. The temperature for the hardening operation is chosen with a view to obtaining in the steel the properties that are regarded as the most important ones. The strength at

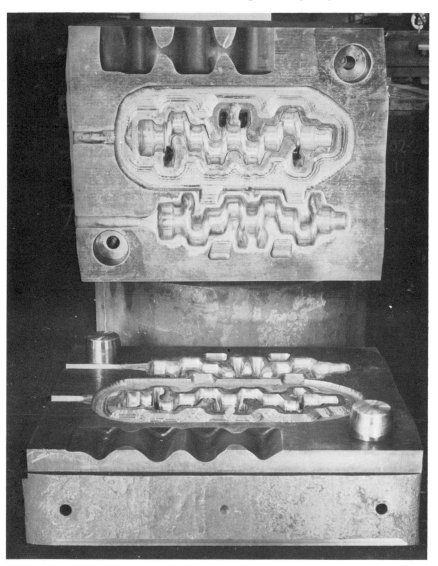

Figure 6.37 Drop-forging dies for motor-car crankshaft. Dies made from
UHB PREGA

elevated temperatures increases as the hardening temperature is raised but
this increase in strength takes place at the expense of the toughness. This
fact is illustrated for steel H 13 in *Figure 6.38* in which the elongation and
reduction of area are the properties taken as a measure of the toughness
(actually ductility). The higher the hardening temperature the more
susceptible is the steel to grain growth. A coarse-grained steel is less tough
than a fine-grained one. The recommended holding times that were given
in a previous section must be adhered to (*see* p. 252). To illustrate this
point, *Figure 6.39* shows the grain size as a function of the hardening

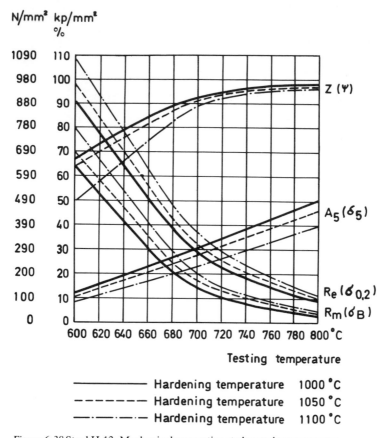

Figure 6.38 Steel H 13. Mechanical properties at elevated temperatures
after hardening from various temperatures followed by double tempering at
600 °C for 2 h

temperature and time of holding. Owing to its high V-content, grade H 13
is not unduly susceptible to grain growth. As has been pointed out before,
the grain size number after hardening should be at least 7 ASTM. When
the heating is carried out in a neutral salt bath the holding time is easily
kept under control. If the tool is pack-heated it is difficult to judge the
actual holding time and therefore when this method of heating is used the
lower temperature of the hardening range is recommended.

As the hardening temperature increases, more and more carbides go
into solution. For grade H 13 the consequences of this are that the higher
the hardening temperature the sooner is the start of the carbide
precipitation that precedes the pearlite and bainite formation when
quenching the steel. This fact is brought out by the
continuous-cooling-transformation (CCT) diagram reproduced in *Figure
6.40*. These diagrams as well as the one in *Figure 6.41* date from a period
earlier than the one from which the diagram shown previously in *Figure
4.49* orginates and therefore they differ somewhat from the latter.
However, from an educational point of view the older diagrams are fully
acceptable.

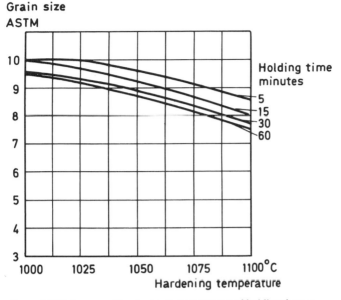

Figure 6.39 Influence of hardening temperature and holding time on
grain size of steel H 13

Impact strength determinations carried out on specimens of steel H 13,
after having been hardened from various temperatures, show that
toughness decreases with increasing hardening temperature. In this
instance grain growth cannot have been the only factor concerned since the
steel is relatively insensitive to grain growth. Tests involving different rates
of cooling have shown that the lowest cooling rate resulted in the lowest
impact strength. It has also been demonstrated that there is some
connexion between low impact values and the presence of precipitated
grain-boundary carbides in the steel. Hence, if the highest possible impact
strength is to be imparted to grade H 13, a moderately high hardening
temperature combined with a high cooling rate should be used.

From the point of view of carbide precipitation *Figures 6.41a, b* and *c*
indicate which dimensions are the crucial ones when the steel, after having
been austenitized at 1030°C for 15 min, is cooled in different cooling
media.

When a 100 mm diameter bar is quenched in oil the curve representing
the rate of cooling at the centre, as traced in the diagram, just touches the
curve of incipient carbide precipitation. This state of affairs will hardly
cause any deterioration in the impact strength of the steel. However, if the
above-mentioned bar is quenched in a salt bath, carbides will be
precipitated both in the core and the surface layers. With air cooling it is
not possible to avoid carbide precipitation in any of the dimensions
indicated in the diagrams. Hence, if high impact strength is required, a
rapid rate of cooling is necessary. Since oil quenching may result in more
distortion than salt-bath quenching or air cooling (*see* Chapter 7) it is
recommended that oil quenching be resorted to only when the
requirements of dimensional stability are not of primary importance.

352

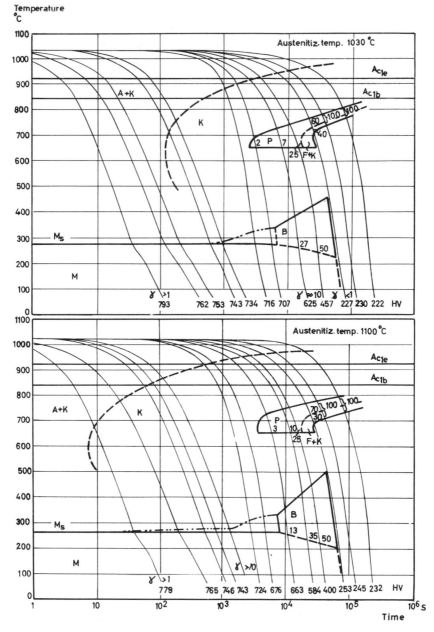

Figure 6.40 Continuous-cooling-transformation (CCT) diagrams for grade
H 13, austenitized at 1030 and at 1100°C, respectively, for 15 min (from
Atlas zur Wärmebehandlung der Stähle).
A = Austenite, B = Bainite, K = Carbide, M = Martensite, F = Ferrite,
γ = Retained austenite, P = Pearlite, M_s = Start of martensite formation

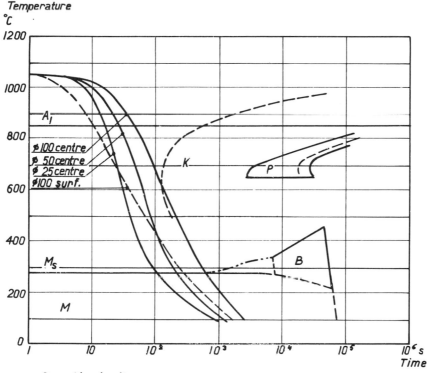

a. Quenching in oil.

Figure 6.41a (b and c are on p. 354) Continuous-cooling-transformation (CCT) diagrams
for grade H 13, austenitized at 1030°C for 15 min. Cooling curves for
specimens of various dimensions and different cooling media superimposed
on diagrams

In practice it has been found that with regard to large and complicated
dies made from grade H 13, such as die-casting dies for aluminium, the
longest service life is obtained if they have been martempered. *Figure 6.42*
shows such a die in the process of being martempered.

On *tempering* hot-work steels in the neighbourhood of 500°C a
secondary hardening is obtained (*see Figure 6.43*). Practical experience has
shown that since there is a reduction in impact strength on tempering the
steel at this temperature, it is better to forgo maximum hardness and only
in exceptional cases to resort to hardnesses greater than 50–52 HRC.
However, since wear resistance increases with hardness, the highest
attainable hardness is usually aimed at. A general rule states that the
smaller the cross-sectional area of the tool the greater is the obtainable
hardness. In an analogous way a tool of simple design may, without
jeopardizing its functioning, be given a hardness near the upper limits of
the recommended hardness range.

The significant influence of time when these steels are tempered above
500°C has been described in detail in the preceding chapter. For the sake
of completeness the tempering diagram for grade H 13 is reproduced in
this chapter also (*see Figure 6.44*). In order to estimate a suitable hardness

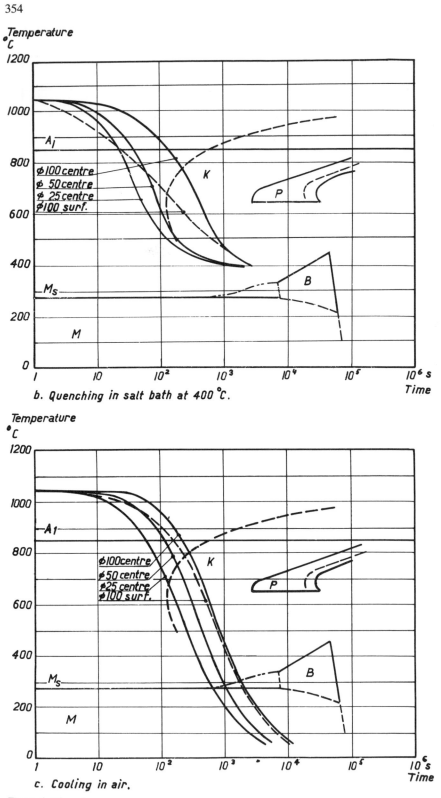

b. Quenching in salt bath at 400 °C.

c. Cooling in air.

Figure 6.41b and c

Figure 6.42 Martempering of die-casting die for aluminium.
Die made from steel H 13

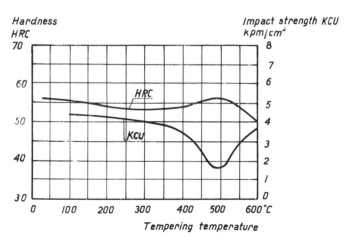

Figure 6.43 Hardness and impact strength of steel H 13 as functions of
tempering temperature

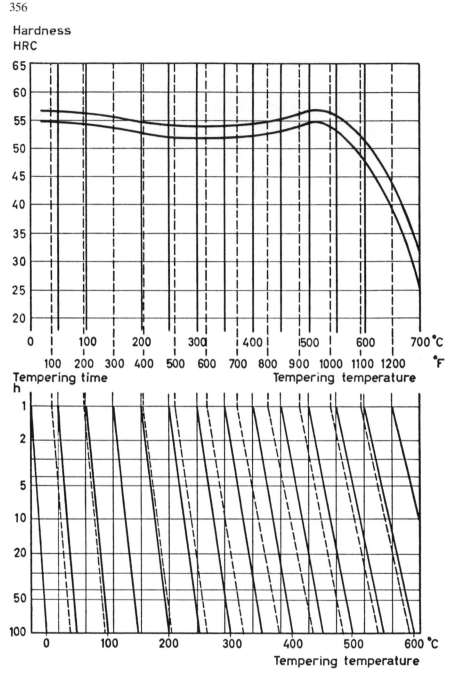

Figure 6.44 Tempering curves showing interdependence of time and temperature for steel H 13. Oil quenched or air cooled from 1050 °C

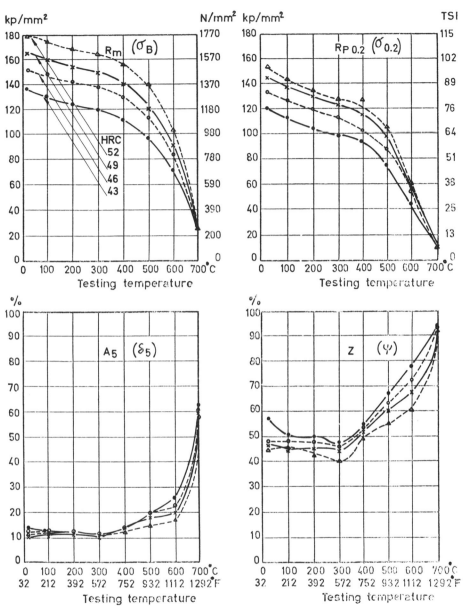

Figure 6.45 Mechanical properties at elevated temperatures of grade H 13.
Specimens air cooled from 1050°C and tempered at various temperatures

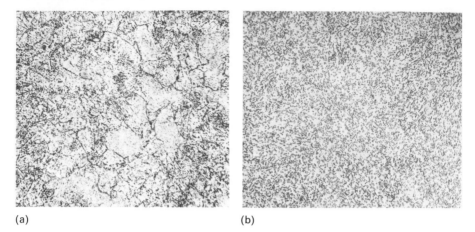

(a)   (b)

Figure 6.46 Microstructure of steel H 13 specimens, 400 mm diameter after
alternative annealing treatments: (a) Conventional annealing;
(b) Structure-treating (microdizing)

for hot-work tools subjected to known pressures and temperatures it is
assumed that the yield strength of the steel at the temperature concerned
should be greater than the working stress. The diagrams in *Figure 6.45*
show the mechancial properties at elevated temperatures of grade H 13
after specimens have been air hardened and tempered to various
hardnesses. For working temperatures up to 500 °C the initial hardness has
a very marked influence on the tensile strength and yield point. As the
temperature increases, this influence diminishes and at 700 °C the values
are the same for all the specimens.

Despite the high ultimate tensile strength of this steel at room
temperature the figures for elongation and reduction of area are also very
high. However, as the dimensions of the steel increase, the impact strength
decreases, particularly in the transverse direction. By remelting the steel,
making use of the V A R or E S R consumable electrode process (*see*
Chapter 3), the transverse impact strength is improved but in many cases,
not enough. For this reason the Bofors Company and the Uddeholm
Company since the middle of the 1960s have been applying a heat
treatment process called 'structure treating' or 'microdizing'. This
treatment consists of an austenitization followed by cooling prior to the
annealing treatment proper. The effect of this treatment on the structure is
shown in *Figures 6.46a* and *b*. The values illustrated by the bar chart in
Figure 6.47 are the result of tests carried out on specimens taken from
200 mm diameter bar stock that had been processed according to different
methods.

An important quality improvement in grade H 13 has been achieved by
Hagfors Steelworks of the Uddeholm group. The modifications to the
technical production process are mainly the following:

SiCa injection, which gives an extremely low sulphur content
(<0·001%);

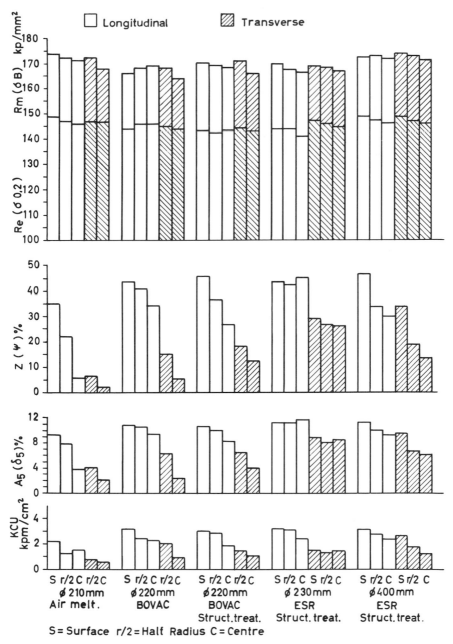

Figure 6.47 Influence of steelmaking process and annealing method on properties of steel H 13. Specimens taken from centre of 8 in diameter bars. Tensile strength 170 kp/mm² (1670 N/mm²)

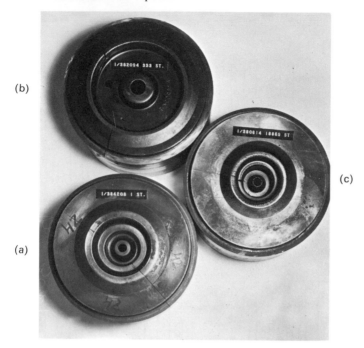

Figure 6.48 Forging dies made from steel H 13. Diameter 200 mm

a modified heat-treating process;
strictly controlled regenerative treatment which produces a consistent 'microdized' result.

H 13, processes in this way, is known by the designation UHB ORVAR M SUPREME. It has the following improved properties:

greater impact-strength level in all directions;
greater hardness may be used while the steel still maintains the same impact-strength level;
increased resistance to thermal shock.

These improvements are regarded as being particularly important for dies used for pressure die-casting of large aluminium castings. The dies shown in *Figure 6.48,* which is an example taken from actual practice, show the importance of a correct initial structure and hardness of a forging die made from grade H 13. The die, marked *a,* cracked under the very first forging blow. Its hardness was 43 HRC which is normal for a die of this size. The microstructure was much inferior to that shown in *Figure 6.46a.* The die, marked *b,* cracked after 333 blows. In this instance the microstructure was ideal. The hardness was 53·5 HRC which is too high for a die of this type. The die, marked *c,* was removed from service owing to its beginning to show heat checking after 18859 blows. The structure was normal and the hardness 45 HRC.

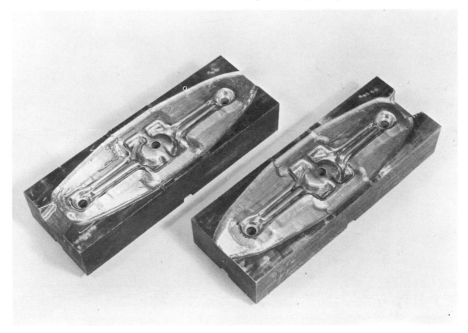

Figure 6.49 Forging dies made from steel H 13 for production of piston rods.
Hardness of die: 38–42 HRC

Further examples of hot-work dies and tools made from grade H 13 are show in *Figures 6.49* and *6.50*.

The excellent mechanical properties of grade H 13 have resulted in the use of the steel for highly stressed structural members operating at room temperature. This field of application is discussed in Section 6.2.6 Ultra high-strength steels. Like the other hot-work steels mentioned above, grade H 13 can be nitrided with excellent results. Data concerning this process are given in Section 6.7.5.

BS BH 10A (UHB QRO 45)

This steel was developed at the Bofors Company and is patented in several countries. The first patent application was lodged in 1956. The composition of the steel corresponds to BS BH 10A. It is a relatively low-alloy steel; the low-alloy content being the result of stringently optimizing the composition with reference to the overall properties of the steel. It possesses good hot-wearing properties, low susceptibility to fatigue cracking and high elevated-temperature strength. In the great majority of instances QRO 45 serves as a wholly satsifactory substitute for the previously very popular 5% and 9% tungsten steels, respectively. The hardening temperature is the same as for grade H 13, viz. 1000–1050°C; 20–30 min is a satisfactory holding time at the hardening temperature. Longer times or temperatures higher than 1050°C tend to make the steel

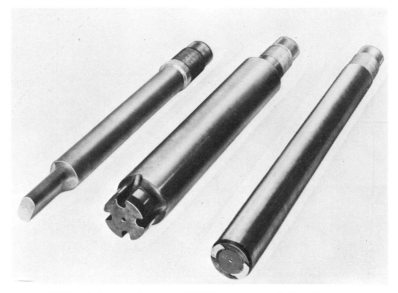

Figure 6.50 Mandrels made from steel H 13 for extrusion of aluminium.
Diameter 30–50 mm, hardness 50–52 HRC

coarse-grained (*see Figure 6.51*). Compared with grade H 13, QRO 45 has a greatly superior resistance to tempering, which can be seen in *Figure 6.52*. At temperatures around 600°C and above, the elevated-temperature strength of QRO 45 is higher than that of grade H 13 (*see Figure 6.53*).

Typical fields of application for QRO 45 are dies for die casting and hot forging of brass, mandrels and dies for the extrusion of copper, tools for the hot-upsetting of steel bolts and nuts (*Figure 6.54*). For this last example

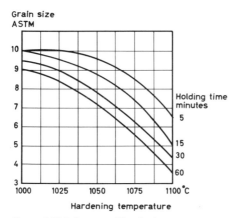

Figure 6.51 Influence of hardening temperature and holding time on grain size of BH 10A (QRO 45)

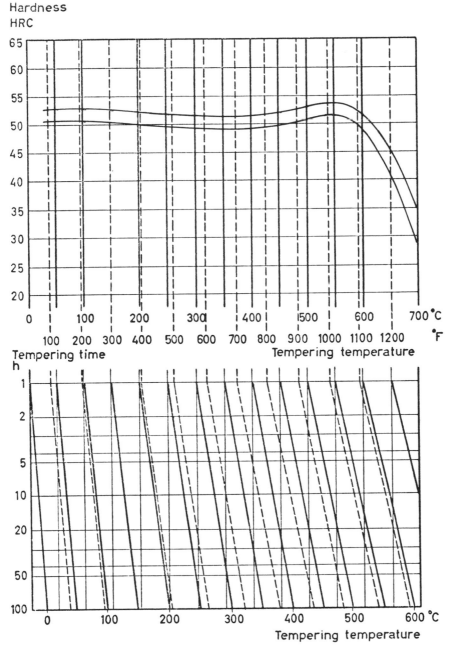

Figure 6.52 Tempering curves showing interdependence of time and temperature for BH 10A (QRO 45). Oil or air hardened from 1050°C

the wear resistance of the steel can be greatly improved by nitriding or by case hardening (*see* Sections 6.5 and 6.7). Owing to the low hardness of QRO 45 as annealed, cold hobbing is a feasible alternative process for shaping tools of suitable design (*see Figure 6.55*).

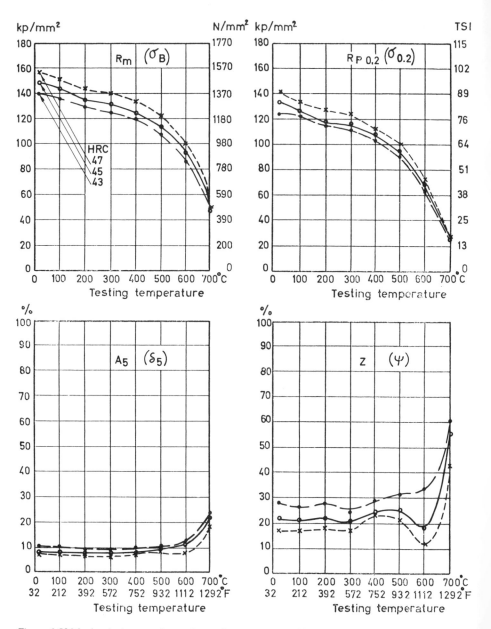

Figure 6.53 Mechanical properties at elevated temperatures of BH 10A (QRO 45), air hardened from 1050 °C and tempered to various hardnesses

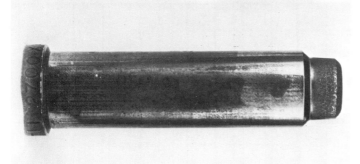

Figure 6.54 Punch made from steel BH 10A (QRO 45) for the hot forming of nuts, 15 mm diameter × 70 mm. Hardness 48–50 HRC

Figure 6.55 Cold hobbed die made from BH 10A (QRO 45). Used for brass die-casting

UHB QRO 80 M

This steel is a further development of QRO 45 (BH 10A), the aim being to achieve improved properties and at the same time to keep down the alloying costs. This objective has been successfully reached by an optimum balancing of the contents of carbon and carbide formers (Cr, Mo and V), the result being a very effective and thermally stable precipitation of mainly MC-carbides[5].

Hardening temperatures up to 1075 °C may be used without any risk that the grain-size number will fall below 9 ASTM. Not until 1100 °C is reached does the grain-size number fall to 8 ASTM, a figure that may still be regarded as being very satisfactory.

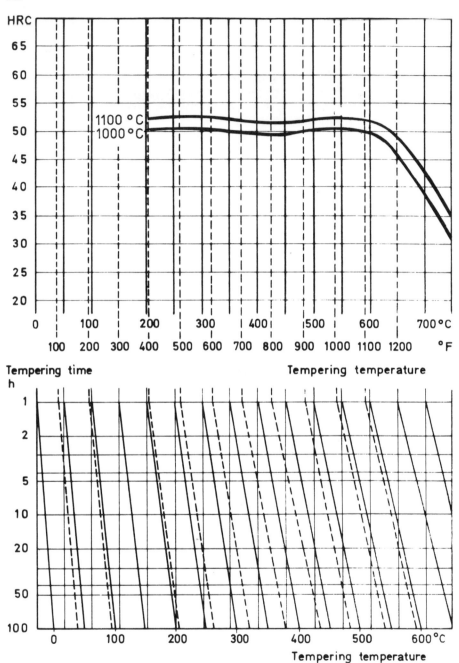

Figure 6.56 Tempering diagram for U H B Q R O 80 M. Austenitizing
temperature 1000 °C and 1100 °C. Holding time at tempering temperature 2 h

Figure 6.56 shows the resistance of QRO 80 M to tempering after sample specimens were hardened from various temperatures. Compared with QRO 45, the hardness values are about the same for both steels up to a tempering temperature of 600°C. At higher temperatures QRO 80 M showed higher hardness values than QRO 45.

Figure 6.57 shows the elevated-temperature proof-stress values for QRO 80 M compared with those for other hot-work tool steels.

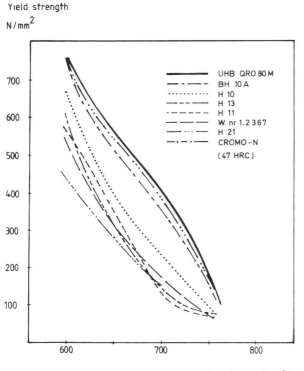

Yield strength

N / mm^2

700

600

500

400

300

200

100

UHB QRO 80 M
BH 10 A
H 10
H 13
H 11
W. nr 1.2 367
H 21
CROMO - N

(47 HRC)

600   700   800

Testing temperature °C

Figure 6.57 Hot yield strength at different testing temperatures after hardening and tempering to 47 HRC (after Norström[5])

Some of the steels investigated were also tested with regard to impact strength after having been hardened and tempered. As may be seen in *Figure 6.58*, after tempering the steels at about 500°C the impact strength is low for all of them except for QRO 80 M.

Tools made from hot-work steel are not only subjected to high temperatures during service but also to considerable temperature fluctuations which cause thermal cracking of a reticular type (heat checking). These cracks will grow successively to such a depth that ultimately the tool becomes unserviceable. Such cracks are called thermal fatigue cracks. The susceptibility of the steel to this type of cracking is evaluated by means of a method based on heating a cylindrical test specimen by means of a high-frequency electric current to 750°C and then

Impact strength KU

J

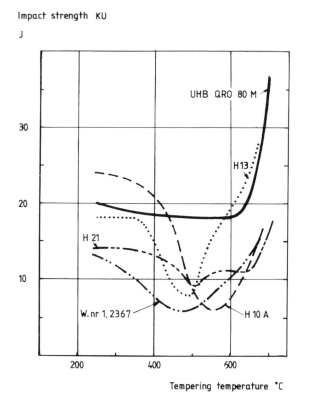

Figure 6.58 Impact strength (KU) at room temperature
after hardening and double-tempering between 200 and
700°C (after Norström[5])

cooling it in water or air. Water quenching takes about one second,
whereas air cooling takes about 50 seconds. The degree of heat checking is
evaluated microscopically and compared with a standard scale. Each test
specimen is given a rating; the higher the numerical value of the rating the
greater the extent of the cracking. The result of such a test is shown in
Figure 6.59 in which the ranking order among the steels is self-evident. An
interesting observation is that the heat-checking frequency of H 13 is
somewhat higher after air cooling; but the opposite state of affairs is
presumed to hold for the other two steels.

A hot-work die may sometimes be exposed to such an extreme thermal
shock that it develops cracks at the first go; cracks so large that the die
becomes unserviceable. The capacity for a steel to withstand such shocks
has been tested and evaluated according to a method developed at the
Uddeholm Company. A notched tensile-test specimen is heated and
clamped between two fixed jaws. The test-piece is cooled very rapidly, thus
causing it to contract and finally to fracture when the tensile stress that is
induced assumes a sufficiently high value. The fall in temperature, being
related to the stress that the test-piece is able to sustain before fracturing, is
a measure of the relative resistance of the steel to thermal shock. After

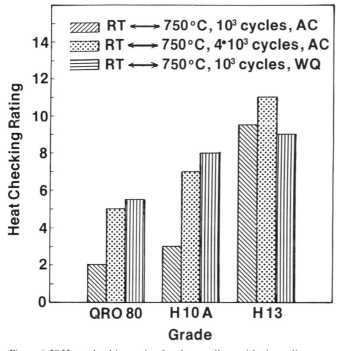

Figure 6.59 Heat-checking rating for slow cyclings with air cooling compared with rapid water quenching after hardening and tempering to 47 HRC. Increasing rating implies increased crack occurrence (after Norström[5])

fracture the stress is calculated and the temperature difference, ΔT, estimated between the highest tempeature and that existing at fracture. The larger the value of ΔT the greater is the resistance of the steel to thermal shock. In *Figure 6.60* there are shown the results of a test with four steels of which QRO 80 M produced the most favourable result, and H 21 the least favourable.

6.1.6   High-speed steels

At the World Exhibition in Paris in 1900, Taylor and White introduced the first high-speed steel, the composition of which was: 1·85% C; 0·15% Si; 0·30% Mn; 3·8% Cr; 8·0% W[6]. This steel was the precursor of the modern high-speed steel T 1 which has kept its position for many years and is still used to some extent. By adding some 5–10% Co and simultaneously increasing the C and V contents the wear resistance is increased. Common to all 18–14–1 steels is their high hardening temperature, viz. 1260–1280°C. Any danger of overheating the steels by using a hardening temperature of 1280°C is therefore out of the question.

As was the case for hot-work steels it has been possible to replace W in the T 1 grade by Mo. This has led to the development of the M 2 type (6–5–4–2) which, for most purposes, can replace T 1. Yet another variant is M 7 which has a higher content of Mo but less W than M 2. For some applications M 7 is said to possess greater toughness and wear resistance than M 2. Like T 1, both M 2 and M 7 can be alloyed with Co which gives them increased hot wear-resistance. Such a variant of M 7 is designated M 42. For the M steels a suitable hardening temperature is 1200–1220°C. Sometimes 1230°C is given as the maximum hardening temperature and under no circumstances should this temperature be exceeded.

A convenient way of characterizing high-speed steels is to set out the sum of the carbide-forming elements (CFE) according to the formula: W + 2Mo + 6V. This is shown in the table below.

Table 6.3 Chemical composition of some commonly occurring high-speed steels

| Designation | | | | Composition | % | | | |
AISI	ISO 4957	C	W	Mo	Cr	V	Co	CFE
T1	HS 18–4–1	0·75	18·0	—	4·0	1·0	—	24
M2	HS 6–5–2	0·85	6·0	5·0	4·0	2·0	—	28
M3:2	HS 6–5–3	1·2	6·0	5·0	4·0	3·0	—	34
M7	HS 2–9–2	1·0	1·8	8·8	4·0	2·0	—	32
M35	HS 6–5–2–5	0·9	6·2	5·0	4·0	2·0	5·0	28
M42	HS 2–9–1–8	1·1	1·5	9·5	3·7	1·2	8·0	28
—	HS 10–4–3–10	1·3	9·0	3·5	4·0	3·3	10·0	36

The seven steels listed in the above table form part of the group of 12 high-speed standardized by ISO. These seven steels cover most of the requirements of the high-speed steel market.

From the heat-treatment angle these steels taken together represent most variants in existence, besides largely covering the fields of application of high-speed steels.

As may be seen from the above, the hardening temperatures for high-speed steels are higher than those for other tool steels. Temperatures of only some tens of degrees below incipient fusion of the steel are used. The chromium carbides go into solution around 1100°C and at the normal hardening temperature for, say grade M 2 there are undissolved carbides left amounting to some 10%, mainly V carbides and double carbides to Mo and W. The high hardening temperatures employed for high-speed steels are conducive to rapid grain growth and hence the holding time must be carefully controlled (see the foregoing chapter). The hardening temperature must also be accommodated to the original dimensions of the steel stock used for the tool since as the stock dimension increases the amount of carbide segregation increases which, in turn, lowers the temperature of incipient fusion. Therefore, the hardening temperature should be kept near the lower limit of the normal hardening temperature range when the dimension of the original steel stock exceeds about 100 mm.

The hardening temperature is chosen to suit the steel in question, always keeping in mind the use to which the tool is to be put. Tools for machining, e.g. turning and planning tools, or for rough milling, should be hardened from the highest temperature in order to be certain that they obtain the best hot-hardness properties since the cutting edges may reach temperatures as high as 600°C.

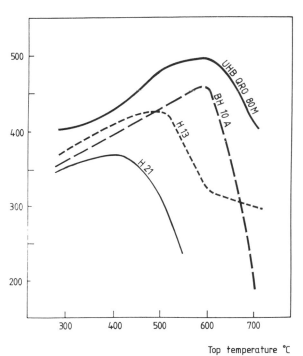

Figure 6.60 Temperature range to fracture (ΔT) versus top temperature after hardening and tempering to a room temperature hardness of 47 HRC (after Norström[5])

Tools to be used at lower temperatures or that require good impact strength, such as cold-upsetting tools, can be hardened from temperatures as low as 1050°C. By this treatment, resistance to tempering is reduced and if the hardening temperature is low enough (below 1000°C) the secondary hardening effect disappears.

As a rule, high-speed steels have good hardenability from which follows that tools made from such steels may be quenched in a salt bath or even air cooled. High-speed steel containing 8–10% Co has a somewhat reduced hardenability and in order to arrive at maximum hardness by means of air cooling, light sections only (less than 30 mm in diameter) can be treated in this way.

Having been quenched from a normal hardening temperature, high-speed steels contain between 20% and 30% retained austenite. As they cool from a tempering temperature of about 575°C or higher there is practically complete transformation to martensite while at the same time the initially formed martensite is tempered. A second tempering treatment is required to give the last-formed martensite its optimum combination of useful properties. Some high-speed steels that tend to have a large amount of retained austenite, e.g. steels containing more than 1% carbon but no cobalt, require a third tempering treatment. In *Figures 6.61a, b* and *c* are shown the microstructures in specimens of grade M 42 after hardening, after a single and after a double tempering treatment, respectively.

Retained austenite can also be transformed by subzero treatment. A prior subzero treatment will not affect the amount of retained austenite after tempering over about 575 °C since this constituent is decomposed at the conventional tempereing temperature used. At tempering temperatures above 550 °C the length of the holding time produces a pronounced effect on the hardness, as may be seen from the curves in *Figures 6.62* and *6.63* which show the interdependence of time and temperature. By increasing the tempering time from 1 to 4 h at 600 °C the hardness of grade M2 falls from 65·5 to 63 HRC.

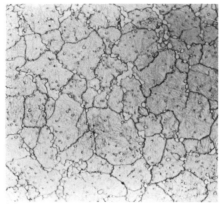

(a)

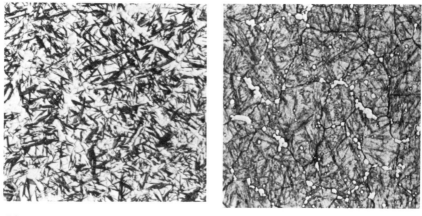

(b)   (c)

Figure 6.61 High-speed steel M 42. Microstructure after oil hardening from 1200 °C. (a) As quenched, 35% retained austenite; (b) Tempered once at 600 °C, 1 h. Tempered martensite and newly formed martensite; (c) Tempered twice at 600 °C, 1 h each time. Tempered martensite. All 700 ×

Cutting tools for which the highest hardness is required, are tempered at 550°C. However, the hardness and tempering temperature must be adjusted to the toughness requirements. The impact strength is highest when the steel is tempered in the range 250–450°C and lowest at the temperature that gives maximum hardness. As the steel continues to be tempered at increasing temperatures the toughness starts to increase again.

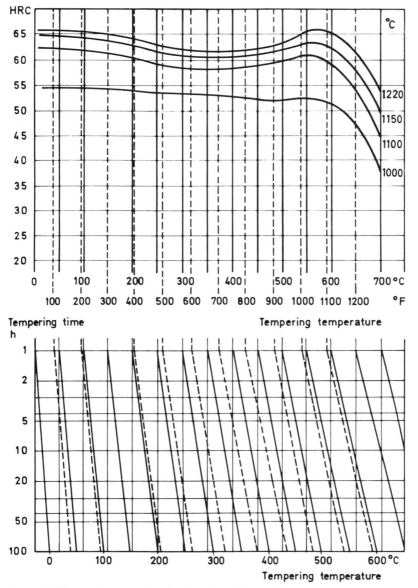

Figure 6.62 Tempering curves showing interdependence of time and temperature for steel M 2

Tools that in service are subjected to high pressures give best results if they have been tempered at about 600°C; the austenite being completely transformed at this temperature.

The torsion-impact test with unnotched specimens yields results that in many cases are in agreement with those from impact strength tests. According to *Figure 6.64* the torsion–impact test indicates a minimum

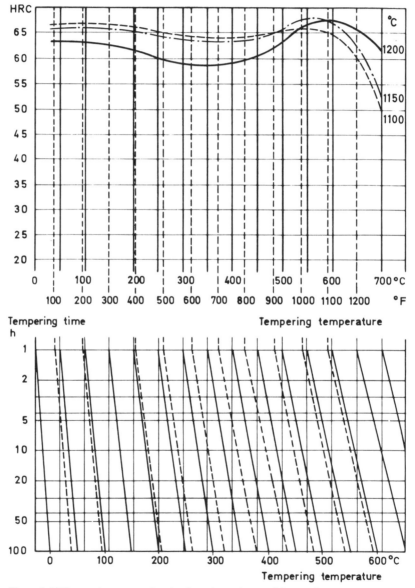

Figure 6.63 Tempering curves showing interdependence of time and temperature for steel M 42

value at 650°C (1200°F) whereas according to *Figure 6.65* the unnotched specimen indicates a minimum value at 510°C (950°F). The diagrams in the latter figure also show that grade M2 has higher impact strength and hardness than grade T1.

The hardness of a high-speed steel, quenched and tempered according to current practice, stays at a reasonably high level up to about 500°C (*see Figure 6.66*). This hot hardness property is of very great importance to cutting tools, the cutting-edge temperature of which may rise to about 600°C. As was mentioned earlier, hot hardness increases as the hardening temperature increases, but the composition of the steel is the determining factor. Co and V are the alloying elements that have the most marked influence, which may be inferred from *Figure 6.66*.

High-speed steels find their most usual field of application as tools for machining. The most highly alloyed grades, e.g. M42, are best able to cope with the high cutting speeds and high cutting-edge temperatures which machining tools, such as lathe and planing tools, are subjected to. Unfortunately the superior wear resistance of this steel cannot always be utilized in milling cutters, for which toughness is imperative. Although the hardness of a turning tool may be about 68 HRC, a figure of 65 HRC should be the maximum hardness for milling cutters.

The large amount of carbides in high-speed steels ensures that they have good wear resistance at room temperature also. Hence high-speed steels find use to some extent as punching and blanking tools for which is required a greater degree of wear resistance than that available in grades

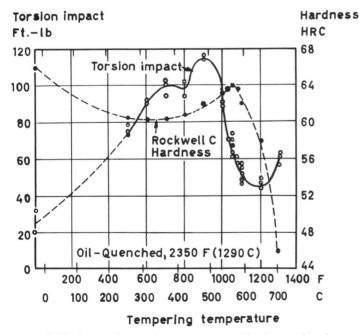

Figure 6.64 Influence of tempering temperature on hardness and torsion impact properties of grade T 1 high-speed steel, oil hardened from 1290°C (2350°F) (after Luerssen and Greene[7])

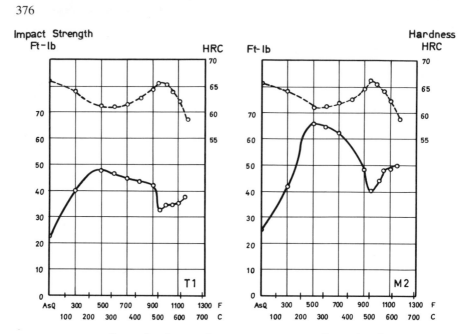

Figure 6.65 Influence of tempering temperature on hardness and unnotched Izod impact strength of grade T 1 high-speed steel, austenitized at 1200°C (2350°F), and grade M 2, austenitized at 1220°C (2225°F), respectively (after Grobe and Roberts[8])

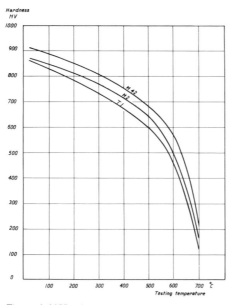

Figure 6.66 Hot-hardness of high-speed steels. Load applied for 7 s

D6 or D2. However, the higher price and the more complex heat-treating process are factors that also should be taken into account when high-speed steels are being considered for such uses.

Owing to their high strength at elevated temperatures, high-speed steels are also used for hot-work tools, usually for such as perform a shearing operation, for example hot shears and punches in hot-punching machines that produce nuts. The hardness of such tools should be 55–60 HRC. The wear resistance of high-speed steels may be enhanced by carburizing or by nitriding. These processes are discussed further on in this chapter.

Since 1970 a type of high-speed steel called powder high-speed steel has been produced by a powder-metallurgical method called the ASEA-STORA Process (ASP)[9]. The molten steel is atomized in a protective gas, the resulting product being a fine-grained metallic powder. This powder is then isostatically compacted in two stages. The first stage takes place at room temperature at a pressure of 4000 atmospheres; the second stage at 1150°C and 1000 atmospheres. The result is a compact and homogeneous steel product which is subsequently wrought to the required dimensions by conventional forging or rolling. This metallurgical process endows the steel with enhanced properties with regard to grindability, toughness and dimensional stability in hardening.

The most usual types of high-speed steels produced in this way are AISI M3:2, designated ASP 23, and a cobalt-alloyed variant designated ASP 30.

The ASP steels are generally more costly than their conventionally produced equivalents and hence they are used only for high-duty industrial tools in which the advantages cited are wholly utilized. Foremost applications are tools of complicated design, tools possessing exceptional machining properties, tools for machining difficult materials and tools used for cold work.

Those interested in studying further the subject of high-speed steels can find literature reviews and references covering the years 1969 to 1973 in reference[10] and for the years 1973 to 1977 in reference[11]. Literature connected with high-speed steel conferences held during the years 1979 to 1981 may be found in references[12, 13, 14, 15].

6.2 Hardening and tempering of conventional constructional steels

6.2.1 Definitions

A constructional steel, quenched from the austenitizing temperature and subsequently tempered in the range 500–700°C, obtains useful mechanical properties, e.g. a high ratio of yield strength to ultimate tensile strength, high elongation, reduction of area and impact strength. The usual hardness range is 210–400 HB, which corresponds to an ultimate tensile strength of 700–1280 N/mm² (70–130 kp/mm²). This heat treatment—in German called *Vergütung*—may be described as tough-hardening. Some decades ago a hardness of 330 HB was considered the upper hardness limit that cutting tools could cope with. This limit has been successively raised and at the same time the division between 'tough-hardened' steels and what might be termed hard-hardened steels has been erased. 'Hard-hardened' constructional steels are such as have been quenched and tempered at approximately 200 °C.

The boundary between the classic 'tough-hardening' steels and other steels has also been bridged. The designation 'Ultra high-strength steels', which covers several types of steel including some tool steels, is currently becoming popular and in the cases under discussion it is applied to steels that have a higher tensile strength than the conventional 'tough-hardening' steels.

If it is possible to start out from prehardened steel it is usually more economic in the long run to purchase this than to buy annealed steel and carry out the necessary heat treatment after the machining operations. When the steel is bought as forgings or if mechanical properties are required greater than those available in material from stock the quenching and tempering may sometimes be carried out by the purchaser.

In the chapter on hardening, the interrelationships of composition, dimensions and hardness of the steel have been discussed in detail. This information is now being put to good use. Formerly it was customary to use rather highly alloyed grades of steel even in light dimensions and this may sometimes be justified when it is expedient to stock as few different grades as possible. However, the trend of the times is to use the least expensive steel for each function. This calls for detailed knowledge of the hardenability of each steel concerned. For the rest of this chapter the steels are classed in groups, based partly on their chemical composition and partly on their fields of application.

6.2.2   Plain carbon steels

According to IS O/R 683/I–1968 there are eight heat-treatable plain carbon steels. The lowest carbon content is 0·25% and the highest, 0·60%—see *Table 6.4*.

Table 6.5 contains the mechanical properties attainable on quenching and tempering the steels. Even carbon-steel parts greater than 100 mm in diameter—or equivalent size—are sometimes subjected to 'tough-hardening'. The purpose of giving the steel such a heat treatment is to give it a fine-grained ferritic-pearlitic structure which will result in higher strength and toughness values than what a simple normalizing treatment will give. The steels designated C 35 and C 45 have the most general application. In light dimensions steel C 45 in particular is used for parts heat treated to high hardness for high toughness, e.g. axles and bolts. Oil quenching may be used, but as shown in *Figure 6.67* the steel exhibits a high degree of section sensitivity although the chemical composition lies above the upper limit for C 45. Tests with steel C 50, which has a somewhat higher Mn content than C 45, show that this steel also exhibits section sensitivity even if the Mn-content is high. The steel in question has the composition: 0·52% C; 0·24% Si; 0·90% Mn; 0·06% Cr. *Figures 6.68* to *6.71* contain the results of a series of hardening tests with this steel. After cooling in conventional quench oil a bar specimen 30 mm in diameter showed only 40 H RC at the surface. When specimens of the same diameter were quenched in fast-quenching oil the hardness was

Table 6.4 ISO/R 683/I–1968. Types of steel and chemical composition guaranteed

ISO Type of steel	C%	Si%	Mn%	P% max	S% max
C 25 C 25 e	0·22–0·29	0·15–0·40	0·40–0·70	0·050 0·035	0·050 0·035
C 30 C 30 e	0·27–0·34	0·15–0·40	0·50–0·80	0·050 0·035	0·050 0·035
C 35 C 35 e	0·32–0·39	0·15–0·40	0·50–0·80	0·050 0·035	0·050 0·035
C 40 C 40 e	0·37–0·44	0·15–0·40	0·50–0·80	0·050 0·035	0·050 0·035
C 45 C 45 e	0·42–0·50	0·15–0·40	0·50–0·80	0·050 0·035	0·050 0·035
C 50 C 50 e	0·47–0·55	0·15–0·40	0·60–0·90	0·050 0·035	0·050 0·035
C 55 C 55 e	0·52–0·60	0·15–0·40	0·60–0·90	0·050 0·035	0·050 0·035
C 60 C 60 e	0·57–0·65	0·15–0·40	0·60–0·90	0·050 0·035	0·050 0·035

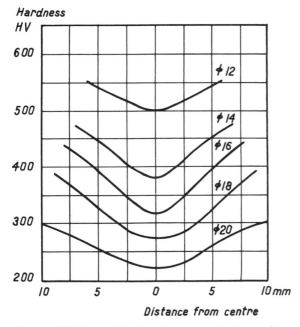

Figure 6.67 Influence of section diameter on hardness, after quenching, of a plain-carbon steel containing 0·45% C, 0·30% Si and 0·88% Mn. Hardening temperature: 850°C, quenching medium: still oil

Table 6.5 ISO/R 683/1–1968. Mechanical properties for quenched and tempered condition*

Type of steel	$\phi \leqslant$ mm (0·63 in)				16 mm (0·63 in) $< \phi \leqslant$ 40 mm (1·58 in)				40 mm (1·58 in) $< \phi \leqslant$ 100 mm (3·94 in)			
	R_e min kgf/mm² N/mm²	R_m kgf/mm² N/mm²	A min %	KCU min kgf.m/cm² /J	R_e min kgf/mm² N/mm²	R_m kgf/mm² N/mm²	A min %	KCU min kgf.m/cm² /J	R_e kgf/mm² N/mm²	R_m kgf/mm² N/mm²	A min %	KCU min kgf.m/cm² /J
C 25 and C 25 e	37 360	55–70 590–690	19	— 7/34	31 300	50–65 490–640	21	— 7/34	—	—	—	—
C 30 and C 30 e	40 390	59–74 580–730	18	— 6/29	34 330	55–70 540–690	20	— 6/29	30** 290	50–65** 490–640	21**	— 6/29**
C 35 and C 35 e	43 420	63–78 620–770	17	— 5/25	37 360	59–74 580–730	19	— 5/25	33 320	55–70 540–690	20	— 5/25
C 40 and C 40 e	46 450	67–82 660–800	16	— 4/20	40 390	63–78 620–770	18	— 4/20	35 340	59–74 580–730	19	— 4/20
C 45 and C 45 e	49 480	71–86 700–840	14	— 3/15	42 410	67–82 660–800	16	— 3/15	38 370	63–78 620–770	17	— 3/15
C 50 and C 50 e	52 510	75–90 740–880	13	—	45 440	71–86 700–840	15	—	41 400	67–82 660–800	16	—
C 55 and C 55 e	55 539	80–95 780–930	12	—	47 460	75–90 740–880	14	—	43 420	71–86 700–840	15	—
C 60 and C 60 e	58 570	85–100 830–980	11	—	50 490	80–95 780–930	13	—	46 450	75–90 740–880	14	—

* R_e = yield stress (0·2% proof stress)
R_m = tensile strength
A = percentage elongation after fracture ($L_0 = 5\ d_0$)
KCU = impact strength with U-notch
** Up to 63 mm (2·5 in) maximum diameter
ISO gives the values of mechanical properties as kgf.m/cm² and tonf/in². In the table tonf/in² has been replaced by N/mm² and kgf.m/cm² has been supplemented with J

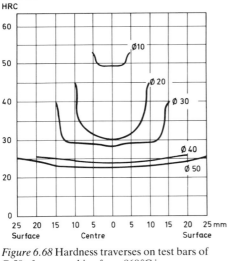

Figure 6.68 Hardness traverses on test bars of
C 50 after quenching from 860°C in
conventional hardening oil. Chemical
composition: 0·52% C, 0·24% Si, 0·90% Mn,
0·06% Cr

45 HRC; when quenched in 10% Aquaquench solution the hardness was
56 HRC; and when quenched in water, 58 HRC. Note that also the depth
of hardening in the last two tests was very shallow. The 50 mm diameter
test specimen, after being quenched in water, showed a surface hardness of
60 HRC, which is higher than that found on the other specimens. This is
due to the presence in the outer layer of the 50 mm diameter specimen, of

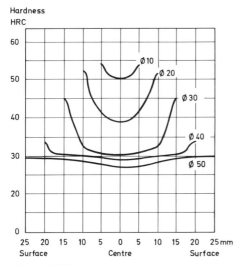

Figure 6.69 Harndess traverses on test bars of
C 50 after quenching from 860°C in
fast-quenching oil. Chemical composition:
0·52% C. 0·24% Si, 0·90% Mn, 0·06% Cr

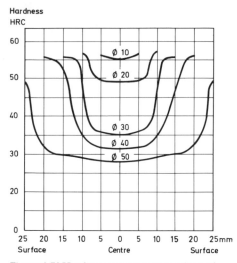

Figure 6.70 Hardness traverses on test bars of
C 50 after quenching from 840°C in water with
10% Aquaquench. Chemical composition:
0·52% C, 0·24% Si, 0·90% Mn, 0·06% Cr

compressive stresses that are sufficiently high to give rise to an additional
hardness increment. *Figure 6.72* contains a Jominy end-quench
hardenability curve which gives a more complete presentation of the
hardenability picture of the steel being examined.

The large hardness differences obtained when hardening carbon steels of
different dimensions are evened out by tempering at 500 to 700°C. This
point is illustrated in *Figure 6.73* which shows the mechanical properties

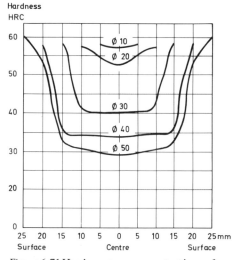

Figure 6.71 Hardness traverses on test bars of
C 50 after quenching from 840°C in water
containing 5% Na₂CO₃. Chemical composition:
0·52% C, 0·24% Si, 0·90% Mn, 0·06% Cr

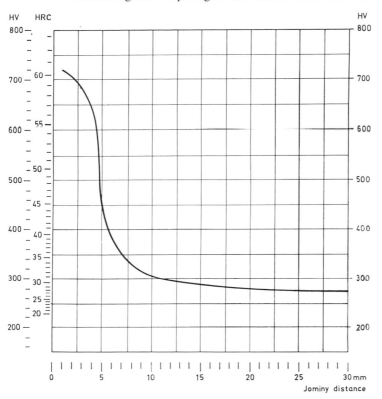

Figure 6.72 Jominy end-quench hardenability diagram for the investigated steel C 50. Chemical composition: 0·52% C, 0·24% Si, 0·90% Mn, 0·06% Cr

obtained after tempering at the stated temperatures.

Water quenching must be used for heavy sections. This treatment is used for parts such as crankshafts for motor-car engines. The limits of the composition range for such steels must be set very close and the heat treatment carried out with the utmost care. From many points of view water quenching is a convenient method of cooling. Water is a cheap quenching medium and the quenched parts need not be put through any extra cleaning operations as is the case when oil quenching is resorted to. Besides, water quenching produces maximum hardness and maximum depth of hardening. However, owing to the large temperature differences set up in the steel during the water quenching and which may cause cracking and distortion, this treatment is limited to parts of simple design and preferably to such as possess rotational symmetry.

6.2.3 Alloy steels

With regard to the standardization of alloy constructional steels it has been possible to reach agreement, even internationally, on a reduction in the number of steel types. So-called tough-hardening steels containing about 1% Cr have long been standardized in several countries. From an economic point of view they constitute in many instances good alternatives to the

Cr-Mo steels. They are standardized according to ISO/R 683/VII–1970 and their chemical composition is given in *Table 6.6*. A survey of corresponding national standards is given in *Table 6.7*. The mechanical properties after hardening and tempering of the steels are shown in *Table 6.8*.

Table 6.6. Chemical composition of steels standardized according to ISO/R 683/VII–1970

Type of steel	C%	Si%	Mn%	P% max	S% max	Cr%
1	0·30–0·37 ⎫					
2	0·34–0·41 ⎬ 0·15–0·40	0·60–0·90	0·035	0·035	0·90–1·20
3	0·38–0·45 ⎭					

There are two alternatives for the S-content: (a) 0·020–0·035%; (b) 0·030–0·050%.

Table 6.7 Desginations of Cr-steels corresponding to ISO/R 683/VII–1970

ISO	AISI	BS	En	DIN Euronorm	NF
1	5132	530A32	18B	34 Cr 4	32 C 4
2	5135	530A36	18C	37 Cr 4	38 C 4
3	5140	530A40	18D	41 Cr 4	42 C 4

The above listed chrome steels have a higher hardenability than the carbon steels, which is apparent by comparing the Jominy end-quench hardenability curves for 41 Cr 4 (*Figure 6.74*) and the plain carbon steel (*Figure 6.72*). The chrome steels are generally quenched in water; only light sections may be quenched in oil. This state of affairs is illustrated in the case of 41 Cr 4 in *Figures 6.75* and *6.76*.

Among the low-alloy steels there are also the Cr–Mo steels, as specified by ISO/R 683/II–1968. Their chemical composition is shown in *Table 6.9*. Some equivalent national standards are given in *Table 6.10*. *Table 6.11* includes the mechanical properties attainable on quenching and tempering the steels. The NF steels have slightly different chemical compositional limits. The steels are generally oil quenched but sometimes water quenching is resorted to. Steel No. 1 gives the best results after this treatment. *Figure 6.77* shows the surface hardness obtained in various dimensions of steel No. 1 quenched in oil and water, respectively. When quenched in water and then tempered at 200°C the steel shows a hardness difference of about 150 HB units between surface and core in bar stock 100 mm in diameter (*see Figure 6.78*). When the steel is tempered at 600°C, which is the temperature used for 'tough-hardening', the hardness difference is reduced to some 50 HB units only. A point in favour of steel No. 1 is its good weldability.

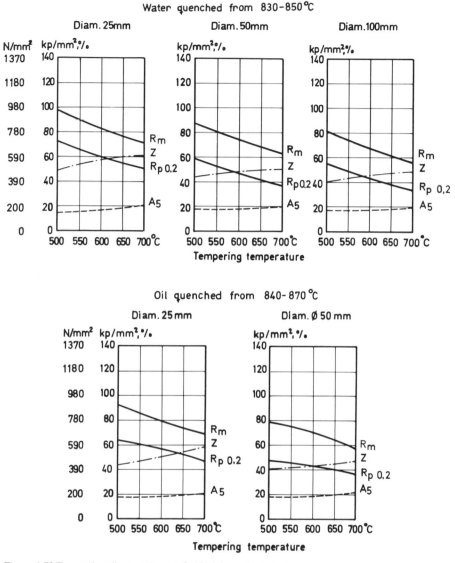

Figure 6.73 Tempering diagram for the C 45 plain-carbon steel

Figure 6.79 shows a tempering diagram for steel No. 3. Note the distinct reduction in impact strength after tempering the specimen around 300°C. To a large extent steels Nos. 2 and 3 are both used for parts that are either flame or induction hardened.

If high mechanical properties in heavy sections are stipulated, higher-alloy steels are used. Examples of such are given in *Table 6.12,* viz. a Cr–Ni–Mo steel and a Cr–Mo steel.

Table 6.8 ISO/R 683/VII–1970. Mechanical properties for the quenched and tempered condition.*

Type of steel	$\phi \leqslant 16$ mm				16 mm $< \phi \leqslant 40$ mm				40 mm $< \phi \leqslant 100$ mm			
	R_e min kgf/mm² N/mm²	R_m kgf/mm² N/mm²	A_5 min %	KCU min[1] kgf. m/cm² J	R_e min kgf/mm² N/mm²	R_m kgf/mm² N/mm²	A_5 min %	KCU min[1] kgf. m/cm² J	R_e min kgf/mm² N/mm²	R_m kgf/mm² N/mm²	A_5 min %	KCU min[1] kgf. m/cm² J
1, 1a	70	90–110	12	5 / 25	60	80–95	14	6 / 29	47	70–85	15	6 / 29
1b	690	880–1080		—	590	780–930		—	460	690–830		—
2, 2a	75	95–115	11	4 / 20	64	85–100	13	5 / 25	52	75–90	14	5 / 25
2b	740	930–1130		—	630	830–980		—	510	740–880		—
3, 3a	80	100–120	11	4 / 20	68	90–110	12	5 / 25	57	80–95	14	5 / 25
3b	780	980–1180		—	670	880–1080		—	560	780–930		—

1. The impact strength is not specified for steels Nos. 1b, 2b and 3b.
2. ISO gives the values of mechanical properties as kgf. m/cm² and tonf/in². In the table tonf/in² has been replaced by the N/mm² and kgf. m/cm² has been supplemented with J.

R_e = yield stress (0·2% proof stress)
R_m = tensile strength
A = percentage elongation after fracture ($L_0 = 5\ d_0$)
KCU = impact strength with U-notch

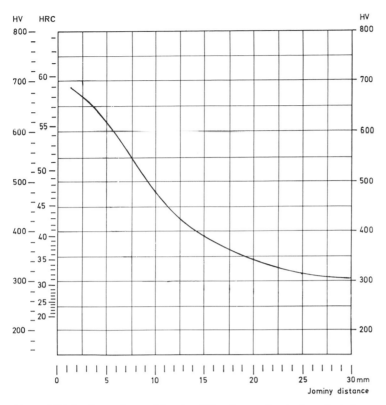

Figure 6.74. Jominy end-quench hardenability diagram for the investigated steel 41 Cr 4. Chemical composition: 0·41% C, 0·27% Si, 0·75% Mn, 0·90% Cr, 0·19% Ni, 0·020% Al

Table 6.9 ISO/R 683/II–1968. Types of steel and chemical composition guaranteed

Type of steel	C %	Si %	Mn %	P % max	S % max	Cr %	Mo %
1	0·22–0·29		0·50–0·80	0·035	0·035		
2	0·30–0·37	0·15–0·40	0·50–0·80	0·035	0·035	0·90–1·20	0·15–0·30
3	0·38–0·45		0·50–1·00	0·035	0·035		

Table 6.10 Designations for Cr–Mo Steels

ISO	AISI	BS	En	DIN Euronorm	SS	NF
1	4130	—	—	25 CrMo 4	2225	25 CD 4
2	4135	708A37	19B	34 CrMo 4	2234	30 CD 4
3	4142	708A42	19C	42 CrMo 4	2244	40 CD 4

Table 6.11 ISO/R 683/II–1968. Mechanical properties for the quenched and tempered condition*

Type of steel	$\phi \leqslant 16$ mm (0·63 in)				16 mm (0·63 in) $< \phi \leqslant 40$ mm (1·58 in)				40 mm (1·58 in) $< \phi \leqslant 100$ mm (3·94 in)			
	R_e min kgf/mm² N/mm²	R_m kgf/mm² N/mm²	A min %	KCU min kgf. m/cm² J	R_e min kgf/mm² N/mm²	R_m kgf/mm² N/mm²	A min %	KCU min kgf. m/cm² J	R_e min kgf/mm² N/mm²	R_m kgf/mm² N/mm²	A min %	KCU min kgf. m/cm² J
1	70 / 690	90–110 / 880–1080	12	6 / 29	60 / 590	80–95 / 780–930	14	7 / 34	47 / 460	70–85 / 690–830	15	7 / 34
2	80 / 780	100–120 / 980–1180	11	5 / 25	68 / 670	90–110 / 880–1080	12	6 / 29	57 / 560	80–95 / 780–930	14	6 / 29
3	90 / 880	110–130 / 1080–1280	10	4 / 20	78 / 770	100–120 / 980–1180	11	5 / 25	65 / 640	90–110 / 880–1080	12	5 / 25

Type of steel	100 mm (3·94 in) $< \phi \leqslant 160$ mm (6·30 in)				160 mm (6·30 in) $< \phi \leqslant 250$ mm (9·85 in)			
	R_e min kgf/mm² N/mm²	R_m kgf/mm² N/mm²	A min %	KCU min kgf. m/cm² J	R_e min kgf/mm² N/mm²	R_m kgf/mm² N/mm²	A min %	KCU min kgf. m/cm² J
1	—	—	—	—	—	—	—	—
2	52 / 510	75–90 / 740–880	15	6 / 29	57 / 460	70–85 / 690–830	15	6 / 29
3	57 / 560	80–95 / 780–930	13	5 / 25	52 / 510	75–90 / 740–880	14	5 / 25

* R_e = yield stress (0·2% proof stress)
R_m = tensile strength
A = percentage elongation after fracture ($L_0 = 5\,d_0$)
KCU = impact strength with U-notch

ISO gives the values of mechanical properties as kgf. m/cm² and tonf/in². In the table tonf/in² has been replaced by N/mm² and kgf. m/cm² has been supplemented with J.

Hardness

HRC

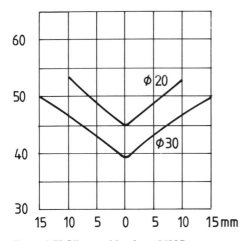

Figure 6.75 Oil quenching from 860°C.
Hardness traverses on test bars of steel 41 Cr 4
after hardening in oil. Chemical composition:
0·41% Cr, 0·27% Si, 0·75% Mn, 0·90% Cr,
0·19% Ni, 0·020% Al

Hardness

HRC

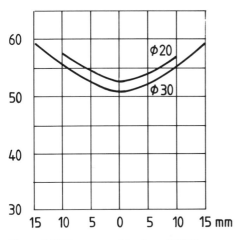

Figure 6.76 Water quenching from 850°C.
Hardness traverses on test bars of steel 41 Cr 4
after hardening in water. Chemical composition:
0·41% Cr, 0·27% Si, 0·75% Mn, 0·90% Cr,
0·19% Ni, 0·020% Al

390

Table 6.12 High-alloy heat-treatable high-strength steels

C	Cr	Ni	Mo	ISO/R 683	BS	En	DIN	SS
0·35	1·4	1·4	0·2	VIII–3	817M40	24	34 Cr–Ni–Mo 6	2541
0·30	3·0	—	0·4	VI–1	—	29B	32 Cr–Mo 12	2240

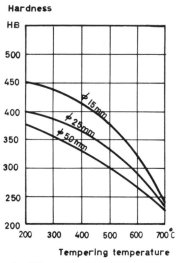
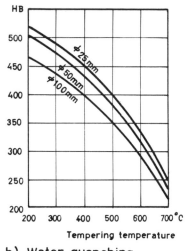

a) Oil quenching

b) Water quenching

Figure 6.77 Surface hardness of SS 2225 (25 CrMo 4) after quenching and tempering. Quenching temperature 850°C

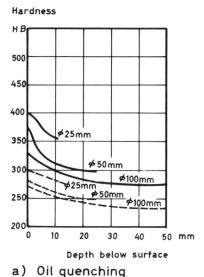
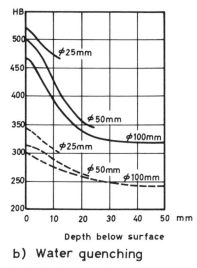

a) Oil quenching

b) Water quenching

Figure 6.78 Curves showing depth of hardening of SS 2225 (25 CrMo 4). Quenching temperature: 850°C.

 Tempered at 200°C
– – – – – – – – Tempered at 600°C

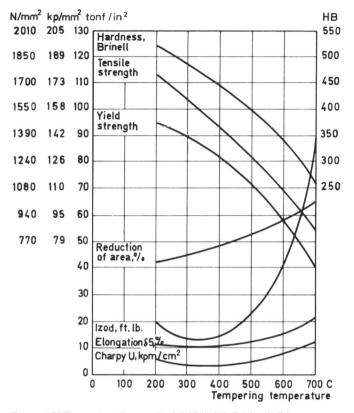

Figure 6.79 Tempering diagram for SS 2244 (42 CrMo 4). Diagram applicable to bar stock 30 mm (1³⁄₁₆ in) in diameter, oil quenched from 850°C

The mechanical properties according to ISO are given in *Table 6.13*.

Figure 6.80 shows the tempering diagram for the Cr–Ni–Mo steel. Although this steel is essentially a prehardening steel it has begun to be increasingly used in a condition of high hardness, i.e. quenched and tempered at about 200°C. With this treatment a hardness of 500–550 HB is obtainable in oil-quenched sections up to about 75 mm diameter.

Some examples of quenched and tempered parts are shown in *Figure 6.81*. All of them are subjected to high stresses, both static and dynamic and therefore the fillets between the changes of section have been burnished. This finish treatment increases the fatigue strength by 25–50%.

6.2.4 Stainless steels

Included among the so-called stainless steels there is a group of martensitic or heat-treatable steels containing about 13% Cr. *Table 6.14* gives the carbon content and designations of some of the most frequently occurring steels of this group. *Table 6.15* gives the hardness obtainable on quenching from 1000°C in oil or air, followed by tempering at different temperatures.

Table 6.13 ISO/R 683/VI-1970 and ISO/R 683/VIII-1970. Mechanical properties for the quenched and tempered condition*

Type of steel	φ ≤ 16 mm (0.63 in)				16 mm (0.63 in) < φ ≤ 40 mm (1.5 in)				40 mm (1.5 in) < φ ≤ 100 mm (3.94 in)			
	R_e min kgf/mm² N/mm²	R_m kgf/mm² N/mm²	A min %	KCU min kgf.m/cm² J	R_e min kgf/mm² N/mm²	R_m kgf/mm² N/mm²	A min %	KCU min kgf.m/cm² J	R_e min kgf/mm² N/mm²	R_m kgf/mm² N/mm²	A min %	KCU min kgf.m/cm² J
VI-1	90 880	110–130 1080–1280	10	5 25	85 830	105–125 1030–1280	10	6 29	80 780	100–120 980–1180	11	6 29
VIII-3	100 980	120–140 1180–1370	9	40 20	90 880	110–130 1080–1280	10	5 25	80 780	100–120 980–1180	11	6 29

Type of steel	100 mm (3.94 in) < φ ≤ 160 mm (6.3 in)				160 mm (6.3 in) < φ ≤ 250 mm (9.85 in)			
	R_e min kgf/mm² N/mm²	R_m kgf/mm² N/mm²	A min %	KCU min kgf.m/cm² J	R_e min kgf/mm² N/mm²	R_m kgf/mm² N/mm²	A min %	KCU min kgf.m/cm² J
VI-1	75 740	95–115 930–1130	12	6 29	70 690	90–110 880–1080	12	6 29
VIII-3	70 690	90–110 880–1080	12	6 29	65 640	85–100 830–980	12	6 29

* R_e = yield stress (0.2% proof stress)
R_m = tensile strength
A = percentage elongation after fracture ($L_0 = 5\,d_0$)
KCU = impact strength with U-notch

ISO gives the values of mechanical properties in kgf.m/cm² and tonf/in². In the table tonf/in² has been replaced by N/mm² and kgf.m/cm² has been supplemented with J

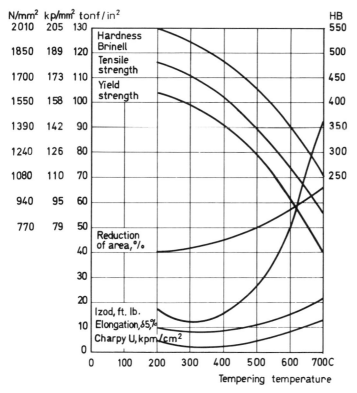

Figure 6.80 Tempering diagram for SS 2541 (34 CrNiMo 6). Diagram applicable to bar stock 50 mm (2 in) in diameter, oil quenched from 850°C

Table 6.14 Martensitic stainless steels containing 13% Cr

% C	ISO/R 683/ XIII–1972	AISI	BS	En	DIN	SS
0·10	3	410	410 S 21	56A	X 10 Cr 13	2302
0·20	4	420	420 S 37	56C	X 20 Cr 13	2303
0·30	5	—	420 S 45	56D	—	2304

Table 6.15 Hardness of 13% Cr steels on quenching and tempering

ISO	% C	Hardness HB after quenching and tempering at °C					
		200	500	550	600	650	700
3	0·10	350	350	325	250	225	200
4	0·20	450	450	405	320	275	250
5	0·33	510	515	445	345	300	280

Figure 6.81 Quenched and tempered machine parts. Fillet
surfaces containing changes of cross-section have been
burnished

The values hold for 50 mm diameter bar stock. A complete tempering
diagram for AISI 420 is shown in *Figure 6.82*. This steel is generally
stocked quenched and tempered to a tensile strength of 690–830 N/mm^2
(70–85 kp/mm^2) but it may also be supplied heat treated to a tensile
strength of 880–1080 N/mm^2 (90–110 kp/mm^2) in dimensions up to
200 mm. One of the uses of this steel is for structural members in the
chemical and cellulose industries; another for components in steam and gas
turbines. It is also used for propeller shafts in ships sailing in fresh water.

Grade AISI 420 and steels containing higher carbon contents are used
when high hardness is required. A popular field of application for such
steels is stainless steel cutlery.

The current 13% Cr steels are not readily weldable. During the 1960s,
steels were developed that have the same mechanical properties and
corrosion resistance as these steels and that, in addition, can be welded
cold and need no subsequent heat treatment. The chemical composition
and mechanical properties of two such steels are given in *Table 6.16*. The
figures given in the table are obtained in specimens, air cooled from
1000 °C and tempered at 600 °C.

In the hardened, untempered condition the steel is martensitic. On
tempering there is a partial retransformation to austenite, *see Figure 6.83*,
and this is accompanied by an increase in impact strength. When the steels
are welded in this tempered condition they still retain about 25% austenite
after completion of the welding operation and it is for this reason they can
be welded cold without danger of cracking. This valuable property is of
great importance when water turbines are being repaired (*see Figure 6.84*).

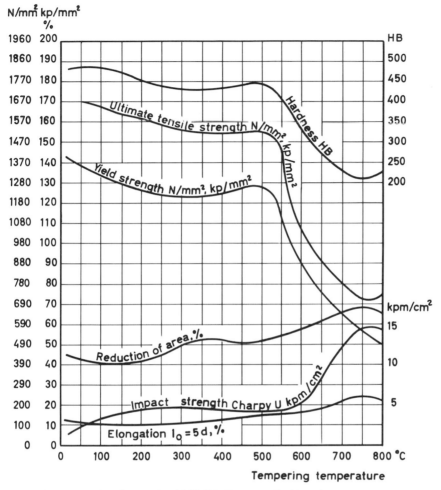

Figure 6.82 Tempering diagram for AISI 420. Curves applicable to bar stock 50 mm diameter, oil quenched or air cooled from 1000°C

Besides water turbines, feed screws in the cellulose industry and compressor components are made from this steel. For the last application the high impact strength of the steel at low temperatures is most useful. This type of steel has excellent hardenability and hence it is also suitable in

Table 6.16 Nominal composition and mechanical properties of weldable martensitic stainless steels

	Nominal composition %				*Mechanical properties*						
					$R_{p0.2}$ min		R_m min		A_5 min	Hard- ness	ISO-V min
Bofors	C	Cr	Ni	Mo	kp/ mm²	N/ mm²	kp/ mm²	N/ mm²	%	HB	J
2 RM2	0·05	13	6	—	60	590	80	780	15	250–300	60
2 RMO	0·05	13	6	1·5	62	610	85	830	15	250–300	60

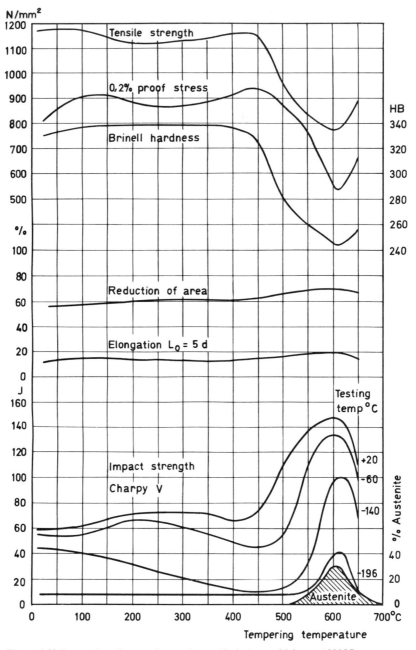

Figure 6.83 Tempering diagram for steel type 13–6. Austenitizing at 1000°C. Cooling in air

heavily dimensioned machine and structural members, such as pump axles in nuclear plant. The two steels, 2 RM 2 and 2 RMO are patented by the Bofors Company[16, 17].

6.2.5   Spring steels

The chemical composition of these steels varies from that of plain carbon steels to alloy steels, depending on the dimensions and fields of application. *Table 6.17* gives the chemical composition of the 14 spring steels that are standardized according to ISO/R 683/XIV–1973. *Table 6.18* gives the national designations of some of the commonest spring steels, the chemical compositions of which either wholly or most nearly match the ISO steels.

Table 6.17 Chemical composition of steels according to ISO/R 683/XIV–1973

Steel No.	% C	Si	Mn	Cr	Mo	V
1	0·72–0·85	0·15–0·40	0·50–0·80	—	—	—
2	0·72–0·85	0·15–0·40	0·50–0·80	—	—	—
3	0·43–0·50	1·50–2·00	0·50–0·80	—	—	—
4	0·47–0·55	1·50–2·00	0·50–0·80	—	—	—
5	0·52–0·60	1·50–2·00	0·60–0·90	—	—	—
6	0·57–0·64	1·70–2·20	0·70–1·00	—	—	—
7	0·57–0·64	1·70–2·20	0·70–1·00	0·25–0·40	—	—
8	0·52–0·59	0·15–0·40	0·70–1·00	0·60–0·90	—	—
9	0·56–0·64	0·15–0·40	0·70–1·00	0·60–0·90	—	—
10	0·56–0·64	0·15–0·40	0·70–1·00	0·60–0·90	—	—
11	0·42–0·50	1·30–1·70	0·50–0·80	0·50–0·75	0·15–0·30	—
12	0·56–0·64	0·15–0·40	0·70–1·00	0·70–0·90	0·25–0·35	—
13	0·48–0·55	0·15–0·40	0·70–1·00	0·90–1·20	—	0·10–0·20
14	0·48–0·56	0·15–0·40	0·70–1·00	0·90–1·20	0·15–0·25	0·07–0·12

N.B. For steel No. 1, max P % and max S % are 0·050% each.
    For steel No. 2 and Nos 8–14, max P and S are 0·035% each.
    For steels Nos. 3–7, max P and S are 0·040% each.
    For steel No. 10, min. B is 0·0005%.

Table 6.18 Spring steel designations

ISO	AISO	BS	En	DIN	Euronorm	SS
2	1074	070A78	42	C 75	—	1770
3	—	—	—	—	45 Si 7	—
4	—	—	—	51 Si 7	50 Si 7	—
5	9255	250A53	45	55 Si 7	55 Si 7	2090
6	9260	250A61	45A	—	60 Si 7	—
7	—	—	—	60 SiCr 7	60 SiCr 8	—
8	5155	—	—	55 Cr 3	—	—
9	5160	527A60	48	—	—	—
13	6150	735A50	47	50 CrV 4	50 CrV 4	2230

Some of the steels may show slight deviations from the chemical compositions of the ISO steels.

Figure 6.84 Francis and diagonal turbine runner of cast steel Bofors 2 RM 2

Plain carbon-steel springs are made from steel containing between 0·60% and 0·90% C. On account of their low hardenability carbon steels are used for light springs only, i.e. springs made from material up to 4 mm in thickness are oil quenched and up to 12 mm, water quenched. Hardenability is dependent to a large extent on the Mn-content of the steel. Helical springs constitute a large field of application for plain carbon steels. The starting spring-wire material is first hardened by patenting and then cold drawn to the required strength. A tempering treatment at 200–250°C after the cold drawing will increase the elastic limit still more. In some cases a tempering treatment at about 400°C is carried out which enhances toughness and ductility.

Among the alloyed spring steels the Si-alloyed steels make up a large group. The flat sections range from 3 mm to 15 mm. The type of steel and its hardenability must be carefully chosen with regard to the section dimensions and the cooling medium used for hardening. Very often flat spring steel stock is ordered in the as-rolled condition and a maximum hardness is stipulated, usually about 300 HB. This sets very close requirements when choosing suitable heats of each steel type. Silicon spring-steel is used for, e.g. flat motor-car springs and harrow springs and also for helical springs for various applications. (*See Figure 6.85*.) An advanced spring-steel design for rail anchors, made from Si-steel, is shown in *Figure 6.86*. Cr–Mn alloyed spring steels, equivalent to AISI 5150–5160, are increasingly being used for the same purposes and in

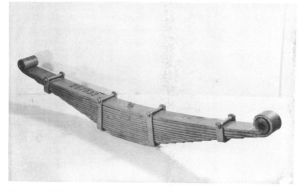

Figure 6.85 Leaf spring set for lorries

the same dimensions as Si spring-steel. If the as-rolled steel sections are to be cold sheared the contents of C, Mn and Cr must lie near the lower compositional limits as stipulated for the dimensions concerned. Cr–Mn steels, which are always quenched in oil, are used in thicknesses up to about 20 mm. For thicknesses up to about 30 mm Cr–Mn–V steels of type 50 CrV 4 are also resorted to. This steel is also used for valve springs for which a high fatigue strength is stipulated. The so-called parabolic springs make up another field of application for this steel.

Table 6.19 gives the maximum diameter or section thickness recommended for the spring steels listed in the ISO standard ISO/R 683/XIV.

Figure 6.86 Rail anchors made from spring steel, Lesjöfors AB

Table 6.19 Recommended maximum diameter or section thickness for the steels listed in ISO/R 68/XIV. The table is supplemented with hardening temperatures and quenching media

Steel No.	Thickness max mm	Diameter max mm	Quenching medium	Hardening temperature °C
1	8	12	Water	820–850
2	8	12	Water	820–850
3	14	20	Water	850–880
4	16	24	Water	845–875
5	8	12	Oil	840–870
6	14	20	Oil	830–860
7	16	24	Oil	830–860
8	18	28	Oil	830–860
9	22	33	Oil	830–860
10	24	35	Oil	830–860
11	30	45	Oil	830–860
12	47	70	Oil	830–860
13	27	40	Oil	850–880
14	40	60	Oil	850–880

The hardness of a spring must be matched to its dimensions. In principle, the smaller the dimensions of the spring the higher is the hardness. Watch springs of only a few tenths of a millimetre in thickness are tempered at 160–300°C after being quenched. Leaf springs of thicknesses of 1–3 mm are tempered at 300–400°C. Tempering in the 300°C range is not considered to be harmful to the serviceability of springs, provided they are not subjected to sharp blows; on the contrary, tempering in this temperature range sometimes produces beneficial results since the yield point often shows a maximum when the steel is tempered at about 300°C. Springs of heavier sections are usually tempered at 420–500°C which results in hardnesses of 400–460 HB. In addition this treatment gives the steel optimum mechanical properties.

The diagrams in *Figure 6.87* should be studied when suitable tempering temperatures for steels 55 Si 7 and 50 CrV 4 are being discussed. Springs working at temperatures up to 250°C should be made from grade H13. For springs to be used at elevated temperatures (up to 550°C) grade A286 should be resorted to.

6.2.6 Ultra high-strength steels

The designation ultra high-strength steel applies to steels having a yield point of at least 1380 N/mm² (200 ksi), according to *Metals Handbook Vol. 4, Heat Treating*. Also included in this group are steels of type H 13 and 34 CrNiMo 6. As may be seen from *Figures 6.79* and *6.80*, which apply to these steels, the stipulated yield point may be attained by tempering the steels at about 200°C. The present section will, however, treat only the two steels having the chemical compositions and typical mechanical properties as given in *Tables 6.20* and *6.21*.

The heat treatment and properties of grade H 13 have been discussed in detail in Section 6.1.5. In its role as a structural steel it covers a wide spectrum of uses both with regard to fields of application and component

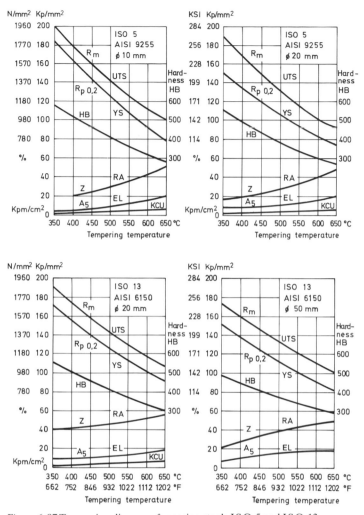

Figure 6.87 Tempering diagrams for spring steels ISO 5 and ISO 13. Hardening temperature 850–870°C. Quenching in oil

size. For example, *Figure 6.88* shows a pawl, 55 mm in length, which was heat treated to 50–52 HRC. The tip of the pawl is subjected to both wear and impact. At the other extreme, *Figure 6.89* shows a component made from the same steel. It forms part of the largest press (80 000 tonf) in the world. This member has the same hardness as the pawl.

Table 6.20 Nominal composition of high-strength steels

| Type of steel | Nominal composition % | | | | | | | |
	C	Si	Cr	Ni	Mo	V	Co	Ti
H 13	0·40	1·0	5·3	—	1·4	1·0	—	—
Maraging steel, grade 250	0·02	—	—	18	5·0	—	7·5	0·4

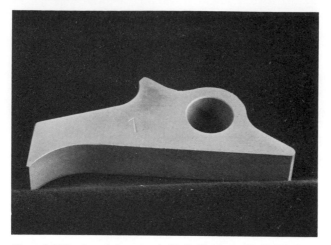

Figure 6.88 Pawl made from grade H 13. Hardness 50–52 HRC, weight 47 g

There is also a steel designated grade H 11 modified, which is a variant of grade H 13 with a somewhat modified chemical composition. This grade is used largely in the aircraft industry which in addition stipulates high cleanness requirements for this steel. *Figure 6.90* shows a structural member for aircraft landing gear.

Figure 6.89 Blank for pressure stamp of steel H 13 and delivered to Messrs A B Carbox, Ystad, Sweden. Component is part of the most powerful press in the world (80 000 tonf). Weight of component 1553 kg; largest diameter 1172 mm; hardness 48–52 HRC

Table 6.21 Mechanical properties for high-strength steels after hardening and tempering

Type of steel	$R_{p0.2}$		R_m		A_5	Z	Hardness	Tempering
	kp/ mm²	N/ mm²	kp/ mm²	N/ mm²	%	%	HB	temperature °C
H 13	155	1520	180	1770	10	45	520	570
Maraging steel, grade 250	170	1670	180	1770	10	50	470	480

The second steel in the table above belongs to what are known as maraging steels. The heat treatment consists of heating the steel to 820°C from which temperature it forms martensite on cooling. However, this martensite is soft and thus the heat treatment is often referred to as annealing. On subseqently ageing the steel at 480°C for 3 h, very high mechanical properties are obtained. There are four types of maraging steel, viz. grade 200, 250, 300 and 350; the number indicating the ultimate tensile strengths in ksi (kilopound, square inch). The tensile strength depends on the Ti content which varies between 0·1% and 1·5%.

The maraging steels are less susceptible to hydrogen embrittlement than are steels of type H 13. They were developed in the USA where they are used for structural members in aircraft and spacecraft designs, principally owing to their high impact properties at low temperatures.

Fatigue tests with unnotched specimens give a somewhat higher endurance limit for grade H 13 than for the 250-grade maraging steel. Tests with notched specimens having a stress concentration factor of 2·8 show that the endurance limit is dependent on the required service life. These observations are set out in diagram form in *Figure 6.91* which holds for vacuum-remelted steel.

Figure 6.90 Aircraft landing gear member made from modified grade H 11 vacuum remelted. Weight 40 kg

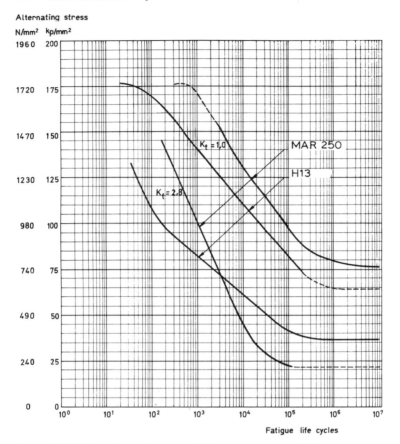

Alternating stress

Figure 6.91 Wöhler diagram for grade H 13 and maraging steel 250 vacuum remelted. Ultimate tensile strength 180 kp/mm² (1770 N/mm²). Tests carried out with unnotched specimen ($K_t = 1 \cdot 0$) and with notched specimen ($K_t = 2 \cdot 8$)

6.2.7   Hadfield steel

This rather special steel is named after its inventor, Sir Robert Hadfield. In its quench-annealed condition the steel is completely austenitic. Its hardness is about 220 HB and its composition $1 \cdot 25\%$ C, 13% Mn and $0 \cdot 5\%$ Cr. Owing to its high Mn-content the steel is often merely referred to as a manganese steel. On being subjected to cold work the austenite is transformed to martensite and the hardness rises to about 600 HB. According to Schumann[18] the transformation mechanism is as follows:

Austenite → disclocations → stacking faults → ε-martensite → α-martensite.

In the mining industry Hadfield steel is used for crusher plates and jaw crushers. In such service the degree of cold work or deformation is sufficiently high to promote the transformation of austenite to martensite. Another use is for prison gratings, (*see Figure 6.92*). The degree of cold

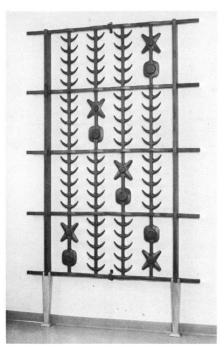

Figure 6.92. Cast prison grating, made from Hadfield steel

work resulting when an attempt is made to saw through the bars is sufficient to harden them so that they cannot be sawn through.

The Hadfield steel has been dealt with earlier in this book, *see* Sections 5.1.5 and 5.5.

6.3 Hardening and tempering of boron-alloyed steels

6.3.1   Boron steels in general

As a result of the never-ending quest for steels which are lower in cost than those of conventional types, the boron-alloyed steels, already known since the beginning of the 1920s, were reintroduced during the 1970s after a trial period of varying success. Doubts still prevail in some quarters regarding whether the use of boron steels may be justified, but nowadays the known advantages of these steels are such that it may be stated without serious doubt that this time the boron steels are here to stay. However, these steels also have their limitations and it is just as important to be aware of them as it is to recognize the very great usefulness of boron steels. These points are discussed in the present section.

6.3.2   Chemical composition of boron steels

Boron steels have been standardized by ISO, AISI, AFNOR and DIN, and there are some hundred types in existence. The majority of boron

Table 6.22 Chemical composition of boron steels (in ascending order of carbon content)

Type of steel	C %	Si %	Mn %	Cr %	Ni %	D_i (in)[1] with B	D_i (in)[1] without B
Mn steel							
A 1	0·15–0·20	0·10–0·30	0·70–0·90	—	—	0·5–1·0	1·1–2·1
2	0·15–0·20	0·10–0·30	0·90–1·10	—	—	0·6–1·2	1·3–2·5
3	0·18–0·23	0·10–0·30	0·70–0·90	—	—	0·5–1·1	1·1–2·2
4	0·18–0·23	0·10–0·30	0·90–1·10	—	—	0·6–1·3	1·3–2·6
5	0·18–0·23	0·10–0·30	1·10–1·30	—	—	0·7–1·5	1·6–3·0
6	0·20–0·25	0·10–0·30	0·70–0·90	—	—	0·5–1·1	1·1–2·2
7	0·20–0·25	0·10–0·30	0·90–1·10	—	—	0·7–1·3	1·4–2·6
8	0·20–0·25	0·10–0·30	1·10–1·30	—	—	0·8–1·5	1·6–3·1
Ni–Mn Steel							
B 1	0·18–0·23	0·15–0·35	0·70–0·90	—	0·40–0·60	0·6–1·3	1·3–2·7
2	0·18–0·23	0·15–0·35	0·90–1·10	—	0·40–0·60	0·7–1·5	1·6–3·0
3	0·20–0·25	0·15–0·35	0·90–1·10	—	0·40–0·60	0·8–1·6	1·6–3·1
4	0·25–0·30	0·15–0·35	0·90–1·10	—	0·40–0·60	0·9–1·7	1·7–3·2
Cr–Mn steel							
C 1	0·18–0·23	0·15–0·35	1·10–1·30	0·10–0·20	—	0·9–1·8	2·0–3·7
2	0·18–0·23	0·15–0·35	1·10–1·30	0·40–0·60	—	1·4–2·9	3·0–5·8
3	0·20–0·25	0·15–0·35	1·10–1·30	0·10–0·20	—	1·0–1·9	2·0–3·7
4	0·20–0·25	0·15–0·35	1·10–1·30	0·40–0·60	—	1·4–3·0	3·0–5·9
5	0·25–0·30	0·15–0·35	1·10–1·30	0·10–0·20	—	1·1–2·0	2·1–3·8
6	0·25–0·30	0·15–0·35	1·10–1·30	0·40–0·60	—	1·6–3·3	3·2–6·2
7	0·30–0·37	0·15–0·35	1·10–1·30	0·10–0·20	—	1·2–2·3	2·2–4·0
8	0·30–0·37	0·15–0·35	1·10–1·30	0·40–0·60	—	1·8–3·5	3·3–6·2
9	0·38–0·45	0·15–0·35	1·10–1·30	0·10–0·20	—	1·3–2·5	2·3–4·0
10	0·38–0·45	0·15–0·35	1·10–1·30	0·40–0·60	—	2·0–4·0	3·3–6·4

For all steels: P and S max 0·035% each
For other alloying elements, generally:
$B_{soluble}$ 0·0010–0·0030%
B_{total} max 0·0050%
Al   0·020–0·050%
Ti   0·020–0·050%. Ti shall be about 5 × N
1. *See* Section 4.2.1 and 6.3.4

steels are alloyed with Mn (around 1%) and Cr (up to about 0·5%) and in some cases with Ni and Mo. Carbon ranges from 0·15% to 0·45%. The steels are mainly used in the hardened and tempered condition, but in some instances also case-hardened. Common to boron steels is that the content of acid-soluble boron is 0·0010–0·0030% and the total boron context, which also includes insoluble boron compounds, should not exceed 0·0060%. Boron steels are deoxidized mainly by means of Al giving a residual content of about 0·030%. They are also denitrided, usually with about 0·030% Ti.

Up till now there has been no standardization of boron steels in Sweden but MNC (the Swedish Committee of Standardization, Metals Division) has prepared a publication—MNC 1204—which lists the common boron steels currently available to Scandinavia and the rest of Europe. These steels are listed in *Table 6.22*.

Table 6.23 gives the standardized Continental boron steels that most closely correspond to some of the boron steels shown in *Table 6.22*. Some of the standardized boron steels have $B_{min} = 0.0005\%$ other have $B_{min} = 0.0008\%$; by which is meant the content of acid-soluble boron. P and S are generally 0·035% max each.

Table 6.23 MNC-designated boron steels most closely equivalent to European and US standards

MNC	Steel standard	Chemical composition (B, Ti Al not stated)			
		C %	Si %	Mn %	Cr %
A3	ISO 4954/E1	0·17–0·23	0·15–0·35	0·50–0·80	—
	DIN 1654 part 4/22B2	0·19–0·25	0·15–0·40	0·50–0·80	—
	NF A 35–551/21B3*	0·18–0·24	0·10–0·40	0·60–0·90	—
A4	ISO 4954/E2	0·17–0·23	0·15–0·35	0·80–1·10	—
	NF A 35–566/20MB4	0·17–0·23	0·10–0·35	0·90–1·20	—
	SAE (AISI)/15B21	0·17–0·24	0·15–0·30	0·70–1·20	—
	BS 3111 part 1/Type 9	0·17–0·23	0·15–0·35	0·80–1·10	—
A5	ISO 4954/E3	0·17–0·23	0·15–0·35	1·10–0·40	—
	NF A 35–551/20MB5*	0·16–0·22	0·10–0·40	1·10–1·40	—
C2	SAE/EX19	0·18–0·23	0·20–0·35	0·90–1·20	0·40–0·60
C7	ISO 4954/E7	0·32–0·39	0·15–0·35	1·10–1·40	—
	ISO 4954/E10	0·34–0·41	0·15–0·35	0·50–0·80	0·20–0·40
	NF 35–551/38CB1*	0·34–0·40	0·10–0·40	0·60–0·90	0·20–0·40
	NF 35–551/38MB5*	0·34–0·40	0·10–0·40	1·10–1·40	—
C8	ISO 4954/E10	0·34–0·41	0·15–0·35	0·50–0·80	0·20–0·40
C10	SAE (AISI)/50B40	0·38–0·43	0·15–0·30	0·75–1·00	0·40–0·60

P and S generally max 0·035% each
*Also included under 35–557

ISO:	International Organization for Standardization
DIN:	Deutsches Institut für Normung
NF:	Normes Françaises
SAE:	Society of Automotive Engineers (USA)
AISI:	American Iron and Steel Institute
BS:	British Standard (British Standards Institution)

6.3.3   Hardening mechanism of boron steels

The considerable interest shown in boron steels is based on the fact that the quite minute amount of boron, about 0·0020%, may increase twofold the effect on hardenability of the other alloying elements and thereby contribute by an amount that is by no means negligible to the reduction in costs of producing steel to a specified hardenability. Boron is present in steel both as acid-soluble and non-soluble forms. Most of the predominant boron compound, the so-called effective compound $Fe_{23}(BC)_6$, is soluble in sulphuric acid, for example. It also goes into solution in the steel at the usual austenitizing temperatures. This implies that, like carbon atoms, free boron atoms are able to diffuse throughout the steel and a large number of boron atoms diffuse towards the grain boundaries.

There are several theories for the hardening mechanism of boron steels. Morral and Cameron[19] have reviewed four of the most credible mechanisms. All theories presuppose that the hardenability depends directly or indirectly on the fact that boron is concentrated at the grain boundaries of the steel.

When a boron steel is cooled from the hardening temperature the solubility of boron is reduced, which results in a still greater concentration of boron at the grain boundaries. Minute grains of boron carbide $Fe_{23}(BC)_6$ are formed there and to some extent they assume an orientation coherent with one of the two austenite grains between them which separate out. Atomic contact is thereby established between $Fe_{23}(BC)_6$ and austenite, resulting in a reduction in the surface tension and grain-boundary energy. The presence of boron in solid solution and coherent boron carbide in the grain boundaries delays the formation of ferrite and pearlite and also to some extent, bainite; hence increasing the hardenability of the steel.

6.3.4   Evaluation of the effect of boron on hardenability

The hardenability of boron steels is most readily determined by the Jominy end-quench test. An austenitizing temperature of 900°C is recommended for steels containing between 0·15% and 0·30% C. *Figure 6.93* shows the difference in hardenability as determined by such a test involving steels with and without boron. Further examples of Jominy end-quench hardenability curves for boron steels are reproduced in *Figures 6.105, 6.106* and *6.107*.

In the case of such boron steels as are currently used, the increase in hardenability caused by boron is most obvious at Jominy distances between 5 mm and 20 mm. The hardness measured in this Jominy range is approximately that attained at the centre of oil-quenched bars of diameters ranging from 15–50 mm. However, the hardness of the quenched face of the test specimen depends only on the carbon content and is not influenced by the boron. At large Jominy distances the influence of boron is relatively small. This implies that a slow cool, such as that resulting from air cooling from the rolling or forging operations, does not cause any appreciable hardness increase in boron steels when compared with a corresponding non-boron-containing steel. Light sections, however, usually cool so rapidly that they will experience some hardness increase.

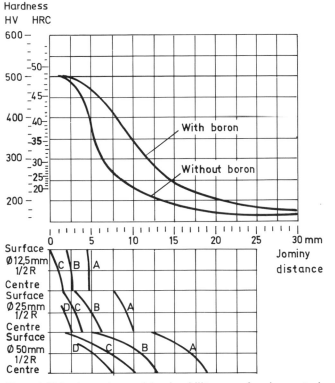

Figure 6.93 Jominy end-quench hardenability curves for a boron steel
and for one without boron.
Base composition: 0·27% C, 0·30% Si, 1·25% Mn, 0·20% Cr
Compare also Figure 4.40

The effect of boron may be expressed quantitatively as the Boron
Factor, which is the ratio of the Ideal Critical Diameters, D_i, (according to
Grossmann) for the steel with and without boron, viz.

$$B_F = \frac{D_i \text{ with boron}}{D_i \text{ without boron}}$$

D_i (with boron) is derived from the Jominy end-quench hardenability
curve; D_i (without boron) is calculated from the chemical composition of
the steel.

The Boron Factor decreases as the carbon content increases and is about
unity for C = 0·90% (i.e. around the euctectoid carbon content). There
are several empirical equations covering this relationship, one of which is:

$$B_F = 1 + 1·5 (0·90 - \% \text{ C}) \tag{6.1}$$

For carbon contents around 0·25%, B_F is about 2, which implies that boron
increases the hardenability effect by twofold of the other alloying
elements. When carbon contents are around 0·90%, which are usual for
case-hardened steels, boron does not have any enhancing effect on the

hardness. As a consequence case-hardening steels are seldom alloyed with boron. The Boron Factor also decreases with increasing alloy content, probably owing to the lower carbon content of the eutectoid composition. If the steel base has a D_i value greater than approximately 5 in it is quite meaningless to introduce boron as an alloying element. This also indicates that boron steels have a limited hardenability, which means that in practice they should be used only in dimensions up to about 100 mm diameter or equivalent sections.

6.3.5　Effect of titanium on the properties of boron steels

A large number of papers report that there is a maximum boron effect when the steel contains about 0·0010% acid-soluble boron. However, from investigations on some hundred boron-steel heats, made according to three different metallurgical processes, it has been found that the variations in the boron effect, as recorded in steels containing from 0·0005% to 0·0030% soluble boron, depend in the majority of cases on the ratio of the contents of titanium and nitrogen in the steel. Since Ti by means of another mechanicsm also affects the properties of boron steels, its influence on some of the properties will now be discussed.

Hardenability

Kapadia, Brown and Murphy[20] investigated how the effective boron content is dependent on Ti. From these observations the equation below has been developed to cover such Ti/N ratios as will admit of 'normal' Boron Factors[21].

$$\% \ Ti = 5 \ (\% \ N-0·003) \tag{6.2}$$

The constant, 0·003, applies to B_{sol} contents of approximately 0·0015% and Al_{tot} approximately 0·035%. In the case of higher contents of B or Al the constant may increase to about 0·005.

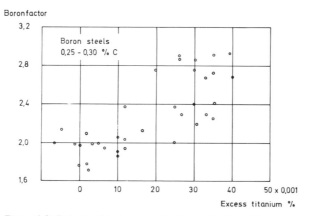

Figure 6.94 Relationship between the Boron Factor and excess titanium in boron steels containing 0·25–0·30% C. Austenitizing temperature 900 °C

For a nitrogen contentof 0·009%, Equation (6.2) gives a Ti/N ratio close to 3.42 which is the stoichiometric ratio for Ti/N. Higher Ti contents than those arrived at in the equation are designated 'excess titanium' in the discussion below. According to *Figure 6.94* the Boron Factor is about 2 for a titanium excess up to 0·020% but beyond this, B_F increases rapidly. A well-defined example of such a drastic effect is shown in *Figure 6.95*. Both steels have very similar base compositions and the D_i values without boron are about the same. The difference in the amounts of excess titanium give rise to large variations in the Boron Factor. These differences are readily observable direct from the Jominy end-quench hardenability curves. The evidence seems to point clearly to the fact that titanium has an inherent capacity for enhancing hardenability and this fact has been observed a number of times, e.g.[22]. It was even noticed when Ti was added direct to conventional heat-treatable steels[23].

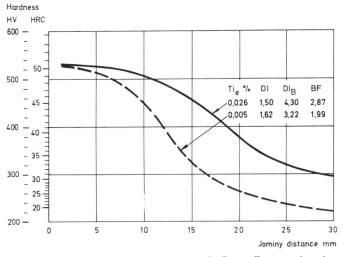

Figure 6.95 Influence of excess titanium on the Boron Factor and on the Jominy end-quench hardenability for steel of type MNC C5. Austenitizing temperature 900°C

Hardness in the untreated condition

One of the great advantages of boron steels is that the as-hot-rolled or as-forged hardness is about the same as that of a non-boron containing steel of the same composition. Also, the hardness is considerably lower than in conventional steels of the same hardenability. An investigation embracing some thirty heats having the approximate composition 0·27% C; 0·30% Si; 1·20% Mn and 0·50% Cr has shown very clearly the effect of excess titanium (*see Figure 6.96*). The hardness increase in the untreated condition is due to the solution-hardening effect of Ti and to the dispersed-phase hardening effect of titanium carbo-nitride.

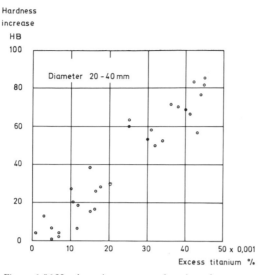

Figure 6.96 Hardness increase as a function of excess titanium in steel type MNC C6, hot rolled to round bars 20–40 mm diameter

Impact strength in the hardened and tempered condition

During the course of determining the Charpy V impact strength it was noted that there was a fall in the impact values when the excess titanium increased to about 0·020% or more.

Machinability

Nakasato and Takahashi[24] pointed out that high titanium contents in boron steels impair machinability. They carried out research work aimed at making titanium-free boron steels. The result with regard to hardenability is summarized in *Figure 6.97* and shows that very low nitrogen contents are a prerequisite to maintaining hardenability in titanium-free steels. The range of 'excellent hardenability' is in very good agreement with Equation (6.2).

6.3.6   Heat treatment of boron steels

In principle the same heat-treatment processes apply both to boron steels and to conventional steels of similar hardenability and carbon content. There are, however, certain differences and therefore the boron steels will now be discussed in greater detail.

Hardening temperature

Practical tests show that the temperature ranges given below produce optimum properties of the steels.

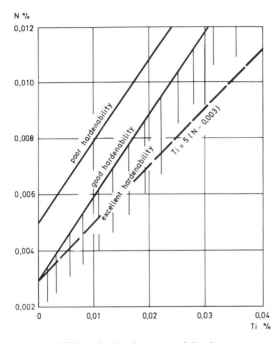

Figure 6.97 Variation in nitrogen and titanium contents for various degrees of boron hardenability for low C–Mn–B steels with 0·036–0·050% Al (after Nakasato and Takahashi[24]). A line representing Equation (6.2) is drawn in the diagram

% C	Hardening temperature °C
0·15–0·20	900–920
0·20–0·30	880–900
0·30–0·45	870–890

At several plants it was found that it is also feasible to harden boron steels direct from the forging temperature provided that this is not too high.

Figures 6.98 and *6.99* show Jominy end-quench hardenability test results from two heats of steel, type M N C C6. For both steels the hardenability was only slightly affected when the hardening temperature was raised from 900 °C to 1000 °C. By heating the steel at a higher temperature and increasing the time of dwell a marked reduction in hardenability was observed. At a Jominy distance of 20 mm the differences in excess titanium affect the hardenability very noticeably.

The steel showing the higher hardenability was studied with regard to ASTM grain size and mechanical properties including impact strength, after oil hardening from various temperatures and subsequent tempering at 500 °C. The impact strength was determined at four temperatures, three test-pieces for each temperature. Two tensile tests were carried out for each hardening temperature. The mean values are recorded in

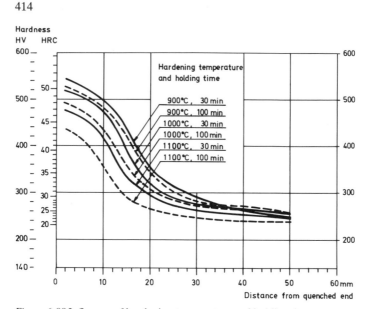

Figure 6.98 Influence of hardening temperature and holding time at temperature on the Jominy end-quench hardenability of a boron steel of composition:

%C	Si	Mn	P	S	Cr	B_{sol}	Ti	N	Al
0·27	0·21	1·16	0·017	0·035	0·51	0·0017	0·035	0·009	0·032

Ti excess 0·005

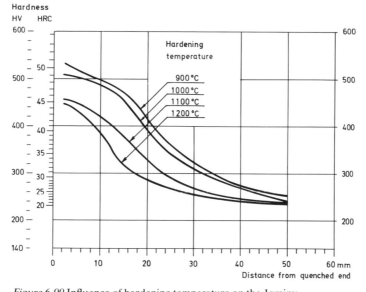

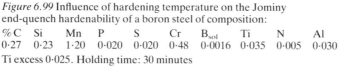

Figure 6.99 Influence of hardening temperature on the Jominy end-quench hardenability of a boron steel of composition:

%C	Si	Mn	P	S	Cr	B_{sol}	Ti	N	Al
0·27	0·23	1·20	0·020	0·020	0·48	0·0016	0·035	0·005	0·030

Ti excess 0·025. Holding time: 30 minutes

Figures 6.100 and *6.101* and from these it is seen that acceptable values of grain size, impact strength and tensile properties result on hardening the steel from 1000°C. Even, after hardening from 1100°C the results are surprisingly favourable.

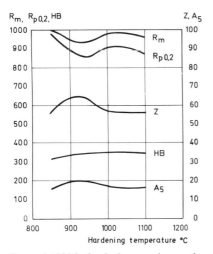

Figure 6.100 Mechanical properties as a function of hardening temperatue of steel, type MNC C6, of composition:

%C	Si	Mn	P	S	Cr	B$_{sol}$	Ti	N	Al
0·27	0·23	1·20	0·020	0·020	0·48	0·0016	0·035	0·005	0·030

Oil quenched. Tempered at 500°C for 1 h. Bar 30 mm diameter

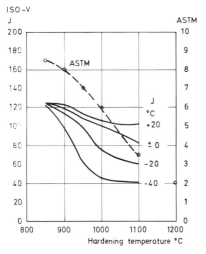

Figure 6.101 Grain size and impact strength as a function of hardening temperature of steel, type MNC C6, of composition:

%C	Si	Mn	P	S	Cr	B$_{sol}$	Ti	N	Al
0·27	0·23	1·20	0·020	0·020	0·48	0·0016	0·035	0·005	0·030

In order further to check the hardenability, test-pieces measuring 30 mm in diameter by 100 mm were cooled in fast-quenching oil from the above-listed temperatures. The hardness was determined at the centre of a cross-section through the middle of the test-piece. *Figure 6.102* shows that the through-hardening is fairly satisfactory even after quenching from from 1100°C. This, however, does not quite agree with the Jominy end-quench hardenability curves for this steel. When quenching from 1200°C the hardenability is markedly reduced.

The hardening temperatures recommended in the opening paragraph are valid if optimum properties are required. However, hardening temperatures up to 1000°C may be used without deleterious effect on the properties, but only if the steel has been satisfactorily fine-grain treated.

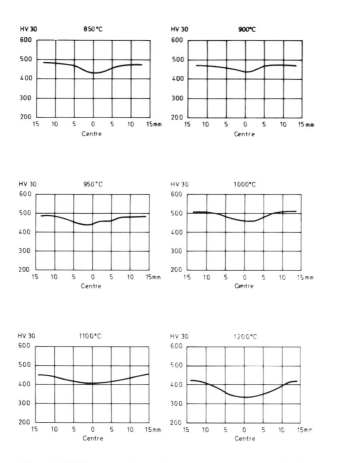

Figure 6.102 Diagrams showing hardness traverse curves in a boron steel after being hardened from various temperatures in fast-quenching oil. Test specimens 30 mm diameter × 100 mm. Chemical composition:

%C	Si	Mn	P	S	Cr	B_sol	Ti	N	Al
0·27	0·23	1·20	0·020	0·020	0·48	0·0016	0·035	0·005	0·030

Cooling media

Boron steels, like other steels, require that the cooling medium be adapted to the hardenability and dimension of the part being treated. Since boron steels in many cases are used for machine part where the requirements of dimensional and shape permanence are not particularly severe, e.g. certain components for agricultural machinery, it is more usual to quench these steels in water than other steels. Another reason for water quenching is that boron steels for the most part have a carbon content below 0·30%. It has been shown that this carbon level consitutes quite a sharp limit below which hardening cracks are rather infrequent even when water quenching is used. *Figure 6.103* shows the difference in depth of hardness for the boron steel C6 in various dimensions after quenching in water and in oil. The chemical composition of the steel in question lies near the upper limit of the range.

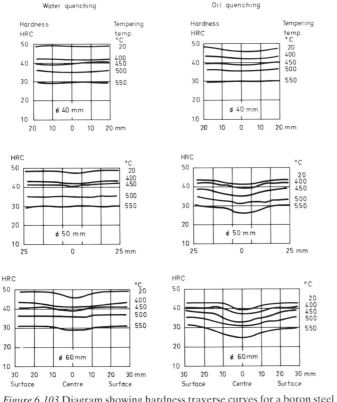

Figure 6.103 Diagram showing hardness traverse curves for a boron steel after quenching test specimens in water and in fast-quenching oil from 890°C. Chemical composition of steel:

%C	Si	Mn	P	S	Cr	Ni	Mo	B_{sol}	Ti	N
0·27	0·25	1·24	0·021	0·023	0·62	0·19	0·03	0·0013	0·039	0·010

Tempering

In the case of boron steels containing less than 0·25% carbon tempering is quite often omitted. Since the M_s at this carbon level is rather high some auto-tempering of the first formed martensite takes place, and this tempering is often adequate to render the steel sufficiently tough. To attain optimum values of mechanical properties, including impact strength, a tempering at about 200°C is recommended (*see Figure 6.104*). Like other heat-treatable steels, boron steels have an impact-strength minimum at about 300°C and hence tempering temperatures between 250 and 400°C should be avoided. If the steel is required to possess a reasonably high toughness and still retain good mechanical properties a tempering treatment at about 500°C is recommended. Boron as an alloying element in steel does not enhance in any way its resistance to tempering, and therefore its tempering temperature is 50–100°C lower than that for a Cr–Mo steel, for example, possessing the same hardenability.

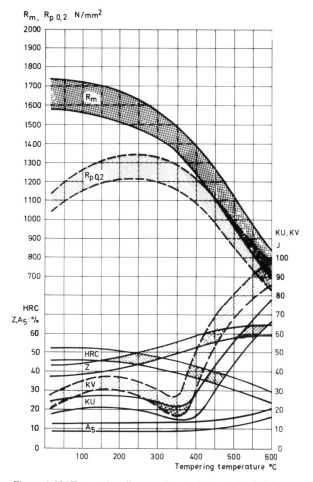

Figure 6.104 Tempering diagram for steel, type MNC C6. Hardening in oil. Test bar diameter 30 mm

6.3.7   Properties of boron steels and Cr–Mo steels—a comparison

The substitution of boron steels for certain Cr–Mo steels is an economically interesting topic for discussion, the significant points being the lower cost of the boron steels and their low hardness in their untreated condition. However, a prerequisite is that the properties of the two steels are similar in other respects. Below are discussed briefly a number of interesting properties.

Hardenability

Hardenability may most simply be judged by comparing the Jominy end-quench hardenability diagrams for the steels concerned. This is done in *Figures 6.105*, *6.106* and *6.107*. As is explained in *Figure 4.40*, it may be

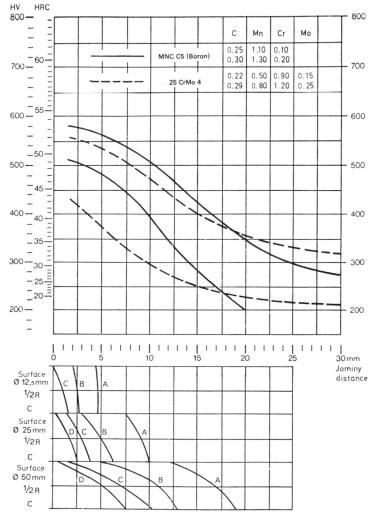

Figure 6.105 Jominy diagrams and hardening conversion diagrams for boron steel MNC C5 and steel 25 CrMo 4

deduced from the lower part of the diagrams, for example, that the three Cr–Mo steels and the corresponding boron steel attain acceptable hardness values in the dimensions given below, after quenching in strongly-agitated oil (curve B).

Diameter mm	CrMo grade	MNC designation
25	25 CrMo 4	C5
50	34 CrMo 4	C6
75	42 CrMo 4	C8

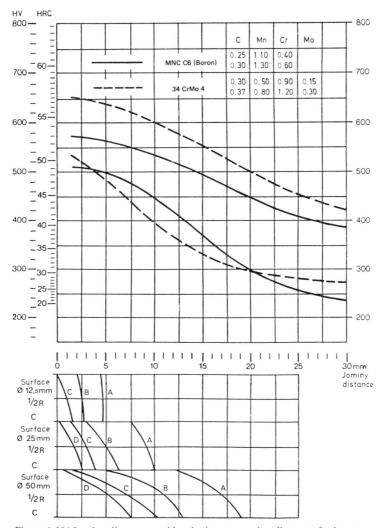

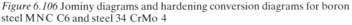

Figure 6.106 Jominy diagrams and hardening conversion diagrams for boron steel MNC C6 and steel 34 CrMo 4

Figure 6.108 shows some CCT diagrams of the new type that has been previously described, and in *Figure 6.109* a diagram for 25 CrMo 4 is compared with one for an equivalent boron steel. The chemical composition of the latter lies in the middle of the composition range while that of the Cr–Mo steel lies near the upper limit of the range. Both steels attain the same hardness in the bars of 25 mm diameter, and in the 50 mm diameter bars the hardnesses are almost the same. It is also interesting to compare the hardness values of a 200 mm diameter section which corresponds approximately to a 40 mm diameter section cooled in air (*See Figure 4.52*). For 25 CrMo 4 the hardness is 285 HB; for the boron steel, 215 HB. This point will be discussed in more detail in the next section.

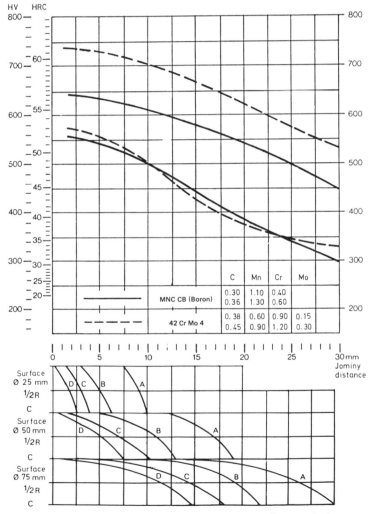

	C	Mn	Cr	Mo
MNC CB (Boron)	0.30	1.10	0.40	
	0.36	1.30	0.60	
42 Cr Mo 4	0.38	0.60	0.90	0.15
	0.45	0.90	1.20	0.30

Figure 6.107 Jominy diagrams and hardening conversion diagrams for boron steel MNC C8 and steel 42 CrMo 4

Hardness in the as-rolled condition

Both from the Jominy end-quench hardenability curves and the CCT diagrams it can be deduced that in heavy sections the boron steels have inferior hardenability compared with Cr–Mo steels, i.e. the boron steels are softer after slow cooling. This fact is of great practical importance when, after cooling from the hot rolling, boron steels are appreciably softer than Cr–Mo steels of similar hardenability. Actually the hardness difference is greater than what can be deduced from the Jominy curves, since during rolling or forging the steels are heated to considerably higher temperatures than those used for hardening. On being heated to high temperatures Cr–Mo steels tend to increase in hardness more than boron steels, which is due to the complete dissolution of Cr and Mo. *Table 6.24*

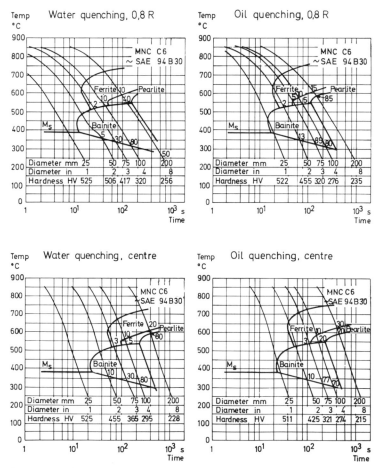

Figure 6.108 CCT diagram for steel, type MNC C6. Specimens water quenched and oil quenched from 900°C. Test area located at 0·8R and at centre of cross-section. Chemical composition:

%C	Si	Mn	Cr	B_{sol}	Ti	N
0·27	0·31	1·17	0·48	0·0013	0·035	0·007

shows the likely hardness values of the steels under discussion, after hot rolling, in the dimensional range 20–50 mm diameter.

Table 6.24 Hardness of boron and Cr–Mo steels, hot rolled

Type of steel	Hardness H B
MNC C5	170–200
25 CrMo 4	250–300
MNC C6	180–220
34 CrMo 4	280–330
MNC C8	200–240
42 CrMo 4	300–400

These hardness results tend to imply that boron steels may, in fact, be supplied in the as-rolled condition. Even after forging, the hardness is the same as in the as-rolled condition.

Most of the Cr–Mo steels must be delivered annealed in order to enable them to be sheared or machined. After forging, machining usually follows, but before this operation the Cr–Mo steels must be annealed once again. This involves increased costs and longer delivery times.

Mechanical properites of the hardened and tempered steels

In *Table 6.25*, which is set out in principle in the style of the Swedish Standards, similar boron steels have been grouped along with equivalent

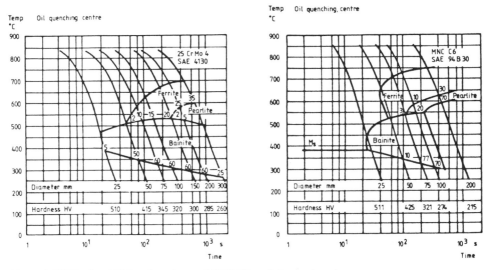

Figure 6.109 Hardenabilities of steel, type MNC C6, and standard grade 25 CrMo 4—a comparison.
Compositions:

Type of steel	% C	Si	Mn	Cr	Ni	Mo	B_{sol}	Ti	N
25 CrMo 4	0·26	0·33	0·86	1·17	0·30	0·23	—	—	0·010
MNC C6	0·27	0·31	1·17	0·48	—	—	0·0013	0·035	0·007

Cr–Mo steels. The stipulations determining the choice of heats having a composition suited to given section sizes are more stringent for boron steels than for Cr–Mo steels; one reason for this being the desire to avoid the necessity of annealing the boron steels.

Boron steels are used also heat-treated to high hardness, i.e. quenched and low-temperature tempered (about 200 °C). Cr–Mo steels, too, may be used in a similarly hard condition and, to a greater extent than what is indicated in the table, they are in fact used in the quenched and tempered condition to an ultimate tensile strength (R_m) of 1000 and 1100 N/mm^2. After being quenched and tempered to the required tensile strength the listed boron steels also fulfil the other stipulations of the Swedish Standard.

Table 6.25 Mechanical properties of Cr–Mo and boron steels

Diameter mm	Ultimate tensile strength (R_m) min N/mm^2				
	700	800	900	1000	1100
⩽25	25 CrMo 4	25 CrMo 4	25 CrMo 4	34 CrMo 4	34 CrMo 4 42 CrMo 4
	C5	C5	C5	C5, C7	C7, C9
>25⩽40	25 CrMo 4	25 CrMo 4	25 CrMo 4 34 CrMo 4 42 CrMo 4		
	C5, C6	C5, C6	C6, C8	C6, C8	C6, C8
>40⩽100	25 CrMo 4 34 CrMo 4	25 CrMo 4 34 CrMo 4	42 CrMo 4		
	C6, C8	C6, C8	C8, C10	C8, C10	C8, C10

From the point of view of mechanical properties and section size the boron steel designated MNC C6 is most closely equivalent to 34 CrMo 4 but its carbon content corresponds to that of 25 CrMo 4. MNC C6 may also replace 42 CrMo 4; in fact this has already occurred. High-strength bolting steel in dimensions up to 50 mm diameter has replaced 42 CrMo 4 in practice and on a large scale. A case in point is shown in *Table 6.26*, which gives a typical test result on a 25 mm diameter fastener. The specimen was quenched in fast-cooling oil and tempered at 380 °C.

Table 6.26 Stipulated mechanical properties for fasteners according to 12·9 and test results for boron steel C6 with diameter 25 mm

		Values stipulated according to 12·9	Test result
$R_{p0·2}$	N/mm^2	min 1100	1290
R_m	N/mm^2	min 1220	1400
A_5	%	min 8	11
Z	%	——	51
KCU	J	min 15	20
Hardness HB		353–409	405

For 50 mm diameter fasteners with other stipulated properties the test results of steel C6 after being quenched in fast-quenching oil and tempered at 520°C are shown in *Table 6.27*. The test bars were cut out after heat treatment from the half-radius location on the cross-section. The steel had the following chemical composition:

C	Si	Mn	P	S	Cr	B_{sol}	Al	Ti	N
0·27	0·25	1·40	0·025	0·020	0·52	0·0015	0·030	0·025	0·005

Table 6.27 Stipulated properties of 50 mm diameter fastener and test results

		Stipulated values	*Test result*
$R_{p0·2}$	N/mm²	800–880	900
R_m	N/mm²	900–1040	990
A_5	%	—	17
Hardness	HB	266–342	312
KV at −20°C	J	22	32

Impact strength in the hardened and tempered condition

Previously the impact strength of heat-treated alloy steels was tested according to the Charpy K U (I S O-U test specimen) test. Since this method does not yield sufficient information about the behaviour of steels at cryogenic temperatures, a number of investigations have been carried out with K V-test bars (I S O-V test specimen). Comparisons between the C6 boron steel and 25 CrMo 4 showed that, provided the steels had the same hardness and similar P- and S-content, these steels had very similar impact-strength curves. *Table 6.28* shows a typical test result.

Table 6.28 Impact strength (K V), at various temperatures, of boron steel C6 and of the standard S S 2225. Bar diameter 25 mm, quenched in 5% Aquaquench solution and tempered to 325 H V

Type of steel	*Impact strength (Charpy K V J) at various testing temperatures*				
	+20°C	±0°C	−20°C	−40°C	−60°C
Boron steel C6	104	97	96	86	69
25 CrMo 4	103	91	72	47	29

Type of steel	% C	Si	*Chemical composition* Mn	P	S	Cr	Mo	Ti	B_{sol}
Boron steel C6	0·27	0·28	1·27	0·018	0·027	0·41	—	0·025	0·0013
25 CrMo 4	0·25	0·39	0·71	0·026	0·015	1·08	0·21	—	—

A comparison was also made between boron steels produced by the L D-process (designated L) and by electric-arc melting (designated E), also the standard grade 25 CrMo 4 electric-arc melting (designated S)[25]. The chemical composition of the steels is given below.

Type of steel	% C	Si	Mn	P	S	Cr	Ni	Mo	Ti	Al	B	N
Boron steel L	0·27	0·24	1·18	0·020	0·020	0·48	0·03	0·01	0·026	0·023	0·0015	0·005
Boron steel E	0·28	0·24	1·15	0·019	0·030	0·42	0·10	0·03	0·050	0·040	0·0025	0·009
Cr–Mo steel S	0·42	0·29	0·75	0·008	0·030	1·05	0·10	0·16	—	0·012	—	0·009

The steels were heated to $R_m = 950$ N/mm^2 (300 HV) and to $R_m = 1250$ N/mm^2 (400 HV)—all figures approximate—in 30 mm diameter rounds. The Charpy K V test specimens were taken from the half-radius location on the cross-section and tested at three different temperatures. The results given in *Figure 6.110* constitute the mean of three tests and may be regarded as the same for all three steels.

Fatigue strength in the heat-treated condition

The fatigue strength of a steel is just as important to its usefulness as its impact strength. Hence the steels discussed in the preceding section (steels L, E and S) were tested in fatigue with reversed tension and compression, zero mean stress. The test bars were 14 mm in diameter and furnished with a diamond-paste polished stress-raising notch, having a bottom radius of

1·0 mm. The shape factor $\alpha = \dfrac{\sigma \max}{\sigma \text{ nom}}$ was about 2·3

Table 6.29 shows the fatigue strength expressed as σ_D and σ_D / R_m for the two tensile-strength levels of the steels being investigated.

Table 6.29 Ultimate tensile strength and fatigue strength expressed as the ratio σ_D / R_m

Steel designation	UTS $(R_m) \approx 950$ N/mm^2			UTS $(R_m) \approx 1250$ N/mm^2		
	R_m N/mm^2	σ_D N/mm^2	$\dfrac{\sigma_D}{R_m}$	R_m N/mm^2	σ_D N/mm^2	$\dfrac{\sigma_D}{R_m}$
L	953	225	0·236	1339	295	0·220
E	994	230	0·231	1229	268	0·218
S	926	210	0·227	1271	253	0·199

Figures 6.111 and *6.112* show 'normalized' Wöhler curves for both strength levels of the steels being investigated. These curves show that the boron steel and the Cr–Mo steel are quite equal as regards fatigue strength.

Fracture toughness in the heat-treated condition

The determination of fracture toughness, using conventional methods, is rather costly. Hence, several research workers, Barsom and Rolfe[26], *inter alia*, have tried to find other and simpler ways of determining fracture toughness. Sandberg and Åkerman[27] and Sinha and Sandberg[28] investigated four Cr–Mo steels, one Cr–Ni–Mo steel and one B–steel. The test specimens were cut out from bars of cross-section 100 mm square and

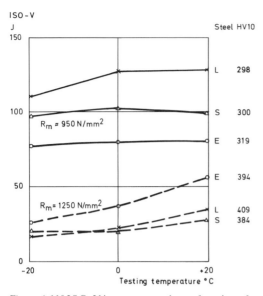

Figure 6.110 ISO–V impact strength as a function of
the testing temperature[25]

Steel L = Boron steel, MNC C6, LD process
Steel E = Boron steel, MNC C6, arc-furnace process
Steel S = Standard grade, 42 CrMo 4, arc-furnace
      process

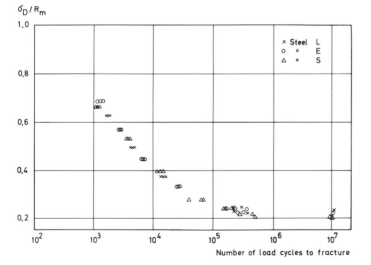

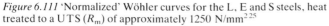

Figure 6.111 'Normalized' Wöhler curves for the L, E and S steels, heat
treated to a UTS (R_m) of approximately 1250 N/mm²[25]

428

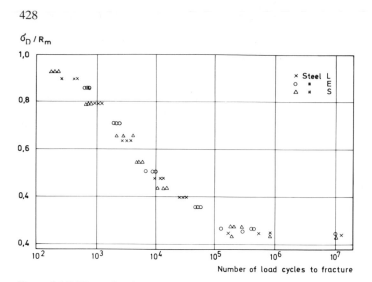

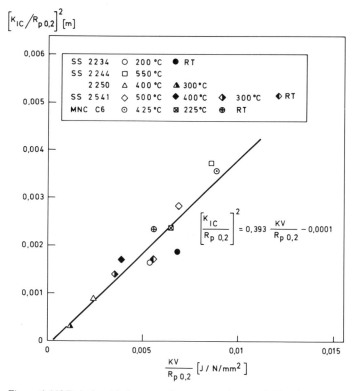

Figure 6.112 'Normalized' Wöhler curves for the L, E and S steels, heat treated to a UTS (R_m) of approximately 950 N/mm² [25]

$$\left[K_{IC} \middle/ R_{p\,0,2} \right]^2 = 0,393 \frac{KV}{R_{p\,0,2}} - 0,0001$$

Figure 6.113 Relationship between fracture toughness, yield point and upper-shelf impact strength of Cr–Mo steel, Cr–Ni–Mo steel and B-steel after hardening and tempering up to 550°C. Test specimens taken at right angles to direction of rolling (after Sandberg[27, 28])

at right angles to the direction of rolling. An empirical relationship was established between the upper-shelf values of the Charpy KV impact strength and the fracture toughness of heat-treated test specimens, i.e. specimens that were quenched and tempered at various temperatures. The relationship holds only in the transverse direction which yields the lowest impact values on account of the anistropic character of the steel. In order to 'normalize' the values the term $R_{p0\cdot 2}$ is included in the equation. *See Figure 6.113*, which shows the results as they vary with the tempering temperature up to 425 °C for the B–steel and up to 550 °C for the other steels. At higher temperatures the relationship is somewhat different, as shown in *Figure 6.114*.

According to this investigation all three steel types are governed by the same formulae and hence it should be possible to determine the fracture toughness of these and similar steels from the value of the yield point and the upper-shelf value of the impact strength.

Figure 6.115 shows by the more familiar approach the relationship between fracture toughness and ultimate tensile strength of the investigated steels.

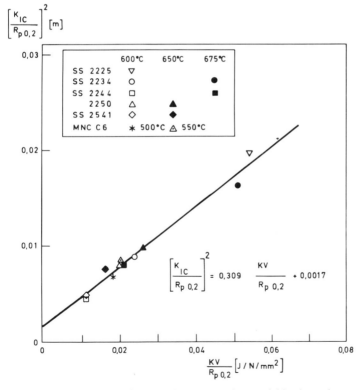

Figure 6.114 Relationship between fracture toughness, yield point and upper-shelf impact strength of Cr–Mo steel, Cr–Ni–Mo steel and B-steel after hardening and tempering at 500 °C and higher. Test specimens taken at right angles to direction of rolling (after Sandberg[27, 28])

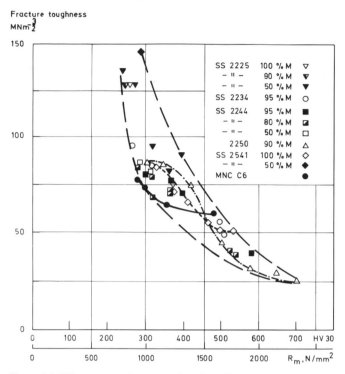

Figure 6.115 Fracture toughness as a function of hardness and UTS (after Sandberg[27, 28])

6.3.8   Fields of application for boron steels

Boron steels are used in the same fields of application as low-alloy heat-treatable steels. The hardenability of these steels sets the limit for the maximum useful dimension, which is about 90 mm diameter or equivalent flat sections, e.g. 160 × 60 mm. Examples of application are: bolts, lifting hooks, connecting rods and forks for fork-lift trucks. Boron steels are suitable for use under abrasive conditions where previously spring steels were used, e.g. as grader blades and plough constructions. *Figure 6.116* shows part of a plough in which the frame, the beam, the mould board extension, the ploughshare and the share point are all made from boron steel.

6.4 Heat treatment of high-strength low-alloy constructional (HSLA) steels

6.4.1   High-strength low-alloy steels in general

The micro-alloyed high-strength low-alloy steels—HSLA steels—were developed during the 1960s and 1970s. During the 1980s a substantial growth in production is predicted for these steels, which will therefore increasingly replace conventional-type constructional steels.

Figure 6.116 In a plough the share points and the shares are the parts which are most subjected to wear as well as impacts. For these parts the boron steels MNC C5 and C6 are often chosen

Owing to their higher yield point it is possible to effect a considerable weight reduction by changing over to HSLA steels. Their employment as a substitute for the usual heat-treatable steels will also increase since the heat treatment of HSLA steels can be carried out by less energy consuming and hence less costly methods compared with the conventional quenching and tempering. In fact, this treatment may be omitted altogether.

The term 'micro-alloyed' implies that small amounts of niobium, vanadium or titanium have been added to the steel. These additions contribute to an increase in strength, partly by grain refinement and partly by precipitation hardening due to carbides, nitrides or carbo-nitrides. Grain refinement which also contributes to enhanced impact strength is brought about by normalizing and also by controlled rolling, the latter process being a less costly one.

A consequence of the small alloying additions is an increase in the yield-point: tensile-strength ratio compared with a corresponding steel that does not have these additions. When HSLA steels are being graded it is usual to base the classification on their yield point. This property is increasingly being used in calculations involving the strength of materials.

HSLA-designed steels are capable of being grouped in several different ways and there is a very large number of steels that are incorporated under this designation.

In the three following sections there will be discussed:

Weldable HSLA steels
HSLA dual-phase steels
HSLA pearlitic steels

6.4.2   Weldable HSLA steels. Steel grades and quality classes

The upper limit for carbon in weldable constructional steels is usually set at 0·2%. Some standards tolerate somewhat higher carbon and others stipulate lower carbon contents. The majority of steels in this group are classed according to IIW (the International Institute of Welding) in various quality classes based on impact strength at the testing temperatures given below. The most usual value is Charpy KV (ISO–V) 27 J minimum. The Swedish Standard SS designations have a terminal digit which refers to the quality class.

Table 6.30. Quality classes for weldable constructional steels

Quality class	Terminal digit according to SS	27 J at testing temperature, °C
A	1	No stipulation
B	2	No stipulation
C	3	± 0
D	4	−20
E	5	−40

The maximum chemical composition of the SS steels shown in *Table 6.31* is:

% C	Si	Mn	P	S	Nb	V	N
0·20	0·5	1·6	0·035	0·035	0·050	0·15	0·02

The steels shall also fulfil certain values of the carbon equivalent (E_C), viz;

$$E_C = C + \frac{Mn}{6} + \frac{Ni + Cu}{15} + \frac{Cr + Mo + V}{5}$$

Table 6.31 Mechanical properties of weldable HSLA steels

SS steel grade	Impact strength min 27 J at °C	Thickness mm	Yield point R_{eL} N/mm^2 min	R_{eH} N/mm^2 min	UTS R_m N/mm^2 min	Elong. A_5 % min	E_C min
2132–01	–	–16	350	360			
2134–01	–20	(16)–35	340	350	470–630	20	0·41
2135–01	–40	(35)–50	330	340			
		(50)–70	320	330			
2142–01	–	–16	390	390			
2144–01	–20	(16)–35	380	380	490–650	20	0·45
2145–01	–40	(35)–50	370	370			
		(50)–70	360	360			

The condition of the finished material is as normalized, but plate and bar may, if nothing to the contrary has been stipulated, also be delivered in an equivalent condition obtained in the rolling process (controlled-rolled).

Table 6.32 gives the designations of approximately equivalent international and national standards.

Table 6.32 Designations of weldable, micro-alloyed steels

Sweden SS	ISO 4950/2	Euro- norm 113–72	Great Britain BS 4360: 1972	Germany S-E-W 089–70	USA ASTM[1]
21 32	E355CC	FeE355KG	50C		—
21 34	E355DD	FeE355KW	—	StE39[2]	—
21 35	E355E	FeE355KT	—		—
21 42	E390CC	FeE390KG	55C		A 572 Grade 60
21 44	E390DD	FeE390KW	—	StE43[2]	A 572 Grade 60[2]
21 45	E390E	FeE390KT	55E		A 572 Grade 60[2]

N.B.
1. Note that the maximum carbon content of the majority of these American steels varies from more than 0·20% to 0·35% and therefore they are not directly comparable with other steels specified according to European practice.
2. Requirements of impact properties shall be stipulated along with the order.

Besides the steels listed in *Table 6.31* there are steels complying with foreign standards with higher yield points. This usually implies that the carbon is in excess of 0·20%, which, however, diminishes weldability and impact strength.

The steels listed in *Table 6.33* are graded according to SAE J 710 and have a carbon content of about 0·25%.

Table 6.33 Mechanical properties of SAE steels, J 710

Steel grade SAE	Yield point N/mm^2 min	UTS N/mm^2 min	Elongation on 50 mm % min
950 X	345	448	22
955 X	379	483	20
960 X	414	517	18
965 X	448	552	16
970 X	483	586	14
980 X	552	665	12

Heat treatment

After conventional rolling, correctly alloyed steels obtain the yield point and ultimate tensile strength as stated in the above tables. On the other hand, it is not possible to guarantee as a matter of course a minimum impact strength of 27 J at testing temperatures below 20°C, but by introducing a normalizing treatment this is feasible. It is found, though, that after such a treatment the vanadium-alloyed steels, which on the whole possess the highest mechanical properties after hot rolling, undergo a reduction in the yield point by about 100 N/mm². For the niobium-alloyed steels the reduction is about 30 N/mm² but the strength of these steels after hot rolling is lower than that of the vandium-alloyed steels.

By means of the so-called controlled-rolling technique it is possible to combine high strength with high toughness. Controlled rolling implies that the finishing temperature of rolling is phased, so that after the final pass the temperature is too low for any recrystallization to take place or, if any does take place, there is no appreciable grain growth. The effect produced by micro-alloying is to raise the lower limiting temperature. *Figure 6.117* shows three hot-rolling and recrystallization sequences, after Le Bon and de Saint-Martin[28], in which austenite is first deformed by hot rolling then recrystallizes afterwards in various ways.

Sequence I. This sequence takes place during an ordinary rolling operation at which the finishing temperature is above 1000°C. The deformed austenite grains recrystallize very rapidly and grain growth may become quite considerable.

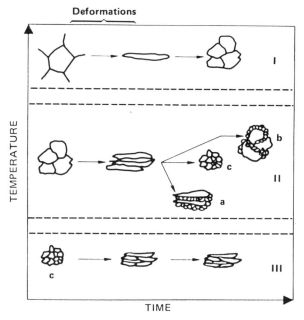

Figure 6.117 Modifications of the austenite grains during controlled rolling (after le Bon and de Saint-Martin[29])

Sequence II. The progress of the recrystallization may follow one of the alternative courses described below when the finishing temperature is around 1000 °C.

Alternative (a): The recrystallization is incomplete and takes place only at the grain boundaries.

Alternative (b): The recrystallization is complete but simultaneously a partial grain growth takes place.

Alternative (c): The recrystallization is complete and gives rise to a fine-grained structure. The subsequent transformation from austenite to ferrite results in the finest and most even grain size which imparts the best properties to the material.

Sequence III. At temperatures around 900 °C and lower no recrystallization takes place and the final structure consists of fine-grained polygonal ferrite with low dislocation density.

It is also possible to influence the mechanical properties by controlling the cooling rate after the finishing pass. *See Figure 6.118,* which also shows the reduction in strength as a result of normalizing.

The strength of steel can be increased by cold working, but at the expense of toughness and ductility (cf. *Figure 1.46*). The strength is further

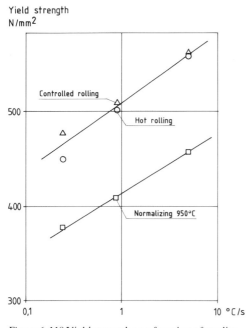

Figure 6.118 Yield strength as a function of cooling rate after controlled rolling, hot rolling and normalizing.
Chemical composition: 0·12% C, 0·35% Si,
1·35% Mn, 0·09% V,
0·013% N
(after Roberts[30])

increased by heating the steel to about 200°C, but at the same time toughness and ductility are reduced. This process has been discussed in Section 5.7—Strain ageing. On heating a cold-worked HSLA steel to a temperature higher than the ageing temperature a substantial increase in strength is obtained compared with the as-cold-worked condition. Also, toughness and ductility attain acceptable values. This is illustrated in *Figure 6.119* which is applicable to an HSLA steel, reduced 15% by cold drawing before tempering.

6.4.3   Dual-phase steels

General: dual-phase steels and their heat treatment

The iron-carbon diagram and similar equilibrium diagrams are mainly used to obtain information on what austenitizing temperatures are required for hardening or normalizing. Such information is obtained by consulting *Figure 5.1* and it is seen that the austenitizing temperature should be about 50°C above A_3. But if a steel containing, for example, 0·20% C is heat treated at a temperature 50°C below A_3, i.e. at 800°C, the steel, after equilibrium has set in, will consist of 50% ferrite and 50% austenite containing 0·40% carbon (*see Figure 6.120*). On quenching in water the steel will contain 50% ferrite and about 50% martensite. A typical microstructure after such a treatment is reproduced in *Figure 6.121*.

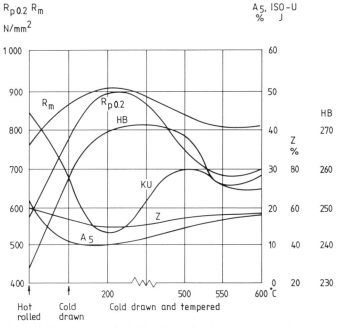

Figure 6.119 Influence of cold drawing and subsequent tempering at 200–600°C for ½ h on the mechanical properties of a hot-rolled steel having a yield strength of 575 N/mm². Bar diameter 30 mm: cold drawn reduction of 15%.
Chemical composition:

%C	Si	Mn	P	S	Cr	V
0·21	0·40	1·36	0·035	0·024	0·15	0·09

Steel that has been heat treated by this method or by some similar one so that it contains two phases, is called a Dual Phase or DP steel. The ferritic–martensitic structure may also be obtained if the steel, after the finishing pass, is allowed to cool to a temperature in the two-phase region and then quenched from that temperature.

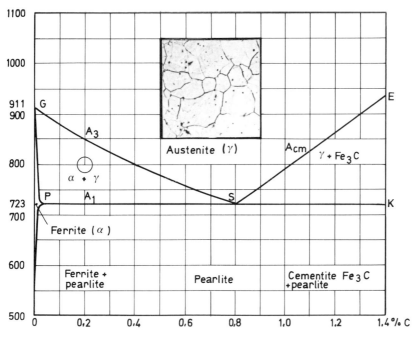

Figure 6.120 At equilibrium at 800 °C a 0·20% C steel contains 50% ferrite (α) and 50% austenite (γ)

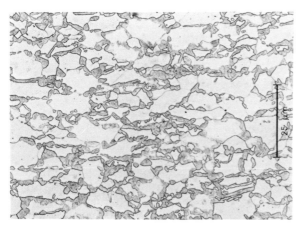

Figure 6.121 Micrograph of dual-phase steel with ferrite and about 30% martensite. Magnification 1000 ×

Rashid[31], who was the first to test the properties of DP steel, found that the stress-strain curve had a different shape and that the DP steel had a considerably larger elongation than conventional hot-rolled steel of approximately the same composition. *Figure 6.122* contains Rashid's diagrams for three ferritic-pearlitic steels and a DP steel designated GM 980 X. The ferritic-pearlitic steels all have a pronounced yield point and show stretcher strains, the so-called Lüder's Lines. As is well known, that the elongation decreases as the UTS increases. The DP steel has a relatively low yield strength (as determined by the 0·2% offset method) but the elongation is 50% greater than that of the ferritic-pearlitic steel with the same UTS.

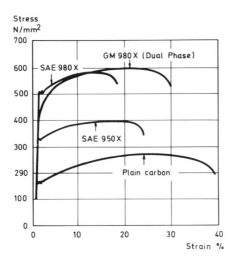

Figure 6.122 Stress-strain diagram of a dual-phase steel and three other cold-pressing steels (after Rashid[31])

The ferrite grains that are formed during the dual-phase treatment are polygonal (i.e. having many sides) in contrast to acicular ferrite. The polygonal ferrite configuration has a greater capacity than the acicular one for containing deformation. The high strength of DP steels is due to the martensite, which contains about 25% austenite as well, an observation that went unnoticed at first.

Thus, dual-phase steels contain three phases but the original designation is still retained. For the first generation of DP steels the aim was to produce about 20% martensite and 80% ferrite, but later on the martensite was reduced to about 10% in order to enhance the cold-pressing properties of the steels. Sometimes amounts in excess of 20% may also be found. DP steels workharden rapidly up to a reduction of about 5% and then more slowly.

DP steels are now produced commercially, mainly as sheet or strip, and are used for stretch-pressed parts such as motor-car wheel hubs. The amount of stretch required for such parts may be carried much further in DP steel than in ordinary hot-rolled steel. The DP steel acquires a high hardness provided that all sections of the part have been deformed by at least 3%.

Chemical composition of D P steels

Both plain unalloyed steels and conventional HSLA steels are capable of being treated so that the required phases are obtained. However, the composition must relate to the cooling rate subsequently applied to the partial austenitization treatment and to the required mechanical properties. Silicon, besides increasing the hardenability, gives rise to a favourable combination of UTS and elongation. Manganese is well known in its capacity for increasing hardenability, and vanadium gives a fine-grained structure which enhances the mechanical properties, including the toughness. A hot-rolled DP steel that acquires the desired micro-structure in light sections after being cooled with compressed air has the following typical composition:

C %	Si %	Mn %	Al %	V %	N %
0·10	0·60	1·5	0·025	0·06	0·007

Mechanical properties of D P steels

As stated above, the mechanical properties after the dual-phase treatment are dependent on the chemical composition of the steel and on the cooling rate. Öström, Lönnberg and Lindgren[32] have studied two Si–Mn steels of the composition mentioned above—one with vandium and one without. Test specimens, $10 \times 2·5$ mm cross-section were annealed for 10 minutes at 840°C and cooled in different ways—see below. Optimum values of UTS and elongation were obtained in the vandium-alloyed steel when cooled in compressed air (FAC) and in the unalloyed steel when quenched in brine at 99°C. The cooling methods used are given below:

		Rate of cooling
QC	Specimen enclosed in quartz sheath	3 K/s
AC	Air cooling	5 K/s
FAC	Forced air cooling	30 K/s
BQ 99	Quenching in brine at 99°C	80 K/s
BQ 88	Quenching in brine at 88°C	180 K/s

Figures 6.123 and *6.124* show dramatically the results of these investigations.

Very extensive research work on HSLA steels and DP steels has been carried out at the Institute for Metals Research in Stockholm[33]. The literature covering this work and other research results is particularly comprehensive. The book entitled, *Micro Alloying 75*, running to about 750 pages, contains the collected knowledge presented by distinguished research workers in this field[34].

6.4.4   High-strength low-alloy pearlitic steels

The weldable HSLA steels have a microstructure consisting mainly of ferrite which imparts a fairly good toughness to the steel. An increase in carbon increases the amount of pearlite, which in turn causes a decrease in toughness. This fact is illustrated by Pickering's classic figure (*see Figure 6.125*).

The micro-alloying of pearlitic steels enhances their toughness only if they are treated to give a fine-grained structure. Precipitation hardening, which is a logical follow-up to micro-alloying, involves a treatment at a high solution temperature however; but this may result in increased grain size, a natural consequence of which is a reduction in toughness. In spite of this fact, pearlitic steels containing about 0·50% carbon and alloyed with approximately 0·1% V have been used with good results since the end of the 1960s for highly stressed machine parts such as motor-car engine crankshafts and connecting rods. These parts are produced by forging at about 1200°C and are then allowed to cool at a rate such that the required mechanical properties are obtained as a consequence of the precipitation of vanadium nitrides or vanadium carbo-nitrides. This precipitation hardening affects the yield point in the first instance.

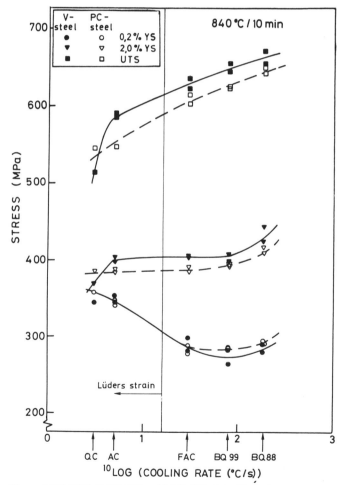

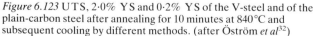

Figure 6.123 UTS, 2·0% YS and 0·2% YS of the V-steel and of the plain-carbon steel after annealing for 10 minutes at 840°C and subsequent cooling by different methods. (after Öström et al[32])

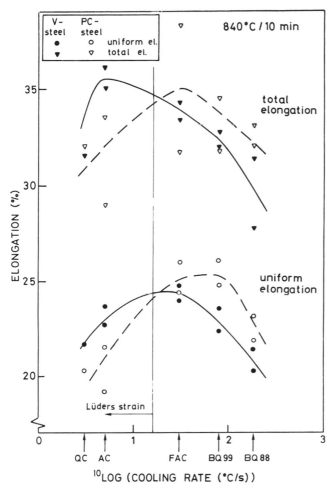

Figure 6.124 Total and uniform elongation of the V-steel and of the plain-carbon steel after annealing for 10 minutes at 840°C and subsequent cooling by different methods (after Öström *et al*[32])

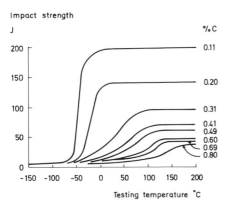

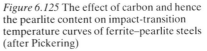

Figure 6.125 The effect of carbon and hence the pearlite content on impact-transition temperature curves of ferrite–pearlite steels (after Pickering)

The most popular steel was developed in Germany and is designated 49 MnVS 3. Its composition is:

C %	Si %	Mn %	S %	V %
0·44	—	0·6	0·045	0·08
0·54	0·6	1·0	0·065	0·13

Controlled cooling from the forging temperature produced the following mechanical properties in the steel:

Yield point	min 500 N/mm^2
Ultimate tensile strength	800–900 N/mm^2
Elongation	approx 12%
Reduction of area	approx 30%
Impact strength DVM	15–20 J

Engineer and von den Steinen[35] investigated the influence of the cooling rate on the mechanical properties of the above-mentioned steel after it was solution treated at 1250°C for 0·5 h. Test bars, 60 mm in diameter, gave the results shown below.

Table 6.34 Influence of the cooling rate on the mechanical properties of steel 49 MnVS 3

Cooling rate	$R_{p0·2}$	R_m	$\dfrac{R_{p0·2}}{R_m}$	A_5	Z
°C/min	N/mm^2	N/mm^2	%	%	%
2	385	701	55	16	31
10	522	830	63	12	28
30	563	868	65	10	16
50	609	912	67	10	15

N.B. 10°C/min equivalent to free cooling in air

In order to obtain better values of impact strength and ductility a steel of the composition given below was tested.

C %	Si %	Mn %	S %	Cr %	V %	Al %	N
0·40	0·78	0·72	0·06	0·08	0·15	0·06	0·02

After heating to 1225°C and holding for 0·5 h, followed by free cooling in air, a fine-grained structure and the mechanical properties given below were obtained.

$R_{p0·2}$ N/mm^2	R_m N/mm^2	$R_{p0·2}/R_m$ %	A_5 %	Z %	DVM J
560	840	67	16	40	28–33

Since the late 1960s AB Bofors-Kilsta have been making forged crankshafts from micro-alloyed pearlitic steel and this type of steel is being developed along roughly the same lines as at Smedjebacken-Boxholm Stål AB.

Figure 6.126 shows the forging of a crankshaft at AB Bofors-Kilsta in a 16000 tonne (160 MN) eccentric press. This press can forge crankshafts weighing up to 250 kg and lengths up to 1900 mm.

Figure 6.127 gives a general view of the five production units that are integrated with the wholly automated press line.

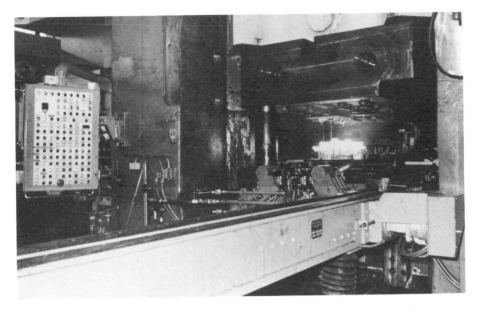

Figure 6.126 Forging a crankshaft in a 16000 tonne (160 MN) eccentric press at A B Bofors-Kilsta

Figure 6.127 General view of the five production units that are integrated with the wholly automated press line at A B Bofors-Kilsta

6.5 Case hardening

6.5.1   Definitions

Low-carbon steels, which as a rule are readily machinable, can have their surface layers carburized and subsequently hardenend. This treatment, called case hardening, gives the steel a hard and wear-resisting surface or case. By virtue of the fact that the core remains comparatively soft and tough the component as a whole shows high impact strength. Owing to the development of compressive stresses in the surface layers during the case-hardening treatment the fatigue strength of the steel is also increased.

Carburization is effected either in solid media (carburizing compounds), in salt baths or in gases at temperatures normally between 825°C and 925°C. The transport of carbon from the carburizing medium always takes place via a gaseous phase, usually carbon monoxide (CO).

The hardening treatment may be carried out as follows:

(a) Direct, i.e. quenching straight from the carburizing medium.
(b) Single quenching, i.e. heating and quenching the carburized parts after first allowing them to cool to room temperature from the carburizing treatment.
(c) Double quenching which usually consists of a direct quench and then a re-quench from a lower temperature. Double quenching may also imply that after carburization the parts are allowed to cool to room temperature and are then subjected to their first quenching or normalizing treatment from 850–900°C. The second quenching is carried out in the same way as in the single quenching, usually from a temperature range between 780°C and 820°C, which is the proper quenching range for the case.

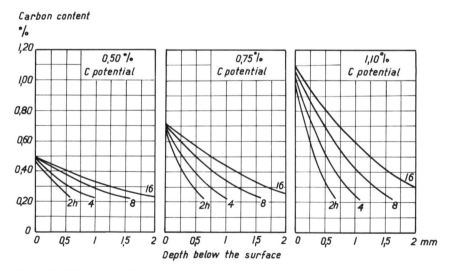

Figure 6.128 Carbon profile of carburized case versus carbon potential and duration of carburization for a plain-carbon steel carburized at 920°C (after Heyn)

Double quenching, when carried out according to the latter method is applied to steels that have not been fine-grain treated. The other methods, however, are applied mainly to fine-grained steels. The various methods of quenching and tempering will be described in more detail in Section 6.5.3.

By the term *depth of carburization* is meant either the distance below the surface to a definite *carbon concentration*, or the total depth of *carbon penetration*. In addition to time and temperature, the depth of carburization depends on the carbon potential of the carburizing medium and on the composition of the steel. The higher the carbon potential the higher the carbon concentration at the surface of the steel, when equilibrium has been established, and the deeper the carburizing depth.

The *carbon potential* may be defined as that carbon content which a specimen of a carbon-steel foil acquires when equilibrium conditions have been established between the carbon potential of the carburizing medium and the carbon content of the foil. The dependence of the depth of carburization on the carbon potential and carburizing time is shown in *Figure 6.128*.

A mathematical model for calculating the carburized concentration profile for carbon has been advanced by Collin, Gunnarson and Thulin[36]. The highest carbon content that the austenite of plain carbon steel can hold may be deduced from the iron-carbon equilibrium diagram. If the carbon concentration is higher, carbides will have formed. This formation is facilitated if the steel contains carbide-forming elements or if it has a composition that favours carbide formation (*see* Section 3.1.4).

The *depth of case hardening,* DC, is defined, according to SS 11 70 08, as the distance from the surface to a plane at which the hardness is 550 HV. This definition is also adopted by ISO (ISO 2639–1973). These Standards cover depth of hardening of 0·3 mm and more. For smaller depth the Standards SS 11 70 10 or ISO 4970–1979 apply. The depth of case hardening is *determined* by HV measurements, using a load of 1 kp, on a section through a case-hardened surface and at right angles to it. *Figure 6.129* shows the hardness profile across a section of a case-hardened steel. With the aid of a magnifying lens the depth of case hardening, viewed at a magnification of about 10 times, may be approximately *assessed* on samples or on actual parts fractured after case hardening. The appearance of such fractured test specimens is shown in *Figure 6.130*.

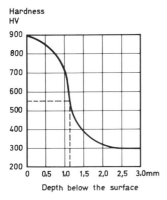

Hardness
HV

Depth below the surface

Figure 6.129 Schematic diagram showing hardness profile of case in a case-hardened steel. Depth of case hardening, DC, as shown in figure, is 1·15 mm

(a)   (b)

Figure 6.130 Test specimens of free cutting steel SIS 1922, quenched and fractured in order to assess depth of case hardening. Water quenched from (a) 800 and (b) 850 °C

There exists a fair measure of agreement between the determined and the assessed depths of case hardening, and the latter is widely used for control and check of the carburizing process. Should an exact measure of the depth of case hardening be required, hardness measurements must be resorted to.

Figure 6.131 has been constructed to serve as an aid to heat-treatment operators who handle steel parts of various grades and dimensions. The curves are based on a large number of case-hardening tests and are sufficiently accurate for practical carburizing work when using a solid compound, a salt bath or a gas. The curves apply to alloy case-hardening steels in dimensions up to about 50 mm in thickness, oil quenched; also to unalloyed case-hardening steels in sections up to about 100 mm, water quenched. The diagram is in good agreement with documented data when the depth of case hardening has been determincd in the same way as that described in ISO 2639–1973.

The validity of the simple diffusion equation $x = k\sqrt{t}$ has been checked against the diagrams and the following values have been obtained for the constant k when $x = DC$ in mm and $t = $ time in hours.

Temperature °C	875	900	925
Constant k	0·34	0·41	0·52

Example: What is the depth of case hardening obtained when carburizing for 12 h at 900 °C followed by hardening?

$$DC = 0·41\sqrt{12} = 1·42 \text{ mm}$$

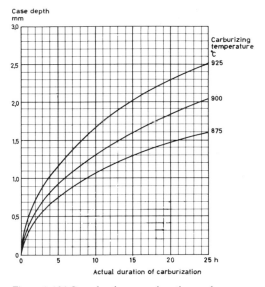

Figure 6.131 Case depth versus duration and
temperature of carburization

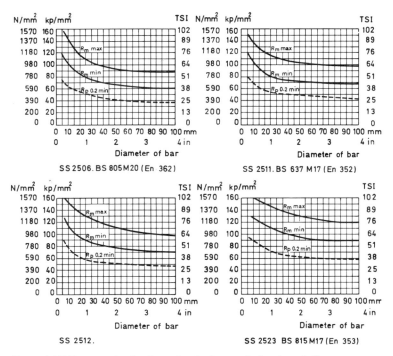

Figure 6.132 Four standardized case-hardening steels showing relation
between bar diameter and core strength after case-hardening with water
quenching

6.5.2 Grades of steel

ISO has issued a recommendation, ISO R/683/XI–1970, covering 15 case-hardening steels, the composition of each being given in *Table 6.35*. *Table 6.36* lists ten of the ISO steels and their nearest equivalent national designations. *Table 6.37* gives the mechanical properties of five steels that are standardized by BS and SS. *Figure 6.132* shows by means of a series of diagrams the guaranteed mechanical properties of the core of the alloy steels, after case-hardening with water quenching, in dimensions up to 100 mm. The diagrams are intended to serve as a guide in the selection of steels of various dimensions and for various requirements of mechanical properties.

Table 6.35 Chemical composition of steels as covered by ISO/R 683/XI–1970

Steel No.	C	Mn	Cr	Mo	Ni
1	0·07–0·13	0·30–0·60	—	—	—
2	0·12–0·18	0·30–0·60	—	—	—
3	0·12–0·18	0·60–0·90	—	—	—
4	0·17–0·23	0·60–0·90	0·70–1·00	—	—
5	0·13–0·19	1·0 –1·30	0·80–1·10	—	—
6	0·17–0·23	0·40–0·70	—	0·20–0·30	1·60–2·00
7	0·15–0·21	0·60–0·90	0·85–1·15	0·15–0·25	—
8	0·17–0·23	0·60–0·90	0·30–0·50	0·40–0·50	—
9	0·12–0·18	0·60–0·90	0·80–1·10	—	1·3 –1·7
10	0·11–0·17	0·35–0·65	1·4 –1·7	—	1·3 –1·7
11	0·10–0·16	0·35–0·65	0·60–0·90	—	2·75–3·25
12	0·17–0·23	0·60–0·90	0·35–0·65	0·15–0·25	0·40–0·70
13	0·14–0·20	0·60–0·90	0·80–1·1	0·15–0·25	1·2 –1·6
14	0·11–0·17	0·30–0·60	0·80–1·1	0·20–0·30	3·0 –3·5
15	0·12–0·18	0·25–0·55	1·1 –1·4	0·20–0·30	3·8 –4·3

N.B.
For all steels: P % = 0·035 max and Si % = 0·15–0·40
For steels 1–3: S % = 0·035 max, 0·020–0·035 or 0·030–0·050
For steels 4–15: S % = 0·035 max or 0·020–0·035

Table 6.36 A selection of equivalent case-hardening steels with their national designations

ISO	AISI	BS	En	DIN	NF	SS
2	1015	—	—	C 15	(CC10)	—
3	1016	080M15	32C	—	XC18	1370
4	5120	527M20	—	(15 Cr 3)	—	—
5	(5115)	—	—	16 MnCr 5	16MC6	—
8	—	—	—	20 MoCr 4	—	—
(9)	—	637M17	352	—	16NCD6	2511
10	—	—	—	15 CrNi 6	—	—
11	—	655M13	36A	—	—	(2514)
12	8620	805M20	362	—	20NCD2	2506
(13)	—	815M17	353	—	18NCD6	2523

Table 6.37 Guaranteed properties of case-hardened steels. The mechanical properties are valid for the test-bar dimensions recommended by ISO. The values refer to the material of the core

BS	SS	Dia mm	$R_{p0.2}$ min kp/ mm^2	N/ mm^2	R_m kp/ mm^2	N/ mm^2	A_5 % min	Hard- ness HB	KCU kpm/cm^2
080M15	1370[1]	11	40	390	70–110	690–1080	9	205–330	5
		30	25	250	50–70	490–690	16	140–205	5
080M15	1370[2]	11	60	590	90–135	880–1320	7	270–405	5
		30	35	340	60–80	590–780	14	170–240	
805M20	2506[3]	11	65	640	100–140	980–1370	8	300–420	
		30	50	490	75–105	740–1030	11	220–315	6
		63	40	390	65–90	640–880	13	185–270	6
637M17	2511[3]	11	65	640	100–135	980–1320	8	300–405	
		30	50	490	75–110	740–1080	10	220–330	6
		63	45	440	70–100	690–980	11	200–300	6
—	2512[3]	11	75	740	110–150	1080–1470	8	330–450	
		30	55	540	85–125	830–1230	9	255–375	5
		63	50	490	75–105	740–1030	10	220–315	5
815M17	2523[3]	11	85	830	125–160	1230–1570	7	375–480	
		30	70	690	105–145	1030–1420	8	315–435	5
		63	60	590	90–125	880–1230	9	270–375	5

1. Water quenched from 780°C ± 10°C, tempered for 1 h at 180°C.
2. Water quenched from 890°C ± 10°C, tempered for 1 h at 180°C.
3. Oil quenched from 820°C ± 10°C, tempered for 1 h at 180°C.

6.5.3 Methods of carburizing

Regardless of what method is used, carburization always takes place via a gaseous phase. However, each method has its own intrinsic characteristics which produce differing case-hardening results.

Carburizing by means of solid substances (granular compounds, pack carburizing)

Carburization with charcoal alone (without activating agents or energizers) can take place thanks to the presence of the atmospheric oxygen enclosed in the carburizing box. The oxygen reacts with the charcoal and during the heating-up period a CO_2-rich mixture is produced which then continues to react with the charcoal as follows:

$$CO_2 + C \rightarrow 2CO \tag{6.3}$$

As the temperature rises the equilibrium is displaced to the right, i.e. the gas mixture becomes progressively richer in CO.

At the steel surface the CO breaks down as follows:

$$2\,CO \rightarrow CO_2 + C \tag{6.4}$$

The atomic or nascent carbon thus liberated is readily dissolved by the austenite phase of the steel and diffuses into the body of the steel, even to the extent of forming carbides (cementite). The CO_2 produced reacts again with the charcoal according to (6.3) above and the cycle of reactions is repeated.

Since the amount of atmospheric oxygen in a charcoal packing can vary, and may be insufficient to produce the carburizing gas it is current practice to mix the charcoal with an energizer, usually barium carbonate, which reacts during the heating-up period as follows:

$$BaCO_3 \rightarrow BaO + CO_2 \qquad (6.5)$$

$$CO_2 + C \rightarrow 2\ CO \qquad (6.6)$$

The CO_2 originally formed then reacts with the carbon in the charcoal, producing the active CO.

Since all the reactions mentioned above can proceed in both directions, i.e. they are reversible, the correct way of representing reaction (6.6) is

$$CO_2 + C \rightleftarrows 2\ CO \qquad (6.7)$$

If the temperature is increased but the pressure kept constant the reaction proceeds from left to right, i.e. more CO is produced. On lowering the temperature the proportion of CO_2 increases at the expense of the CO. *Figure 6.133* shows the equilibrium relationships existing between the constituents in the gaseous mixture expressed as percentages by volume, at a total pressure of one atmosphere. At 900°C, which is a customary carburizing temperature, equilibrium is at about 96% CO and 4% CO_2.

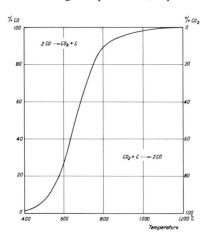

Figure 6.133 Equilibrium diagram for reaction $2\ CO \rightleftarrows C + CO_2$ at pressure of one atmosphere

This equilibrium curve is also included in *Figure 6.134*, which is part of the equilibrium diagram for the Fe–O–C system. Given that at 900°C the mixed gases are in equilibrium according to Equation (6.7) the carbon concentration at the surface of the steel will adjust itself to some value between 0·70% and 1·2%, which is the equilibrium carbon concentration at the phase-field boundary A_{cm} at this temperature. If the CO_2 content falls to 2% some Fe_3C may be precipitated; conversely, should for some reason the CO_2 content increase, owing to say, the admittance of air to the carburizing compound, the carbon content of the steel will fall correspondingly. For example: at 900°C, 6% CO_2 results in about 0·70% C and 14% CO_2 in about 0·30% C.

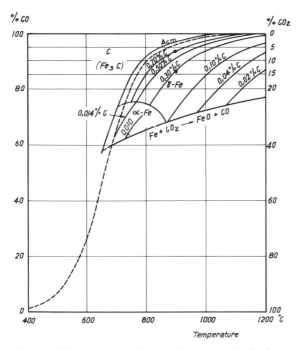

Figure 6.134 Section of equilibrium diagram for Fe–O–C at pressure of one atmosphere

When steel parts are being pack carburized in a carburizing compound continuously or regularly, boxes should be made from heat-resisting steel, type 25% Cr, 20% Ni being suitable for this purpose. A box for occasional work may be made from ordinary mild-steel sheet.

After being cleaned the parts are packed into the boxes on the bottom of which is first laid a layer of the compound, about 25 mm thick. The purpose of this layer is to keep the parts from coming into contact with the bottom itself of the box since they might otherwise become deformed during the carburizing process. When a layer of the parts has been placed in the box more compound is added to make a somewhat thinner layer, then more parts are packed and so on. When the top layer of the parts has been packed in the box there should be room at the top for a 50 mm layer of compound before the lid is placed in position.

It is important to make the lid airtight with clay or similar material as the carburizing process may be prevented by air ingress. During the heating-up period the clay dries and may crack and hence its sealing properties are not wholly reliable. A good way of preventing the ingress of air is to make a rimmed lid about 25 mm larger than the box. When the lid has been placed in position the box is turned upside-down and the space between it and the rim of the lid is sealed with refractory compound (*see Figure 6.135*). Round boxes made from heat-resisting steel sheet have adequate rigidity and there is generally no need to use any sealing material.

Conventional carburizing compounds contain between 6 and 20% of energizer. This substance is usually evenly mixed with the charcoal and in principle, the carbon of the compound can be fully utilized. A batch of fresh carburizing compound will shrink a good deal in volume after its first use, a reduction by as much as a fifth of its original volume being not unusual. Hence, before starting another treatment fresh compound is added to make up the original volume. In this way a mixture is obtained which shrinks less and, more important, which shrinks by a known amount.

Carburizing compounds have relatively poor heat conductivity. Consequently when large carburizing boxes are used, workpieces lying in the middle of the box are carburized considerably less than those lying near the sides. The smaller the boxes used the more uniform is the carburizing effect likely to be. The boxes themselves must not be placed so close together that heat radiation or the free circulation of the furnace atmosphere is impeded. It is good practice to make the boxes cylindrical whereby more space between them is automatically created. The service life of cylindrical boxes is also longer and the risk of cracking smaller. From the point of view of both production economics and uniformity of carburization, optimum reults are obtained if the diameter or width of the box is about 150 mm. Boxes having a diameter or width exceeding 300 mm should not be used unless the workpiece dictates this size.

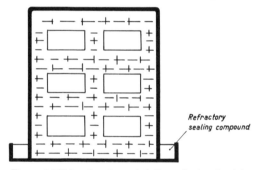

Refractory
sealing compound

Figure 6.135 Section through fully packed carburizing box

The *case depth* is kept under control by inspecting test specimens or, if large series are concerned, by checking actual parts. The curves in *Figure 6.131* can serve as a guide when estimating the time necessary to obtain a stipulated case depth. Since the curves refer to holding times at the temperatures concerned, the heating-up time must be assessed. The dimensions of the boxes, the size of the charge and the heating capacity of the furnace are all important factors. A simple way of judging the temperature in the middle of the box is to insert a steel rod through a hole in the box and then pull it out when it is thought that the preset temperature has been reached. The colour of the red-hot bar will give an indication of the temperature inside the box. Alternatively, a tube closed at one end and protruding into the box may then be welded to one side and at right angles to it. The temperature may be determined by means of a thermocouple inserted into the tube through a hole in the box.

In order to assess heating time, tests have been carried out with boxes, 130 mm diameter × 320 mm. Specimens 10 mm diameter × 50 mm, were packed in the boxes at different distances from the wall. In one test they were placed immediately next to larger specimens as sketched in *Figure 6.136*. Thermocouples were attached to the specimens and were connected to a temperature recorder. The furnace measured $378 \times 500 \times 800$ mm and was rated at 24 kW.

During the heating-up period the temperature increased at the rate shown in the diagrams. Specimen No. 1 in box *A* reached the furnace temperature only slightly ahead of specimen No. 2. Specimen No. 3 required an additional 10 min to reach 900°C. The specimens in box *B* required about twice as long to reach the pre-set temperature but they all reached it in about the same time. At first the temperature of specimen No. 6 increased more slowly than that of the other (*see* specimen No. 3). At the higher temperatures the rate of heating was faster compared with the other specimens. This was due to the radiation absorbed at the top of the box and also to the lower rate of heating of the larger specimens.

In the test just described it is estimated that the specimens in box *A* obtained about one hour's longer carburizing time than the specimens in box *B*. Assuming a total furnace time of say, 5 h at 900°C, this would imply 4 h and 3 h of actual carburizing time, respectively, which, in turn, would result in a carburized depth of 0·85 mm and 0·70 mm for the specimens in boxes *A* and *B* respectively (according to *Figure 6.131*).

Salt-bath carburizing

The active carburizing agent in a salt bath is sodium cyanide (NaCN) or potassium cyanide (KCN). It is believed that the carburization proceeds via a gaseous phase according to the following reaction:

$$2\,NaCN + O_2 \rightarrow 2\,NaCNO \tag{6.8}$$

$$4\,NaCNO \rightarrow 2\,NaCN + Na_2CO_3 + CO + 2\,N \tag{6.9}$$

$$3\,Fe + 2\,CO \rightarrow Fe_3C + CO_2 \tag{6.10}$$

The first reaction takes place at the interface between the salt bath and the atmosphere; the other two reactions take place at the interface between the salt bath and the steel. Some of the nitrogen liberated by reaction (6.9) is also taken up by the steel.

The conventional cyanide baths that give a case depth up to 0·8 mm have a cyanide content (NaCN) of 40–50%. When a fresh cyanide bath is being prepared it is customary to start with pure NaCN which is melted down in the furnace. NaCN decomposes quite rapidly at the beginning of the process whereby the cyanide content assumes the value stated above.

A salt bath containing about 20% NaCN may be used to treat very thin parts requiring a case depth of only a few tenths of a millimetre. The parts are quenched direct into water after being carburized. Such a bath is prepared by melting together equal parts of sodium cyanide and anhydrous sodium carbonate.

To obtain case depths up to 1·5 mm, so-called activated baths are employed, the NaCN content of which is about 10%. These baths are also used in conjunction with baths for the martempering of steel at about 200°C, which contain sodium nitrite and potassium nitrate. When the parts are transferred from the cyanide bath to the martempering bath some

454

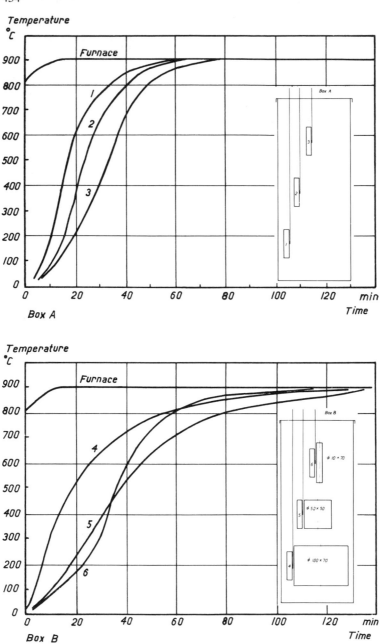

Figure 6.136 Rate of temperature increase during heating-up period for boxes A and B and experimental carburizing set-up in order to assess heating-up time at various distances from box walls. Box dimensions: 130 mm diameter × 320 mm

cyanide salt will be carried over. If the content of the bath is more than 10% NaCN the reaction between the carbon in the cyanide salt being carried over and the oxygen in the martempering bath will become explosively violent.

The NaCN content of the cyanide bath should be checked daily and fresh salt added as required. Generally the NaCN content is satisfactorily maintained simply by replacing the volume of salt carried out with an equal amount of fresh salt. The surface of the bath should be kept covered with a layer of graphite fines which prevents radiation loss and gas evolution from the bath.

The amount of carbon and nitrogen picked up by the steel depends mainly on the cyanide content of the bath and on its temperature. *Figure 6.137* shows that the carbon concentration increases whereas the nitrogen decreases as the cyanide content increases from 10 to 50%.

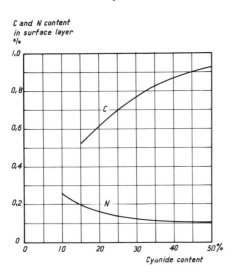

Figure 6.137 Surface concentrations of carbon and nitrogen as functions of the NaCN content when carburizing unalloyed case-hardening steel for 2·5 h at 950°C in a liquid carburizing bath under a protective cover of graphite. The carbon and nitrogen estimations were carried out on a 0·075 mm thick surface layer on the specimen (after Waterfall[37])

With rising temperature, the NaCN content remaining constant at about 50%, the concentration of carbon increases and that of nitrogen decreases as shown in *Figure 6.138*. Both diagrams have common coordinates at 950°C and 50% NaCN. The differences existing between the curves at this point are due to the somewhat different conditions under which the tests were carried out[37]. Obtained from the same source, *Figure 6.139* gives the hardness profiles of a mild steel, specimens of which were carburized in a cyanide bath at different temperatures, oil quenched, heated to 780°C and water quenched.

The diagram in *Figure 6.131* can also be used to assess the case depth. Leaflets and brochures describing the various liquid bath carburizing salts generally give curves showing the total carburized depth but this information is usually of only minor interest.

Salt-bath carburizing is mainly used for small parts that in general require a case depth less than 0·5 mm. The smaller the case depth required, the greater is the economy achieved by using a salt bath since the

C and N content
in surface layer
%

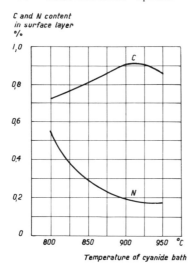

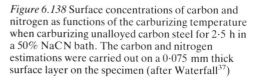

Temperature of cyanide bath

Figure 6.138 Surface concentrations of carbon and nitrogen as functions of the carburizing temperature when carburizing unalloyed carbon steel for 2·5 h in a 50% NaCN bath. The carbon and nitrogen estimations were carried out on a 0·075 mm thick surface layer on the specimen (after Waterfall[37])

rate of heating is very high compared with that in solid carburizing media. Before the parts are immersed in the salt bath they should be preheated to between 100°C and 400°C, partly to remove any traces of moisture and partly to utilize more efficiently the capacity of the salt bath. For further information on carburizing salts the reader is recommended to contact the suppliers concerned.

It may happen than even when parts are being carburized in a cyanide bath the surface carbon concentration becomes too high, with the result that the quenched hardness may show somewhat low values on account of there being too much retained austenite present. The carbon content increases as the time of treatment increases and as the temperature rises. Therefore, when the parts are to be quenched direct from the cyanide bath a moderate carburizing temperature should be chosen. The result of a trial hardening treatment, using BS 655M13 (En 36 A) is shown in *Figure 6.140*. From these curves it can be seen how the thickness of the soft surface zone increases as the treatment time increases. The surface hardness of the specimens was also checked, using HRC, and from this it could be seen that in this instance the case depth should be about 0·4 mm if the hardness is to exceed 60 HRC. By tempering the specimens at 180°C the surface hardness is reduced but the case-hardened depth remains largely the same (*see Figures 6.140 and 6.141*).

If a case depth exceeding 0·5 mm is required it is more economical to use higher carburizing temperatures, e.g. 900–925°C. When alloy steels are quenched direct the surface hardness is rather low and consequently they are usually rehardened by a requenching treatment from some lower temperature. If such double quenching is resorted to, the first quench takes place in oil or in a salt bath. If the parts are allowed to cool in air oxidation of the surface may occur.

Bungardt, Brandis and Kroy[38] have studied the carburizing process in a liquid salt bath at temperatures between 900°C and 1000°C in great detail and found that a normal NaCN salt bath will give a soft surface due to high

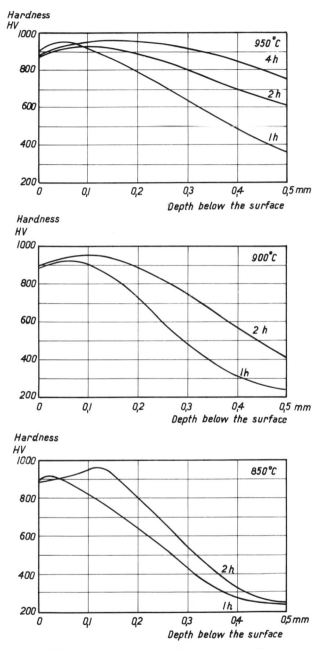

Figure 6.139 Curves showing depth of hardening in mild-steel
test bars, 14 mm diameter, carburized in a cyanide bath at
various temperatures and oil quenched. Bars reheated to 780 °C
and water quenched (after Waterfall[37])

458

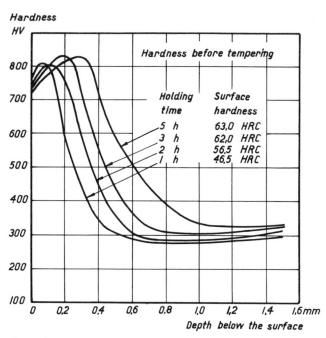

Figure 6.140 Steel BS 655M13 (En 36 A). Hardness as
quenched after cyaniding at 870°C for various times. (No
tempering.)

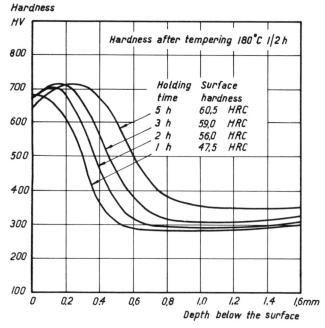

Figure 6.141 Steel BS 655M13 (En 36 A). Hardness after
cyanide hardening at 870°C for various times. Tempering for
½ h at 180°C

content of retained austenite. There are, however, now available on the market salt baths that will build up a definite surface carbon concentration, say 0·80% and a tolerable content of retained austenite.

N.B. Sodium cyanide is extremely poisonous if it gets into the human system. There are special instructions to be followed when work with cyanide salts is in progress and workers should make themselves familiar with these and the procedure to be adopted in case of accidents. Municipalities using bacteriological water purification, i.e. purifying sewage by means of bacteriological action, will require the installation of a detoxification unit since even very low cyanide concentrations will destroy the bacteria. Detoxification is usually carried out by 'neutralizing' the salt by means of sodium hypochlorite. The disposal of waste salt must be rigorously controlled in conformity with official regulations.

There are now available on the market a number of cyanide-free carburizing salt baths, the residues of which need not be detoxified.

Gas carburizing

During the last few decades carburizing by means of gas has become the most popular method of case hardening. The furnaces are very dependable and relatively simple to operate. Since hydrocarbons enter into the process several reactions take place simultaneously. In addition to the reaction

$$2 CO \rightleftharpoons C + CO_2 \tag{6.11}$$

there is also the methane reaction

$$CH_4 \rightleftharpoons C + 2 H_2 \tag{6.12}$$

and the water-gas reaction

$$CO + H_2 \rightleftharpoons C + H_2O \tag{6.13}$$

The composition of the furnace gas can be determined by analysis and hence the carbon potential may be estimated. The moisture content of the gas has a great influence on the carbon potential as indicated by reaction (6.13) and may be measured by simply determining the dew point of the gas. The dew point is that temperature at which water droplets (dew) are deposited from the gas (saturation). The relationship between moisture content and dew point is shown in *Figure 6.142*.

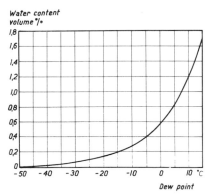

Figure 6.142 Relation between dew point and moisture content

F. E. Harris[39] has described a method of calculating the carbon potential from the composition of the gas. In the following are discussed only a few examples that show, by controlling the dew point, what carbon concentrations are obtainable at the surface by varying the temperature and gas composition. *Figure 6.143* shows how the surface concentration of carbon varies with the dew point and the percentage of H_2 at a temperature of 925°C when the CO content is kept constant at 20%. If a carbon concentration of, say, 0·80% is aimed at, the dew point should be −3°C with 60% H_2 or −17°C with 20% H_2.

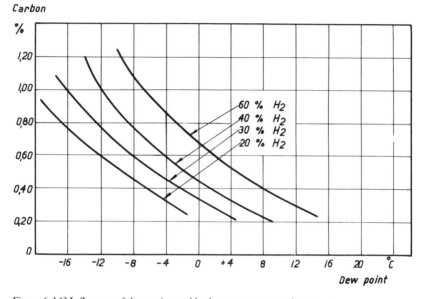

Figure 6.143 Influence of dew point and hydrogen content and constant 20% CO at 925°C on concentration of surface carbon (after Cullen)

Figure 6.144 shows how the temperature and the dew point of a gas containing 40% H_2 and 20% CO influence the concentration of carbon at the surface. At 925°C, 0·80% C is obtainable if the dew point is about −6°C. If the temperature is only 815°C the same carbon concentration would require a dew point of about +6°C. The calculations are valid only when equilibrium has been established. In a retort furnace this occurs after a few hours' carburization. If a gas of known composition is used it is possible to follow the progress of the process from dew-point determinations.

There are a number of methods in existence for the continuous measurement and control of the carbon potential. One such method is based on infrared rays which are used to register the concentration of CO_2 in the gas. This value gives a measure of the CO concentration which, in turn, relates to the carbon potential. By means of a special measuring unit the resulting impulse can be converted direct to a numerical value of the carbon potential. The control may also be carried out by measuring the resistance of a steel wire which is calibrated for different carbon

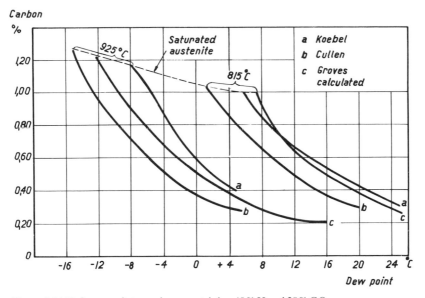

Figure 6.144 Influence of atmosphere containing 40% H and 20% C O on carbon concentration at surface

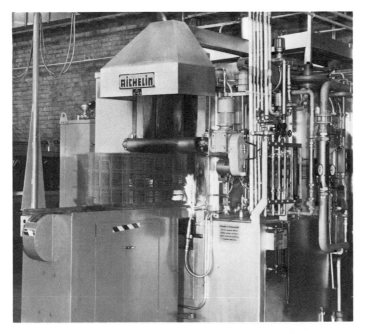

Figure 6.145 Gas carburizing furnace. A charging basket containing a batch to be carburized is being pushed into the furnace. The three 'tubes' to the right of the picture are flowmeters for carrier gas, propane and air, respectively

concentrations. The resistance measured is proportional to the carbon concentration. Indirect measurement of the carbon content can take place by means of an oxygen probe.

When gas carburizing is being employed, a primary requirement is the availability of furnaces sufficiently gas tight to prevent air ingress, since this would interfere with the process.

In principle there are four different ways of producing the gas, viz.:

1. Dripping carburizing liquid into the furnace. This liquid generally contains ethyl alcohol and white spirit or only isopropyl alcohol. It drops on to a plate inside the furnace where it is dispersed and vaporized. By this method it is possible to employ a liquid that produces a carrier gas having a relatively low carbon potential. This potential can then be adjusted by dripping a more active liquid into the furnace. In practice, however, only one liquid is used and the carbon potential is adjusted by varying the volume of liquid added.
2. Admission of gas from a gas-producing unit. The gas is produced by endothermic combustion. The raw gas, usually propane, is mixed with air in carefully balanced proportions and burnt in the gas-producing unit to produce a carrier gas with a carbon potential of 0·35%–0·50%. Before it enters the furnace the carrier gas is given an extra addition of propane, thereby adjusting the carbon potential to the required level which is usually around 0·8%.
3. Direct admission of gas and air into the furnace. The gas actually doing the carburization is produced in the furnace chamber by the reaction between the gas and air which are admitted in balanced amounts.
4. Carburizing in a vacuum furnace. Only hydrocarbons are admitted into the furnace from which air is evacuated. Since no oxygen takes part in the reaction there is no internal oxidation. The process can take place at a temperature higher than that normally used, thus allowing the carburization to proceed more quickly[40, 41].

Further information about gas carburizing is given in the reference [42, 43] and [44]. Production of protective gases has been treated in Section 5.3.1.

6.5.4 Influence of heat treatment and steel composition on case depth, surface hardness, core hardness and microstructure

As a result of the carburizing treatment the steel has obtained a certain carbon content in the surface zone. On subjecting the steel to further heat treatment this zone can give it the required surface properties. Case hardening is, however, not merely a matter of surface hardness and depth of penetration. Stipulated requirements may call for a certain correlation between the hardness of the case and the core, which will be discussed further in the text.

Depth of hardness penetration (case depth)

In the first instance the depth of hardness penetration depends on the carbon content of the carburized layer. Provided that only martensite is formed in the case on quenching, the depth of case hardening would be equivalent to a depth of carbon penetration down to 0·40% C. This would agree well for small parts but as the section dimensions increase, the rate of cooling decreases and hence the conditions necessary for the formation of martensite are changed. This implies that steel components of different sizes but which have the same depth of carbon penetration obtain a case-hardening depth that is dependent on the dimensions of the piece. *Figure 6.146* illustrates this fact as applied to BS 637M17. The test specimens were all carburized together and were placed in such positions as would ensure their reaching the carburizing temperature approximately at the same time. A complete set of test specimens was case hardened at each of six different heat-treatment shops. In two instances the hardness was lower than 550 HV in the 145 mm diameter specimen. Results from practical experience also tend to verify that the hardenability of BS 637M17 may be insufficient when the dimensions of a solid component exceeds 100 mm. Note that this limitation does not apply to gears since the rate of cooling of a gear tooth is greater than that of a solid shaft.

Obviously the depth of case hardening is also dependent on the quenching medium. Water quenching gives a greater depth of case hardening than oil quenching, particularly in heavy sections. The differences in case-hardening depths for different dimensions are evened out if the quenching takes place in water (*see Figure 6.147*).

Since the hardenability is influenced by the quenching temperature the latter is also a factor affecting the depth of case hardening. This may be seen in *Figure 6.130* by comparing the depth of case hardening in the two specimens which were carburized at the same time but were quenched from different austenitizing temperatures. When the quenching is

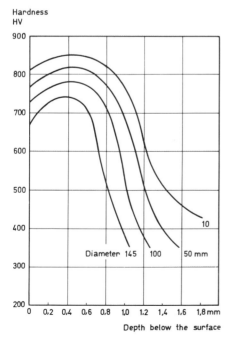

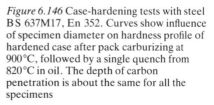

Figure 6.146 Case-hardening tests with steel BS 637M17, En 352. Curves show influence of specimen diameter on hardness profile of hardened case after pack carburizing at 900°C, followed by a single quench from 820°C in oil. The depth of carbon penetration is about the same for all the specimens

performed direct from the carburizing temperature the depth of case hardening is usually somewhat greater than if the rehardening takes place from a lower temperature.

The course of the carburizing process may conveniently be followed by means of test bars of case-hardening steel which are taken out of the furnace and quenched in water. The bars are fractured and the fractured surface is inspected at a magnification of about × 10 to assess the depth of case hardening. (*See Figure 6.130.*)

For practical reasons it is not always possible to use test specimens of the same dimensions as the parts being case hardened. In such cases *Figures 6.146* and *6.147* may serve as a guide. For example, if the test specimen has a diameter of 10 mm and the workpiece, which is to be given a case-hardening depth of about 1 mm, has a diameter of 100 mm then, assuming that the conditions prevailing when *Figure 6.146* was drawn also apply in the present instance, the depth of case hardening in the test specimen must be 1·35 mm.

The fact is often lost sight of that the chemical composition of the steel, and hence its hardenability, does exert a great influence on the case-hardened depth; also that with low-alloy case-hardening steels in heavy dimensions it is not possible as a rule to obtain large depths of case hardening on quenching in oil even if the carburizing times have been very long. It should also be borne in mind that since the carbide-forming elements tend to bind the carbon, hardenability may be reduced.

With a view to illustrating these circumstances an investigation was undertaken with four grades of steel, as specified below.

Table 6.38 Chemical composition and D_i-value of the investigated steels

SS steel grade	BS steel grade	C %	Si %	Mn %	P %	S %	Cr %	Ni %	Mo %	D_i without C
2506	805M20	0·20	0·29	0·88	0·022	0·044	0·50	0·58	0·18	1·85
2511	637M17	0·18	0·30	0·79	0·022	0·025	0·78	1·46	0·07	2·2
2514	655M13	0·12	0·25	0·52	0·025	0·007	0·71	2·9	0·04	2·2
2523	815M17	0·20	0·27	0·94	0·022	0·043	1·00	1·17	0·12	3·0

N.B. The composition of SS 2511 was modified a number of yerars ago. S S 2514 is no longer a Swedish standard steel.

Test-pieces of diameters 25, 50, 75 and 100 mm and length 3xd were used for the test. The pieces were gas carburized at 925°C for 5, 15 and 25 h respectively. They were cooled in a martempering bath, then heated to 840°C, quenched is oil and finally tempered at 180°C. For the first two-thirds of the carburizing time the carbon potential was 1·1% and for the last third it was about 0·80%.

The carbon profile was determined on the 50 mm diameter bars after a carburizing time of 15 h. *Figure 6.148* shows that equalization of the carbon content in the case was not satisfactory and that the carbon profile was influenced by the steel chemistry. The maximum carbon content is dependent on the amount of carbide formers and carbide stabilizing

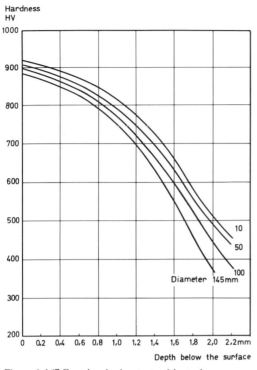

Figure 6.147 Case-hardening tests with steel
BS 637M17, En 352. Curves show influence of
specimen diameter on hardness profile of hardened
case after pack carburizing at 900°C, followed by a
single quench from 780°C in water. The depth of
carbon penetration is about the same for all the
specimens

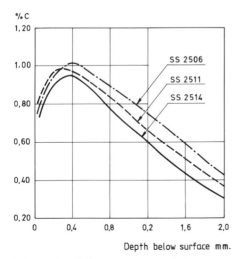

Figure 6.148 Carbon profile across 50 mm test
bars after having been carburized for 15 h

agents. *See* Sections 3.1.4 and 3.1.5, which show that Cr is the most potent carbide former and at the same time has a high carbide stabilizing coefficient. In addition, it is known that in case-hardening steels containing between 3% and 4% Ni the carbon content of the case seldom exceeds 0·9% even if the carbon potential of the carburizing medium is higher.

Figure 6.149 shows the hardness distribution across the case of 25 mm diameter bars of grade SS 2506, carburized for 5, 15 and 25 h respectively. For each bar the case depth was found to be as expected and the microstructure in the case to be mainly martensite. With increasing bar diameter the case depth became smaller and a reduced hardness was found in a zone below the surface even in the 50 mm bar. The hardness profile across the 100 mm diameter bar is shown in *Figure 6.150* which contains an inset micrograph showing the microstructure obtained after a case-hardening treatment comprising a carburizing time of 25 h. The outer surface zone, about 0·2 mm thick, has a hardness of approximately 700 HV and consists mainly of martensite. The hardness of the underlying zone falls to 500 HV and the microstructure contains carbides and pearlite nodules (previously called troostite). At a depth of 1 mm the hardness is nearly 700 HV. The structure now consists of martensite interspersed with bainite which increases in amount with increasing depth from the surface. The thickness of the case is 1·6 mm as compared with 2·8 mm in the 25 mm diameter bar which was treated in the same way as the 100 mm diameter bar.

With regard to the other steels, there was no noticeable reduction in the hardness of the case for bar diameters up to 75 mm. The hardness profiles of the 100 mm diameter bars are shown in *Figures 6.151, 6.152* and *6.153*.

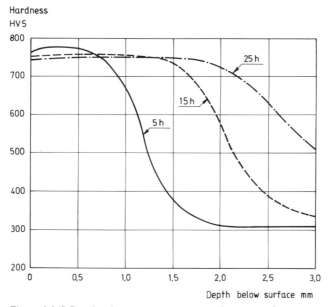

Figure 6.149 Case hardness traverse across surface zone of SS 2506 test bars, 25 mm diameter after having been carburized for 5 h, 15 h and 25 h, respectively, and subsequently hardened

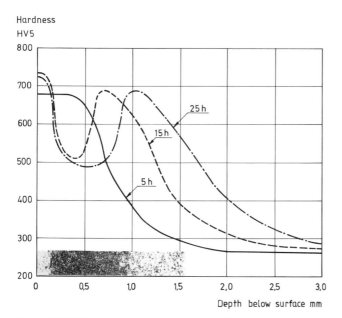

Figure 6.150 Case hardness traverse across surface zone of
SS 2506 test bars, 100 mm diameter, after having been carburized
for 5 h, 15 h and 25 h, respectively, and subsequently hardened.
Inset micrograph shows microstructure after 25 hours carburizing
and hardening

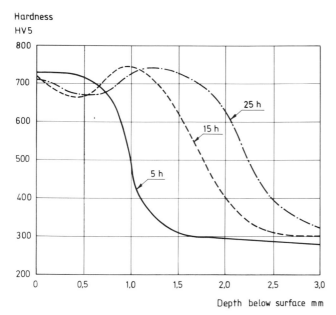

Figure 6.151 Case hardness traverse across surface zone of
SS 2511 test bars, 100 mm diameter after having been carburized
for 5 h, 15 h and 25 h, respectively, and subsequently hardened

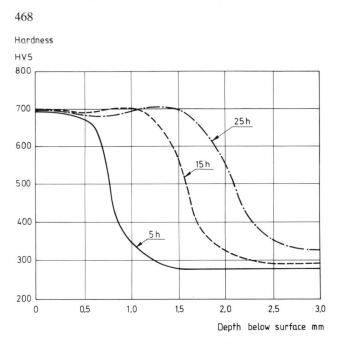

Figure 6.152 Case hardness traverse across surface zone of
SS 2514 test bars, 100 mm diameter after having been carburized
for 5 h, 15 h and 25 h, respectively, and subsequently hardened

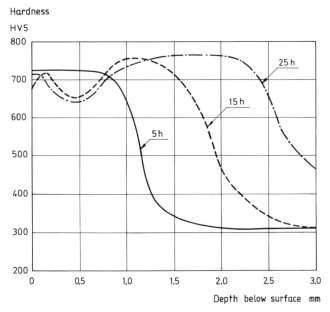

Figure 6.153 Case hardness traverse across surface zone of
SS 2523 test bars, 100 mm diameter after having been carburized
for 5 h, 15 h and 25 h, respectively, and subsequently hardened

It may be observed that also grades SS 2511 and SS 2523 show a hardness decrease in the region situated about 0·5 mm below the surface. Rather better results might have been obtained if the carbon potential had been kept somewhat lower during the first part of the carburizing operation but the carburizing cycle as it was actually carried out corresponds to current case-hardening practice. When carburized for 25 h the steels being studied obtained the case depths shown below.

Grade SS	SD 550 mm
2506	1·6
2511	2·2
2514	2·0
2523	2·7

As may be deduced from the table, grade SS 2523 is the most economic choice of case-hardening steel for large parts that require substantial case depths owing to the short carburizing time required. In addition, this steel is quite reasonably priced.

Grade SS 2506 should not be used for parts larger than about 75 mm in diameter or the equivalent since the surface hardness may be uneven after case hardening. This drawback applies also to some extent to grade SS 2511. It ought to be mentioned, however, that the quenching arrangements prevailing during the investigations reported above were considerably more favourable than those usually prevalent during large-scale heat treatments.

Surface hardness

As before, the carbon dissolved in the austenite is the decisive factor determining surface hardness. When the carbon concentration at the surface of conventional alloy case-hardening steels exceeds 0·70% the M_s temperature falls steeply, the amount of retained austenite after quenching increases and the hardness decreases. If the surface layer of the steel contains the appropriate concentration of carbon for maximum hardness the quenching temperature is of minor importance to the hardness provided that the grain size is not altered. By varying the quenching temperature of an 'over-carburized' steel it is possible to control the amount of carbon going into solution and hence the amount of retained austenite which, in turn, affects the hardness.

During the course of an investigation with steel grades BS 655M13 (En 36 A), BS 637M17 (En 352) and BS 805M20 (En 362), specimens 25 mm in diameter × 200 mm were pack carburized for 9 h at 925 °C and quenched in oil. In one test series of specimens the hardness was determined, on a cross-section, from the surface towards the centre and in another the carbon concentration was estimated at 0·1 mm steps below the surface by machining off stock to the approporiate depth, the specimens having been previously annealed for 1 h at 630°C. The results are shown in *Figure 6.154.*

The carbon concentration at the surface of the steels is about 1·40%. This figure represents a carbon content somewhat higher than soluble in austenite at 925°C. For a given carbon content above approximately 0·90%, the lower the Ni content of the steel the higher is the hardness that

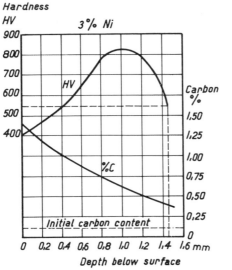

Steel BS 655M 13, En 36 A

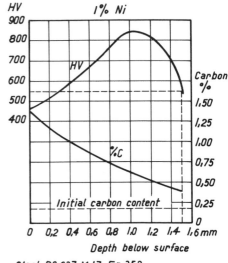

Steel BS 637 M 17, En 352

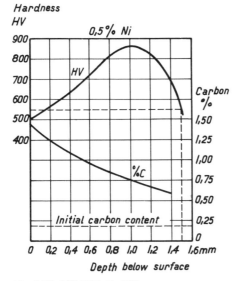

Steel BS 805 M20, En 362

Figure 6.154 Hardness and carbon content at surface of case-hardening steel specimens after pack carburizing for 9 h at 925°C, followed by direct quench from the box in oil

corresponds to this carbon content. Maximum hardness is obtained 1 mm below the surface where the carbon concentration is 0·65%–0·75%.

A similar investigation has been carried out by E. Theis. The curves in *Figure 6.155* show the dependence of the hardness on the carbon content after gas carburizing and direct quenching from 920°C. As the Ni content of the steel increases the figure for the carbon concentration corresponding to maximum hardness is reduced. For the Cr–Mn steel the maximum hardness occurs at 0·70% C and for the Cr–Mo steel (uppermost curve) at 1% C. H. U. Meyer has coined the term 'surface hardenability' to describe the capacity of different steels to attain a high surface hardness, particularly after direct quenching.

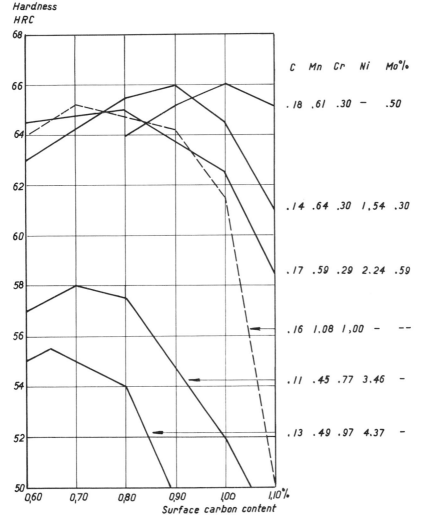

Figure 6.155 Influence of steel composition and surface carbon content on hardness obtained by direct quenching from 920°C, quenching from the box (E. Theis)

A low surface hardness implies a high content of retained austenite. The case of an over-carburized and hardened steel, the surface hardness of which is 500 HV, will have a microstructure like the one shown in *Figure 6.156*. This type of microstructure is definitely unsuitable for machine components and tools and makes them unserviceable for ordinary working conditions. It is therefore necessary to adopt measures that will give a maximum or near-maximum surface hardness.

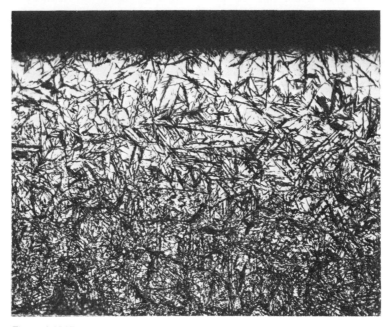

Figure 6.156 Retained austenite in surface layer of case-hardened steel, 100 ×

When pack carburization is used the carbon concentration in the case can be controlled to a certain extent by adjusting the activity of the carburizing compound and the carburizing temperature. A low carbon activity in the compound and a low carburizing temperature result in a low carbon content in the surface. However, the carbon content increases with the carburizing time and this is illustrated by the curves in *Figure 6.157*. If the steel is requenched from a lower temperature, less carbon will be dissolved in the austenite and hardness gradients such as those shown in *Figure 6.158* are obtained. Further lowering of the quenching temperature will result in still less carbon being dissolved in the austenite and still higher hardnesses. For light sections, quenching temperatures as low as 770 °C can be used whereby hardnesses as high as 67 HRC are obtained.

When carburizing in gas it is possible to control the carbon content and keep it at the desired level, generally 0·7–0·8% C. The quenching temperature is not so decisive a factor in this instance. For practical reasons the temperature in the furnace is lowered to about 830 °C before quenching. At this low temperature carbide is formed and hence the

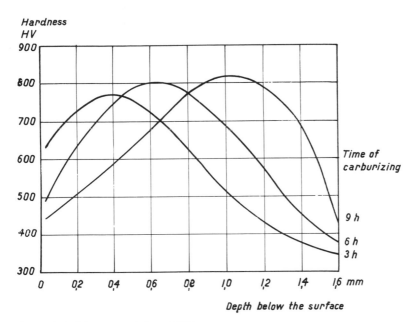

Figure 6.157 Direct quench from 925 °C

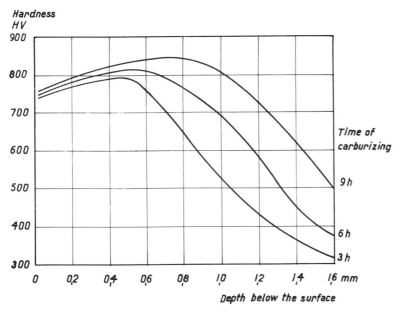

Figure 6.158 Single quench from 830 °C

Figures 6.157 and 6.158 Case hardening of steel B S 637M17, En 352.
Hardness profile of case after pack carburizing at 925 °C, followed by direct
quenching and single quenching from 830 °C in oil, respectively

amount of carbon dissolved in the austenite is somewhat less. In addition the lower quenching temperature reduces distortion.

It is difficult to decide how much retained austenite will produce the best result in each individual instance. For some machine components, such as gears, it has been found, in practice, that retained austenite in larger amounts than previously considered advisable can have beneficial effects. The surface hardness is obviously influenced by the rate of cooling. When quenching into a martempering bath the hardness falls as the quenching bath temperature is raised. To ensure a hardness of at least 60 HRC the temperature of the bath should be about 180 °C. If, because of the presence of too much retained austenite, the hardness has inadvertently become too low, subzero treatment may be resorted to. There is no danger of cracks developing provided that the component concerned has a depth of case hardening that is normal for its size and shape.

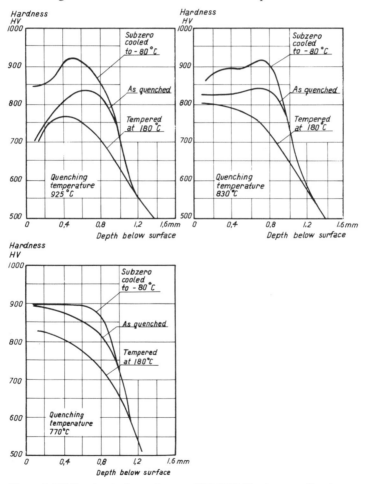

Figure 6.159 Case hardening carbon steel S S 1370. Hardness profile of case after pack carburizing and single water quenching from different temperatures followed by subzero cooling for 1 h at −80 °C, alternatively followed by tempering at 180 °C

It is sometimes difficult to obtain a high surface hardness in alloy case-hardening steels, and a subzero treatment is thus commonly employed. Even unalloyed steels (which are water quenched) may contain appreciable amounts of retained austenite in the surface layers, particularly when the parts are quenched direct after a deep carburizing treatment. As is apparent from *Figure 6.159*, the surface hardness increases as the quenching temperature is lowered. When quenched from 770°C the case has a hardness of about 900 HV and this hardness is not affected by a subzero treatment.

On account of the presence of oxygen or oxygen-containing compounds in most carburizing furnace atmospheres it often happens that certain alloying elements in the surface of the steel are oxidized. This occurrence is called internal oxidation and as the oxygen proceeds to bind the alloying elements the hardenability tends to be reduced resulting in incomplete martensite formation in the outer zone. The oxygen diffuses into the steel, mainly along the grain boundaries, and reacts with those elements that have a greater affinity for oxygen than iron has. Silicon, manganese and chromium are susceptible to oxidation whereas molybdenum and nickel are not [45]. It is not unusual for the oxidation to penetrate as much as about 25 μm. With regard to Cr–Mn steels, particles of Cr–oxide have been found inside grains that lie near the surface of the steel. Further in from the surface oxides of Mn and Si may be found, mainly along the grain boundaries[46]. The degree of internal oxidation is dependent on the CO_2–content and on the time and temperature of the carburization. In *Figure 6.160*, which shows a cross-section through a case-hardened steel, there is a typical instance of internal oxidation in the surface zone.

Internal oxidation gives rise to only a small hardness reduction in the surface zone, but, more importantly, it causes a reduction in the fatigue strength of the material, a fact that has been pointed out by several authors who have demonstrated that a reduction of 20% and more in the figure for the fatigue limit can occur when the depth of internal oxidation is 13 μm or

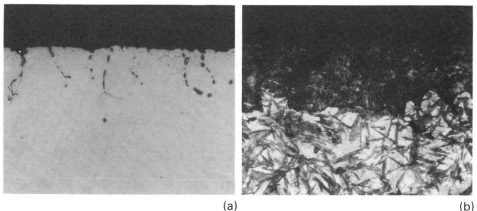

(a)   (b)

Figure 6.160 Typical appearance of the surface layer of a low-alloy case-hardening steel showing internal oxidation after gas carburizing and quenching (after Magnusson[54]). (a) Polished condition; (b) Etched condition

more[47, 48, 49]. When carrying out conventional case hardening it is not possible to avoid some internal oxidation[46]. Since this phenomenon is dependent on both time and temperature the degree of internal oxidation may be kept to a minimum by making the case depth as small as permissible.

Core hardness

When a gear, for example, is being designed with regard to its flexural strength the mechancial properties of the material of the core form the basis for the calculations. *Table 6.37* shows the dependence of core strength on the grade of steel and on the dimension. For example, a comparison between SS 2511 and SS 2512 in the dimension 11 mm shows that the former steel develops a core hardness of 300–405 HB and the latter 330–450 HB. As the dimensions of the sections increase the hardness differences between the two grades are reduced. The only difference between them in chemical composition is 0·05% carbon. However, this is quite large enough to give rise to substantial hardness differences in the core on quenching. Should a Jominy hardenability curve be available for the heat from which the part in question was made, it should be possible to calculate approximately the hardness to be expected on quenching.

The increase in hardness just below the carburized layer, particularly noticeable in light sections, may give rise to machining difficulties when the material of the core is being machined after the hardening treatment. In such cases the core hardness can be controlled by adjusting the temperature of quenching. *Figure 6.161*, which applies to SS 2511 (0·15% C), shows the effect produced by varying the quenching temperature. There is a measured difference of 60 HV between the specimens quenched from the highest and from the lowest temperature. After being quenched from 780°C the specimen contained about 50% ferrite; after being quenched from 830°C the ferrite content was judged to be 5%.

A similar investigation has been carried out on a plain carbon steel, SS 1370 (BS 080M15) the composition in this instance being 0·16% C, 0·35% Si and 0·84% Mn. Test specimens with diameters between 10 mm and 50 mm were carburized at 925°C to a penetration depth of about 1 mm. One set of test specimens was quenched direct into water from the carburizing temperature. Another two sets were requenched in water; one from 830°C and the other from 770°C. The hardness transverse curves are shown in *Figure 6.162*. It can be seen that the hardness is most influenced by the sectional dimension. After being quenched from 850°C the largest and the smallest specimens show at the centre of their sections a hardness difference of about 200 HV.

6.5.5   Fatigue strength of case-hardened steel

By and large the increase in fatigue strength brought about by case hardening is proportional to the hardness increase in the surface zone. It is also dependent on the compressive stresses in this zone. The maximum fatigue strength value is obtained when the carbon is about 0·6%. Most

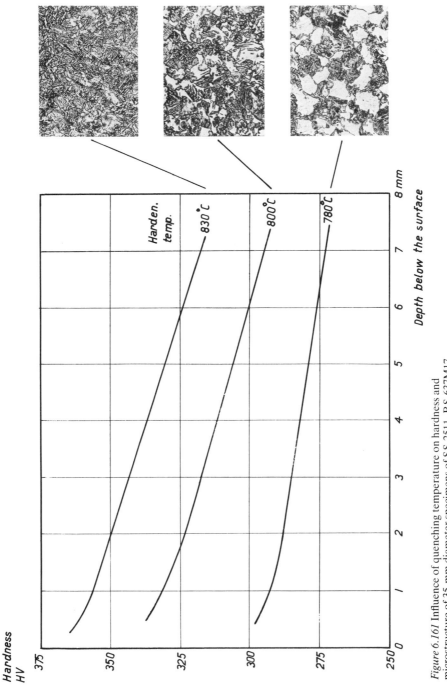

Figure 6.161 Influence of quenching temperature on hardness and microstructure of 35 mm diameter specimens of SS 2511, BS 637M17, En 352. Oil quenching

478

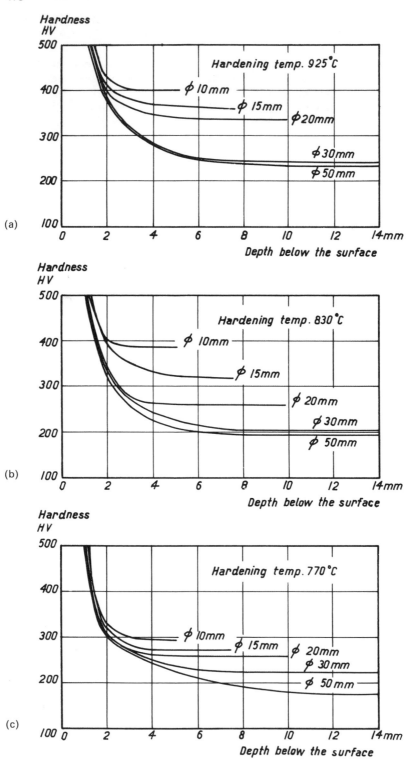

Figure 6.162 Influence of quenching temperature and dimensions on core hardness of case-hardened carbon steel SS 1370 (BS 080M15). Water quenching

fatigue strength investigations of case-hardening steels show that optimum properties are obtained when the surface carbon content is 0·6–0·8% and the surface hardness at least 700 H V[50, 51, 52, 53].

Magnusson[54] has studied the fatigue strength of the case-hardening steels SS 2511 and 2506. After subjecting them to various heat treatments he arrived at results that were in accord with current opinions. His findings are summed up in *Figure 6.163* which shows the fatigue strength scatter bands for the steels concerned, the surface carbon content of which ranged from 0·60–0·95%. Optimum results were obtained when the surface carbon content lay around 0·70%. If subzero treatment was applied to steel with a surface carbon content of 0·85% there was a reduction in the fatigue strength.

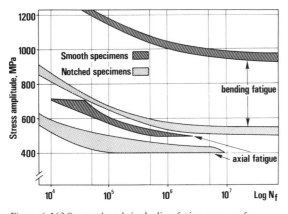

Figure 6.163 Scatter bands including fatigue curves for different groups of case hardening steels tested (after Magnusson[54])

Fatigue fractures that occur in unnotched highly polished test specimens do not start at the surface but often originate in the boundary area between the core and the hardened case. This phenomenon may be explained by means of the schematic stress diagram shown in *Figure 6.164* which represents a cross-section through the plane of the surface and the neutral axis. H represents the strength of the material of the core and of the hardened case. If OP represents the stress acting across the section the strength, H, of the steel is not exceeded. A larger stress, OP', exceeds the strength of the marterial in the outer zone of the core (*see* the black area) and hence a crack appears here. To avoid this a higher core strength H^1 or a larger case depth should be chosen.

The fatigue crack initiated in the boundary area is quickly propagated through the body of the core thus inducing a progressively increasing stress in the case. Cracks begin to appear here and ultimately the test specimen fractures suddenly. From this we see that the strength of the core has a decisive influence on the fatigue limit owing to the fact that the cracks are usually initiated in the core, the fatigue limit of which is lower than that of the carburized case.

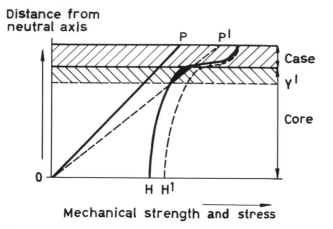

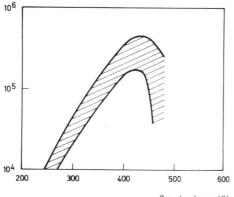

Figure 6.164 Schematic stress diagram for section between surface and neutral axis

There is, however, an optiumum value of the core strength which, if it should be exceeded, will cause a reduction in the fatigue limit. In step with increasing contents of the alloying elements—carbon in particular—there is a successive lowering of the transformation temperature range of the austenite during its cooling from the hardening temperature of the steel. The strength of the core is thereby increased but the longer the time lapse before the transformation takes place in the core the smaller will be the compression stresses in the case. This implies a lower fatigue limit and a greater notch sensitivity of the latter. For maximum resistance to fatigue fracture the optimum tensile strength of the core after case hardening is usually in the region of 1300 N/mm^2 corresponding to approximately 400 HV.

Figure 6.165 Fatigue life of case-hardened gears, gear modulus 3, as a function of core hardness (courtesy Volvo)

This point is illustrated in *Figure 6.165*, which shows the fatigue life of case-hardened gears with a gear modulus of 3, as a function of the core hardness.

6.5.6   Recommendations for case hardening

From practical experience gained in this field the following general heat-treatment recommendations are suggested for case-hardening steels. The treatment is designed to give a surface hardness of at least 60 HRC or 710 HV after quenching.

Carburizing in solid compound (pack carburizing)

Temperature: 900–925 °C

Depth of case hardening mm	Method of hardening
< 0·50	Direct quench
0·50–1·25	Single quench from 800 to 820 °C
< 1·25	Double quench

When direct quenching is employed a carburizing temperature of 900 °C is recommended. If a case-hardening depth of more than 1·25 mm is aimed at it might be good practice to perform the traditional practice of a double quench, i.e. a first quench from about 880°c and then a final quench. During the first heating for quenching, the carbon concentration is reduced by diffusion. In addition it breaks up and dissolves the carbide network that usually forms when pack carburizing is used to give a deep carbon penetration.

Carburizing in a salt bath (liquid bath carburizing)

Temperature: 850–900 °C

Depth of case hardening mm	Method of hardening
< 0·50	Direct quench
0·50–1·0	Requench from 800–820 °C (Double quench)

For direct quenching, a carburizing temperature of 850–870 °C is recommended; for double quenching, 900 °C may be used.

Carburizing in gas (gas carburizing)

Temperature: 900–940 °C

The carburization is carried out to the required depth of case hardening and to a surface carbon concentration of 0·70–0·80%. When the process is completed the temperature in the furnace is lowered to 830 °C and the part is then quenched in a suitable medium.

Tempering

Case-hardened steels are tempered at temperatures generally around 160–220°C. Temperatures below 160°C should not be used, particularly if a grinding operation is to follow, since grinding cracks develop very easily. Tempering is not necessary after a cyaniding treatment that gives a case depth of only a few tenths of a millimetre.

The hardness falls quite rapidly when the steel is tempered between 160°C and 200°C. If a hardness of 60 HRC is required the tempering temperature should not be higher than 180°C.

6.5.7   Case hardening of tool steels

The wear resistance of tools made from certain medium-carbon steels may be increased by case hardening. This treatment is very often applied to simple punches, for instance, made from S 1. When such tools are being hardened they are heated in a cyanide bath at 900°C, i.e. the normal hardening temperature for the steel, for some 10–30 min longer than the

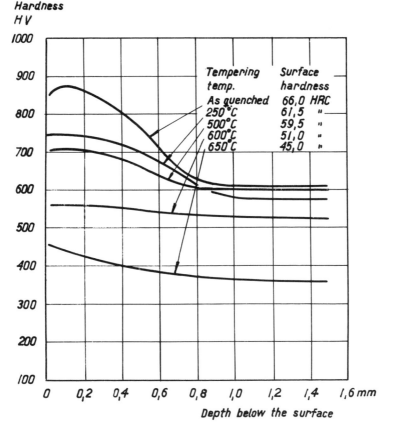

Figure 6.166 Case hardening of steel H 13. Pack carburizing for 3 h at 900°C, followed by cooling in the box. Oil quenched from 1000°C, tempering for 1 h at 250 to 650°C. Specimen 50 mm diameter × 30 mm

usual holding time and then quenched in the customary way in water or oil.

Hot-work tool steels are case hardened for certain applications. First, the tools are carburized at about 900°C, generally in a solid compound, and then allowed to cool in the compound. If the tool is to be put to hot-work service it is quenched from 1000°C. For cold-work applications it is better to use a lower quenching temperature, 980°C for example. This gives the steel a somewhat higher impact strength.

Figure 6.166 shows the results obtained from a case-hardening test with H 13. First, the specimens were pack carburized for 3 h at 900°C. They were hardened by quenching from 1000°C and then tempered at temperatures up to 650°C.

It has been observed both from laboratory tests and in actual practice that H 13 has a definite tendency to give irregular case-hardening results. This might possibly be due to the high Si content of the steel. On the other hand consistent results are always obtained when the steel QRO 45 BH 10A is case hardened. *Figure 6.167* shows the hardness transverse curves in QRO 45 after it had been case hardened and then tempered at various temperatures.

The heat treatment just discussed, involving low quenching temperatures, is applicable if the tool is to work at moderate temperatures. If high working temperatures are intended the tool, after carburization, should be rehardened by quenching from 1050–1100°C.

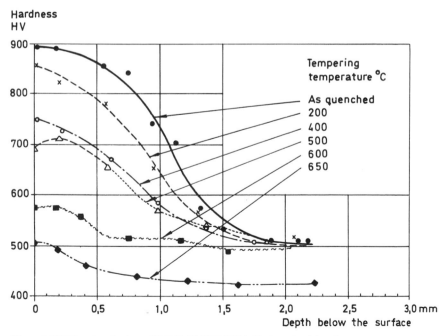

Figure 6.167 Case hardening of QRO 45 (BH 10A). Pack carburizing for 5 h at 950°C. Oil quenching followed by tempering for 1 h at various temperatures

6.5.8 Protecting against carburization (selective carburizing)

For various reasons it may be necessary to perform some machining operation after the carburizing treatment. Hence restricted carburization may be carried out according to some of the methods outlined below.

1. An *extra machining allowance* greater than the total depth of carbon penetration and usually about three times D C, is allowed to remain on those surfaces that are to be left soft. If the parts are pack carburized they should be allowed to cool slowly in the box. After this they are generally quite machinable. If the parts are too hard they may be tempered at 650°C. Should they have been quenched from the carburizing temperature they must be tempered. After machining the parts are hardened in the usual way.

2. *Protective pastes* of 'N O C A S E' type are applied to those surfaces that are to remain soft. These pastes are made up of a powder that contains mainly copper suspended in a varnish-type binder. The pastes are not suitable for use with gas or salt-bath carburizing. Specially compounded pastes are available for gas carburizing.

3. *Electrolytic copper plating* to a thickness of 30–50 μm. It is important to ensure that the parts are very carefully cleaned before being copper plated and that the copper bath itself is clean. An alkaline bath gives a more compact and even coating than an acidic one. Any pores in the copper deposit may be detected by means of a reagent consisting of 10 parts $K_4Fe(CN)_6$, 10 parts of NaCl and 20 parts of gelatine dissolved in distilled water. A filter paper is moistened with the solution and placed on the copper deposit. If there are any pores blue spots will appear on the paper in 3–5 min.

Those surfaces that are to be carburized can be laid bare in different ways:

(a) By machining off the copper deposit. In this case a machining allowance equal to the depth of cut must be planned beforehand. Great care must be exercised so that the copper coating is not damaged on those surfaces that are to be protected by it.

(b) By locally coating the surface with an insulating varnish or by sticking on a special tape before copper plating.

(c) By locally dissolving the copper deposit by chemical means. This method consists of painting the surfaces that are to remain soft with a special varnish (Bonosol B 40) after the copper plating is completed. A red pigment is added to the varnish for detection purposes. When the varnish has dried the copper deposit is dissolved from the unprotected areas by immersing the parts in a solution containing 5 1 water, 2·4 kg chromic acid and 125 ml concentrated sulphuric acid. Before commencing the carburizing treatment the protective varnish is dissolved by exposing it to trichlorethylene vapour for 2 min. *Figure 6.168* shows a component that has been treated according to the above.

Electrolytic copper plating also serves to protect a surface against carburization by a cyanide bath of the 10% NaCN type. Higher concentrations of NaCN will cause the copper deposit to dissolve fairly

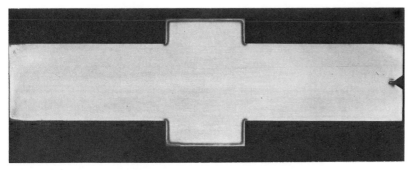

Figure 6.168 Selective carburized part shielded by copper plating

quickly. Even a low concentration of cyanide will gradually dissolve copper. Regular carburization of copper-plated parts in a cyanide bath should be avoided since the copper that is dissolved in the salt will cause the bath gradually to lose its carburizing power.

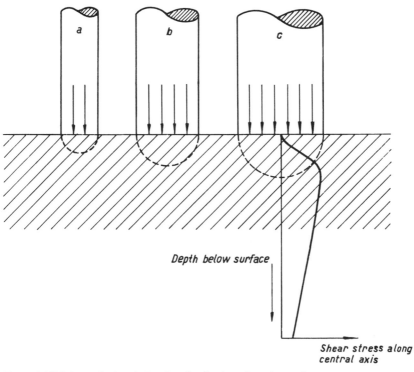

Figure 6.169 Schematic sketch showing distribution of maximum shear stresses created by a force pressing on a surface. Largest value of shear stress exists at points under the surface along an arc of a semicircle that runs from the edges of the contact surfaces. The unit pressure (in N/mm^2) is the same in all three cases. In (c) are drawn the shear stress variations along the central axis

6.5.9   Choice of case-hardening depth

When steel is case hardened it develops increased surface hardness and wear resistance, greater ability to withstand flexural and torsional forces and enhanced fatigue strength. The depth of case hardening is of decisive importance to several of these properties. If the surface of the steel is subjected to adhesive wear at low loads a case depth of some tenths of a millimetre is generally sufficient. As the load increases the case depth must also be increased.

The pressure applied to a case-hardened surface is propagated into the material. The most dangerous of the stresses thus created is considered to be the shear stress. In a section through a stressed surface the largest shear stresses act along a semicircle emanating from the edges of the surface to which the load is applied. If the same unit stress (N/mm^2) is applied, the distance down to the greatest shear stress increases as the area of the surface under load increases (*see Figure 6.169*). If the shear stress directly below the hardened case exceeds the yield strength in shear of the material the latter will deform plastically and the case will have to support a greater load which may cause cracking. Hence the choice lies between, increasing the depth of case hardening or increasing the core strength of the steel.

The choice of case depth is often a matter of experience and is rarely the result of theoretical considerations. In actual practice, calculations are made complicated by the fact that the load is seldom evenly distributed across the surfaces of contact between two bodies. For case-hardened gears the design load is proportional to the gear module. The depth of case hardening is usually taken to be 0·15–0·18 times the module. Case depths of more than 2·0 mm are rarely encountered in gears.

6.6 Carbonitriding

6.6.1   Definition

During carbonitriding, C and N are absorbed simultaneously in the steel, the N increasing the hardness of the carburized layer. This process can be carried out in a salt bath or gas. The treatment temperature is normally 800–900°C, but both lower or higher temperatures can be employed. Basically, carbonitriding in a salt bath is the same as cyanide-bath hardening. In the section on case hardening we have already discussed N pick-up during carburizing in the cyanide bath. However, unless anything to the contrary is stated, by carbonitriding is meant carburizing in gas with simultaneous N pick up. Carburizing takes place in the same way as described previously. Nitrogen, in the form of gaseous ammonia, is supplied direct into the furnace or in the same feed pipe as is used by the carburizing gases. The ammonia content is usually kept between 3 and 8%. The normal value for the N content of the surface layer is about 0·4%. After treatment the steel is hardened in the same way as after conventional gas carburizing.

In some cases ammonia is supplied only during the last 30–60 min of the process. The object of this procedure is to avoid the soft skin which can occur as a result of internal oxidation in the surface layer. As a result of such internal oxidation, the alloying elements are combined with oxygen and thus do not contribute to hardenability.

The description 'carbonitriding' is somewhat misleading, because we are not dealing with nitriding of the lower temperature type described in the following Section 6.7. The main emphasis is on carburizing and more correctly the process should be called nitro-carburizing. The name 'carbonitriding' is however so widely used now that it probably cannot be changed. Since 'correct' carbonitriding, i.e. nitriding with a supply of carbonaceous gas to the ammonia gas, has begun to be used on an increasing scale (*see* Section 6.7.2), we should perhaps speak of carbonitriding above or below A_1.

6.6.2   Theoretical background

At the treatment temperatures normally applied the solubility of N in α-iron is approximately 0·1%. Its rate of diffusion is roughly the same as that of C, as shown in *Table 6.39*. The diffusion of N in α-iron proces $\sim$ 50 times more quickly than it does in γ-iron. The corresponding figure for C is 40. However, the equilibrium diagram for iron–nitrogen resembles the iron–carbon diagram (*see Figure 6.175*) and at the temperatures used for carbonitriding, nitriding can be taken in principle as equivalent to carburizing.

Table 6.39 The diffusion coefficients of nitrogen and carbon for diffusion in alpha iron (according to Fast and Verrijp)

Temperature °C	Diffusion coefficients $cm^2 \cdot s^{-1}$ in α-iron	
	N	C
300	$5·3 \times 10^{-10}$	$4·3 \times 10^{-10}$
500	$3·6 \times 10^{-8}$	$4·1 \times 10^{-8}$
700	$4·4 \times 10^{-7}$	$6·1 \times 10^{-7}$
900 (extrapolated)	$2·3 \times 10^{-6}$	$3·6 \times 10^{-6}$

6.6.3   Practical results

Chatterjee-Fischer and Schaaber[55] studied the relationship between N content on the one hand and treatment temperature, the amount of ammonia supplied and the C content on the other. *Figure 6.170* shows the effect of treatment temperature and the amount of ammonia supplied. The N content increases with increasing quantity of ammonia and decreases with increasing temperature. *Figures 6.171a* and *b* show that the C content has very little influence on the N content with ammonia addition up to 5%.

The investigations were made on iron foils 0·05 mm in thickness. Parallel tests were also made using cylindrical test specimens made from plain-carbon and alloy steels, from which thin surface layers were machined off to enable the N content to be determined. In practically all cases the same N contents were found in the removed 0·1 mm outer layer, as were found in the foils. *Figure 6.172* shows a typical distribution of C and N in the surface layer of a carbonitrided mild carbon steel.

Prěnosil[56] made detailed studies of the effect of N on the hardenability

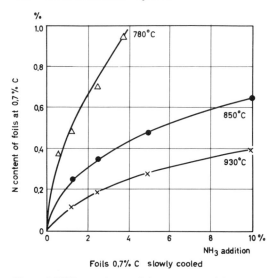

Figure 6.170 Dependence of nitrogen potential at different temperatures on ammonia addition during carbonitriding (after Chatterjee–Fischer and Schaaber[55])

of steel and established that contents of around 0·20% N have a favourable influence. This is illustrated in *Figure 6.173* which is a TTT diagram for a steel with 0·80% C and 0·22% N. For comparison purposes the transformation curves have been drawn for the same steel without N. Apart from a direct increase in hardenability, N results in a lowering of both A_1 and M_s. Contents higher than 0·20% increase hardenability to a moderate extent, but lower A_1 and M_s considerably. The values in *Table 6.40* were obtained on the same steel as above, with higher N contents. Since N lowers M_s substantially the amount of retained austenite in alloy steels can become high when carbonitriding.

Tests have also been made using Jominy test bars which were carburized or carbonitrided using different N activities[56]. The test bars were then austenitized at 820°C for 30 min and quenched in the conventional manner. The hardness was measured after the test bars had been ground down to depths between 0·1 and 0·5 mm below the surface. The contents of C and N in these layers were analysed. The results obtained are shown in *Figure 6.174*. With N contents below 0·20% there is hardly any tendency towards increased hardenability, but this tendency increases noticeably with higher N content. The increase in hardenability can be studied in the

Table 6.40 The effect of nitrogen content on A_1 and M_s for a steel containing 0·80% C

N%	A_1 °C	M_s °C
0·0	720	205
0·39	682	154
0·66	670	108

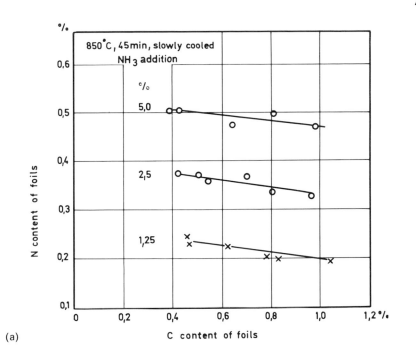

(a)

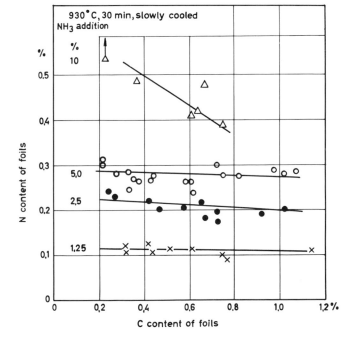

(b)

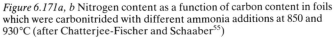

Figure 6.171a, b Nitrogen content as a function of carbon content in foils
which were carbonitrided with different ammonia additions at 850 and
930°C (after Chatterjee-Fischer and Schaaber[55])

diagrams at Jominy distances around 10 mm. If the nitrogen content exceeds about 0·6% there is a risk that pores may appear in the surface zone, which in turn will reduce fatigue strength very considerably, especially if the load tends to exceed the fatigue limit[57].

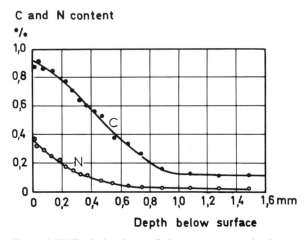

Figure 6.172 Typical carbon and nitrogen concentration in a mild carbon steel after carbonitriding at 860°C (after Prěnosil[56]). ● = C, ○ = N

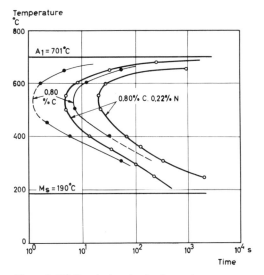

Figure 6.173 Graph showing isothermal transformation of austenite with 0·80% C and 0·22% N after carbonitriding at 860°C (after Prěnosil[56]). For comparison we have included transformation curves of a carbon steel containing 0·80% C

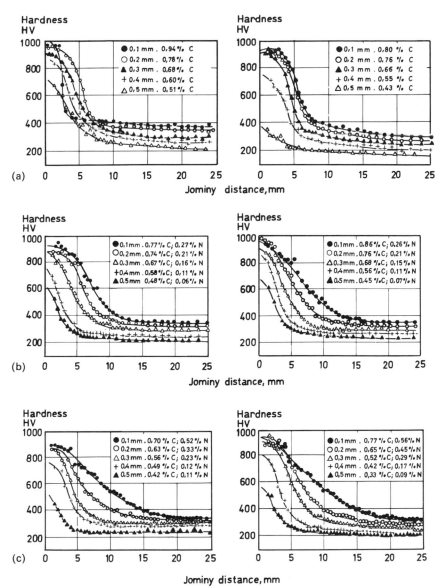

Figure 6.174 Jominy diagrams. After either carburizing or carbonitriding the Jominy test specimen is austenitized at 820°C for 30 min. Base material is 0·08% C, 0·19% Si, 0·40% Mn. The keys in the diagrams give the carbon and nitrogen content at different depths below the surface (after Přenosil[56]): (a) carburized Jominy tests specimen; (b) carbonitrided Jominy test specimen (0·06–0·27% N); (c) carbonitrided Jominy test specimen (0·09–0·56% N)

6.6.4 Conclusions

1. Carbonitriding gives increased hardenability and wear resistance compared with ordinary case hardening.
2. Carbonitriding gives a hard and uniform layer or case to such plain-carbon steels as tend to have soft spots when quenched in water.
3. Plain-carbon and low-alloy steels 'which normally have to be quenched in water may, when carbonitrided, sometimes be quenched in oil. This is applicable to small parts only. The potentiality of nitrocarburizing for larger parts must not be overestimated.

6.7 Nitriding

The carburizing treatments which were discussed in the previous section are austenitic thermochemical treatments, in that they involve the diffusional addition of the interstitial alloying elements into the austenite phase and rely on the subsequent transformation of the austenite to martensite to produce a high surface hardness. The process of nitriding, however, is a ferritic thermochemical treatment and usually involves the introduction of atomic nitrogen into the ferrite phase in the temperature range 500–590°C and consequently no phase transformation occurs on cooling to room temperature. The method was first used at the end of the 1920s and since then its application has continuously spread, due among other things to the fact that the process has been further developed and can now be applied to a much larger number of steels than was originally thought possible.

The properties imparted to steel by nitriding can be summarized as follows:

1. High surface hardness and wear strength, together with reduced risk of galling.
2. High resistance to tempering and high-temperature hardness.
3. High fatigue strength and low fatigue notch sensitivity.
4. Improved corrosion resistance for non-stainless steels.
5. High dimensional stability.

These properties will be dealt with later in this chapter; dimensional stability, however, will not be discussed until Chapter 7. The main factors which govern the depth of the nitride case are: treatment time, temperature, nitrogen activity and steel composition.

When studying the nitriding process we make use of the iron–nitrogen equilibrium diagram (*see Figure 6.175*). At the nitriding temperatures customarily used, the nitrogen will dissolve in iron, but only up to a concentration of 0·1%. When the nitrogen content exceeds this value, γ'-nitride is formed, the chemical formula of which is Fe_4N. If the nitrogen concentration exceeds about 6%, the γ'-nitride starts to change into ε-nitride ($Fe_{2-3}N$). Below 500°C ζ-nitride begins to form. The nitrogen content of this phase is about 11% and its chemical formula Fe_2N.

When making observations in the metallurgical microscope the γ' and ε-nitrides can be seen as a white surface layer, called the 'white layer' or 'compound layer', *Figure 6.176*. Simultaneously with the increase in thickness of the white layer during nitriding, the nitrogen diffuses further into the steel. When the solubility limit is exceeded, nitrides are precipitated at the grain boundaries and along certain crystallographic planes.

Among the alloying elements, A1, Cr and Mo are used as nitride formers. Nitrogen together with carbon, forms carbonitrides. The amount of carbonitride can be increased by simultaneous pick-up of nitrogen and carbon by the steel.

The high surface hardness which is obtained after nitriding is due to the formation of finely dispersed nitrides and carbonitrides which distort the ferrite lattice. Basically, all steels can be nitrided[58] and for further details regarding the different types of steel, see p. 523.

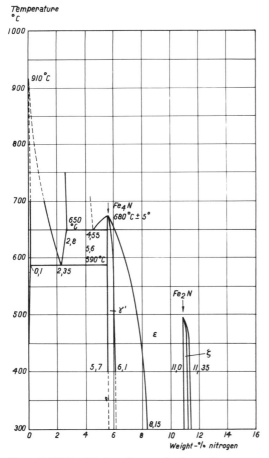

Figure 6.175 Equilibrium diagram for iron–nitrogen (M. Hansen)

Figure 6.176 White layer and diffusion layer with precipitated acicular iron nitride in steel containing 0·15% C after gas nitriding for 10 h at 500°C, 500 ×

6.7.1   Determination of depth of nitriding

The depth of nitriding, designated DN, is determined in accordance with SS 11 70 09 by taking HV measurements on a cross-section through a nitrided surface. The limiting hardness is 400 HV. This figure is applicable only to constructional steels of the type included in *Table 6.41*. Other limiting values should be chosen for other steels. The Standard is applicable to nitriding depths of 0·3 mm and more. For depths smaller than 0·3 mm the Standard SS 11 70 10 applies, which is most nearly equivalent to ISO 4970–1979, *Steel-Determination of Total or Effective Thickness of Thin Surface-Hardened Layers*. However, this International Standard is based on somewhat different norms and specifies a method of measuring the total or effective thickness of thin surface-hardened layers having thicknesses less than or equal to 0·3 mm, obtained, for example, by mechanical (shot-blasting, shot-peening, etc.) or thermochemical (nitrocarburizing, carburizing and hardening, etc.) treatment. The two methods usually selected are the micrographic method and the microhardness measurement method.

The choice of the method and its accuracy depend on the nature of the thin layer and on its presumed thickness. Since the method used also affects the result obtained, the choice has to be made by prior agreement between the parties concerned. Surface-hardened layers with thicknesses of more than 0·3 mm are covered by ISO 2639, *Steel-Determination and Verification of the Effective Depth of Carburized and Hardened Cases*; and ISO 3754, *Steel-Determination of Effective Depth of Hardening After Flame or Induction Hardening*.

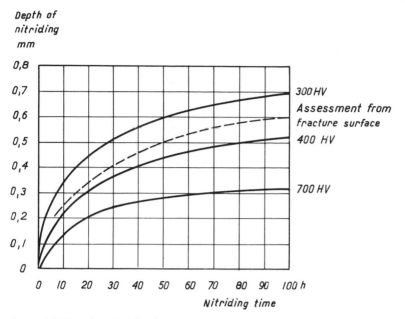

Figure 6.177 Depth of nitriding for B S 905M39, En 41 B as function of
nitriding period at 510°C

The latter of the two methods consists in determining the changes in
Vickers or Knoop microhardness under a 2·94 N (300 gf) load, measured
from the circumference inwards to the centre of the product.

The *total thickness* of the thin surface layer is defined by the distance
from the surface to the limit beyond which the hardness of the unaffected
metal is reached.

The *effective thickness* is defined by the distance from the surface to the
limit beyond which the required hardness (reference hardness) is reached.

The depth of nitriding can be assessed with the aide of a fractured
surface in the same way as that employed for case-hardening steels. *Figure
6.177* shows the agreement obtained between the values assessed and
determined at 400 H V for steel B S 905M39 (En 41 B).

6.7.2   Methods of nitriding

The four main methods of nitriding are gas nitriding, salt-bath nitriding,
plasma nitriding and powder nitriding. There are variations within these
main methods and the commonest variations will be described. Nitriding
by means of ammonia gas, which involves the absorbtion of nitrogen only,
is usually referred to simply as nitriding or gas nitriding. When besides
nitrogen, carbon also is introduced the treatment is called nitrocarburizing,
and this process may take place in a gaseous atmosphere or in a salt bath.

Gas nitriding with ammonia

In the original method of gas nitriding, ammonia is allowed to flow over the parts to be hardened, normally at about 510°C. The ammonia dissociates in accordance with the equation:

$$2NH_3 \rightarrow 2N_{Fe} + 3H_2$$

At the instant of dissociation, nitrogen occurs in the atomic form, and as such it can be absorbed by the steel. *Figure 6.178* gives a schematic representation of the dissociation and nitrogen pick-up in the steel.

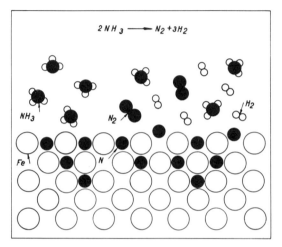

Figure 6.178 Dissociation of ammonia and nitrogen pick-up in steel during gas nitriding

Gas nitriding is used for parts that require a case depth between 0·2 and 0·7 mm. Nitriding is undertaken in an electric furnace equipped for precise temperature control. For large-scale nitriding pit furnaces with specially designed containers provided with tightly closing lids are used (*see Figure 6.179*). For small-scale nitriding operations a conventional muffle furnace with a separate nitriding box, as shown in *Figure 6.180*, can be used. Aluminium sheet or asbestos yarn is used as the caulking material between the flange and cover of the box.

The boxes used for nitriding should be made from a material which does not react with the gases. Nickel, inconel and similar alloys are ideal, but heat-resisting steels with 25% Cr and 20% Ni, particularly, have also proved quite suitable for this purpose. If only a few nitriding operations are to be carried out in an *ad hoc* installation, the nitriding boxes can be made from iron sheet. However, the iron will react with the gas and hence it is difficult to control the processes taking place during the initial nitriding operations in a newly made nitriding box as some of the gas is used up in nitriding the inner surfaces of the box.

The gas always contains a small quantity of water vapour, which must be removed before the gas is introduced into the nitriding box, otherwise the water vapour can cause oxidation of the nitrided parts. This removal is best

Figure 6.179 Various components for nitriding placed in charging basket, ready for lowering into nitriding retort in modern pit furnace at Bofors

Figure 6.180 Standard-type muffle furnace used for nitriding

achieved by passing it through a filter of unslaked lime, which should be replaced or regenerated by heating it at 1000°C before all the lime has become slaked.

To check that the ammonia gas in the nitriding box has decomposed (dissociated) according to the requirements of the process, the composition of the exit gas is determined by means of a dissociation pipette. The sketch in *Figure 6.181* shows the general arrangement of one type of pipette customarily employed. During the course of the nitriding process, or

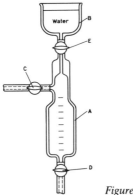

Figure 6.181 Dissociation pipette for nitriding

preparatory to making a measurement, the gas passes through the taps *C* and *D*. During the measurement these taps are closed after which tap *E* is opened. The water contained above at *B* is then sucked into the meter because the undissociated ammonia dissolves in the water, which then takes up precisely the volume occupied previously by the ammonia. The remaining gas, i.e. the dissociated ammonia, which consists of nitrogen and hydrogen, does not dissolve in the water. *Figure 6.182* shows some examples of height of the water column for different degrees of dissociation.

The parts to be nitrided should be well cleaned and degreased. Under no circumstances should there be any traces of rust or mill scale. If certain areas of the part are to remain soft, they can be given an electrolytic

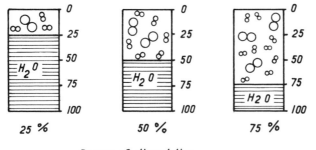

Degree of dissociation

Figure 6.182 Height of water column in meter for different degrees of dissociation

coating, e.g. of copper, tin or nickel, which prevents the nitrogen from diffusing into the steel. 'Stopping-off' agents containing tin are also available, and these are painted on to the surfaces to be protected. Since the layer of tin is molten at the nitriding temperatures, it is most important that the coating should be so thin (not exceeding 0·01 mm) that it does not spread to surfaces that are to be nitrided.

The parts are placed inside the nitriding boxes in such a way that all surfaces come into contact with the gas. After the box has been closed, it is purged with ammonia gas until all the air has been expelled, after which it is placed inside the furnace. The composition of the exit gas is checked regularly, using the dissociation pipette, and for the first 5–10 h it is kept at 15–20% residual gas. If determinations indicate a lower value than this, the gas velocity is too high and should be reduced. If the converse should prevail, the gas velocity must be increased. When, after adjustment, the system has arrived at a steady-state condition, it is enough to check the gas composition two or three times per day, a residual gas content of about 50% being regarded as a suitable value. Formerly a value of about 30% residual gas was regarded as satisfactory, but this can result in a more brittle nitrided layer. For further details on this point *see* Section 6.5.4.

When the nitriding is completed, the box is removed from the furnace without interrupting the flow of gas. After the charge has cooled to 200 °C, the gas supply is cut off and the gases remaining in the box are most conveniently expelled with compressed air before the box is opened. The nitrided parts now normally exhibit a characteristic matt grey colour. If on occasions the colour has shades of yellow, blue or purple, it does not mean that the nitriding procedure has been at fault. The various tints derive from the presence of oxygen in the system, which may have originated from incomplete drying of the gas, or from some leak in the box or in the supply tubing.

Gas nitriding with ammonia and nitrogen or hydrogen

Although this method has very seldom been used it is included here for the sake of completeness. Minkevič and Sorokin[59] recommend a mixture of 20% ammonia and 80% nitrogen. The low ammonia content gives a low nitrogen activity, and this results in a tougher case. However, similar results can be achieved by increasing the degree of dissociation when nitriding with ammonia only or by adding hydrogen.

Gas nitriding with ammonia and hydrocarbons (nitrocarburizing)

When gas nitriding with ammonia, hydrocarbons can also be added, usually in the form of pure propane or endogas, produced from propane and air. Nitrocarburizing takes place in a gaseous atmosphere by several different methods. These methods differ from one another according to the character and composition of the carbonaceous gases. Three of the most popular methods are referred to below:

Designation	Carbonaceous gas
Nitemper	Endothermic gas
Triniding	Exothermic gas + natural gas
Nitroc	Exothermic gas with about 90% N

Nitrocarburizing is performed at a somewhat higher temperature than normal gas nitriding, e.g. 570°C. The carbon diffuses into the steel at the same time as the nitrogen and together they form a carbonitride mainly of the ε-type which is illustrated in *Figure 6.183*[60]. The composition and micro-hardness of the surface layer of pure iron which has been nitrided in an atmosphere of 50% ammonia and 50% propane at 580°C for 4 h is shown in *Figure 6.184*[61]. The high carbon content in the surface layer gives improved wear resistance and less risk of scuffing.

Bell and Lee[62] studied the compound layer formed on pure iron and on En 32 (0·12% C, 0·60% Mn). Specimens were nitrocarburized in an endothermic gas generated from methane and which contained 32% ammonia in the exhaust gas. The treatment was carried out at 570°C for 3 h.

It was found that the compound layer inherits the matrix carbon content as well as obtaining carbon from the gaseous atmosphere. The carbon content of the compound layer was found to be 1·25% for the pure iron specimen and 1·50% for the En 32 specimen. The average N/C ratio of the compound layer on pure-iron samples was 6·6, whereas the N/C ratio with En 32 was 5·4. The average combined interstitial content of the compound layer formed on both materials was found to be within the region of the ε-carbonitride phase field.

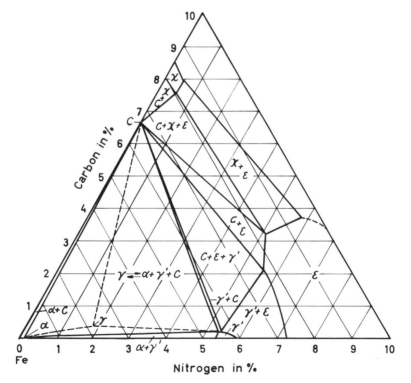

Figure 6.183 Isothermal section of the Fe–C–N phase diagram at 565°C (after Naumann and Langenscheid[60]) C = cementite

X-ray structural analysis confirmed that the compound layer did consist mainly of the cph phase, ε-carbonitride, which has a variable composition in the range $Fe_3(CN)$ and $Fe_2(CN)$. The carbonitride phase is derived from the ε-nitride phase by the replacement of nitrogen atoms by carbon. A small amount of Fe_4N and traces of oxygen were also present in the compound layer.

Plasma or glow-discharge nitriding

The glow-discharge nitriding process was patented by Berghaus at the beginning of the 1930s. For many years this was not used extensively on an industrial scale, because of relatively expensive equipment and the diversified technology which is essential if the process is to be carried out correctly. Thanks to development work during recent years, the process is gaining in popularity[63, 64, 65].

The glow-discharge or plasma nitriding process makes use of an ionized gas that serves as a medium for both heating and nitriding. The parts to be treated are charged into an airtight chamber, which constitutes the anode of the plasma nitriding unit. The parts making up the charge are so arranged that they are in electrical contact with the cathode, either by being placed on a bottom plate or suspended in a suitable manner. After being evacuated the chamber is filled with the process gas which, besides nitrogen, may contain hydrogen and methane. Thus, plasma nitriding may function both as conventional gas nitriding and as nitrocarburizing. The

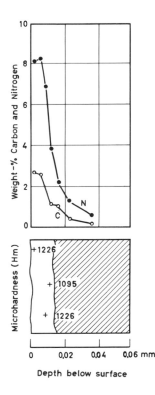

Figure 6.184 Composition and microhardness of pure iron after nitriding for 4 h at 580°C in a mixture of 50% ammonia and 50% propane. Degree of dissociation of ammonia was 20% (after Prěnosil[61])

working pressure lies between 0·1 and 10 mbar. On applying a potential difference of 500–1000 V the gas becomes ionized with the result that positive nitrogen ions having a high kinetic energy bombard the charge. Thereby the parts become heated to the required temperature and at the same time the nitrogen required for the nitriding process is introduced into the surface of the parts. All surfaces of the charge may be subjected to ion bombardment. This causes the formation of a blue-white glow discharge plasma around the components. *See Figure 6.185*. The temperature is controlled by means of thermocouples located in the charge. The heating as well as the cooling after treatment may be controlled without difficulty.

Figure 6.185 During plasma nitriding a blue-white glow discharge plasma appears round the components in the charge

The process may be operated at temperatures between 420 and 700°C which admits of a large degree of flexibility. Maraging steels, for example, may be simultaneously plasma nitrided and precipitation hardened at 420–480°C. As with gas nitriding, the time of plasma nitriding is governed by the required case depth. Plasma nitriding gives a nitriding depth that is about the same as that obtained by conventional gas nitriding at the same temperature. However, in actual practice for treatment times, up to approximately 30 h, the depth of hardening with plasma nitriding is greater than that of conventional gas nitriding since the heating up time for plasma nitriding is appreciably shorter.

The composition of the compound layer can be controlled so that either γ' or ε-nitride is formed or that both phases form simultaneously. This is the most usual case according to Varchoschkow and Toschkow[66]. Other authors claim that the most usual situation involves the formation of the monophased γ', Fe_4N nitride. Stainless steels and other heat-resisting steels (chromium-alloyed) which generally are difficult to nitride by other methods are readily plasma nitrided.

The only factor restricting a more general application of the plasma nitriding process seems to be the high installation costs, but taking into

Figure 6.186 Plasma nitriding unit at Brukens Härdverkstäder (The Bruken
Heat-Treatment Plant) at Älvsjö, Sweden

account that it is possible to control the composition of the compound layer
it should be accepted that at present this is, technically, the most advanced
process. *Figure 6.186* shows a photograph of a modern plasma nitriding
unit.

Salt-bath nitriding

The nitriding of steel in a bath of molten salt is almost as old as the original
gas nitriding method. The salt mixtures initially used contained 60–70% by
weight NaCN and 30–40% by weight KCN. In addition there are a few per
cent of carbonates (Na_2CO_3) and cyanates (NaCNO). By ageing the bath
at 575°C for 12 h (*see* Equation (6.8) on p. 453), the cyanate can be raised
to the desired level, about 45%. Steel parts should not be immersed in the
bath during the ageing stage. The normal working temperature for
salt-bath nitriding is 550–570°C and the holding time seldom exceeds 2 h.
The salt bath gives off carbon and nitrogen in accordance with Equations
(6.9) and (6.10) on p. 453.

Up to the middle of the 1950s, nitriding salt baths were used only on a
very restricted scale and then mainly for short-period nitriding (10–30 min)
of twist drills for example. If longer nitriding periods were used, pitting
occurred on the steel surfaces and in addition the results of the nitriding
were irregular. Both of these phenomena were due to variations in the
cyanate content at different levels in the bath and also to the solution of
iron from the crucible, resulting in oxidation of the surface of the steel
parts comprising the charge. During the last decades the method has

become more popular because, by injecting air into the bath, it has become possible to obtain better control of the cyanate content. A further improvement has been achieved by employing a titanium crucible, thus avoiding the decomposition of the salt which took place previously as a result of contact with the iron crucibles hitherto used. The method which has been developed by DEGUSSA is called the Tufftride process (German: *Teniferbehandlung*). Extensive literature has been published on the Tufftride process[67].

Another variation of salt-bath nitriding is 'Sulfinuz[68] treatment', in which the active constituent in the bath, apart from NaCN and NaCNO, is sodium sulphide (Na_2S). During conventional salt-bath nitriding, nitrogen and carbon are absorbed by the surface of the steel. During the Sulfinuz treatment, sulphur is present and this enhances still further the excellent anti-friction properties of the nitrided layer. *Figure 6.187* shows the composition of the surface layer of a carbon steel containing 0·20% C after Sulfinuz treatment at 570 °C for 2 h.

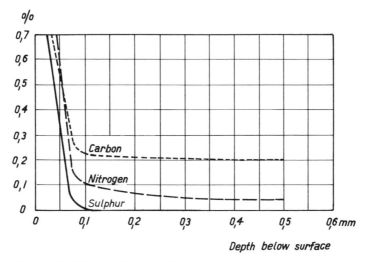

Figure 6.187 Sulphur, nitrogen and carbon contents in surface layer of mild steel after Sulfinuz treatment at 570 °C for 2 h

In some cases of Sulfinuz treatment it is possible for light corrosion of the steel surface to take place, and this can amount to a loss of 0·005 mm from the surface. Furthermore a thin high-sulphur layer can be formed on the outside of the steel parts. This layer should not be ground off, because it has high anti-friction properties. Should it have to be removed for dimensional reasons, this is best done with steel wool. Normally, grinding should not be performed after salt-bath nitriding.

After Tufftride or Sulfinuz the best surface is obtained if the steel parts are quenched in warm water. This rapid cooling results in a supersaturated solid solution of nitrogen in α-iron, which adds an extra contribution to the increase in fatigue strength which is a characteristic feature of nitriding. However, the high internal stresses introduced at the same time reduce toughness. With regard to plain-carbon and low-alloy steels, if these are tempered at 200 °C for 1 h, nitride precipitation takes place which reduces the fatigue strength and increases toughness[69]. During slow-cooling, as with gas nitriding, this precipitation is obtained automatically (*see Figure 6.176*).

It is essential to make a daily check on the cyanate content of the salt bath (also the sulphur content for the Sulfinuz bath). The deposits which form in the bath and which sink to the bottom of the crucible should be collected and removed by means of a scoop. The entire bath should be renewed at regular intervals.

Powder nitriding

In powder nitriding[70] the parts to be treated are packed into boxes in roughly the same way as when pack carburizing. About 15% by weight of an energizer is first placed on the bottom of the box, after which is laid a layer of nitriding powder on which the parts are placed. (The parts must not come into contact with the energizer). More nitriding powder is then added, one layer of powder alternating with one layer of parts. The boxes are then closed as tightly as possible and are then placed in a muffle furnace at a temperature of 520–570 °C. A nitriding period not exceeding 12 h is recommended.

Nitriding tests on steels B S 905M39 (En 41 B), B S 817M40 (En 24) and H 13, gave roughly the same values for hardness and case depth as those obtained during gas nitriding for corresponding times and temperatures. After powder nitriding, the steel surfaces can exhibit some pitting and spalling.

6.7.3 Comparison between gas and salt-bath nitriding

In the preceding section we have described gas and salt-bath nitriding without making any direct comparisons between these methods. To some extent the differences between the two methods have been exaggerated and many statements which must now be regarded as outmoded are still cited in literature references. In this instance gas nitriding implies nitriding by means of ammonia gas only. The statements usually made when comparing the two methods are as follows:

1. Gas nitriding requires 12–120 h nitriding time whereas salt-bath nitriding requires 1–4 h.
2. Gas nitriding makes the steel brittle, but this is not the case with salt-bath nitriding.
3. The reason why steel becomes brittle during gas nitriding is the presence of the brittle γ'-nitride which is not present after nitriding in the salt bath.

The usual reason for the statement that gas nitriding makes steel brittle is that gas-nitrided aluminium-alloyed steel is compared with plain-carbon or low-alloy steel that has been nitrided in a salt bath. It is difficult to find any test results in the literature relating to salt-bath nitrided aluminium-alloyed steel, and it is questionable whether any data on this subject have actually been published.

In other contexts the same steel has been employed for purposes of comparison, in such tests the case thickness of the gas-nitrided steel is considerably greater.

To permit a closer study of the circumstances prevailing during gas and salt-bath nitriding, a number of test discs were made from quenched and tempered Cr–Ni–Mo-steel, BS 817M40 and then nitrided; one test was carried out in gas for 10–40 h at 510°C, with 50% degree of dissociation, and a second test in a Tufftride salt bath, for 2 and 4 h at 570°C.

Surface hardness and depth of hardening

After nitriding, taper sections were ground on a plane specimen surface and the hardness was measured in HV, load 1 kp.

After *gas nitriding* at 510°C, the surface hardness was found to be about 675 HV. Nitriding at 550°C gave a lower surface hardness of about 575 HV, but on the other hand the depth of nitriding was greater. Nitriding for 20 h at 550°C corresponds to 40 h at 510°C.

The specimens after the Tufftride treatment gave somewhat differing results, depending on where they were treated. In one test the resulting difference in case depth between the 2 and 4 h treatment time was negligible, but in another instance there was a marked difference. In both instances the surface hardness was about 650 HV.

As shown in *Figure 6.188*, a comparison between gas-nitrided and Tufftrided specimens shows that 10 h of gas nitriding at 510°C corresponds very closely to 4 h nitriding in a salt bath at 570°C.

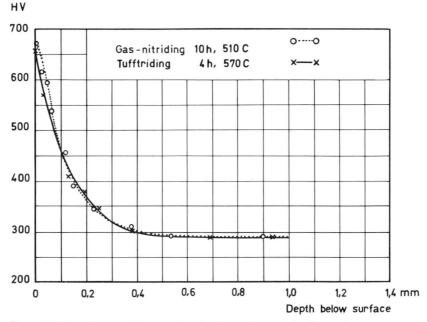

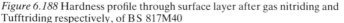

Figure 6.188 Hardness profile through surface layer after gas nitriding and Tufftriding respectively, of BS 817M40

Testing for brittleness

It is difficult to find a reliable method for assessing the brittleness of the white layer. In the present case the assessment was made by means of a metallographic study of a section through the white layer. In addition the hardness (HB 5/750) of the specimens was measured and the appearance of the indentations was studied at 50 × magnification. From this study the following facts emerged:

During *gas nitriding* the thickness of the white layer increased with increasing time and temperature. However, the character of the layer remained homogeneous throughout. On performing the Brinell hardness test on a specimen after it had been nitrided for 10 h at 510°C, both radial and concentric cracks were observed. After nitriding for a longer period or at higher temperatures concentric cracks only were observed, the width of which decreased with increasing nitriding time. This is probably due to the enhanced load-bearing capacity of the nitrided layer.

The *Tufftride* treatment also caused the thickness of the white layer to increase with treatment time. After a treatment of 4 h the layer contained pores which could clearly be distinguished at 500 × magnification. According to Finnern[71] this is a normal phenomenon that occurs even after a 90 min treatment. Brinell hardness testing gave rise to radial and concentric cracks after both a 2 h and 4 h Tufftride treatment.

Figure 6.189 shows the cross-section and test indentations in gas-nitrided and Tufftrided specimens. On the basis of these test methods it is difficult to make any assertions as to which method of nitriding gives the toughest surface layer.

Nitride formation

The composition of the nitride layer in the nitrided specimens was established by means of X-ray diffraction (Cu-radiation). *Figure 6.190a* shows original curves for some of the *gas-nitrided* specimens. After gas nitriding at 510°C for 20 h and 40 h no reflection was obtained for α-Fe. As result of the treatment time there was a reduction in the amount of γ'-nitride and an increase in the amount of ε-nitride. Gas nitriding at 550°C for 10 h and 20 h gave similar results.

After the 2 h Tufftride treatment at 570°C an indication of a α-Fe could be detected. Both γ'-nitride and ε-nitride were also present. As a result of a 4 h treatment at 570°C, the α-Fe disappeared, the γ'-nitride disappeared almost completely, but the ε-nitride increased (*see Figure 6.190b*). The investigation was repeated using Cr-radiation, and basically the same results were obtained.

These investigations have shown clearly that γ'-nitride is present both after gas nitriding and salt-bath nitriding at the temperatures employed. Nevertheless, the amount of γ'-nitride is greater after gas nitriding than after salt-bath nitriding.

Summarizing, the investigations made have shown that none of the three statements cited by way of introduction, and which derive from the published literature, turns out to have any validity whatever. The opinion, often put forward, that low-alloy steel should be nitrided in a salt bath only and not gas-nitrided can thus be disregarded.

508

375×

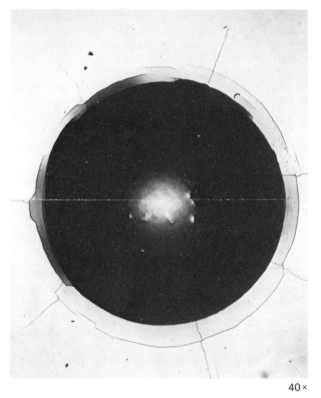

40×

Figure 6.189 Cross-section and test indentation marks on gas-nitrided and Tufftrided specimens of BS 817M40. Gas-nitriding 10 h, 510°C. Tufftriding 4 h, 570°C

TUFFTRIDING

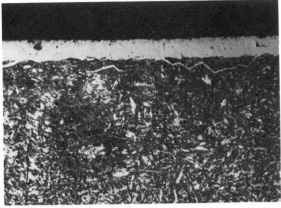

375×

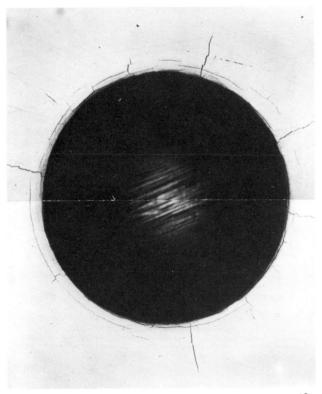

40×

Advantages and disadvantages

The following additional points can be made for the two methods.

1. *Treatment time*
   In this respect, salt-bath nitriding has its obvious advantages when compared with conventional gas nitriding. This applies particularly if a nitriding depth of 0·1 mm (at 400 HV) is adequate, which is normally obtained after salt-bath nitriding for 2 h with the steel investigated.

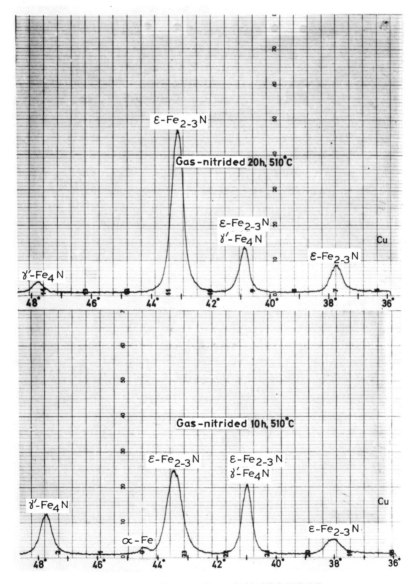

Figure 6.190a X-ray diffraction diagram of gas-nitrided BS 817M40;

2. *Cleanliness*

With gas nitriding there is no soiling of the work premises or of the nitrided components. However, with salt-bath nitriding, it is quite easy to spill salt on the top of the furnace and on the surrounding floor area. Furthermore, parts treated in the salt bath must be cleaned by rinsing in hot water. The cyanide salt is toxic, but cases of poisoning are very rare provided that the correct procedure is

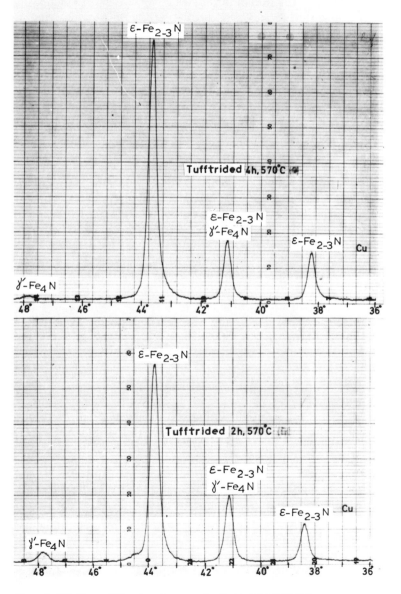

Figure 6.190b X-ray diffraction diagram of tufftrided B S 817M40

followed. Since the rinsing water together with residues of salt can easily be flushed into the drainage system, we can expect that in future the nature conservancy authorities will impose more stringent regulations. Even now, traces of cyanide must be detoxified in districts where the effluent water is subjected to bacteriological purification. The reason for this is that the cyanide destroys the bacteria that are used for purification purposes.

3. *Depth of nitriding*

As a result of the formation of pores, which increases with increasing treatment time during salt-bath nitriding, the nitriding period is restricted to 4 h. No such restrictions are imposed on gas nitriding, but for practical reasons a nitriding time of 90 h should normally not be exceeded. Often a greater depth of nitriding is required than that obtainable with salt-bath nitriding, and this is when gas nitriding comes into its own.

4. *Dimensional stability*

For the same depth of nitriding, the resulting distortion is the same for both methods, if by this we mean the change in dimensions occurring as a result of the compressive stresses induced in the steel during nitriding.

After gas nitriding, the box containing the nitrided parts is removed from the furnace and allowed to cool freely in air, whereas components which have been nitrided in a salt bath are quenched in water or oil after the treatment. As a result of the water quench thermal stresses are created which can result in changes of shape.

5. *Wear resistance*

During salt-bath nitriding a higher content of ε-carbonitride is formed than during conventional gas nitriding In critical cases the higher content of ε-carbonitride provides increased wear resistance and reduced risk of scuffing. By adding hydrocarbons during gas nitriding (nitrocarburizing) the content of ε-carbonitride can be raised to the same level as that obtained during salt-bath nitriding. Nitrocarburizing and salt-bath nitriding yield results that in actual practice are very similar.

6. *Toughness*

As a result of the rapid cooling employed during salt-bath nitriding, the toughness of the nitrided layer is inferior to that obtained by gas nitriding, where cooling takes place slowly. The practical consequences of this have been observed, for instance, when comparisons were made between gas-nitrided and salt-bath nitrided moulds made from quenched and tempered steel. Several instances of spalling were observed after salt-bath nitriding.

What has been said above indicates that the shorter treatment time and the resultant lower costs represent the only definite advantage that salt-bath nitriding can offer *vis-à-vis* gas nitriding, which possesses many advantages within other spheres. Hence the heat treater as well as the user of nitrided material has to select which method is the most favourable one for his special or general case.

6.7.4   Nitridability

The concept of 'nitridability' includes on the one hand the ability of the steel to absorb nitrogen, and on the other the increase in hardness imparted by the nitrogen.

The effect of the *alloying elements* has been described in Chapter 3. For the sake of convenience the diagram shown previously is reproduced in *Figure 6.191*. When deciding on the composition of a nitriding steel for which high surface hardness is required, the choice falls mainly on those elements that form nitrides, such as A1, Cr and Mo. The diagram shows how these elements along with a few others affect the surface hardness after nitriding. A1, closely followed by Ti, has the greatest effect, and then come Cr, Mo and V. In the Ni steels the same surface hardness has been obtained as in plain-carbon steels (about 400 HV). In steels containing several alloying elements higher hardness values are obtainable than if the alloying elements are used separately.

The depth of nitriding decreases with increasing content of alloying element, as shown in *Figure 6.192*. A1 and Ti, which have the greatest effect in increasing the hardness, also have the strongest delaying influence on the diffusion of nitrogen into the steel as their contents increase. An optimum value for hardness and depth of nitriding is obtained at roughly 1% A1, and this is also the content normally used in A1-alloyed nitriding

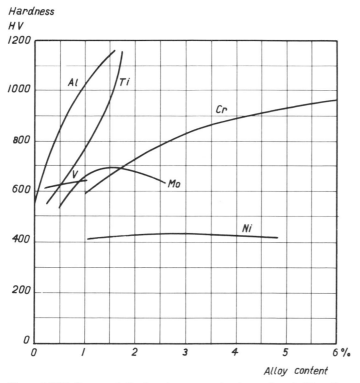

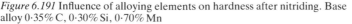

Figure 6.191 Influence of alloying elements on hardness after nitriding. Base alloy 0·35% C, 0·30% Si, 0·70% Mn

514

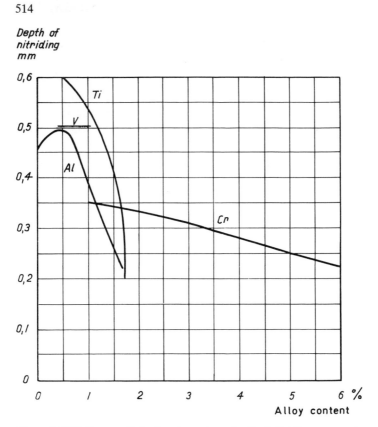

Figure 6.192 Influence of alloying elements on depth of nitriding measured at 400 HV. Nitriding for 8 h at 520°C

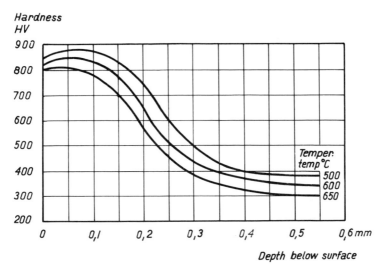

Figure 6.193 Influence of tempering temperature on hardness and depth of nitriding of steel En 29 B. Nitrided for 60 h at 510°C

steels. The reason for the inhibiting effect of the alloying elements is that
they bind the nitrogen as nitrides. Carbon, too, has a strong inhibiting
effect on the diffusion of nitrogen.

The *microstructure* influences nitridability in the following two ways. A
high content of free ferrite favours diffusion of nitrogen. A low carbide
content in the structure favours both the diffusion of nitrogen and
hardness. Usually alloy steels in the heat-treated state are employed for
nitriding, i.e. they have been quenched and tempered at 500–650°C.
Within this temperature range both precipitation and coagulation of
carbides takes place. Since carbide precipitation starts preferentially at the
grain boundaries, a strong barrier is formed at these points, where the
diffusion-retarding effect is accentuated because diffusion—which
normally takes place more rapidly at the grain boundaries than through the
grains—is impeded by the nitrides and carbonitrides formed in the grain
boundaries. The effect of the tempering temperature used during the heat
treatment is illustrated in *Figure 6.193*, which applies to steel En 29 B.

Tests have shown that the nitridability of this steel after (a) induction
hardening and tempering at 180°C, and (b) surface decarburization, is
better (*see Figure 6.194*)—than after quenching and tempering (tough
hardening).

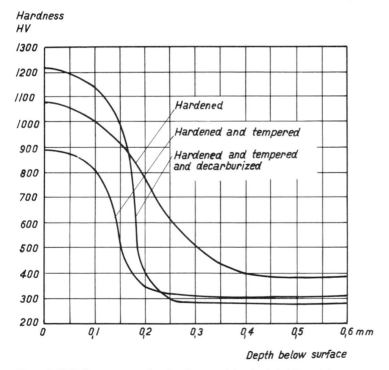

Figure 6.194 Influence on surface hardness and depth of nitriding of the
microstructure of En 29 B after glow discharge nitriding for 8 h at 520°C
(after Noren and Kindbom[72])

Hodgson and Baron[73] have made detailed tests on the nitriding of heat-treatable steels containing approximately 3% Cr and 0·50% Mo, with carbon contents 0·10–0·37%, and subjected to varying tempering temperatures. The results are summarized in *Figure 6.195* which, among other things, shows that the response of the steel to elevated tempering temperatures is dependent on the carbon content.

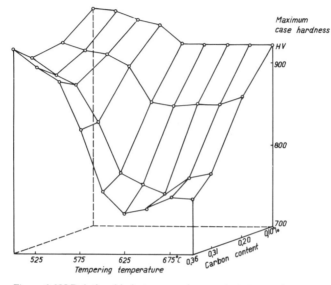

Figure 6.195 Relationship between carbon content, tempering temperature before nitriding and hardness after nitriding. Tempering time 100 h. Base analysis 3% Cr and 0·5% Mo (after Hodgson and Baron[73])

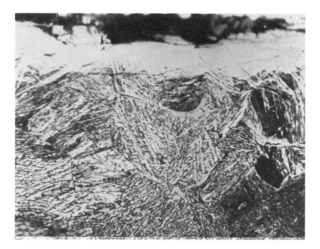

Figure 6.196 Nitrided surface layer with carbonitrides in grain boundaries, ISO steel 3 (En 41 B), 400 ×

The *grain size* and *carbide distribution* in the steel influence the surface smoothness obtained after nitriding. With increasing grain size, the nitrides are precipitated in wide bands along the grain boundaries. If carbides have already been precipitated at the grain boundaries, the nitrogen forms carbonitrides with the carbides, which will hereby increase in size. As a result the carbonitrides can displace individual grains in the surface layer, so that the white layer will become deformed or crack (spall). The photograph in *Figure 6.196* gives an idea of the mechanism involved.

Measurements of surface smoothness on nitrided specimens of SS 2940 (ISO steel 3) show that there is a definite correlation between grain size and surface smoothness, as has been shown by measurements made on material taken from the surface and centre of hardened and tempered rods, roughly 150 mm diameter. As shown in *Figure 6.197*, the surface smoothness of the nitrided specimens taken from the centre is inferior to the smoothness of those taken from the surface. The centre portions contain considerably more grain-boundary precipitates than the sections located near the surface, owing to the slower cooling of the centre during hardening. This investigation resulted in a modification of the composition of steel SS 2940 (ISO steel 3), so that its hardenability improved and the tendency to form grain-boundary precipitates in sections exceeding 100 mm was reduced (ISO steel 4). *See Table 6.39 p. 524.*

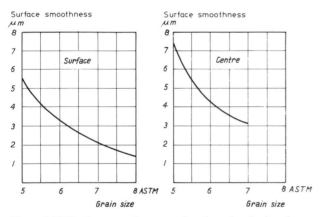

Figure 6.197 Surface smoothness as a function of grain size of nitrided specimen of ISO steel 3 taken from surface and centre

Figure 6.198 shows the appearance of the nitrided surface of SS 2940, unmodified composition (ISO steel 3), with different grain sizes. If high surface smoothness is required in the central sections when nitriding large dimensions of ISO steel 3, it is advantageous to carry out the hardening and tempering after the rough machining.

During hot working or heat treatment of steel SS 2940, the surface layer may become decarburized, and results in grain growth. On nitriding, this will have the consequences outlined above. If the machining allowance has been inadequate, spalling will occur during nitriding (*Figure 6.199*). If stress relieving takes place at too high a temperature, recrystallization with

(a)   (b)

Figure 6.198 Appearance of surface after nitriding of steels with different grain size, ISO steel 3. (a) Grain size 7·5 A STM. Surface smoothness 1·5 μm; (b) Grain size 5 ASTM. Surface smoothness 5 μm

the formation of columnar grains may occur in the surface layer. This type of grain growth, *Figure 6.200*, also causes rough surfaces and spalling.

The effect of *nitriding time* on the depth of nitriding can, for low-alloy steels, be derived from the simple formula for diffusion:

$$D_N = k\sqrt{t}$$

where D_N = depth of nitriding
k = a constant
t = time in hours

Figure 6.199 Spalling of surface layer of machine component made from BS 905M39, En 41 B. Component did not have sufficient machining allowance before nitriding

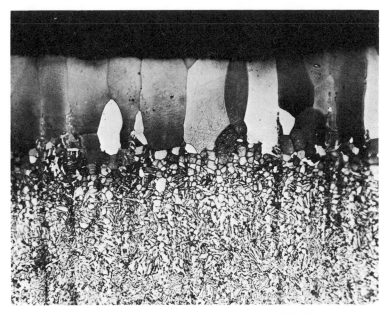

Figure 6.200 Columnar grain growth in surface layer of B S 905M39, En 41 B after stress–relief annealing at 700 °C and subsequent nitriding, 200 ×

The constant is a function of temperature and material.

When studying the nitrogen content at different depths below the surface after salt-bath nitriding of a carbon steel containing 0·15% C, the values shown in *Figure 6.201* were obtained. These confirm the validity of the equation shown above quite well, if the calculations assume a constant nitrogen content. Basically, the effect of the nitriding temperature is such that the lower the temperature the higher is the surface hardness but, at the same time, the smaller is the depth of nitriding. This applies particularly to the alloy steels and is shown by way of example in *Figures 6.202* and *6.203* which apply to steels En 41 B and En 29 B.

Nitrogen activity is another of the controlling factors during nitriding. In accordance with the laws governing diffusion, the degree of nitrogen penetration is governed by the temperature and the nitrogen content which can be built up in the outermost layer of the steel. In the nitriding salt bath, nitrogen activity can be controlled by means of the cyanate content which, as mentioned previously, should be maintained at about 45%. During gas nitriding the nitrogen activity is controlled by the degree of dissociation and the flow rate of the gas and the following equation is assumed to apply:

$$A_N \sim a \times v$$

where A_N = activity of atomic nitrogen
a = degree of dissociation
v = flow rate.

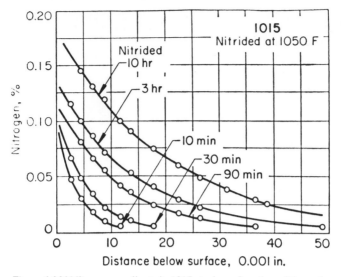

Figure 6.201 Nitrogen gradients in 1015 steel as a function of time of nitriding at 1050 °F (566 °C), using the aerated bath process. (Reproduced by permission, from *Metals Handbook* **2**, American Society for Metals, 1964)

Consequently the nitrogen activity is a function of the number of ammonia molecules dissociated at the steel surface per unit of time. At constant pressure and temperature the degree of dissociation is reduced as the flow rate increases but the product *av* increases. Therefore in actual practice the nitrogen content can be checked by means of the degree of dissociation and controlled by the flow rate.

A high nitrogen activity is thus obtained with high flow rates, which give

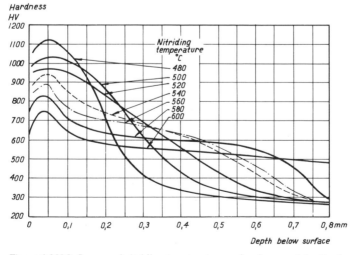

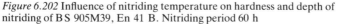

Figure 6.202 Influence of nitriding temperature on hardness and depth of nitriding of B S 905M39, En 41 B. Nitriding period 60 h

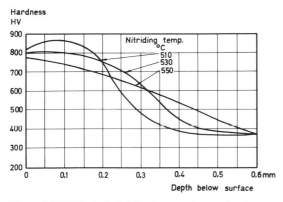

Figure 6.203 Effect of nitriding temperature on hardness and depth of nitriding of En 29 B. Nitriding period 60 h

a low degree of dissociation. A low nitrogen activity is obtained with low flow rates, which give a high degree of dissociation. If the flow rate is too low, the turbulence in the nitriding box is inadequate unless the box is provided with a fan. This can result in uneven nitriding at different locations in the box. The degree of dissociation increases as the temperature increases so that by raising the temperature a constant degree of dissociation can be ensured with a high gas flow rate.

At the commencement of the process it is customary to employ a low degree of dissociation (high gas flow rate), the object being to expel the air present rapidly and to build up a high nitrogen content in the surface layer. The thickness of the white layer with a 30% degree of dissociation increases in the manner shown in *Figures 6.204* and *6.205*. Layer thicknesses around 0·02 mm or more may be unwelcome because of the risk of spalling, and hence it is essential to control the layer thickness so that it assumes a suitable value.

In 1943 C. F. Floe took out a patent for a method of controlling the thickness of the white layer by means of the degree of dissociation. However, this method has not been used on the scale anticipated. A modified method however has been developed by Bofors and has been used by them for some years with good results. During preliminary investigations performed using En 41 B before the method was taken into use, the degree of dissociation and nitriding temperature were varied. Some of the results of the tests are shown in *Figure 6.206*. An almost ideal build-up of the nitriding layer is obtained if, after 5 h, the degree of dissociation is increased to 85%. At still higher values the steel becomes de-nitrided.

In actual practice it has proved feasible to operate with a degree of dissociation of roughly 50% even in furnaces without fans. If higher degrees of dissociation are required, it is essential to have furnaces with fans, or to increase the nitriding temperature. An increase in the degree of dissociation from 30 to 50% at 510°C implies a reduction in the amount of ammonia by *ca* 25%. If the degree of dissociation is increased to 50–60%, considerable increases are achievable in the service life of gas-nitrided components.

522

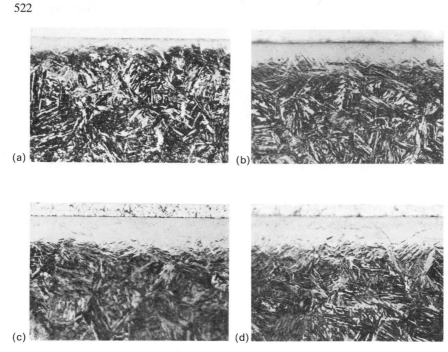

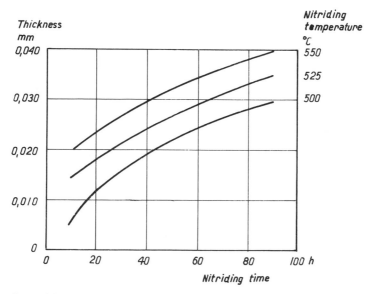

Figure 6.204 Growth of white layer during nitriding of B S 905M39, En 41 B at 500 °C for 10–90 h. Degree of dissociation: 30%, 200 ×. (a) 10; (b) 30; (c) 60; (d) 90 h

Figure 6.205 Thickness of white layer as a function of nitriding period and temperature. B S 905M39, En 41 B. Degree of dissociation 30%

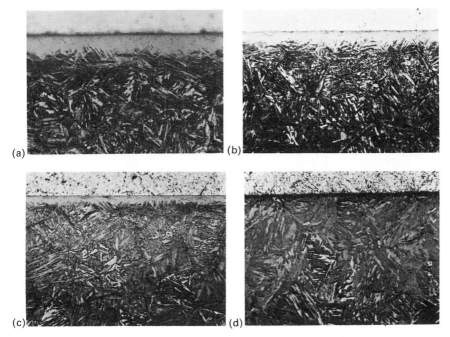

Figure 6.206 Thickness of white layer resulting from different methods of nitriding BS 905M39, En 41 B. (a) 30 h at 500°C and 20% degree of dissociation; (b) 5 h at 500°C and 20% degree of dissociation then 25 h at 525°C and 85% degree of dissociation; (c) 5 h at 500°C and 20% degree of dissociation then 25 h at 500°C and 55% degree of dissociation; (d) 5 h at 500°C and 20% degree of dissociation then 25 h at 550°C and 90% degree of dissociation

6.7.5   Nitriding different types of steel

Some of the types of steel that are not commonly used for nitriding have been discussed in the preceding section, but we will now give details of a few more. ISO/R 683/X–1970 includes the following steels, *Tables 6.41, 6.42* and *6.43*, which are primarily intended for nitriding.

Chromium–molybdenum-alloyed nitriding steels containing approximately 3% Cr and 0·5% Mo are made with carbon contents between 0·20 and 0·40%. One of the more widely used steels is En 29 B (ISO steel 1) which has a hardness profile as shown in *Figure 6.207*. This steel finds its main use for large machine components such as shafts and gear wheels (*see Figure 6.208*).

Aluminium-alloyed nitriding steels are manufactured with varying compositions, depending on hardenability requirements. One of the rather commonly occurring steels is BS 905M39 (En 41 B), which on nitriding achieves a surface a hardness of about 1100 HV and has the hardness profile shown in *Figure 6.209*. A nitriding period shorter than 5 h is not recommended if the very highest surface hardness is required.

Alloy heat-treatable steels of the type BS 708A37 (En 14 B) and

Table 6.41 Nitriding steels in accordance with ISO/R 683/X–1970

Type of steel	C %	Si %	Mn %	P % max	S % max	Al %	Cr %	Mo %	Ni % max	V%
1	0·28–0·35	0·15–0·40	0·40–0·70	0·030	0·035	—	2·80–3·30	0·30–0·50	0·30	—
2	0·35–0·42	0·15–0·40	0·40–0·70	0·030	0·035	—	3·00–3·50	0·80–1·10	—	0·15–0·25
3	0·30–0·37	0·20–0·50	0·50–0·80	0·030	0·035	0·80–1·20	1·00–1·30	0·15–0·25	—	—
4	0·38–0·45	0·20–0·50	0·50–0·80	0·030	0·035	0·80–1·20	1·50–1·80	0·25–0·40	—	—

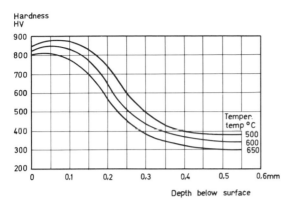

Figure 6.207 Influence of tempering temperature on hardness and depth of nitriding of steel En 29 B. Nitrided for 60 h at 510°C

Figure 6.208 Nitrided gears made from En 29 B

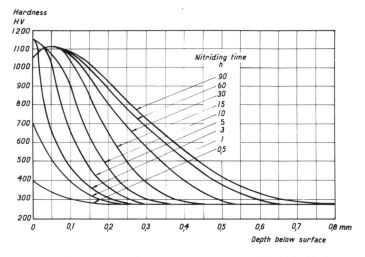

Figure 6.209 Hardness profile through surface layer of BS 905M39 after gas nitriding for 0·5–90 h at 510 °C

BS 817M40 (En 24) exhibit a surface hardness of around 650 HV after gas nitriding at 510 °C, if their initial state was hardened and tempered. If the nitriding temperature is raised to 550 °C, the hardness drops to around 550 HV. Basically the graph in *Figure 6.210* applies also to BS 768A37.

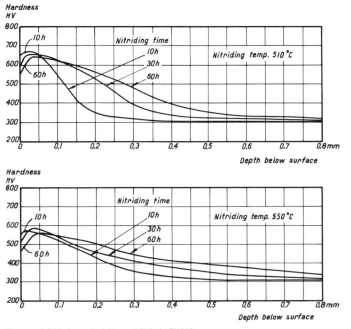

Figure 6.210 Gas nitriding of BS 817M40

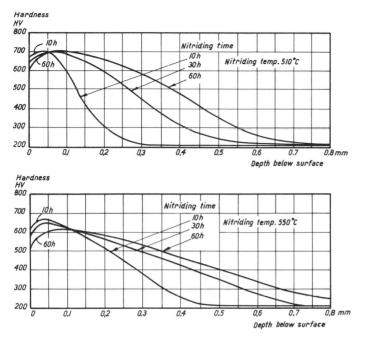

Figure 6.211 Gas nitriding of annealed BS 637M17

Table 6.42 Standardized nitriding steels

ISO	Nitralloy	BS	En	DIN	SS
1	—	(722M24)	29 B	32 CrMo 12	2240
2	—	897M39	40 C	32 CrMoV 1210	—
3	G	—	—	34 CrAlMo 5	2940[1]
4	135 M	905M39	41 B	—	2940

1. Previous composition replaced by steel 4.

Alloy case-hardening steels, e.g. BS 637M17 can be nitrided in the annealed state, whereby the hardness and depth of nitriding obtained are somewhat higher than that achieved with heat-treatable steels (*see Figure 6.211*).

Unalloyed case-hardening steels and plain low-carbon steels show, after gas nitriding the hardness profiles given in *Figure 6.212*. After salt-bath nitriding or nitrocarburizing the hardness distribution obtained is roughly the same. However, in the outermost layer the surface hardness is somewhat greater.

Low-alloy tool steels type NiCrMo can be nitrided either in the annealed or hardened and tempered state. The surface hardness becomes roughly 500 HV.

High-alloy tool steels containing 5–12% Cr of type A 2, D 2 and D 6 follow a common pattern, which suggests that the highest hardness is

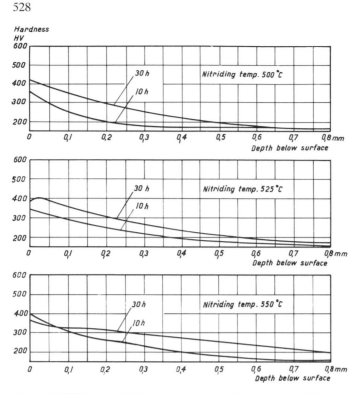

Figure 6.212 Effect of nitriding temperature on the hardness and depth of nitriding of SS 1370 (0·15% C steel)

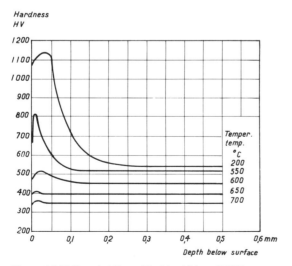

Figure 6.213 Gas nitriding of D 6 for 10 h at 510°C after hardening from 980°C followed by tempering at different temperatures

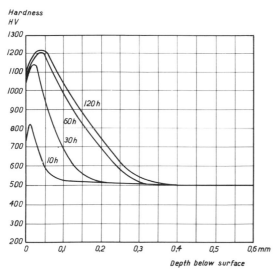

Figure 6.214 Gas nitriding of D 6 for 10–120 h at 510°C after hardening from 980°C followed by tempering at 550°C

obtained if, after hardening the steel is tempered at only 200°C. The surface hardness is 1100–1200 H V (*see Figures 6.213* and *6.214*).

If, for example, the hardening temperature is raised to 1050°C, secondary hardening takes place during nitriding. The hardness of the core obviously drops with increased nitriding time but it is considerably higher than if a normal hardening temperature had been used (*see Figure 6.215*).

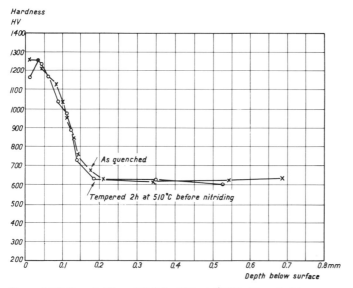

Figure 6.215 Gas nitriding of D 2 for 30 h at 510°C with and without prior tempering for 2 h at 510°C. The specimens were first hardened from 1050°C in oil

Table 6.43 Mechanical properties of nitriding steels in accordance with ISO

Type of steel	Diameter mm	R_e min kgf/mm² N/mm²	R_m kgf/mm² N/mm²	A min %	KCU min kgf.m/cm² J	Hardness of nitrided surface HV min	HR 15N min
	≤ 16	90 880	110–130 1080–1280	10	5 25	800	92
	> 16≤ 40	85 830	105–125 1030–1230	10	6 29	800	92
1	> 40≤ 100	80 780	100–120 980–1180	11	6 29	800	92
	> 100≤ 160	75 740	95–115 930–1130	12	6 29	800	92
	> 160≤ 250	70 690	90–110 880–1080	12	6 29	800	92
2	≤ 70	110 1080	130–150 1280–1470	8	3 15	800	92
3	≤ 70	60 590	80–95 780–930	14	5 25	950	93·5
	≤ 100	75 740	95–115 930–1130	12	4 20	950	93·5
4	> 100≤ 160	65 640	85–100 830–980	14	5 25	950	93·5

* R_e = yield stress (0·2% proof stress)
R_m = tensile strength
A = percentage elongation after fracture ($L_o = 5d_0$)
KCU = impact strength with U-notch
HV = Vickers hardness number
HR 15N = Rockwell superficial (N scale) hardness with 15 kgf load

ISO gives the values of mechanical properties as kgf.m/cm² and tonf/in². In the table tonf/in² has been replaced by N/mm² and kgf.m/cm² has been supplemented with J

A tempering treatment at 510°C prior to nitriding has not affected the depth of nitriding or the surface hardness. Steel D 6, after hardening from 1050°C, behaves in precisely the same way during nitriding as steel D 2.

Hot-work steels of the types normally employed exhibit surface hardness values after nitriding that lie between 1100 and 1300 HV. Owing to the high alloy content, the depth of nitriding for these steels is relatively small as is also the case for the high-alloy chromium steels. Since the steels are normally tempered at temperatures above the nitriding temperature, the hardness of the core is not reduced during nitriding. As shown by *Figures 6.216* and *6.217* the temperature employed during tempering has a marked effect on the depth of nitriding for steel H 13. A high initial hardness tends to have a favourable effect on the depth of nitriding. Nitriding can be carried out between 510 and 525°C without noticeably affecting the surface hardness. Nitriding at 550°C gives a surface hardness of about 1100 HV and a somewhat greater depth of hardening than that obtained at lower temperatures. Tools made from this steel are nitrided on quite a large scale for both cold- and hot-work service. *Figure 6.218* illustrates nitrided pistons made from H 13 which are intended for production of plastics under high pressure.

High-speed steel tools are normally salt-bath nitrided for 10–60 min. With gas nitriding the period should not exceed 10 h. If periods longer than this are employed, the outermost layer becomes soft (*see Figure 6.219*). Hardness values above 1300 HV are obtainable which may be too high for

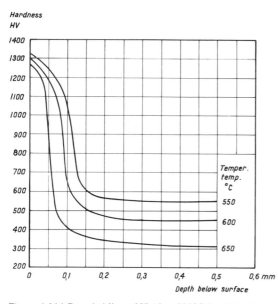

Figure 6.216 Gas nitriding of H 13 at 525 °C for 10 h after hardening followed by tempering at different temperatures

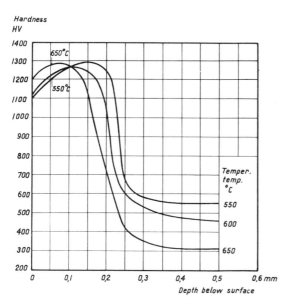

Figure 6.217 Gas nitriding of H 13 at 525 °C for 30 h after hardening followed by tempering at different temperatures

Figure 6.218 Gas nitrided pistons of steel H 13

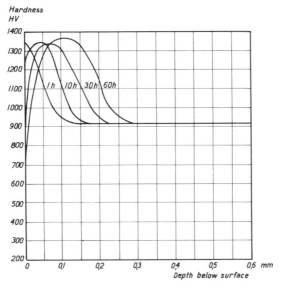

Figure 6.219 Gas nitriding at 510°C for 1–60 h of
high-speed steel M 42. Degree of dissociation: 30%

many purposes. In such cases the tools are tempered at 550–600°C after nitriding, thereby increasing the toughness.

Stainless steels cannot be nitrided without special pre-treatment, because the film of chromium oxide which forms on the surface of the steel after contact with atmospheric oxygen prevents nitrogen absorption. For martensitic stainless steels the oxide film can be removed or thinned down by a light sand blasting immediately before nitriding.

The same also applies to a certain extent to the 18/8 austenitic steels and similar steels. Better results are obtained with these steels if chemical pre-treatment is carried out in accordance with any of the following methods:

1. Pickling in 50% by volume hydrochloric acid at 70°C, with subsequent thorough rinsing in water.
2. Pickling in 1% by volume orthophosphoric acid containing 0·7% primary zinc phosphate at 70°C without subsequent washing.

In both cases nitriding must be carried out directly after the pre-treatment. If, for practical reasons, this should not be possible, the parts can be dealt with in accordance with any of the following methods:

3. Pickling in 10% hot sulphuric acid, rinsing in water, and copper-plating in a copper cyanide bath to give a coating of thickness 0·3 μm.
4. According to Coppola[74] 12% Cr stainless steel can be nitrided after lapping the steel surface followed by passivation in hot nitric acid, or heating in air at 510°C. Heating in air at 260–370°C prior to nitriding gives unsatisfactory results.

When plasma nitriding it is possible to remove the interfering oxide film by a simple sputtering operation carried out in the nitriding chamber immediately before the nitriding proper.

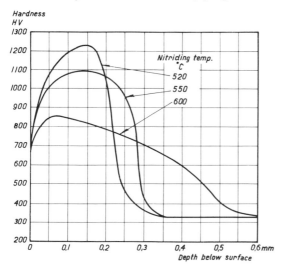

Figure 6.220 Gas nitriding of 13% Cr steel for 60 h at 520–600°C

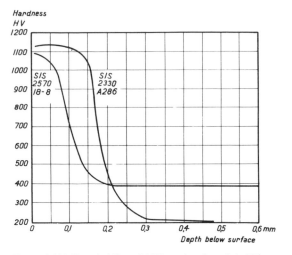

Figure 6.221 Gas nitriding of 18/8 steel and steel A 286 for 60 h at 600 °C

Stainless steels are nitrided at higher temperatures than those used for other steels. For martensite stainless steels 520–550 °C is used, and for the austenitic steels, 570–620 °C. It is important that the air should be purged from the furnace before heating starts.

Figure 6.220 shows results obtained from nitriding tests on creep-resisting 13% Cr steel at temperatures between 520 and 600 °C for 60 h. This graph can also be used for other martensite steels of the type AISI 430. *Figure 6.221* shows hardness curves which have been plotted for an 18/8 steel and steel A 286 after nitriding for 60 h at 600 °C.

As a result of nitriding there is a considerable reduction in resistance to rusting, and hence it is doubtful whether a stainless steel chosen for its corrosion-resisting properies, should be subjected to nitriding. Nitriding of steel A 286 which is employed as a creep-resisting steel can, on the other hand, be quite justifiable. Also, as a result of nitriding, austenitic steels can become slightly magnetic in the nitrided layer, partly because the iron nitride Fe_4N is ferromagnetic, and partly because of partial transformation of austenite to ferrite in the nitrided layer.

6.7.6 Properties of nitrided steels

Wear resistance

The wear resistance of steels increases with increasing hardness. Since the concept of 'wear resistance' is difficult to define, a warning should be given against any generalizations. If the hardness of a cutting edge should exceed a certain value, small fragment of the edge break away; as machining continues, the edge becomes blunt and a pattern reminiscent of wear arises. The nitrided layer is very liable to chip in this way. In the case of abrasive wear of a scratching or eroding nature, it is doubtful whether nitriding will provide any long-term improvement in wear resistance. With

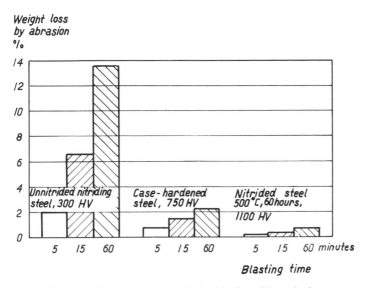

Figure 6.222 Abrasion loss occurring during blasting with steel grit on unnitrided, case-hardened and nitrided test specimens respectively (after Lambert[75])

gouging or grinding abrasion, nitriding is not to be recommended. *Figure 6.222* shows the results obtained from shot-blasting tests with steel grit on steels of different hardness, the least abrasion being reported for the nitrided steel[75].

It is for adhesive wear especially that the main advantages are achieved with nitriding, because the coefficient of friction is reduced, as is also the risk of scuffing. A further improvement is achieved by nitrocarburizing and particularly by the Sulfinuz treatment, a fact confirmed by practical tests and laboratory investigations. During wear tests in an Amsler machine, in which two discs made from a 0·15% carbon steel rotated against each other at a speed of 440 and 400 rpm respectively, the values shown in *Figure 6.223* were obtained. The specimen given the Sulfinuz treatment is superior to both the case-hardened and the nitrided specimens. Scuffing has started at the point where the curves noticeably change direction. In the initial stages the Sulfinuz-treated specimen suffered somewhat more abrasion than the case-hardened or the nitrided specimens because of the wearing away of the black surface coating.

Under certain conditions nitrided steels can function without lubrication, this being particularly marked with Sulfinuz-treated steels. This characteristic has been tested in a Faville-Levally machine, in which a test bar rotates at a speed of 300 rpm between two hardened V-jaws. The pressure is gradually raised until scuffing occurs. When tests were made with oil-lubricated specimens made from 0·15% carbon steel, the untreated bar scuffed at a loading of 380 kp. The Sulfinuz-treated bar exhibited no sign of scuffing whatever at the maximum loading of the machine, i.e. 1140 kp (*see Figure 6.224*). Corresponding tests were made on unlubricated specimens, and the untreated specimen scuffed at a

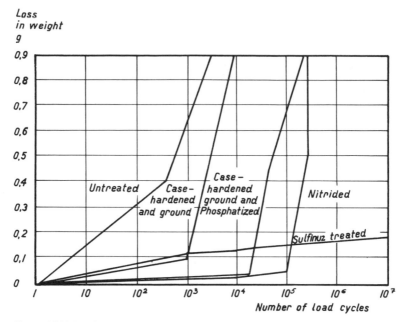

Figure 6.223 Amsler test on discs made from carbon steel with 0·15% C
subjected to different treatments. Loading 20 kp (after Waterfall[68])

loading of 250 kp. The test with the Sulfinuz-treated bar had to be
discontinued at a loading of 630 kp because the specimen lengthened as a
result of the frictional heat (*see Figure 6.225*).

Practical tests were performed involving the Sulfinuz treatment of
case-hardened steel parts that exhibited a tendency to scuffing. As an
example we can mention a bushing made from case-hardened Cr–Ni steel
which started scuffing against the shaft after 7 min at a speed of 4000 rpm.
After Sulfinuz treatment at 550°C for 2 h the bushing could be run for 2 h
without scuffing. Because of the heat of friction generated the machine
then started to operate sluggishly. After being allowed to cool to room
temperature the running test could be resumed.

High resistance to tempering and high hardness at elevated temperatures

These are valuable properties of nitrided steels. After nitriding, the steel
can be heated up to the same temperature as that at which it was nitrided,
and in some cases to an even high temperature, without loss of hardness at
room temperature (*see Figure 6.226*). *Figure 6.227* shows that the hardness
at elevated temperatures is also relatively high. The excellent wear
resistance and elevated-temperature hardness combined ensure that
nitrided steels can be used with good results for hot-work dies. However,
the choice of steel and nitriding time must be adapted to suit working
conditions.

High fatigue strength and low notch sensitivity

Nitrided steels are characterized by high fatigue strength and low notch
sensitivity. The increase in fatigue strength is highest for plain-carbon and

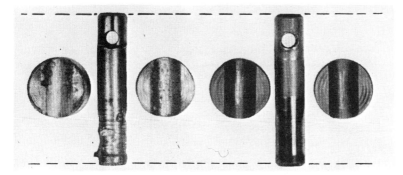

Figure 6.224 Test bars and jaws after Faville–Levally test in oil S A E 30.
Untreated (left): Scuffing at pressure of 380 kp. Sulfinuz treated specimen
(right): No scuffing at pressure of 1140 kp (after Waterfall[68])

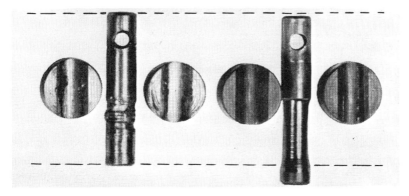

Figure 6.225 Test bars and jaws after Faville–Levally test without oil.
Unteated (left): Scuffing at pressure of 250 kp. Sulfinuz treated (right): No
scuffing at pressure of 630 kp, but test bar lengthened as a result of plastic
deformation (after Waterfall[68])

low-alloy steels, and this generally increases as the nitriding time increases.
The standardized nitriding steels heat treated to an ultimate tensile
strength of 900 N/mm^2 have a fatigue strength of roughly 450 N/mm^2, as
determined during rotary bending fatigue tests on highly finished ground
test bars, 10 mm diameter. By gas nitriding at 510°C for 20 h the fatigue
strength increases to roughly 600 N/mm^2. Nitriding for 60 h gives a fatigue
strength of around 700 N/mm^2 at 10^7 load cycles.

In general the fatigue strength increases as the depth of the nitrided layer
increases. This point is illustrated in the diagrams in *Figure 6.228*, which
apply to the ionitriding of En 40 B where the compound layer was 1 μm
thick. Should a compound layer not form or be ground off the fatigue
strength will be lower[76].

Tests involving gas nitriding and salt-bath nitriding of smooth test bars
made from steel D I N C 35 and 34 Cr 4 show that as regards fatigue
strength in bending gas nitriding is somewhat more advantageous for the

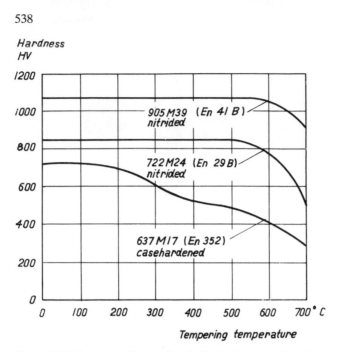

Figure 6.226 Tempering diagram for nitrided and case-hardened steels respectively

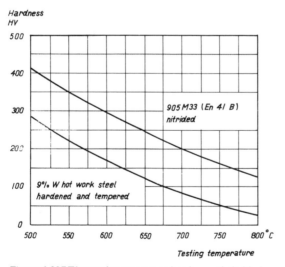

Figure 6.227 Elevated temperature hardness of nitrided steel and hot-work steel

first, and salt-bath nitriding is rather better for the second steel (*see Figure 6.229*)[77]. However, the differences are so slight that that it does not really matter which method of nitriding is employed.

The notch effect on fatigue strength is extremely marked. The effect of this unfavourable factor is lessened by the compressive stresses which are

Bending stress
± N / mm^2

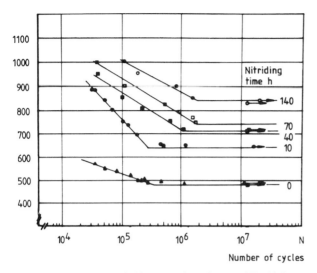

Figure 6.228 Wöhler or S–N curves of specimens of En 40 B
bright plasma nitrided at 480°C for 10, 40, 70 and 140 h
respectively (after Bell and Loh[76])

introduced into the steel during nitriding. Hence nitriding is a very useful
process if the aim is to increase the fatigue strength of machine components
which, because of their ultimate application, have to be provided with
notches, e.g. bolts. *Figure 6.230* shows that the fatigue strength drops by
half as a result of the notch effect in hardened and tempered test bars made
from D I N 14 CrMoV 69. By undertaking a Tufftride treatment for 90 min
at 570°C the fatigue strength increases both for the unnotched and for the
notched bars, the nitrided notched bar reaching the same value as that
achieved by the non-nitrided, unnotched bars[78].

The depth of nitriding which gives the maximum fatigue strength is,
however, dependent on material thickness and shape. This is illustrated in
Figure 6.231 which shows the torsional fatigue strength as a function of
nitriding depth after the gas nitriding of test bars 6·5 mm diameter. One
test bar has a transverse hole 1 mm diameter, which reduces the fatigue
strength. For this bar the maximum fatigue strength is achieved with a
nitriding depth of 0·3 mm. In the case of the undrilled bar the maximum
fatigue strength is obtained with a nitriding depth of 0·4 mm. With
thin-walled parts, salt-bath nitriding is preferable in such cases as
compared with gas nitriding because it is easier to control the depth of
nitriding during bath nitriding when such shallow depths are required.

Should a nitrided part be subjected to such high loads that cracks form in
the nitrided layer the fatigue strength may fall off drastically. Such parts as
have pronounced stress raisers, where stress concentrations are high, are
also particularly, susceptible to fatigue failure after nitriding.

Ishizaki, Fior and Corredor[79] have shown that the fatigue life of nitrided

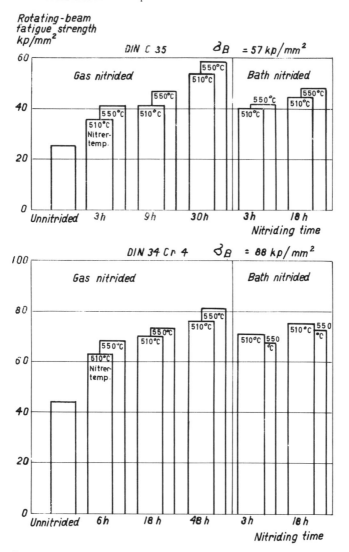

Figure 6.229 Increase in fatigue strength of test bars 5·9 mm diameter made of hardened and tempered steel after gas and salt-bath nitriding at 510 and 550 °C (after Wiegand[77])

SAE 7140 (Nitralloy 135H) is exceptionally short when frequencies of about 5 Hz and smaller are applied. This fact is probably due to longer times of relaxation and increased rate of nitrogen diffusion during the fatigue test.

Corrosion resistance

As a result of nitriding the corrosion resistance of non-stainless steels is increased. From the corrosion viewpoint the white layer is equivalent to a

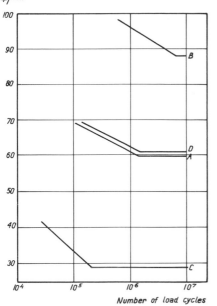

Stress
kp/mm²

Number of load cycles

Figure 6.230 Effect of tufftriding on rotating-beam fatigue strength. Steel DIN 14 CrMoV 69 (0·14% C, 1·5% Cr, 0·09% Mo+V) (Finnern, Vetter and Jesper[78])
  A = smooth test specimen, hardened and tempered at 570 °C, B = smooth test specimen, hardened and tempered at 600 °C then Tufftrided for 90 min at 570 °C, C = notched test specimen, hardened and tempered at 570 °C, D = notched test specimen, hardened and tempered at 600 °C then Tufftrided for 90 min at 570 °C; notched factor $\alpha_k = 2$

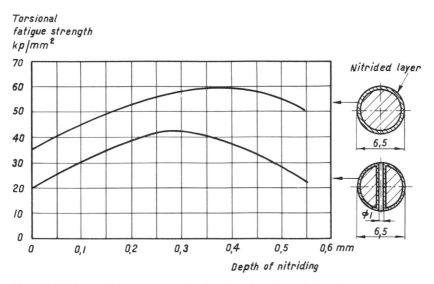

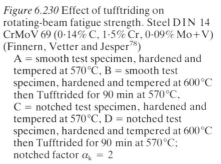

Torsional fatigue strength kp/mm²

Depth of nitriding

Figure 6.231 Torsional fatigue strength as function of depth of nitriding of a smooth test specimen and a drilled test specimen, respectively, made from a Cr–Mo–V steel having the following composition: 0·3% C, 2·5% Cr, 0·2% Mo, 0·25% V; $\sigma_B = 110$ kp/mm² (after Wiegand[77])

Polishing

0·03 mm stock removed

0·06 mm stock removed

B S 905M39

Polishing

0·03 mm stock removed

0·06 mm stock removed

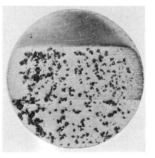

B S 817M40

Polishing

0·03 mm stock removed

0·06 mm stock removed

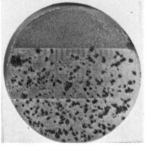

B S 420S37

Figure 6.232 Corrosion test for 16 h using 5% salt-water mist.
Spray period 2 × 30 min. The test specimens were gas-nitrided
for 60 h at 510 °C

13% martensite chromium steel. In stainless steels, corrosion resistance is
reduced because the chromium is bound as nitrides. The temperature used
during nitriding may also render non-stabilized 18/8 stainless steels
susceptible to intercrystalline corrosion during subsequent service.
Corrosion resistance is greatest for nitrided and polished surfaces; however
by grinding off, say 0·03 mm, the white layer is removed and corrosion
resistance is reduced. After grinding 0·03–0·06 mm steel B S 905M39 has
better corrosion resistance than the conventional heat-treatable steels and

B S designation C 50   708A37   817M40   905M39   420S37   302S25

Base material

Sulfinuz-treated

Nitrided

Nitrided+ground

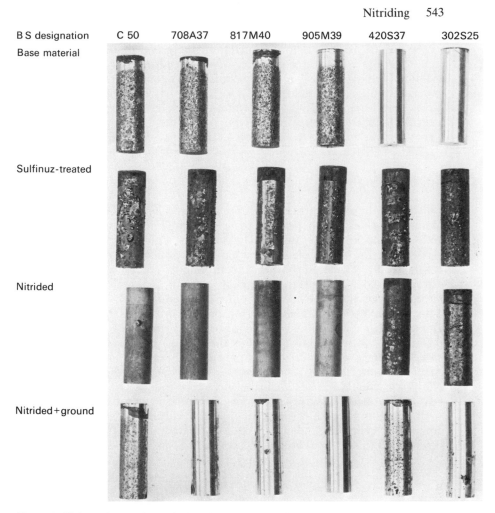

Figure 6.233 Corrosion test in tropical testing cabinet for 16 days on test bars subjected to different surface pre-treatments

BS 420S37. When tested in salt-water spray, the two last mentioned groups exhibited roughly the same corrosion resistance in the nitrided layer (*see Figure 6.232*). From various nitriding tests on 18/8 steels, contradictory results have been obtained, but in any case we must expect a reduction in corrosion resistance, at least down to the same level as that shown by martensitic chromium steels in the hardened and tempered state.

Similar results were obtained during corrosion tests in tropical atmosphere testing chambers. In this instance test specimens were used having dimensions of 15 mm diameter × 50 mm. They were tested in the 'normal state'; after a Sulfinuz treatment for 1 h at 570°C; after gas nitriding for 30 h at 510°C; and after gas nitriding and then having approximately 0·05 mm on the diameter ground off. *Figure 6.233* illustrates the appearance of the specimens after sixteen days in the tropical chamber. The matt surface on the specimens that were nitrided only is due to the film of oxide that forms during nitriding and that tends to reduce somewhat the corrosion resistance. The best results are obtained if, as mentioned above, polishing is undertaken subsequent to nitriding. After grinding, the specimen made from BS 905M39 has withstood the corrosion attack better than have the other grades of steel. The poor results obtained after Sulfinuz treatment were due in part to the fact that the specimens were not polished prior to corrosion testing.

6.8 Boriding

By allowing boron to diffuse into steel a surface layer is formed containing metallic borides which possess extremely high hardness, viz. between 1600 HV and 2000 HV (Vickers Hardness, 0·1 kgf load). Boriding may be carried out by means of gaseous, liquid or solid media. By employing commercially available boriding compounds containing additions of activating agents a surface layer of Fe_2B is formed. This layer has approximately the same coefficient of expansion as steel.

In principle the method used for pack boriding is the same as that employed for conventional pack carburizing. The process takes place at 800–1000°C. After cooling in the box the steel parts may be hardened as in case hardening. However, the surface hardness is the same whether the parts are subjected to a hardening operation (heat treatment) or simply allowed to cool in the box. All steels can be borided. The thickness of the boride layer decreases as the content of alloying elements increases.

Boriding trials with various steels were carried out with a boriding compound called E K A B O R. After a treatment of 850°C for 1–8 h a borided depth of 0·01–0·06 mm, respectively, was obtained at the 550 HV level. The surface hardness was about 1300 HV 1.

Figure 6.234 shows the microstructure in a cross-section through the surface of a steel, type 25 CrMo 4, after being borided at 850°C for 2 h, cooled in air, oil quenched from 850°C and tempered at 600°C. A trial involving direct hardening from the boriding temperature gave a similar microstructure, depth of hardening and hardness profile.

Detailed information on boriding may be found in references[80, 81].

6.9 Induction hardening

6.9.1   Fundamental principles

When electric current passes through a conductor, a magnetic field is created round it. If the conductor consists of a coil, a magentic field is established inside the coil. This field persists even if a metal bar is inserted into the coil, as shown in *Figure 6.235*. If the magnetic flux is created by a high-frequency alternating current, it gives rise to eddy currents in the surface of the metal bar, which consequently becomes heated (*see Figure 6.236*). In iron, hysteresis losses also contribute to some extent to the temperature rise up to the Curie point (768°C), above which iron is non-magnetic.

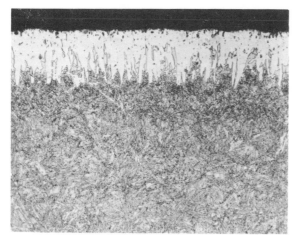

Figure 6.234 Microstructure of cross-section through the surface layer of a borided specimen of steel S S 2225 (25 CrMo 4)

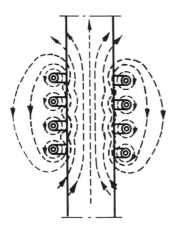

Figure 6.235 Path of magnetic flux through a metal bar inserted in a coil through which an electric current is flowing

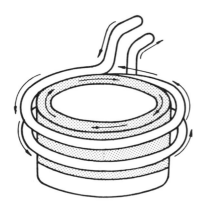

Figure 6.236 Path of the electric current in the coil and metal bar during induction heating

The depth of penetration of the heat is governed mainly by the power and frequency employed. The normal power input is $0 \cdot 1$–2 kW/cm^2 of the heated surface. The relationship between depth of penetration and frequency can be calculated approximately by using the following simplified expressions, which are valid for the temperature rise in steel up to the hardening temperature:

$$d_c = \frac{20}{\sqrt{f}} \text{ cold state } (20°C)$$

$$d_h = \frac{500}{\sqrt{f}} \text{ hot state } (800°C)$$

where d = depth of penetration in mm
$\quad\quad f$ = frequency in cycles per s (Hz)

Owing to heat conduction in the material during heating, the overall depth of penetration is larger. It is possible to calculate the additional penetration due to heat conduction from the following expression.

$$d_l = 0 \cdot 2 \sqrt{t}$$

where d_l = depth of penetration in mm
$\quad\quad t$ = time in s

The total depth of penetration is obviously $d + d_l$. It should be stressed that these expressions give only a rough estimate of the depth of penetration and they have been included here only to show the fundamental effects of frequency and time[82].

6.9.2   Steel grades for induction hardening

By induction hardening we wish to achieve a considerable increase in surface hardness, and steels are used having C contents of $0 \cdot 30$–$0 \cdot 50\%$, which give hardness values of 50–60 HRC. The upper limit is governed by the C content that may risk hardening cracks. From practical tests it has been found that this C content is around $0 \cdot 50\%$. If the heat treatment is well controlled, higher C contents are permissible, e.g. for rolls, that are made from steels having $0 \cdot 80\%$ C, $1 \cdot 8\%$ Cr and optionally $0 \cdot 25\%$ Mo. In accordance with ISO 683/XII–1972, the steels in *Table 6.44* are recommended for flame and induction hardening. Steel 3, with increased Mn content (*ca* $1 \cdot 30\%$) has also proved suitable for this purpose. Steel 9 corresponds to BS 108A42 (En 19 C). Cr–Mo steels with lower carbon contents are also widely used (*see Table 6.10*). The steels are normally quenched in water. In certain cases the alloy steels can be cooled by means of an oil emulsion. Steels 9–11 can also be oil quenched.

The steel, prior to induction hardening, should be in the hardened and tempered or normalized state, and the hardness should be such as is acceptable for the unhardened sections. Fully annealed steel is not so suitable, because normally the time needed for dissolution of the carbides is longer than the normal heating-up time. Consequently induction hardening of fully annealed steel can result in inadequate and irregular values unless the heating time is matched to suit the fully annealed state.

Table 6.44 Steels for flame and induction hardening I S O 683/XII–1972

Type of steel	C %	Si %	Mn %	P % max	S % max	Cr %	Mo %	Ni %
1	0·33–0·39	0·15–0·40	0·50–0·80	0·035	0·035	—	—	—
2	0·38–0·44	0·15–0·40	0·50–0·80	0·035	0·035	—	—	—
3	0·43–0·49	0·15–0·40	0·50–0·80	0·035	0·035	—	—	—
4	0·48–0·55	0·15–0·40	0·60–0·90	0·035	0·035	—	—	—
5	0·50–0·57	0·15–0·40	0·40–0·70	0·035	0·035	—	—	—
6	0·42–0·48	0·15–0·40	0·50–0·80	0·035	0·035	0·40–0·60	—	—
7	0·34–0·40	0·15–0·40	0·60–0·90	0·035	0·035	0·90–1·20	—	—
8	0·38–0·44	0·15–0·40	0·60–0·90	0·035	0·035	0·90–1·20	—	—
9	0·38–0·44	0·15–0·40	0·50–0·80	0·035	0·035	0·90–1·20	0·15–0·30	—
10	0·38–0·44	0·15–0·40	0·70–1·00	0·035	0·035	0·40–0·60	0·15–0·30	0·40–0·70
11	0·37–0·43	0·15–0·40	0·50–0·80	0·035	0·035	0·60–0·90	0·15–0·30	0·70–1·00

6.9.3   Equipment for induction hardening

As the frequency governs the depth of penetration (depth of hardening), equipment should be selected mainly on the basis of the required depth of hardening. Since the latter generally increases with the size of the hardened part, the consequence is that, for large parts, equipment operating at low frequencies and low power input per unit of surface is selected whereas for small parts the best results are obtained using equipment operating at high frequencies and with high power input per unit of surface (*Table 6.45*).

Table 6.45 A guide to frequency selection

Frequency range	Power supply source	Frequency kHz	Minimum depth of hardening obtainable mm	Minimum diameter of workpiece mm
Intermediate frequency	Rotary converter	0·5–10	1·5	*ca* 15
Intermediate frequency	Solid state	0·5–10	1·5	*ca* 15
High frequency	Valve generator	100–500	0·2	*ca* 1·5

As an example it can be stated that for the hardening of large rolls for cold rolling Bofors use three rotary converters operating at 500 Hz with a total power output of 900 kW and a power input per unit surface of 0·1 kW/cm^2, by means of which a hardness depth of up to 20 mm can be obtained. As may be seen in *Table 6.45* rotary converters are used in the frequency range 0·5–10 kHz. The same frequency range is also covered by a new generation of converters called solid state converters, the functioning of which is based in semiconductor techniques. The efficiency of these units lies around 75% as against approximately 60% for the rotary converters. The solid state converters are very reliable in operation but they require servicing by highly qualified personnel. Rotary converters, besides requiring only simple maintenance, also give reliable and trouble-free service. For the hardening of small components, valve generators operating at frequencies around 500 000 Hz and power outputs up to 100 kW are employed.

6.9.4   Working coils and fixtures

The coils, also known as inductors, also become heated by the electric current and by thermal radiation from the heated steel. The coils are often made from copper tubing and are cooled during operation by internal water flow. *Figures 6.237a–c* show three basic types of working coils designed for external and internal heating, and for heating of a flat surface.

The internal diameter of the copper tube is matched to suit generator power output. For the power output normally required, around 50 kW, tubing with an internal diameter of around 5 mm is used. The coils are usually wound with a distance of 2–5 mm between turns and with roughly the same spacing between the coil and the workpiece. By varying this distance it is possible to influence the rate of heating to a very large extent.

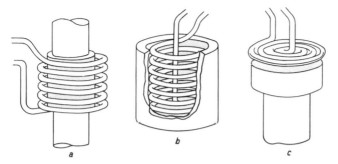

Figure 6.237 Types of coil used for induction heating

Figure 6.238 illustrates a coil that is fixed in a hard plastic mounting. The individual turns of the coil in *Figure 6.239* have been positioned using glass fibre spacers. The first coil is designed for heating rolls for cold rolling, and the second for dealing with shafts having tapered sections.

Parts that are to be hardened must be held in a fixture of some sort so that they are correctly located in the inductor, and small components must be firmly held so that they are not displaced by the magnetic field. A metallic fixture that comes near the coil may become heated by the induced current. To avoid this the fixture and centre washers are normally made from an insulating and heat-resistant material such as 'sindango' or a heat-resistant plastic.

6.9.5 Procedure during induction hardening

According to the grade employed, the steel, after it has been treated in the coil, is cooled with water or oil or, in certain cases, in air. Heating and

Figure 6.238 Coil for spin-hardening of rolls

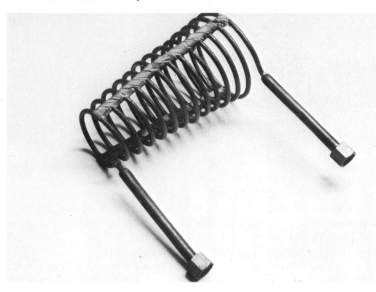

Figure 6.239 Coil for hardening of tapered shafts

cooling can be carried out in accordance with two main methods, these being single-shot hardening and progressive hardening.

Single-shot hardening

During single-shot hardening the component is positioned and heated, after which it can be quenched in various ways as shown in *Figure 6.240*.

   (*a*) The component can either be immersed by hand in the quenching liquid or it can be released from the fixture so that it drops into the liquid.

   (*b*) In hardening machines the heated steel parts can be fed automatically into a quenching spray when heating has terminated.

   (*c*) The coil can also be provided with a quenching spray which starts to operate when the high-frequency current is switched off.

Figure 6.241 shows a gear that has been heated to the hardening temperature in the inductor located above and which is just going to be transferred into the quenching spray.

The parts may also rotate during heating and, if possible, during cooling. This method is employed for parts that have rotational symmetry. This gives a more uniform depth of hardness penetration, because it compensates for any irregularities in the coil.

Progressive hardening

Progressive hardening is used if the power output is not adequate to permit single-shot hardening, or if a particularly shallow depth of hardness is required. The method is employed for very long parts, and normally

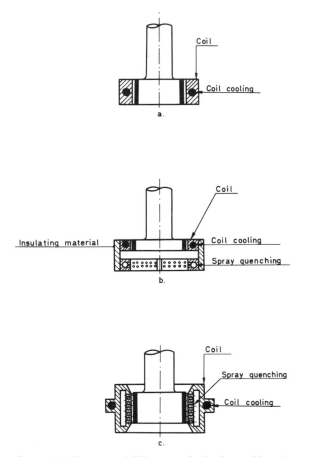

Figure 6.240 Example of different methods of quenching when induction hardening

requires a cross-sectional area that is uniform along the entire length of the hardened surface. During hardening the component should be held in position, e.g. in a rotating chuck or between a pair of jaws. It is fed forward at a certain rate through the coil, so that it is heated and quenched progressively during its passage through the coil and spray. A separate spray as shown in *Figure 6.242a* can be employed; alternatively the working coil can be combined with the spray into one unit, so that the spray ring itself also functions as the inductor, *Figure 6.242b*.

   Figure 6.243 illustrates a unit where the spray coil is brazed to the underlying inductor. (During hardening the spray coil is located underneath the inductor.) *Figure 6.244* shows the progressive hardening of a roll. Progressive hardening can also be carried out on parts that are not axially symmetrical, e.g. as shown in *Figure 6.245*. Here cooling is carried out by means of air.

Figure 6.241 Water quenching of induction-heated gear

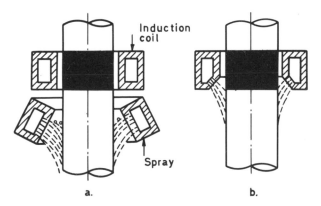

a.   b.

Figure 6.242 Progressive hardening of shafts using (a) separate
spray and (b) integrated working and spray coil

Figure 6.243 Inductor with spray quench ring brazed to it

Figure 6.244 Progressive hardening of a roll for cold rolling

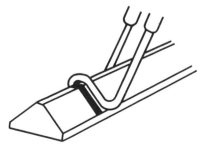

Figure 6.245 Progressive hardening without rotating the workpiece. In this example, air cooling is used

6.9.6   Determination of depth of hardening

The method used for determining the depth of hardening is the same for both induction hardening and flame hardening and is specified according to SS 117007, i.e. by measuring the value of HV 1 on a cross-section through the hardened zone. The limiting hardness value is 400 HV. The letters DI and DF refer to induction hardening and flame hardening, respectively. The corresponding ISO Standard is: ISO 3754–1976, *Steel—Determination of Effective Depth of Hardening After Flame or Induction Hardening*. However, this Specification is set out somewhat differently. This depth of hardening, designated DS, applies to:

(*a*) hardened layers with a depth greater than 0·3 mm;
(*b*) parts which, in the surface-hardened condition, have at a distance 3 × DS from the surface a hardness less than the hardness limit (HV)–100.

Where these conditions are not satisfied the effective depth of hardness after flame or induction hardening should be defined by agreement between the parties concerned. For steels where the hardness of a part at a distance 3 × DS from the surface is above the hardness limit (HV)–100, the criterion may still be used on condition that a higher hardness limit is chosen for the assessment of the effective depth of hardening.

The effective depth of hardening after flame or induction hardening is the distance between the surface of the product and the layer where the Vickers hardness (HV) under a load of 9·8 N (1 kgf) is equal to the value specified by the term 'hardness limit' which is a function of the minimum hardness required for the part, given by the following equation:

hardness limit (HV) = 0·8 × minimum surface hardness (HV)

6.9.7   The influence of various factors on hardness and depth of hardening

As mentioned previously, the C content has a decisive influence on hardness after hardening. As the C content increases up to about 0·80% the hardenability also increases, particularly when in combination with other alloying elements. Both the hardening temperature and holding time are decisive factors that govern the hardness and depth of hardening. For induction hardening with conditions otherwise remaining the same the depth of hardening is governed by the frequency of the inductive current. Since the hardening temperature, heating time and holding time can be controlled relatively easily during induction hardening, it is possible with this method to determine in advance the required hardness depth with a relatively high degree of certainty by using calculated and empirical values[83, 84, 85].

The hardening temperatures employed for induction hardening are normally some 50°C higher than those used for conventional hardening. The holding times are extremely short. In many cases there is no holding time, i.e. cooling starts directly when the desired temperature has been reached. Temperatures are normally controlled by means of optical pyrometers.

The pressure of the cooling medium and the angle of incidence of the jet as well as its width are of paramount importance to the intensity of cooling. Hence these factors influence the depth of hardening and the hardening itself. Kegel and Pennekamp[86] have arrived at the following recommendations based on experimental tests and theoretical calculations of the coefficient of heat transfer through the vapour film:

Pressure	1–3 bar
Angle of incidence	20–30°
Width of slit	0·8–1·2 mm, preferably
	0·9–1·0 mm

Usually the spray nozzle contains a number of holes, the diameters of which are taken as equivalent to the slit width of the above recommendations.

Figures 6.246–249 show examples of hardness and depth of hardening of steels having a base composition of 0·7% Mn, 1·1% Cr, 0·25% Mo and varying C contents. Induction hardening was carried out at 100 000 Hz. In this case the power output employed was the same throughout. Hence the factor governing the hardening temperatue was the heating-up time. The test specimens, 50 mm in diameter, were made from hardened and tempered steel with a hardness of *ca* 300 HB. Quenching took place in water directly when the specimens had reached the desired temperature.

Figures 6.250 and *6.251* show the results from tests using a frequency of 400 000 Hz. In this case the power output was regulated so that the heating-up times to the desired hardening temperature were different. Here too the specimens had a diameter of 50 mm and they were quenched in water directly as the current was cut off. Tests at temperatures of 875 and 900 °C gave almost identical results. Use of the high frequency resulted

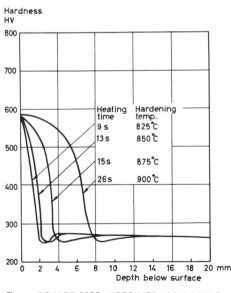

Hardness
HV

Heating time	Hardening temp.
9 s	825 °C
13 s	850 °C
15 s	875 °C
26 s	900 °C

Depth below surface

Figure 6.246 SS 2225, AISI 4130 with 0·24% C, 1·1% Cr, 0·25% Mo

556

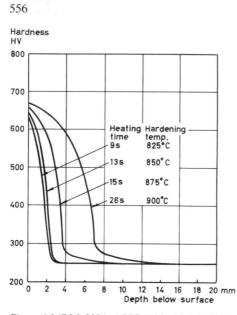

Figure 6.247 S S 2234, A I S I 4135 with 0·33% C, 1·1% Cr, 0·25% Mo

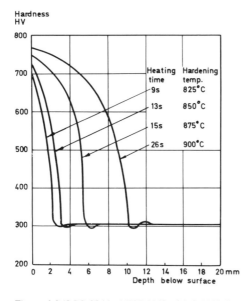

Figure 6.248 S S 2244, A I S I 4142 with 0·40% C, 1·1% Cr, 0·25% Mo

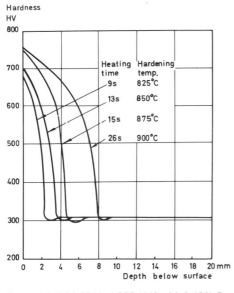

Figure 6.249 SS 2244, AISI 4142 with 0·45% C, 1·1% Cr, 0·25% Mo

Figures 6.246–6.249 Curves showing depth of hardening for induction-hardened Cr–Mo steels with different carbon contents. Frequency 10 000 Hz

in the depth of hardening being just about half the value obtained at 10 000 Hz; compare *Figure 6.251* with *Figure 6.247*.

All the experiments described above were made on steels that had been normalized or alternatively hardened and tempered. It was not possible to observe whether the mode of pretreatment had made any difference. When hardening rolls, relatively deep hardening is required. This can be achieved by employing a low frequency, e.g. 500 Hz, and by pre-heating the rolls. *Figure 6.252* shows the temperature as a function of the time during the heating-up and cooling of a pre-heated roll with a diameter of 300 mm. *Figure 6.253* shows the hardness distribution and the microstructure in the surface layer.

After being induction hardened the steel is usually tempered at 150–200°C, or higher if a lower hardness is desired. Parts that are to be ground should be tempered at at least 160°C. If the parts are tempered at lower temperatures there is considerable risk of cracking during grinding. *Figures 6.254* and *6.255* illustrate how tempering influences hardness.

Tempering can also be performed by induction heating, this being particularly suitable for localized tempering. *Figure 6.256* illustrates a coil design employed for the tempering of flanges on car axles. During tempering two axles are placed in a fixture with the flanges facing each other. The axles rotate during heating.

After tempering at 180°C the hardness values given in *Table 6.46* may be expected for the steels included in ISO 683/XII–1972.

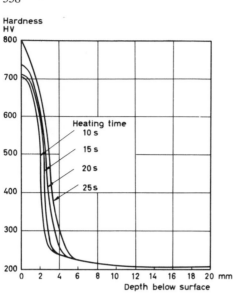

Figure 6.250 C 50

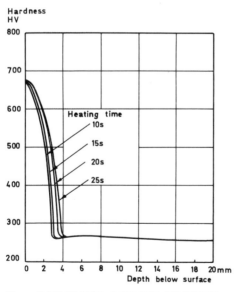

Figure 6.251 SS 2234, AISI 4135, BS 708A37, En 19 B

Figures 6.250 and 6.251 Curves showing depth of hardening for induction-hardened test bars of carbon steel C 50 and SS 2234. Hardening temperature 900°C, frequency 400000 Hz

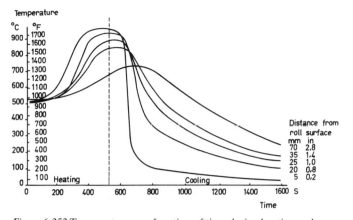

Figure 6.252 Temperature as a function of time during heating and cooling of a roll for cold rolling, 300 mm diameter, preheated to 500 °C (after Zetterlund and Björnander)

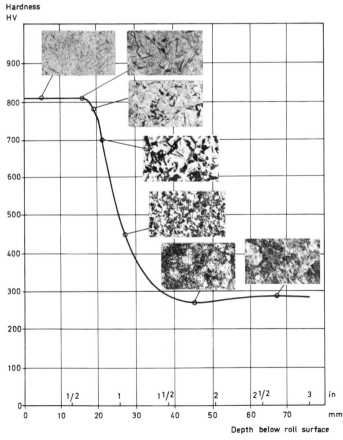

Figure 6.253 Hardness and microstructure in the surface zone of induction-hardened roll for cold rolling (after Zetterlund and Björnander)

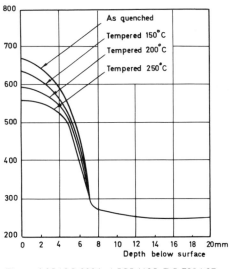

Figure 6.254 SS 2234, AISI 4135, BS 708A37, En 19 B

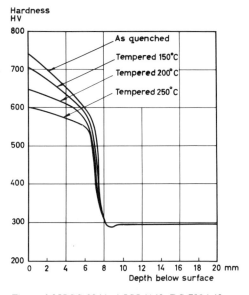

Figure 6.255 SS 2244, AISI 4142, BS 708A42, En 19 C

Figures 6.254 and 6.255 Curves showing hardening for SS 2234 and SS 2244 after induction heating to 900 °C, heating time 25 s, and subsequent tempering at different temperatures. Frequency 10000 Hz

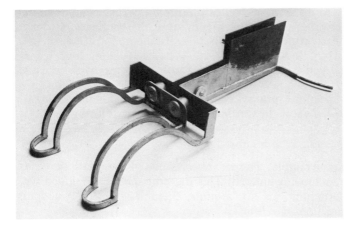

Figure 6.256 Coil for tempering of shaft flanges

Table 6.46 Hardness after induction heating and tempering 180 °C

Type of steel	Hardness HRC
1	50–55
2	53–57
3	56–61
4	57–62
5	58–63
6	56–61
7	52–57
8	54–58
9	55–60
10	55–60
11	55–60

6.9.8   Examples of induction-hardened machine components

Machine components which are induction-hardened are mainly bodies having rotational symmetry, such as gear wheels and shafts. Gears can be hardened in accordance with any of the methods described in *Figure 6.240*, but the gear should preferably rotate. The power input and the temperature employed should be so matched that the central portions of the gear are not heated to temperatures beyond 200 °C, which can be checked from the tempering colours. This is particularly important in the case of gears with finish-machined internal splines, which may be deformed during heating. Rotating the part during induction hardening, also called spin hardening, is generally suitable for gears with modules up to 5.

Acceptable contour hardening of a gear by means of spin hardening can be obtained if a high energy concentration, *ca* 2 kW/cm^2, and high frequency are employed. There is, however, a risk of overheating the tips of the gears and inadequate hardening at the tooth roots. When using intermediate frequencies the teeth first become hot at the roots, after which the heat diffuses towards the tops. In this case the teeth become through-hardened. More uniform contour hardening is achieved if heating is first started at an intermediate frequency until the tooth roots show a slight red heat, after which the tooth tips are heated by means of a high-frequency current for a short period. When hardening large gears, machines are employed which harden one tooth at a time by progressive heating.

In the following, sections summarize methods of induction hardening for gears. *Figure 6.257a–d* show the hardness zones for the differing methods.

(a) Single-shot hardening of tooth tips

The hardening is quite simple. The wear resistance of the teeth increases, but the strength remains unchanged. This method can be employed with high-frequency hardening up to module 3, and with intermediate-frequency hardening up to module 5.

(b) Single-shot hardening of complete teeth

Method (*b*) requires somewhat more experience than method (*a*). When using method (*b*) not only the wear resistance but also the bending strength

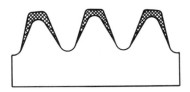

a. Spin hardening of tooth tips.

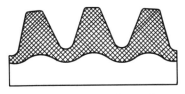

b. Spin hardening of entire teeth.

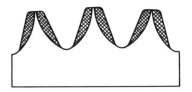

c. Flank hardening.

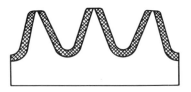

d. Tooth-gap hardening.

Figure 6.257 Appearance of hardened zones after hardening by different methods

of the teeth is increased. This method is recommended for modules up to 5. *Figure 6.241* shows an example of this method of hardening.

(c) Flank hardening (progressive hardening)

This method is used for modules ≥ 2 using high-frequency hardening and for modules ≥ 5 using intermediate frequency hardening. In principle the properties obtained are the same as by method (*a*). The hardening process is relatively simple.

(d) Tooth gap hardening (progressive hardening)

The sphere of application is the same as for method (*b*). A considerable, increase in bending and fatigue strength is achieved. This 'contour hardening' is actually the ideal method for gears, but it requires considerable experience. Trial hardening heat treatment and metallographical investigations are necessary before good results can be expected.

Induction hardening is very advantageous for long shafts on which only the bearing sections need hardening. Thanks to induction hardening we avoid deformation and distortion that could occur if the entire shaft were hardened. *Figure 6.258* shows a grinding-machine spindle that has been induction hardened on two bearing surfaces and on the flange. Shafts fitted with splines made from hardened and tempered steels often crack at the splines owing to overloading and fatigue. This susceptibility to cracking can easily be overcome by induction hardening. The depth of hardening below the base of the spline should at least be equal to height of the spline ribs. In the case of the part shown in *Figure 6.259* the optimal strength was obtained when the depth of hardening was 7 mm.

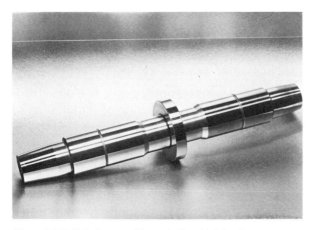

Figure 6.258 Grinding-machine spindle which has been induction-hardened at the cylindrical and tapered bearing surfaces and on the flange at the middle of the spindle. Steel BS 708A42

Figure 6.259 Machine component which has been
induction-hardened on the splines. Steel B S 708A42

6.9.9   Advantages and disadvantages of induction hardening

From the preceding section it is easy to understand that induction
hardening can offer considerable advantages when compared with
conventional methods of hardening. The major advantages are:

1. Restricted localized hardening.
2. Short heating-up periods.
3. Minimum surface decarburization and oxidation.
4. Only slight deformation.
5. Increased fatigue strength.
6. Any straightening required can be carried out on the unhardened
   surfaces and—to a certain extent—on the hardened surfaces.
7. The process can be incorporated in a production line.
8. Low operating costs.

The following are among the disadvantages:

1. The high capital costs necessitate high degree of equipment
   utilization.
2. The method is restricted to components having a shape suitable for
   induction hardening.
3. Only a limited number of grades of steel can be induction hardened.

Induction heating can be used for heat-treatment processes other than
those mentioned above, e.g. for continuous annealing or normalizing of
steel tubing or for brazing and welding.

6.10 Flame hardening

Flame hardening and induction hardening are two methods which, in many
respects, give equivalent results. The advantages of induction hardening
enumerated in the preceding section also apply, by and large, to flame
hardening. The capital investment costs for flame hardening are lower than
for induction hardening, but the operating costs are higher. Flame
hardening in its simplest form consists of heating the steel to the hardening

temperature by a flame from a welding torch and then quenching it in water or oil.

The grades of steel which were recommended for induction hardening are also suitable for flame hardening, which, moreover, can be used for the partial hardening of tools from alloy tool steels. If the steels exhibit sufficiently high hardenability, the requisite hardness can be achieved by cooling in air. In certain cases flame hardening is preferred to induction hardening, because it is easier to direct the heat to selected surface areas.

6.10.1   Methods of flame hardening

The gas used is a mixture of oxygen and a combustible gas, usually acetylene, natural gas or propane. It is mixed in a burner unit which has a shape to suit the part to be hardened. The appearance of the part also governs the method of hardening employed. We can distinguish between the following three main types.

1 Manual hardening

The designation indicates that the steel part is heated by hand with a welding torch or some other burner to the hardening temperature and is then cooled in water, oil or air. This method is suitable if a small surface, e.g. the tip of a screw, is to be hardened. Larger surfaces can also be hardened by hand, but this requires larger burners. By making sweeping movements with the burner the operator ensures that the temperature is uniform over the entire surface to be heated.

When hardening a large number of parts it is customary to keep the burner stationary and to hold the parts by hand or in a fixture. Hardening can also be automated so that the parts are held in a fixture which travels past the burner, where they are heated for some seconds and then dropped into the quenching liquid.

2 Spin hardening

This method is employed for bodies having rotational symmetry, the bodies being placed on a rotating table or in a chuck. The speed of rotation is relatively low, normally about one revolution per second. The number of torches used depends on the size and shape of the part. The method is also applicable to parts that are not absolutely symmetrical. Since the heating operation may take several minutes, the temperature tends to be equalized by heat conduction into the steel. Gears with modules up to 5 can be spin hardened. After they have been heated to hardeneing temperature the parts are either cooled by spray or by immersion into a quench tank. *Figure 6.260* illustrates the spin hardening of a trolley wheel.

3 Progressive hardening

This can be carried out on a flat surface or in combination with spin hardening. In the first instance a burner combined with a cooling spray is passed over the surface which is to be heated (*see Figure 6.261*). The rate of

566

Figure 6.260 Spin-hardening

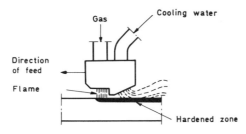

Gas

Cooling water

Direction
of feed

Flame

Hardened zone

Figure 6.261 Principle illustrating progressive
flame hardening

Direction of feed

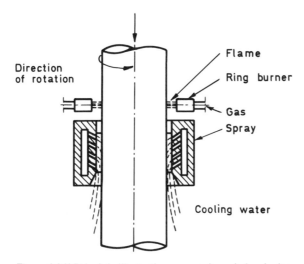

Direction
of rotation

Flame

Ring burner

Gas

Spray

Cooling water

Figure 6.262 Principle illustrating progressive spin hardening

movement is relatively low, 50–200 mm/min, and naturally this has to be adapted to suit the size of burner and the required depth of hardening. The tip of the inner cone of the flame is kept only a few mm away from the workpiece.

If progressive hardening is combined with spin hardening (*see Figure 6.262*), the speed of rotation and feed are of considerable importance to the uniformity of the hardened layer. As a result of the rotary movement the cooling water is thrown off tangentially. At higher rotational speeds this can interfere with the heating-up process and can also become a nuisance owing to water splashing into the workshop. At low rotational speeds, with a high feed and only a few torches, it is possible for 'spirals' to appear around the workpiece which contain alternate soft and hard bands (*see Figure 6.263*). In such a case ring burners should be used.

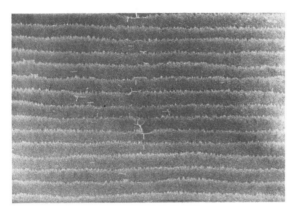

Figure 6.263 Spiral lines on flame-hardened shaft

6.10.2   Hardness and depth of hardening

As is the case with conventional hardening, the hardness is governed by the composition of the steel (mainly the C content), by the hardening temperature and by the rate of cooling. The hardness values given for induction hardening apply also to flame hardening. The depth of hardening, also, is affected by the factors mentioned above. For a given steel cooled at a certain rate, the heating-up time and intensity of heating are the decisive factors. It is very difficult to illustrate the effect of these factors by means of a table or diagram. General knowledge of heat treatment as well as good equipment and sound judgement are required if the desired results are to be achieved.

Irrespective of the method of hardening, the hardening temperature is reached more rapidly with acetylene–oxygen than with propane–oxygen. Hence, with acetylene–oxygen gas a much more sharply defined depth of hardening can be achieved.

With flame hardening there is a risk of oxidation of the steel surface, particularly if the steel should be overheated. At temperatures of around 1100 °C, which can be reached quite easily, and under oxidizing conditions, the steel can become so oxidized that copper—which is always present in steel and which does not oxidize to the same extent as iron—becomes concentrated and melts. If the steel surface is subjected to tensile stresses copper diffuses into the steel along the grain boundaries and causes surface cracking.

The Swedish Association For Metal Working Industries (Sveriges Mekanförbund) undertook a research project on flame hardening to study

Figure 6.264 Gas-cutting machined adapted for flame hardening

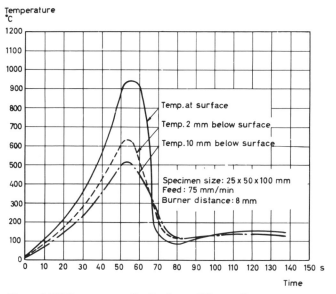

Figure 6.265 Temperature distribution at different distances below the surface during flame hardening followed by water quenching

the extent of hardening and depth of hardening during progressive flame hardening treatments[87]. For the flame-hardening trials the modified flame-cutting machine shown in *Figure 6.264* was used, which incorporated the burner units illustrated in *Figure 6.261*. The temperature distribution in the steel surface was determined on test bars size $25 \times 50 \times 100$ mm, the 25×100 mm surface being subjected to the hardening trials. The temperature distribution with water cooling and air cooling respectively is shown in *Figures 6.265* and *6.266*. By burner distance is meant the distance between the burner orifice and the steel surface being hardened.

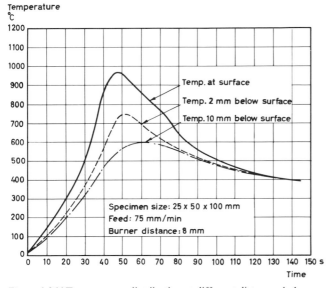

Figure 6.266 Temperature distribution at different distances below the surface during flame hardening followed by cooling in air

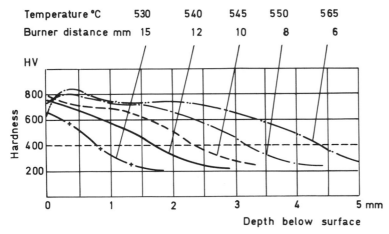

Figure 6.267 Relationship between hardness profile and burner distance during flame hardening of a carbon steel with 0·50% C. Test specimen dimensions: $25 \times 75 \times 100$ mm. Feed 50 mm/min. Water spray quenching. Temperature measured 10 mm below the surface

Figure 6.267 shows the hardness distribution in test bars made from carbon steel C 50 after trial hardening when using a feed rate of 50 mm/min and different burner distances.

It is also possible for tool steels of type O 1 and D 2 to be flame hardened without cracking. With these steels it is sufficient to cool them in air in order to obtain surface hardness values of around 800 H V. *Figure 6.268* shows results from a trial steel O 1. During a hardening run, with the minimum burner distance, the steel became coarse grained in the surface layer.

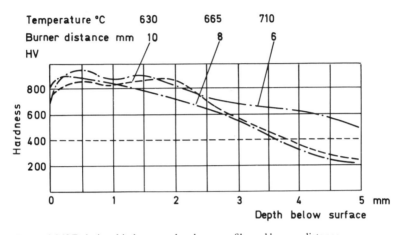

Figure 6.268 Relationship between hardness profile and burner distance during flame hardening of steel O 1. Test specimen dimensions: 25 × 75 × 100 mm. Feed 50 mm/min. Air cooling. Temperature measured 10 mm below surface

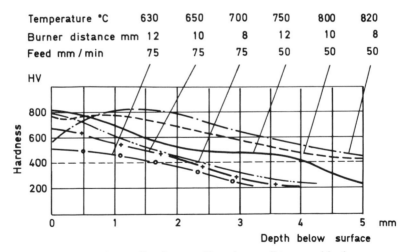

Figure 6.269 Dependence of hardness profile on burner distance and feed during flame hardening of steel D 2. Test specimen dimensions: 25 × 75 × 100 mm. Air cooling. Temperature measured 10 mm below surface

Steel D 2 is unaffected by overheating as regards grain size, but it is liable to contain retained austenite and be prone to melting. During the tests it was clear that this effect was due, among other things, to the rather low heat conductivity of the steel, which resulted in a steep temperature gradient. *Figure 6.269* also shows that the temperatures measured 10 mm below the surface are considerably higher than in the preceding case. When hardening the edges only of the test bars, and with only a slight change in the feed rate, both normal and overheated structures were obtained, as shown in *Figure 6.270*.

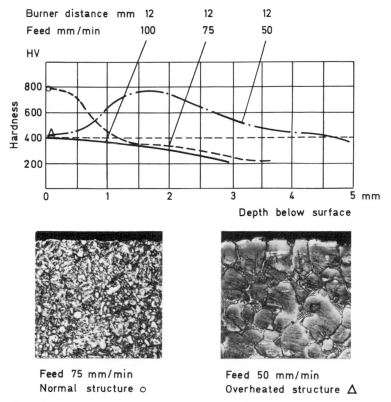

Feed 75 mm/min
Normal structure o

Feed 50 mm/min
Overheated structure Δ

Figure 6.270 Dependence of hardness profile and microstructure on the rate of feed during flame hardening of steel D 2. Test specimen dimensions: 25 × 75 × 100 mm. Edge hardening. Air cooling

6.10.3   Examples of flame-hardened machine components and tools

Gears can be flame-hardend by several methods. Up to module 5, spin hardening is used. If only the tooth tips are to be hardened, gears of larger modules can be spin hardened. With gears of module 4 or more, one tooth at a time can be progressively hardened. This can be undertaken in such a way that either the flanks on one tooth are hardened simultaneously (flank hardening) or the flanks on two teeth are hardened simultaneously (tooth-gap hardening). From module 6 and up it is also possible to harden the space between the teeth (tooth bases), so that the fatigue strength is increased. The principles involved resemble very closely those described for the induction heating of the examples shown in *Figure 6.257*

Progressive hardening can be carried out with a quenching spray directly after heating, this being essential for gears made from carbon steel. In the case of alloy steel gears similar to the one shown in *Figure 6.271*, each tooth is heated progressively, after which the gear wheel is rotated 15–60° and the tooth allowed to cool in air. The other parts of the gear are kept cool by means of cooling water.

Figure 6.271 Flame hardening of a bevel gear made of CrNiMo steel (courtesy Nohab)

A similar method is employed for spur gears and sprocket wheels which are placed vertically, with roughly one quarter of the wheel immersed in water. When the tooth located at the highest position has been heated to the hardening temperature, the wheel is rotated through 180°, so that the heated tooth is quenched. For single-tooth hardening pre-indexed machines are used which give symmetrical stresses in the gear wheel, and thereby only slight deformation is created.

Figure 6.272 Flame hardening of lathe guides made of steel C 45

Long objects

Long objects, such as lathe guides, are hardened progressively, whereby a flame-cutting machine can be used for the feed. *Figure 6.272* shows the way in which hardening progresses. The depth of hardening is regulated by the amount of heat supplied (gas pressure) and the feed rate. *Figure 6.273* illustrates lathe guides that have been hardened by using the same gas-pressure setting but different feed rates.

Figure 6.273 Depth of hardening for different feed rates. For the left-hand lathe guide the feed rate was 250 mm/min and for the right-hand guide 200 mm/min

One special case is illustrated in *Figure 6.274,* where a cam shaft for a printing press is hardened simultaneously along four sections, length about 100 mm each. The larger part of the cam shaft is immersed in water, which provides sufficient cooling to ensure that satisfactory hardening is achieved. A similar procedure can be used when hardening long edges, for example. The object is placed in water with the edge just level with the surface. As the flame sweeps towards the edge, the water is blown away and the edge is exposed and heated. As the burner travels onwards, the water flows back and quenches the edge.

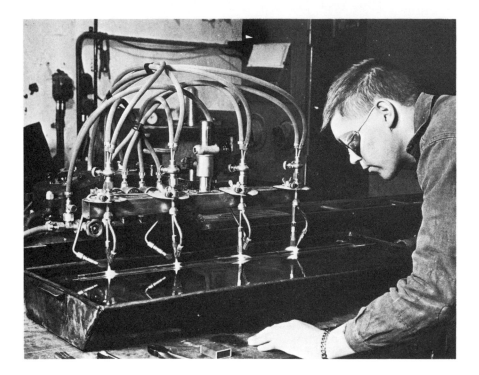

During the flame hardening of long and relatively thin objects it is difficult to avoid deformation. When one side of a straight edge is being hardened it might be presumed that this will become convex, since the martensite formed has a greater volume than the other structural constitutents. This does in fact occur quite frequently and is advantageous when straightening is called for. The straight edge may also become concave, which is due to the fact that the upsetting occurring during heating exceeds the increase in volume that has taken place during the martensite formation. The degree of contraction can be regulated by means of the feed rate.

Plates

Plates can be successfully hardened with the aid of wide burners. Large plates can be hardened in several passes, but a soft zone between the passes must be allowed for. *Figure 6.275* illustrates a cross-section of a flame-hardened plate made from Cr–Ni–Mo steel.

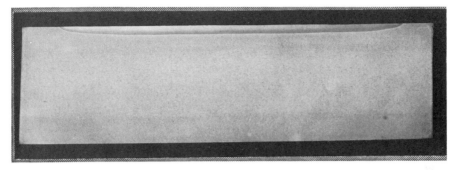

Figure 6.275 Cross-section of flame-hardened plate of Cr–Ni–Mo steel P 20. Plate thickness 72 mm

Simple blanking tools

Simple blanking tools (punching tools), as mentioned previously, can be given a wear-resistant cutting edge by means of flame hardening. This method is used mainly for tools employed in blanking large plates. Such a tool can either be built up of shear plates which are bolted firmly to a parent tool, or it can be manufactured from plate in which the desired contour has been cut to shape by flame cutting and then machined. In the former instance the shear plates can be dismantled and flame hardened separately if so required. In the latter case hardening must be carried out using a torch which is guided around the entire cutting edge. When the torch has returned to the point which was hardened first, a narrow, softer zone forms, the location of which should be at a point where minimum wear is anticipated.

Carbon steel containing 0·50% C, as well as steel W 1, can be used for simple tools. Quenching is carried out by water spraying. It may also suffice to keep the tools immersed in water, with the cutting edge just above the surface of the water.

Moulds for plastics

Moulds for plastics are largely made from steel P 20 hardened and tempered. If the plastic is very abrasive or if the pressure at the mating surfaces is extremely high, flame hardening of these mating surfaces can impart a considerable improvement in wear resistance. After flame hardening and cooling in air the hardness of P 20 is 50–53 HRC.

General recommendations

The following recommendations are given for flame hardening the edges of moulds and other tools made from alloy steels when using a conventional welding torch or burner which is passed over the edges of the tool as shown in *Figure 6.276.*

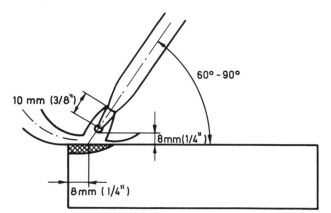

a. View from side.

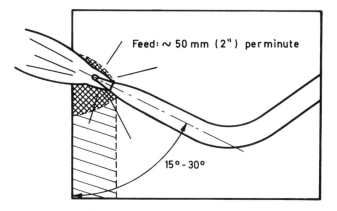

b. View from top.

Figure 6.276 Flame hardening of the edge of a tool made from steel P 20

1. The area to be hardened should be clean and free from scale.
2. An oxy-acetylene flame is preferable, initially with slightly more acetylene than oxygen—a 'blue or green cone' will be noticed in the flame.
3. Oxygen should be increased until the 'blue or green cone' disappears.
4. A slight pre-heat by movement of the flame over the beginning of the area to be hardened should be given.
5. The tip of the core flame should then be held about 8 mm (¼ in) back from the edge of the material, approximately 8 mm (¼ in) above the

material and at an angle of 60–90° to the horizontal plane and 15–30° to the vertical.

6. Once a dull-red hot spot becomes visible, the flame must be moved slowly along, so that the red spot follows the flame at a constant speed and remains a constant size. Speed is approximately 50 mm (2 in) per minute.
7. The flame should not be allowed to dwell or the process to discontinue and then be restarted. If the flame dwells, sparks may appear, which indicates burning of the metal. If the process is discontinued it is virtually impossible to start where the flame left off and the localized uneven heating gives rise to stressed or soft areas.
8. Tempering at 180–200°C is recommended, but not necessary, after flame hardening. Local heating with the flame is used.

A suitable nozzle for flame hardening is AGA Welding nozzle No. 4. Alternatively BOC nozzle No. 18. Saffire 3 range.

References

1. ZMESKAL, O., and COHEN, M., 'The Tempering of Two High-Carbon High-chromium Steels', *Trans. ASM*, **31**, 380 (1943)
2. RAPATZ, F., KRAINER, H. and SWOBODA, K., Hardness Manifestation and Cutting Properties of Steels Containing 9–13% Cr, *Archiv für das Eisenhüttenwesen*, No. 3/4, 115–123 (1949) (in German)
3. BÜHLER, H., POLLMAR, F., and ROSE, A., 'Influence of Tool Material and its Heat Treatment on the Cutting of Thin Steel', *Archiv für das Eisenhüttenwesen*, No. 10, 989–996 (1970) (in German)
4. THELNING, K-E., 'Hot and Cold Hobbing of Hot-work Tool Steel', *Schweiz. Arch. angew. Wiss. Techn.*, **27**, 503–510 (1961) (in German)
5. NORSTRÖM, L-Å., 'Performance of Hot-work Tool Steels', *Scand. J. Metallurgy*, **11**, 33–38 (1982)
6. GLEDHILL, I. M., 'High-speed Steel, its Development and Use', *Iron and Steel*, (October 1904)
7. LUERSSEN, G. V., and GREENE, O. V., 'The Torsion Impact Test', *Trans. ASTM*, **33**, 315 (1933)
8. GROBE, A. H., and ROBERTS, G. A., 'Unnotched Impact Strength of High Speed Steels', *Trans. ASM*, **42**, 686 (1950)
9. HELLMAN, P., *et al.*, 'The ASEA-STORA-Process, Modern Development in Powder Metallurgy', **4**, 573–582 (1970)
10. LEIDEL, B., and SCHÖNBAUER, G., 'New Developments in the Field of High-speed Steels' (Review of the literature since 1969), *Stahl u. Eisen*, **93**, 1266–1270 (1973) (in German)
11. LEIDEL, B., 'New Developments in the Field of High-speed Steel' (Review of the literature since 1973), *Stahl u. Eisen*, **99**, 477–482 (1979) (in German)
12. *Bulletin of the Metals Study Circle*, International Conference on Cutting of Metals, St. Etienne (1979) (in French)
13. *Processing and Properties of High Speed Tool Steels*, Las Vegas Conf. Proc. Metallurgical Society of AIME (1980)
14. *Cutting Tool Material*, Ft. Mitchell Conf. Proc., American Society for Metals (1980)
15. *Standard Specification for High Speed Tool Steels, A 600*, American Society for Testing and Materials
16. BAGGSTRÖM, G., 'New Steel for Turbine Runners', *Water Power* (December 1964)
17. ARWIDSON, S., BAGGSTRÖM, G., and HELLNER, L., *New Steels for Power Industry*, Publication 117, Iron and Steel Institute (1968)
18. SCHUMANN, H., 'On the Causes of the High Degree of Work-hardening of High-manganese (Hadfield) Steel', *Neue Hütte*, **12**, No. 4, 220–226 (1967) (in German)

19. MORRAL, J. E., and CAMERON, T. B., 'Boron Hardenability Mechanisms'. In *Boron in Steel,* The Metallurgical Society of AIME (1980)
20. KAPADIA, B. M., BROWN, R. M., and MURPHY, W. J., 'The Influence of Nitrogen, Titanium and Zirconium on the Boron Hardenability Effect in Constructional Alloy Steels', *Trans. AIME,* **242**, 1689–1694 (1968)
21. THELNING, K-E., 'Evaluation and Practical Application of the Hardenability of Boron Steels'. In *Boron in Steel,* The Metallurgical Society of AIME, 127–146 (1980)
22. BROWNRIGG, A., and BROWN, G. G., 'The Effect of Boron on Transformation and Properties of Low-Carbon Fe–Mn Alloys', *Australian Inst. Met,* **J17**, 192–204 (1972)
23. SHACYAN, G., SANDBERG, A., and LAGNEBORG, R., *The Effect of Micro-alloying Additions on the Hardenability of Quenched and Tempered Constructional Steels,* 23rd Mechanical Working and Steel Processing Conference, Pittsburg, USA, (October 1981)
24. NAKASATO, F., and TAKAHASHI, M., 'Effects of Boron, Titanium and Nitrogen on the Hardenability of Boron-treated Steels for Heavy Machinery', *Metals Technology,* 102–105 (1979)
25. KARLSSON, L., SANDBERG, O., and THELNING, K-E., 'Comparison of the Properties of Steels, of Types Cr–Mn–B and Cr–Mo for Hardening and Tempering', *Research Publication D 394* of Jernkontoret (the Swedish Association of Ironmasters) (in Swedish) (1981)
26. BARSOM, I. M., and ROLFE, S. T., 'Conditions between K_{IC} and Charpy V-notch Test Results in the Transition Temperature Range, Impact Testing of Metals', *ASTM STP* **466**, 281–302 (1970)
27. SANDBERG, O., and ÅKERMAN, J., *Influence of Tempering Temperature, Microstructure and Carbon Content on the Toughness of Cr–Mo–steel for Hardening and Tempering,* Swedish Institute for Metals Research IM 1487, (June 1980) (in Swedish)
28. SINHA, T. K., and SANDBERG, O., *Fracture Toughness of a Quenched and Tempered Boron Steel,* Swedish Institute for Metals Research IM-1526, (January 1981)
29. LE BON, A. B., and SAINT-MARTIN, L. N., 'Using Laboratory Simulations to Improve Rolling Schedules and Equipment', *Micro Alloying,* **75**, 90–98
30. ROBERTS, M. S., 'Micro-alloyed Constructional Steels (HSLA-steels)-Relationship between Microstructure and Properties', *Scanmetall* 1981 (in Swedish)
31. RASHID, M. S., *GM 980 X, Potential Applications and Review,* SAE Conference, Detroit, Michigan (February 1977)
32. ÖSTRÖM, P., LÖNNBERG, B., and LINDGREN, I., 'Role of Vanadium in Dual-Phase Steels', *Metals Technology,* 81–93 (1981) or IM-1404 (1979)
33. ÖSTRÖM, P., HOLMSTRÖM, V., and LAGNEBORG, R., 'The Mechanical Properties and Microstructures of Some Commercially Produced Dual-Phase Steels', *Scandinavian Journal of Metallurgy,* **10**, 163–169 (1981)
34. *Micro Alloying 75,* Proceedings of an International Symposium on High-Strength Low-alloy Steels, Union Carbide Corporation, New York (1977)
35. ENGINEER, S., and VON DEN STEINEN, A., 'Micro-alloyed Pearlitic Steels with Improved Toughness', *Thyssen Edelst. Techn. Ber.* 85–89 (1980) (in German)
36. COLLIN, R., GUNNARSON S., and THULIN, D., 'A Mathematical Model for Predicting Carbon Concentration Profiles of Gas-carburized Steel', *J. Iron Steel Inst.,* **210**, 10, 785–789 (1972)
37. WATERFALL, F. D., 'Case-hardening Steels in Cyanide-containing Salt Baths', *Metallurgia,* 29–36 (May 1949)
38. BUNGARDT, K., BRANDIS, H., and KROY, P., 'Salt-bath Carburizing of Case-hardening Steels at 900–1000°C', *Härterei-Techn. Mitt.,* **19**, No. 3, 146–153 (1964) (in German)
39. HARRIS, F. E., Case Depth—an Attempt at Practical Definition, *Met. Prog.* (August 1943)
40. WESTEREN, H. W., 'Vacuum Furnace Carburizes Faster, More Efficiently', *Met. Prog,* 101–107 (October 1972)
41. GROSCH, J., LIEDTKE, D., KALLHARDT, K., TACKE, D., HOFFMANN, R., LUITEN, C. H., and EYSELL, F. W., 'Gas Carburizing at Temperatures above 950°C in Conventional Furnaces and in Vacuum Furnaces' (Discussion, 36 Participants: Conference on Hardening), *Härterei-Techn. Mitt.,* **36**, 262–269 (September/October 1981) (in German)
42. STRIGL, R., 'New Methods of Gas Carburizing', *Z. f. wirtsch. Fertig.,* **76**, No. 9, 415–417 (1981) (in German)
43. WILLIAMS, M. E., 'A Practical Approach to Carburizing Process Determination', *J. of Heat Treating,* **2**, No. 1, 28–34 (1981)
44. WÜNNING, J., 'Development Trends in Gas Carburizing Technique', *Z. f. wirtsch. Fertig.,* **76**, No. 9, 411–414 (1981) (in German)

45. KOZLOVSKII, J. S., KALININ, A. T., NOVIKOVA, A. YA., LEBEDEVA, E. A., and FEOFANOVA, A. J., 'Internal Oxidation During Case-hardening of Steels in Endothermic Atmospheres', *Metals Sciences and Heat Treatment*, No. 3, 157–161 (1967)

46. CHATTERJEE-FISCHER, R., 'Internal Oxidation During Carburizing and Heat Treating', *Met. Trans. A.,* **9A**, No. 11, 1553–1560 (1978)

47. ARKHIPOV, YA, J., BATYREV, V. A., and POLOTSKII, M. S., 'Internal Oxidation of the Case on Carburized Alloy Steels', *Metal Science and Heat Treatment,* **14**, No. 6, 508–512 (1972)

48. ROBINSON, G. H., *The Effect of Surface Condition on the Fatigue Resistance of Hardened Steel. Fatigue Durability of Carburized Steel,* American Society for Metals, Cleveland, Ohio, 11–46 (1957)

49. FUNATANI, K., 'Influence of Case-hardening Depth and Core Hardness on Bending Fatigue Strength of Carburized Gear Teeth', *Härterei-Techn. Mitt.,* **25**, No. 2, 92–97 (1970) (in German)

50. BEUMELBURG, W., *The Behaviour of Case-hardened Specimens of Various Surface Finish and Surface Carbon Content when Subjected to Rotating Bend Test, Static Bend Test and Impact Bend Test,* Thesis, University of Karlsruhe, W. Germany (1964) (in German)

51. BIERWIRTH, G., 'Retained Austenite in the Surface-hardened Layers of Case-hardened Steel, Grade 25 MoCr 4', *Schweizer Archiv.,* 104–112 (April 1964) (in German)

52. FINNERN, B., 'Effect of Case Hardening on Reversed-stress Fatigue Strength', *Härterei-Techn. Mitt.,* **3**, No. 1, 1–7 (1957) (in German)

53. DIESBURG, D. E., and ELDIS, G. T., 'Fracture Resistance of Various Carburized Steels', *Met. Trans.,* **9A**, No. 11, 1561–1570 (1978)

54. MAGNUSSON, L., *Cyclic Behaviour of Carburized Steel,* Linköping Studies in Science and Technology. Dissertations, No. 56 (1980)

55. CHATTERJEE-FISCHER, R., and SCHAABER, O., *Some Observations on Carbonitriding,* Publication 124, Iron and Steel Institute, London (1970)

56. PŘENOSIL, B., 'Properties of the Carbonitriding Layers Forming as a Result of Carbon Diffusion in Austenite', *Härterei-Techn. Mitt.,* **21**, No. 1, 24–33 (1966) (in German)

57. KIESSLING, L., *Fatigue Strength of a Carbo-nitrided Steel,* IVF Report No. 77630 (1977) (in Swedish)

58. THELNING, K-E., 'The Nitridability of Some Constructional and Tool Steels', Paper read at the *7th International Conference on Heat-treatment Problems,* Vienna (8 June 1960) (in German)

59. MINKEVIČ, A. N., and SOROKIN, JU. V., 'The Nitriding of Steels in a Mixture of Nitrogen and Ammonia', *Härterei-Techn. Mitt.,* **25**, No. 1, 10–16 (1970) (in German)

60. NAUMANN, K., and LANGENSCHEID, G., 'An Investigation of the Fe–C–N System', *Archiv, für das Eisenhüttenwesen,* **9**, 677–682 (1965) (in German)

61. PŘENOSIL, B., Structure of Layers Produced by Salt-bath Nitriding and by Ammonia Atmosphere with Added Hydrocarbons', *Härterei-Techn. Mitt.,* **20**, No. 1, 41–49 (1965) (in German)

62. BELL, T, and LEE, S. Y., *Gaseous Atmospheric Nitrocarburizing in Heat Treatment '73,* The Metals Society, London, 99–107

63. KELLER, K., 'Structure of Layer on Glow-discharge Nitrided Ferrous Materials', *Härterei-Techn. Mitt.,* **26**, No. 2, 120–130 (1971) (in German)

64. KOROTCHENKO, V., and BELL, T., 'Applications of Plasma Nitriding in UK Manufacturing Industries: 1978', *Heat Treatment of Metals,* **4**, 88–94 (1978)

65. STAINES, A. M., and BELL, T., 'Technological Importance of Plasma-Induced Nitrided and Carburized Layers on Steel', *Thin Solid Films,* **86**, 201–211 (1981)

66. VARCHOSCHKOW, A., and TOSCHKOW, W. 'Influence of the Nitriding Treatments on the Formation of the ε-phase in the Plasma State of the Glow Discharge', *Härterei-Techn. Mitt.,* **37**, No. 6, 270–274 (1982) (in German)

67. *Tenifier-Mitteilungen,* Degusssa, Frankfurt am Main

68. WATERFALL, F. D., 'Reducing Scuffing and Wear of Ferrous Metals', *Engineering,* (23 January 1959)

69. WIEST, P., and KRZYMINSKI, H., 'Effects Governed by Age-hardening on the Mechanical Properties of Nitrided, Plain-carbon Steels', *Bänder Blech Rohre,* **7**, 351–358 (1970) (in German)

70. BIRK, P., 'Powder Nitriding, Theory and Practice', *Microtechnics,* **25**, 1 (1970) (in German)

71. FINNERN, B., 'The Testing of Bath-nitrided Components', *Härterei-Techn. Mitt.,* **20**, No. 1, 50–57 (1965) (in German)

72. NOREN, T. M., and KINDBOM, L., 'The Structure of Nitrided Heat-treatable Steels', *Stahl und Eisen*, **78**, No. 26, 1881–1891 (1958)

73. HODGSON, C. C., and BARON, H. G., 'The Tempering and Nitriding of Some 3% Chromium Steels', *J. Iron Steel Inst.*, **182**, 256–265 (1956)

74. COPPOLA, V. J., 'Gas Nitriding of Stainless Steel', *Metal Progress*, 83–84 (July 1961)

75. LAMBERT, R., 'The Surface-hardening of Steel Parts and Progress Achieved with Controlled Nitriding', *Revue de metallurgie.* **11**, No. 7, 553–558 (1955) (in French)

76. BELL, T., and LOH, N. L., 'The Fatigue Characteristics of Plasma Nitrided Three Pct CrMo Steel', *J. Heat Treating*, **2**, No. 3, 232–237 (1982)

77. WIEGAND, H., 'The Present State of Nitriding', *Härterei-Techn. Mitt.*, **21**, No. 4. 263–270 (1966) (in German)

78. FINNERN, B., VETTER, K., and JESPER, H., 'Fatigue Strength and Heat Treatment of High-strength High-grade Constructional Steels', *Zeitschrift für Wirtschaftliche Fertigung*, **60**, No. 8, 381–387. No. 9, 444–448 (1965) (in German)

79. ISHIZAKI, K., FIOR, G., and CORREDOR, L., 'On an Abnormal Load Frequency Dependence in Fatigue Endurance of Nitrided Steel', *Trans. ISIJ*, **20**, 707–709 (1980)

80. FICHT, W., 'New Findings on the Technique of Surface Boriding', *Härterei-Techn. Mitt.*, **9**, No. 2, 113–118 (1974) (in German)

81. GRAF VON MATUSCHKA, A., *Boriding. Monographs on the Techniques of Surface Treatment*, Delta-Verlag AG, CH-1800, Vevey (1975) (in German)

82. OSBORN, H. B., 'Surface Hardening by Induction Heat', *Metal Progress*, 105–109 (December 1955)

83. GEISEL, H., 'Experimental Determination of the Depth of Hardening Obtained by Progressive Induction Hardening, using a Frequency of 10 kHz', *Elektrowärme*, **18**, No. 10 (1960) (in German)

84. HEGEWALDT, F., 'Induction Surface Hardening: Predicting the Depth of Hardening', *Härterei-Techn. Mitt.*, **17**, No 2, 75–81 (1962) (in German)

85. SIEDEL, W., and NETZ, W., 'Predicting, by Calculations, the Result of Induction Heating', *Härterei-Techn. Mitt.*, **37**, No. 5, 211–219 (1982) (in German)

86. KEGEL, K., and PENNEKAMP, H., 'Quenching Trials to Determine the Optimum Angle of Incidence for Jet Quenching, Using Medium and High Frequency Induction Heating', *Härterei-Techn. Mitt*, **29**, No. 2, 67–70 (1974) (in German)

87. THELNING, K-E., 'Flame hardening of Punching and Deep-drawing Tools made of Steel and Cast Iron', *Sveriges Mekanförbund*, Stockholm (1972) (in Swedish with explanation in English)

7
Dimensional changes during hardening and tempering

One of the most difficult problems facing the heat treater is the dimensional changes that occur in the steel during hardening and tempering. By dimensional changes we mean changes in both shape and size.

7.1 Dimensional changes during hardening

One of the main causes of dimensional changes is the stresses which occur as a consequence of the contraction of the material during cooling, i.e. thermal stresses. The other main cause is the transformation stresses which occur as a result of the martensite formation.

7.1.1 Thermal stresses

When a body cools, the outer layer cools more quickly and contracts. The inner, softer parts try, during this process, to assume a spherical shape, this being the shape to which they offer the least resistance during deformation. Hence the main rule is that all bodies with shape deviating from the spherical one attempt to assume this shape during rapid cooling.

The effect of thermal stresses can best be studied in a low-carbon steel, in which we can disregard the likelihood of martensite formation. Austenitic steel is also a suitable object for study. Frehser and and Lowitzer[1] performed a series of investigations which show the effect of cooling rate on slabs size 200 × 200 × 20 mm made from mild steel. Slab *a* in *Figure 7.1* is solid whereas slab *b* has an inner square hole size 100 × 100 mm. To illustrate the dimensional changes more clearly, these have been drawn to a larger scale. *Figure 7.1* shows clearly that the more drastic the quench, the greater are the changes. *Figure 7.2* shows that the greater the temperature drop during cooling, the greater is the deformation.

The high-temperature strength of the steel also has some effect. As shown by *Figure 7.3*, the steel with the greatest high-temperature strength (18/8 steel) exhibits the highest dimensional stability.

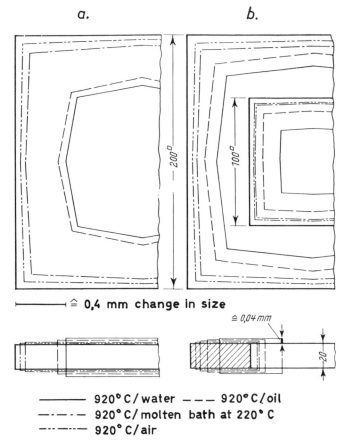

Figure 7.1 Dimensional changes of plates made from mild carbon steel (0·10% C) after cooling in water, oil, molten quenching bath and air, respectively (after Frehser and Lowitzer[1])

7.1.2   Transformation stresses

During heating and cooling steels pass through a series of structural transformations. The various structural constituents possess different densities and hence differing values for specific volume, as shown by *Table 7.1*.

The amount of carbon dissolved in the austenite or martensite has a relatively high effect on the specific volume. When calculating the changes in volume which take place during the transformations in conjunction with hardening, due regard must be paid to the carbon content, as is also shown by *Table 7.2*.

The changes in length can be studied by using a dilatometer, in which a bar of steel is placed between two quartz bars, the changes in length being measured during heating and cooling. If cooling takes place at such a high rate that martensite is formed, we obtain a diagram, as shown in *Figure 7.4,* where the continuous curve is particularly significant. During heating a continuous increase in length occurs up to A_{c1}, where the steel shrinks as it transforms to austenite. After the austenite formation is completed, the length increases again. However, the coefficient of longitudinal expansion is not the same for austenite as for ferrite.

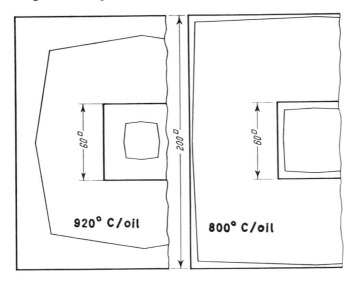

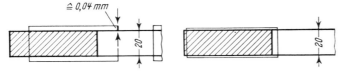

Figure 7.2 Effect of amplitude of heating temperature on dimensional changes occuring during cooling of plates of mild carbon steel in oil (after Frehser and Lowitzer[1])

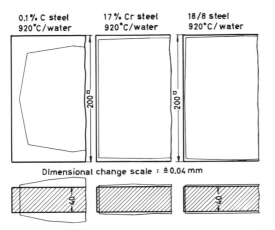

Figure 7.3 Dimensional changes in steel plates of different compositions after cooling in water from 920°C (after Frehser and Lowitzer[1])

Table 7.1 Specific volume of phases present in carbon tool steels[2]

Phase or phase mixture	Range of carbon %	Calculated specific volume at 20°C cm³/g
Austenite	0–2	$0 \cdot 1212 + 0 \cdot 0033 \cdot (\%C)$
Martensite	0–2	$0 \cdot 1271 + 0 \cdot 0025 \cdot (\%C)$
Ferrite	0–0·02	$0 \cdot 1271$
Cementite	6·7 ± 0·2	$0 \cdot 130 \pm 0 \cdot 001$
Epsilon carbide	8·5 ± 0·7	$0 \cdot 140 \pm 0 \cdot 002$
Graphite	100	$0 \cdot 451$
Ferrite + cementite	0–2	$0 \cdot 271 + 0 \cdot 0005 \cdot (\%C)$
Low-carbon content martensite epsilon carbide	0·25–2	$0 \cdot 1277 + 0 \cdot 0015 \cdot (\%C—0 \cdot 25)$
Ferrite + epsilon carbide	0–2	$0 \cdot 1271 + 0 \cdot 0015 \cdot (\%C)$

Table 7.2 Changes in volume during transformation to different phases[2]

Transformation	Change in volume %
Spheroidized pearlite → austenite	$-4 \cdot 64 + 2 \cdot 21 \cdot (\%C)$
Austenite → martensite	$4 \cdot 64 - 0 \cdot 53 \cdot (\%C)$
Spheroidized pearlite → martensite	$1 \cdot 68 \cdot (\%C)$
Austenite → lower bainite	$4 \cdot 64 - 1 \cdot 43 \cdot (\%C)$
Spheroidized pearlite → lower bainite	$0 \cdot 78 \cdot (\%C)$
Austenite → upper bainite	$4 \cdot 64 - 2 \cdot 21 \cdot (\%C)$
Spheroidized pearlite → upper bainite	0

On cooling, thermal contraction takes place and during martensite formation the length of the steel increases. After cooling to room temperature most martensitic steels contain some retained austenite, the amount increasing with the amount of alloying elements dissolved during austenitization. The larger the quantity of retained austenite contained in the steel after hardening the smaller is the increase in length or increase in volume. If the retained austenite content is sufficiently high, we generally obtain a reduction in volume, *see* curve T_3 in *Figure 7.4*. However, he volume of the austenite increases with the quantity of dissolved carbon, and this has to be taken into account when calculating the changes in volume.

Taking as a basis the proportions of martensite and austenite, together with the amount of carbon dissolved therein, it is possible with the aid of the data in *Table 7.2* to calculate the changes in volume occurring during hardening. If the steel contains undissolved cementite this volume has to be deducted during the calculations. Consequently the equation for use in the theoretical calculations takes the following form:

$$\frac{\Delta V}{V} = \left(\frac{100 - V_c - V_a}{100}\right)(1 \cdot 68 \times C) + \frac{V_a}{100}(-4 \cdot 64 + 2 \cdot 21 \times C)$$

where $\Delta V/V$ = change in volume in %
V_c = % by volume undissolved cementite
V_a = % by volume austenite
$100 - V_c - V_a$ = by volume martensite
C = % by weight of carbon dissolved in austenite and martensite respectively.

Change
in length

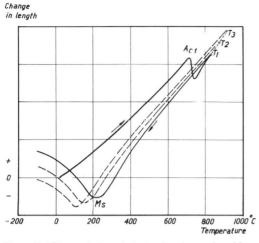

Figure 7.4 Change in length during heating and rapid
cooling of a eutectoid steel

In the case of a carbon steel containing 1% C it should be possible in
theory, to avoid changes in volume during hardening if, for example, the
steel contains 10% undissolved cementite and 13% retained austenite. In
this case the carbon content in the martensite is *ca* 0·38%. However, with
such a low carbon content we do not achieve a retained austenite content
as high as 13% (*see Figure 1.18*). If all the cementite were to be dissolved,
we would require a retained austenite content of *ca* 40% to avoid any
changes in volume occuring in this steel. This situation, however, is
impossible with this carbon steel (*see Figure 1.18*). Hence we must always
expect an increase in volume when hardening a 1% carbon steel, assuming
that full hardening is achieved. Since such a steel becomes fully hardened
only in dimensions less than approximately 10 mm, the increase in volume
in the larger dimensions is quite moderate, because in such cases the
hardened layer comprises only a few per cent of the total volume.

Assuming that the contents of martensite and retained austenite remain
the same as in the carbon steel, we might expect steels with higher
hardenabilities to show larger volume increases. However, since the
amount of retained austenite is greater in the high-alloy steels this
compensates for the increase in volume which would result from the
increased hardenability. In medium-high and high-alloy tool steels the
quantity of retained austenite and hence the increase in volume can be
controlled by means of the hardening temperature, as shown schematically
in *Figure 7.4*. In this case the hardening temperature T_2 does not result in
any increase in volume after cooling to room temperature. If there is a
volume diminution because a high hardening temperature (T_3) is used, the
original volume can be restored by means of subzero cooling.

In a material with the same properties in all directions (isotropic
material) the relative changes in length. Engineering steels are, however,
anisotropic which means that the linear change occurring during hardening
will not be the same in the direction of rolling as in the directions at right
angles to it.

Figure 7.5 Carbides elongated in the direction of rolling of the steel, 300 ×

In, for example, high alloy content chromium steels of type D 2 and D 6 the carbides are elongated in the direction of rolling in the manner shown in *Figure 7.5*. During hardening these steels expand more in the direction of rolling than in other directions. According to Frehser[3] this is because during the heating-up period to the austenitizing temperature the carbides do not expand so much as the matrix. During cooling the carbides impede the thermal contraction of the matrix in the longitudinal direction. No strict scientific proof has been put forward in support of this theory, but the model is simple and easy to grasp and provides a good means of remembering the phenomenon. Other grades of steel also exhibit marked anisotropy.

7.2 Dimensional changes during tempering

Before this section is read, it is recommended that Section 1.4 be read through once more. During tempering structural transformations occur which change the volume of the steel and its state of stress. There is a certain correlation between the tempering temperature, volume and state of stress. From the educational viewpoint it might be advisable here to distinguish between changes in volumes and changes in stress state.

7.2.1   Changes in volume

During tempering the martensite decomposes to form ferrite and cementite which implies that there is a continuous decrease in volume. As a result of tempering at high temperatures, the volume reverts by and large to its original value prior to hardening, provided that we can disregard plastic deformation (*see Figure 7.6*). Since we cannot count on an initial value of 100% martensite prior to tempering, the continuous curve in

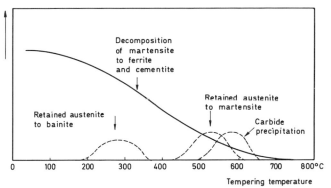

Figure 7.6 Schematic representation of the effect of changes in the structural constitution on volume changes during tempering of hardened steel

Figure 7.6 is not representative of the changes in volume occurring in the grades of steel which can be employed for engineering purposes.

The dashed curves plotted in *Figure 7.6*, which represent increases in volume during different tempering stages, are only schematic, as is the entire diagram which is representative of many different types of steel.

The retained austenite which in plain-carbon and low-alloy steels is transformed to bainite in the second stage of tempering at about 300 °C results in an increase in volume. Because of the relationship between martensite and retained austenite, taken in conjunction with the tempering temperature employed, either an increase or a decrease in volume may take place within this temperature range.

When tempering high-alloy tool steels in the temperature range 500–600 °C, very finely distributed carbides are precipitated. This gives rise to a stress condition which results in increased hardness and greater volume. Simultaneously with the precipitation of carbides the alloy content of the matrix is reduced which implies that the M_s of the retained austenite will be raised. During cooling down from the tempering temperature the retained austenite will consequently be transformed which also results in an increase in volume.

7.2.2 Changes in stress conditions

The stress condition which prevails in the steel after hardening is, as mentioned previously, governed by the thermal stresses and transformation stresses which have occurred during hardening. The continuous decomposition of martensite during tempering causes at the same time a continuous reduction in the state of stress which also facilitates the transformation from austenite to martensite. With the aid of the split-ring test, Brown and Cohen[4] studied the stress changes occurring in the hardened ball-bearing steel of the type AISI 52100. After hardening

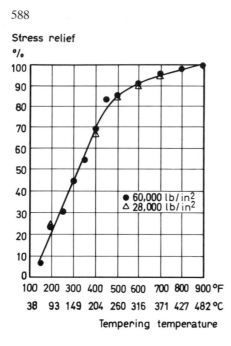

Stress relief
%

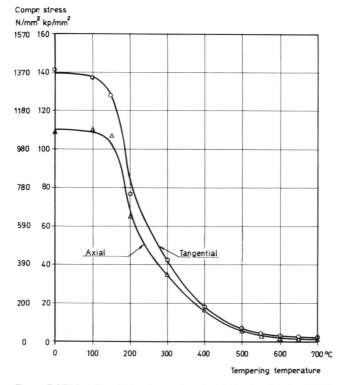

Figure 7.7 Relaxation during tempering of hardened rings made from ball-bearing steel (after Brown and Cohen[4])

Figure 7.8 Relaxation during tempering of an induction-hardened roll for cold rolling

the rings were stressed to 42·2 kp/mm^2 (415 N/mm^2) and 19·7 kp/mm^2 (194 N/mm^2), and then tempered for 1 h at different temperatures.

After relaxation the rings were measured and the changes in stress conditions were calculated. Even after tempering at 260°C (500°F) 85% stress relaxation was measured, *see Figure 7.7.*

With the aid of X-ray diffraction, stress measurements were made by Bofors on an induction-hardened cold roll after tempering at different temperatures. The results of the measurements are shown in *Figure 7.8.* The very high initial stresses were due to the extra stress additions which are created during surface hardening, since the annular hardened zone cannot expand freely during martensite formation. After tempering at 260°C the stress reduction was about 60%. This figure may be considerably less in asymmetrical bodies that have been drastically quenched. In such cases appreciable stress relieving is not obtained before a tempering temperature of around 600°C is attained (*see Figure 5.15*).

7.3 Examples of dimensional changes during the hardening and tempering of tool steels

Theoretical calculations which are based on the specific volumes of the different structural phases can never provide values which are of practical application, because these presuppose a knowledge of the amount of each individual constituent present in the steel after hardening. Furthermore the calculations do not allow for the dimensional changes which occur as a result of thermal stresses, or allow for the anisotropy of the steel. Hence we nearly always have to rely on empirical values. Comprehensive experiments have been made by Bofors aimed at determining the dimensional changes which occur during the hardening and tempering of the most commonly used types of tool and constructional steels. With the theoretical background as outlined above, and using the practical experience obtained both during the tests described above and from close co-operation with heat-treatment shops it has proved possible to assess the dimensional changes occurring even for bodies whose shape and size differ from those employed during the experiments. Each grade of steel was found to have its own characteristics which are not applicable to other grades. Nevertheless, it is possible to assign the steels to groups in the manner shown below. The dimensional changes as a percentage, mentioned in the graphs, represent the average values of several tests and we must expect a scatter of up to ±0·05%. In the following by longitudinal direction we mean the direction of rolling.

7.3.1   Plain-carbon steels

The low hardenability of these steels makes the total increase in volume highly dependent on the material thickness of the tool. The steels obtain a high martensite content, and sections completely hardened through, i.e. material thicknesses not exceeding 10 mm, exhibit the largest possible increase in volume on hardening. In dies of considerable material thickness on the other hand the proportion of martensite by volume is so small that it is possible almost to disregard the increase in volume.

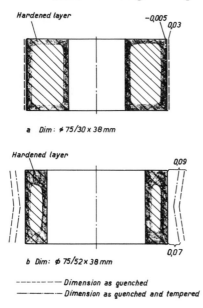

Hardened layer   −0,005
                        0,03

a   Dim: ⌀ 75/30 × 38 mm

Hardened layer   0,09

b   Dim: ⌀ 75/52 × 38 mm   0,07

- - - - - - - Dimension as quenched
— · — · — · — Dimension as quenched and tempered

Figure 7.9 Dimensional changes in rings made from steel W 1 after quenching from 800 °C in water and tempering at 180 °C. The figures have been drawn in section. The martensitic areas are shaded

Example 1. Figure 7.9 shows the results obtained from experiments using embossing rolls made of steel W 1 which could not be ground on the engraved outer diameter after heat treatment. In the roll having the largest material thickness the increase in diameter remained relatively small after hardening. After tempering at 180 °C the dimensions shrank below the original values. In the roll made of thinner material the depth of hardness was greater, so that there was a greater increase in volume. Since the roll hardened through at the ends, the largest increase in diameter occurred there.

7.3.2   Low-alloy steels

The majority of the low-alloy tool steels are employed in dimensions such that they are completely or almost completely fully hardened. In the through-hardening section sizes the dimensional changes occurring are, viewed as a percentage, relatively uniform, even in the case of different material thicknesses. In the event of different degrees of partial through-hardening, however, the dimensional changes occurring are irregular. In those sizes where the steels are surface-hardening, the volume increases least.

Example 2. Steel O 1 is a grade normally used for dies for blanking tools, the thicknesses employed for this purpose ranging usually up to 50 mm. *Figure 7.10* shows how through-hardened plates of this grade change in dimensions on quenching in oil or in a martempering bath. Quenching in oil results in an increase of roughly 0·1% in dimensions in all directions. During martempering the longitudinal dimension increases to a greater extent, mainly at the expense of thickness. Undoubtedly the reason for this is the less drastic thermal shock occurring during martempering which does

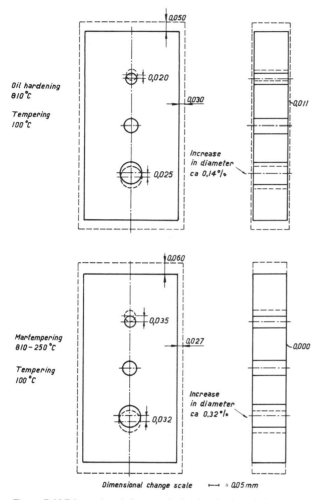

Oil hardening
810°C

Tempering
100°C

Martempering
810–250°C

Tempering
100°C

Figure 7.10 Dimensional changes during hardening of plates made from steel O 1 after quenching in oil and martempering bath, respectively. Specimen size: $18 \times 50 \times 100$ mm. Longitudinal direction = direction of rolling

not result in such a considerable contraction in the longitudinal direction as does the more severe shock imposed during oil quenching. This experiment was undertaken on steels from different heats and at different heat-treatment shops and unequivocal results were obtained. If the considerable risk of cracking involved in oil quenching can be disregarded, this method of quenching, as applied to plates described here as well as to similar plates, is to be preferred to martempering when dimensional stability in longitudianal direction is required. During this experiment the dimensional deviations occurring as regards the planeness of the sides were relatively slight and these have not been included in the diagrams.

Example 3. During one series of experiments plates made from steel O 1

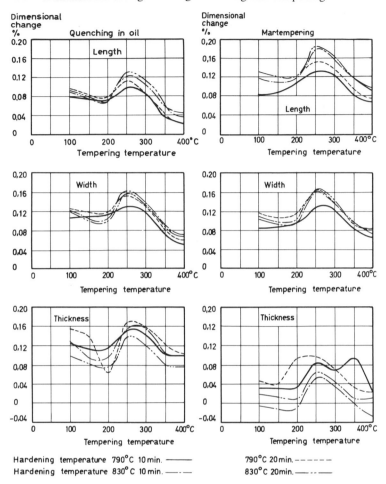

Figure 7.11 Dimensional changes during temperature of steel O 1 after different quenching methods. Specimen dimensions: 18 × 50 × 100 mm. Longitudinal dimension = direction of rolling

were hardened from three different hardening temperatures, involving two different holding times. These variables had only negligible influence, as can also be deduced from *Figure 7.11* which includes the highest and lowest hardening temperatures. On tempering up to 200 °C a slight contraction occurs in all directions of the plate. At higher temperatures there is an increase in dimensions, with a maximum at 300 °C, after which the dimensions once again decrease. The increased volume at 300 °C is occasioned by the transformation of retained austenite. At 700 °C the dimensions have reverted almost completely to their original values (this point has not been included in the diagram).

Example 4. When hardening a die made from steel O 1 having dimensions as shown in *Figure 7.12,* the manufacturer had relied on there being no change in the dimension of 375·00 mm between the holes. Owing

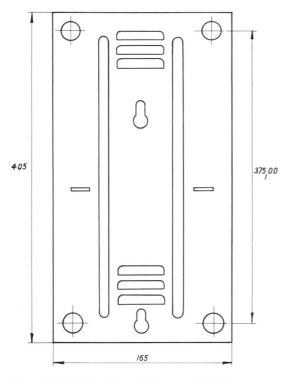

Figure 7.12 Blanking die made from steel O 1, size
30 × 165 × 405 mm

to the shape of the die, it was hardened by quenching in a martempering
bath. After tempering for 2 h at 200 °C the length measurement was found
to be 375·33 m, this being an increase of 0·09%. Further heat treatments
were performed with the following results:

Tempering   4 h 220 °C   375·75 mm

Tempering   3 h 400 °C   375·30 mm

Annealing   10 h 700 °C   374·90 mm

Tempering at 220 °C for 4 h corresponds approximately to tempering at
300 °C for 1 h, which gives the maximum increase in length. By carrying
out a tempering at 400 °C, the length dimension was reduced to somewhat
below the value obtained after tempering at 200 °C. After annealing, the
die reverted to what was probably the original dimension.
    The die was re-hardened by quenching in oil and tempered as follows:

Tempering 4 h 180 °C   374·90 mm

Tempering 6 h 180 °C   374·99 mm

Tempering 6 h 180 °C   375·01 mm

It may seem odd that no increase in length occurred after quenching in
oil. Regarded from the viewpoint of cooling intensity, the dimension

30×165 mm is equivalent to 50 mm diameter. In this dimension the steel is not fully through-hardening so that the increase in volume is less. Tempering for 4 h at 180°C corresponds approximately to the lowest value on the length variation curve. After tempering for 6 h at 180°C the dimension has increased. The long tempering time at 180°C (4 + 6 h) is

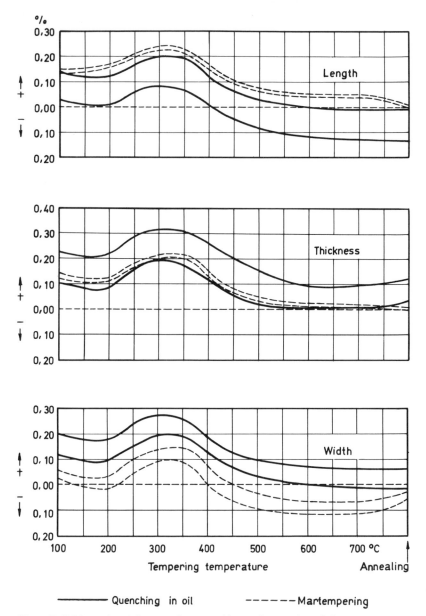

Figure 7.13 Dimensional changes after quenching and tempering of plates made from SS 2092. Specimen dimensions: $18 \times 50 \times 100$ mm. Hardening temperature 870°C. Martempering at 230°C. Tempering for 1 h

equivalent to tempering for 1 h at 205°C, and at this temperature the transformation of the retained austenite starts, which results in increased volume.

Example 5. By subzero cooling, e.g. at −80°C, the dimensions increase, provided that tempering has not been undertaken at excessively high temperatures. In one case a hardened and tempered bushing of steel O 1 was to be shrunk into a cast iron stand. Since it was not possible to heat the stand; the bushing was subzero cooled but as a result it became so large that it could not be shrunk into the stand. In such cases it is advisable to carry out a subzero treatment and then grind the workpiece before cooling for shrinkage is carried out.

Example 6. SS 2092 differs only slightly from SS 2140 (O 1) as regards change in shape. Hardening experiments were undertaken with SS 2092 on test bars as shown in *Figure 7.10*. Hardening temperatures of 850, 870 and 890°C gave very similar results. To illustrate the difference which is encountered as regards warping during hardening in oil and martempering, *Figure 7.13* shows the largest and smallest dimensional deviations for two test bars.

Example 7. When hardening *thin-walled rings* made from SS 2092 and O 1, we must expect an increase in both the outer and inner dimameters. During the subsequent tempering, the rings follow in principle the same course of dimensional variations as illustrated previously. The rings in *Figure 7.14* after quenching were tempered at 180°C only, so that the inner diameter shrank until it reached almost the original value.

Example 8. As the wall thickness of rings made from these types of steel increases, the picture changes. When the inner diameter is reduced so that

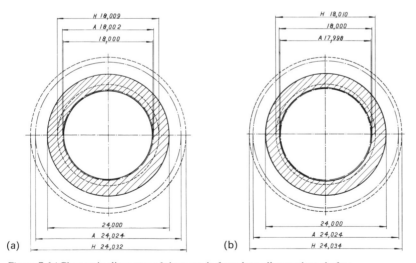

(a)   (b)

Figure 7.14 Change in diameter of rings made from low-alloy tool steel after quenching in oil and tempering at 180°C. Ring dimensions: diameter 24/18 × 35 mm. H = dimensions as quenched. A = dimensions after tempering. (a) Steel O 1. Hardening temperature 810°C. Tempering temperature 180°C; (b) Steel SS 2092. Hardening temperature 860°C. Tempering temperature 180°C

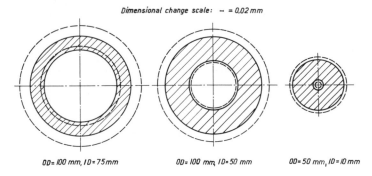

Figure 7.15 Change in diameters of rings made from SS 2092, after quenching in oil from 870°C

it is roughly half the outside diameter, the inner diameter will shrink during hardening (*see Figure 7.15*).

Example 9. When the thickness of rings made from SS 2092 and O 1 increases to cross-sectional areas which are equivalent to a bar *ca* 50 mm diameter, the depth of hardening is reduced and the increase in volume is negligible. As the wall thickness increases still further, the depth of hardening is reduced further and both inner and outer diameters shrink, which is because the thermal stresses dominate.

During measurements made on blanking die rings made from SS 2092 with cross-sectional areas within the critical range, the following variations in outer and inner diameters were observed after quenching in oil and subsequently tempering at 200°C (*Table 7.3*).

Table 7.3

OD	ID	H	Cross-section	ΔOD	ΔID
283	193	65	65 × 45	−0·2	−0·2
376	270	65	65 × 53	−0·5	−0·8
439	281	65	65 × 79	−1·1	−1·4

OD = outer diameter mm
ID = inner diameter mm
H = height mm

Example 10. With bar diameters exceeding about 80 mm, or equivalent rectangular cross-sections (*see Figure 4.41*), the depth of hardening for steel SS 2092 is often inadequate. In such cases it is possible, during hardening, first to quench the steel in water for several minutes and then in oil.

For such a test, rings having dimensions 495/335 diameter × 175 mm were used. The cross-sectional area 80 × 175 mm is equivalent to a diameter of 115 mm. Quenching took place first in strongly agitated oil only, and then in water for 1 min after which the quenching was finished in oil. The following diameter changes were observed (*Table 7.4*).

Table 7.4

Method of hardening	ΔOD mm	ΔID mm	ΔH mm	Depth of hardening mm
Quenching in oil	$-0{\cdot}46 \pm 0{\cdot}14$	$-0{\cdot}81 \pm 0{\cdot}06$	$+0{\cdot}09 \pm 0{\cdot}05$	5
Quenching in water/oil	$-1{\cdot}12 \pm 0{\cdot}42$	$-1{\cdot}86 \pm 0{\cdot}22$	$\pm0{\cdot}00 \pm 0{\cdot}05$	10

In spite of the fact that the depth of hardening was greater for the rings which were quenched in water/oil, the reduction in diameter was greater, this being due to the fact that the thermal stresses exceed the transformation stresses.

Example 11. As mentioned previously, and as shown in *Figure 7.13*, the steel often reverts to its original dimensions after a full anneal. If the heat treatment which has been performed is considered incorrect, as regards dimensional variations, there is consequently the possibility that, by performing a full anneal, the original dimensions will be obtained. Details will be given here of such a case which concerned a die made from SS 2092.

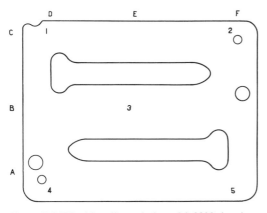

Figure 7.16 Blanking die made from SS 2092 showing dimensional changes after different heat treatments

The length of the die, *Figure 7.16*, was the factor governing the serviceability of the tool. According to information received the length should have been 123·00 mm. After hardening this had increased to 123·38 mm, which is 0·3%. Since this increase in length appeared to be unreasonable, the plate was thoroughly annealed, whereupon it shrank to 123·17 mm, which is 0·17% and which can be regarded as a more reasonable value. The plate was subsequently rehardened in exactly the same manner as employed on the first occasion, whereupon the length increased by 0·17%. By tempering it at 200°C the length was reduced somewhat. It is highly probable that the original length dimension was greater than that stated. Details are given of the results of the measurements and of the complete heat treatment process in *Table 7.5*.

Table 7.5

Heat treatment	Length mm			Width mm			Thickness mm				
	A	B	C	D	E	F	1	2	3	4	5
1	123·38	123·38	123·39	94·85	94·81	94·81	9·98	9·98	9·97	9·98	9·97
2	123·17	123·18	123·16	94·70	94·66	94·67	9·97	9·97	9·95	9·97	9·95
3	123·38	123·36	123·37	94·88	94·81	94·84	9·98	9·98	9·97	9·97	9·96
4	123·31	123·31	123·30	94·85	94·76	94·80	9·97	9·98	9·96	9·96	9·96

The conventional ball-bearing steel AISI 52100 has approximately the same characteristics as the two steels mentioned above. Consequently it

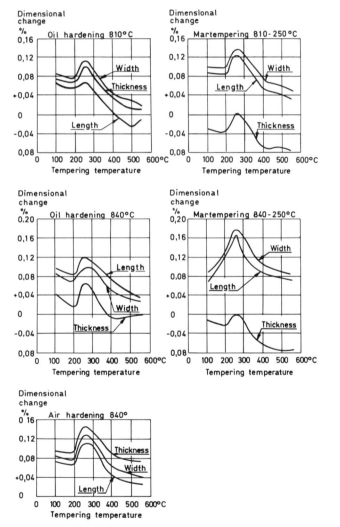

Figure 7.17 Percentage dimensional changes after hardening and tempering of plates of SS 2550. Specimen dimensions: 25 × 70 × 100 mm. Longitudinal direction = direction of rolling

has been possible in several cases to apply experience gained with these steels to the ball-bearing steel.

Example 12. A hardness of 50–55 HRC had been requested for a ring-shaped die diameter 300/210 × 30 mm made from AISI 52100. After hardening in oil the die was tempered at 300°C. Its hardness was found to be 55 HRC. Both the external and internal dimensions had increased by approximately 0·35 mm and the die was unserviceable, because the inner diameter was the ruling dimension. It was then tempered at 400°C so that the hardness became 52 HRC. The inner diameter was reduced by 0·28 mm, which was then sufficient to permit the die to be employed.

SS 2550 varies dimensionally in the same way as steels O 1 and SS 2092, which means that the dimensions increase roughly by 0·1% in all directions on quenching in oil. When quenching in a martempering bath the length and width increase somewhat more at the expense of the thickness which decreases. On tempering, the largest dimension is encountered after heating for 1 h at 250°C, i.e in the temperature range where the retained austenite transforms to bainite.

Example 13. Figure 7.17 shows the values measured on plates size 25 × 70 × 100 mm after different heat treatment processes. The lowest diagram in the figure which shows the dimensional changes occurring after cooling in air indicates that the percentage increase in thickness is somewhat greater than for the other dimensions. Since the absolute dimensional changes as regards thickness are very slight it is probable that the dimensional increase was caused by the thin film of oxide which formed during cooling in air.

Figure 7.18 Mould made from SS 2550 for producing plastic transport boxes

Figure 7.19 Tool made from steel of the following composition: 0·55% C, 1·5% Cr, 3% Ni

As SS 2550 exhibits very good hardenability we can expect that the dimensional changes outlined above should be representative at least for material thicknesses up to roughly 200 mm. The percentage changes in dimension should then decrease as the dimensions increase.

Example 14. In *Figure 7.18* a photograph of the inner part of a mould for pressing plastic transportation boxes weighing 9 kg is shown. The dimensions are 300 × 400 × 800 mm. The part is made in two halves. Hardening was performed from 840°C in compressed air to 240°C and then in still air after which the die was tempered at 400°C twice to give a hardness of 360–390 HB. The length which in this particular case was the vital dimension increased by 0·4 mm, which is 0·05%. Compare this with the lower portion of *Figure 7.17* where the increase in length after hardening in air and tempering at 400°C is 0·04%.

Example 15. Steel DIN 50 NiCr 13 which is very similar to SS 2550 generally exhibits the same dimensional variations during hardening. A clear proof of this is provided by the die in *Figure 7.19* the external dimensions of which are 650 × 590 × 245 mm. During hardening the steel was heated to 840°C and quenched in a martempering bath at a temperature of 250°C. Tempering took place at 180°C. The following dimensional changes were measured:

Length   + 0·30 mm = 0·05%
Width   + 0·65 mm = 0·11%
Thickness + 0·10 mm = 0·04%

There is a surprisingly good agreement with the values obtained after corresponding heat treatment, as shown by *Figure 7.17*. However, it is not advisable to calculate and design too close, dimensionally, because numerous factors can affect dimensional changes.

Steel A 2 occupies an intermediate position between the low-alloy and the high-alloy chromium steels. The hardening temperature used is moderate but is sufficiently high to give rise to changes in shape as governed by the cooling rate. During the hardening of this steel, quenching

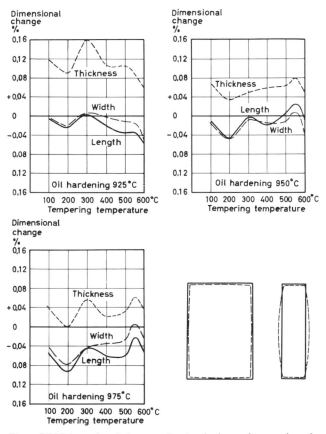

Figure 7.20 Dimensional changes after hardening and tempering of plates made from steel A 2. Specimen dimensions: 25 × 70 × 100 mm. Holding time at austenitizing temperature 25 min. Hardening temperatures 925, 950 and 975°C. Quenching in oil. Holding time during tempering 1 h. Longitudinal direction = direction of rolling

in oil can give rise to much greater deformation than quenching in a martempering bath or cooling in air. This change in shape can, for example, result in a plate becoming thicker in the centre than at the sides, so that there is a change in parallelism between the sides. Hence, this steel should not be quenched in oil. For the sake of completeness we should, however, include the following description of values obtained from hardening tests made using all the above-mentioned cooling media.

Example 16. Figures 7.20, 7.21 and 7.22 show the average values for the dimensional changes occuring in plates, size $25 \times 70 \times 100$ mm after hardening in oil, martempering bath and air respectively. On quenching in oil there is a reduction in the length and width but the thickness increases. A sketch has been incuded in *Figure 7.20* which shows the relatively large deviations involved as regards parallelism between the sides. On cooling in the martempering bath and in air, the length and width increase, but the thickness decreases. As regards the various cooling media, this steel

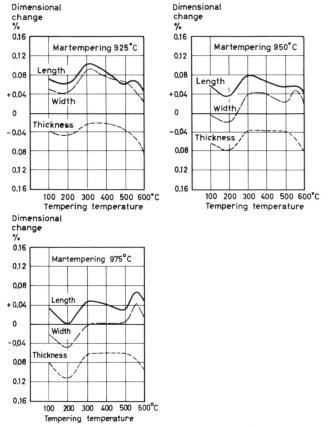

Figure 7.21 Dimensional changes after hardening and tempering of plates made from steel A 2. Specimen dimensions: $25 \times 70 \times 100$ mm. Holding time at austenitizing temperature 25 min. Quenched from 925, 950 and 975 °C into martempering bath at 250°C. Holding time during tempering 1 h. Longitudinal direction = direction of rolling

consequently behaves in fundamentally the same way as the grades mentioned previously. The dimensional changes occurring after martempering are considerably less for this steel, than for the other tool steels.

The effect of the hardening temperature is quite obvious. The volume of the specimen decreases with increasing hardening temperature, this being due to the increased amount of retained austenite.

During tempering, steel A 2 passes through all the transformation stages which were sketched in *Figure 7.6*. Together with the continuous shrinkage which is occasioned by the decomposition of the martensite, there is first at roughly 300°C an increase in volume due to the transformation of the retained austenite to bainite. After tempering at 450°C or higher carbide

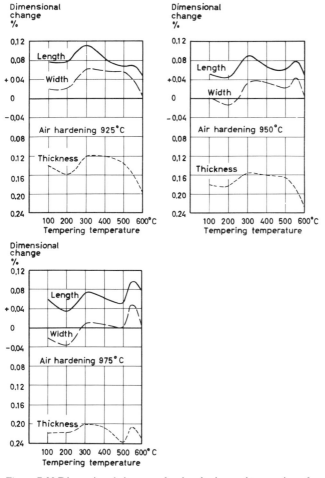

Figure 7.22 Dimensional changes after hardening and tempering of plates made from steel A 2. Specimen dimensions: 25 × 70 × 100 mm. Holding time at austenitizing temperatures 25 min. Air cooling from 925, 950 and 975 °C. Holding time during tempering 1 h. Longitudinal direction = direction of rolling

precipitation takes place and during cooling the retained austenite is transformed to martensite. The increase in volume occurring during tempering is greatest for those specimens which are hardened from the highest temperature. Note also the TTT diagrams in *Figure 1.22* which illustrate the transformation of the retained austenite during the tempering of this steel. In the lower temperature range the transformation takes place isothermally to bainite, and in the upper temperature range martensite is formed when the steel cools from the tempering temperature.

7.3.3   High-alloy cold-work steels

Steels D 2 and D 6 are often called non-deforming steels. However, they have a marked tendency on hardening to increase their length (direction of

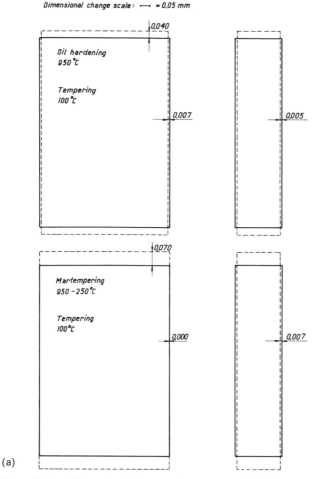

(a)

Figure 7.23a, b Dimensional changes of plates made from steel D 6 after quenching in oil and martempering respectively. Specimen dimensions: 25 × 70 × 100 mm. Holding time at hardening temperature 25 min.

rolling) and to shrink in the other directions. This was previously explained to be due to the carbide stringers in the steels. By selecting suitable hardening and tempering temperatures it is possible for the dimensional changes to be kept in check to some extent.

Example 17. The dimensional changes whch occurred during the hardening of plates made from steel D 6, size 25 × 70 × 100 mm are shown in *Figure 7.23*. The parallelism between certain of the surfaces of the plates deviated somewhat from the extent shown in the figure. On the other hand planeness was extremely good. As in the case of the steels mentioned previously, the length increased to a greater extent after martempering than after hardening in oil. Similarly, the volume decreased with increasing hardening temperature.

During tempering all dimensions decreased with increasing temperature (*see Figure 7.24*). At about 300°C the retained austenite was transformed to bainite which tended to impede the reduction in dimensions, which is at its maximum at 400°C. At higher temperatures the volume increased because of the formation of new martensite and because of carbide precipitation. After tempering at temperatures above 600°C the steel contracted.

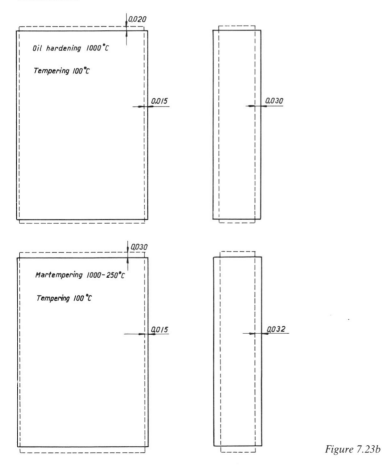

Figure 7.23b

606

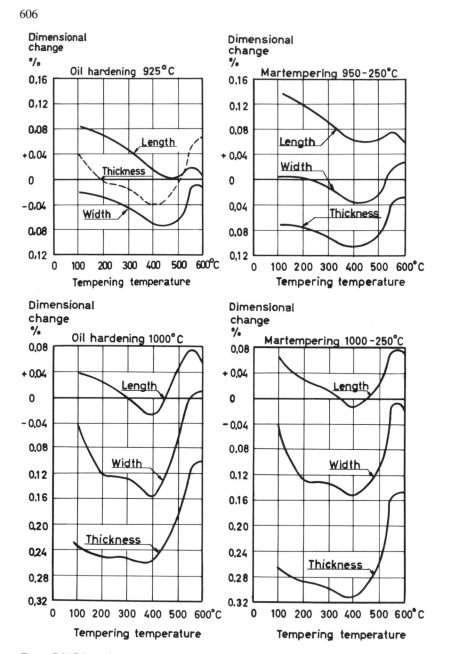

Figure 7.24 Dimensional changes during temperature after hardening of
steel D 6. Specimen dimensions: 25 × 70 × 100 mm. Holding time at
tempering temperature 1 h. Longitudinal direction = direction of rolling

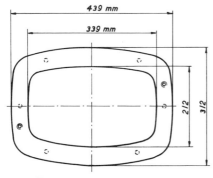

Thickness: 52 mm

Figure 7.25 Blanking die ring made from D 6
which contracted during the first hardening
operation by *ca* 1 mm along the dimensions
shown in the sketch

When assessing the dimensional variations of ring-shaped dies made
from D 6, as regards the change in diameter we can assume as an average
value the mean of the width and thickness variations.

Example 18. A blanking die made from D 6 having dimensions and
appearance as shown in *Figure 7.25* shrank approximately 1 mm on the
specified dimensions during hardening. The die was unserviceable because
the outer dimensions were the decisive factor. As the hardness was only
ca 55 H R C, a high retained austenite content was suspected due to an
excessively high hardening temperature. After annealing the die
measurements indicated the dimensions as shown in the diagram.
Hardening then took place from 950°C whereupon the outer diameter
increased by approximately 0·1 mm. Tempering was undertaken at the
lowest practicable temperature (160°C), after which the die was found to
be exactly to size and could be used in spite of the fact that no grinding
tolerance had been included. In this particular case it would undoubtedly
have been possible to achieve an increase in size had the tool been
tempered at 550°C or if it had been subzero cooled after the first hardening

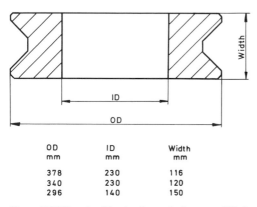

OD mm	ID mm	Width mm
378	230	116
340	230	120
296	140	150

Figure 7.26 Sketch of feed rolls made from steel D 6

operation. Subzero treatment is recommended only in exceptional cases, because of the risk of cracking.

Example 19. Twelve feeder rollers of three different sizes made from D 6 were to be hardened to a minimum of 58 HRC. After hardening and tempering there should still be an allowance left for grinding in the hole, i.e. the internal diameter should not increase. Hardening took place from 980°C in a martempering bath at a temperature of 220°C. The initial tempering was done at 170°C for 1 h, after which the dimensions of the rollers were measured. Tempering was then carried out at 490°C for 1 h followed by measurement. *Figure 7.26* shows a schematic diagram of the rollers and *Figure 7.27* shows the dimensions and dimensional changes occurring after hardening and tempering. With only one exception, all dimensions reduced after hardening and tempering at 170°C. Nearly all dimensions were increased as a result of tempering at 490°C. The final

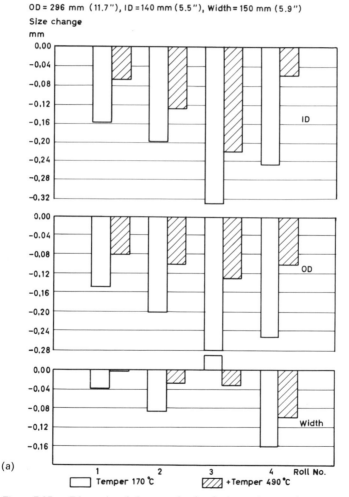

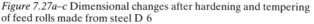

Figure 7.27a–c Dimensional changes after hardening and tempering of feed rolls made from steel D 6

result was that there was a small increase in the inner diameter (0·02 mm) of only one of the twelve rollers.

In this case the tempering temperature of 490°C must be regarded as the maximum value, because a higher temperature would have resulted in an increase in the inner diameter, *see Figure 7.24*. This investigation also gives a good idea of the scatter range to be expected when hardening several tools of the same size.

Example 20. Tools having an elongated shape made from D 2 which were to operate at a somewhat elevated temperature were hardened from 1030°C and then tempered at 560°C. The length increased by *ca* 0·10%, which is 1·5 mm for the length concerned, i.e. 1500 mm. The reason for this relatively considerable increase in length was that the tools had been hardened from the highest temperature within the specified range of steel D 2. Since this excess length has to be ground off, the tool was thenceforth made 1·5 mm shorter in length which—with the heat treatment remaining the same—resulted in a considerable reduction in grinding costs.

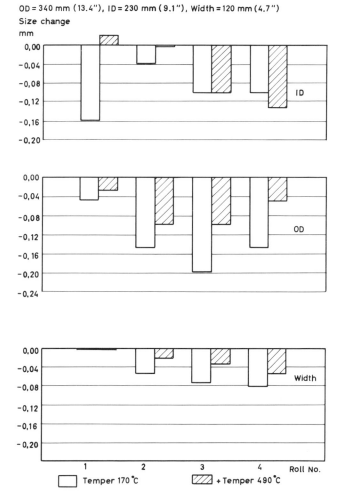

OD = 340 mm (13.4"), ID = 230 mm (9.1"), Width = 120 mm (4.7")

Figure 7.27b

OD = 378 mm (14.9"), ID = 230 mm (9.1"), Width = 116 (4.6")

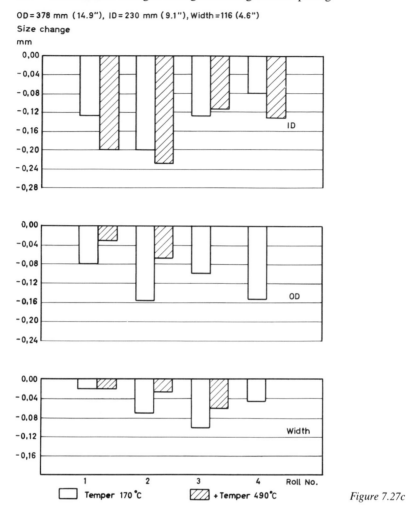

Figure 7.27c

7.3.4 Hot-work steels

Steels BH 13 and BH 10A (UHB QRO 45) differ considerably from other steels as regards dimensional changes because during hardening they always contract in the longitudinal direction (direction of rolling). Normally the width varies least, but the thickness can increase or decrease depending upon the method of quenching and material dimensions. When quenching flat workpieces in oil the deformation can be relatively large, and this is reflected in the thickening of the plates in the centre. Round tools, e.g. dies, change only negligibly in diameter. On cooling in oil it is possible for relatively severe swelling to occur on the plane surfaces.

Example 21. The hardening of straight-edge dies made from H 13, size 25 × 50 × 150 mm, was undertaken by quenching in oil, martempering bath and air respectively. There was a relatively large amount of deformation after quenching in oil as shown in *Figure 7.28* but there was

Dimensional change scale ⊢⊣ = 0,05 mm

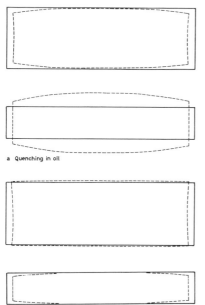

a Quenching in oil

b Martempering

Figure 7.28 Deformation of tools made from H 13 after hardening from 1050°C in oil and martempering bath respectively, and then tempering at 550°C. Specimen dimensions: 25 × 50 × 150 mm

only a negligible change on quenching in the martempering bath (*see Figure 7.28b*). *Figure 7.29* shows the dimensional changes occurring during hardening and tempering, along with the scatter band.

Example 22. Large dies made from H 13 for the forging of steel are usually hardened and tempered to about 400 HB before machining and the change in dimensions of the rough-machined blocks is of minor importance. Hardening in oil is not recommended for finish-machined workpieces where stringent requirements are imposed regarding dimensional stability. Apart from the unavoidable shrinkage in the longitudinal direction, some deformation of the sides also occurs during hardening in air. As shown in *Figure 7.30* these can become slightly concave, but convex deformation is more normally encountered.

Example 23. During hardening of the die-casting die shown in *Figure 5.50* measurements were made before and after heat treatment. The hardening temperature was 1030°C and quenching was carried out in a martempering bath at 300°C. The die was tempered at 590°C for 4 h then at 580°C for 4 h. The changes in dimensions are shown in *Figure 7.31*. The length decreased by 0·08%, which corresponds well with *Figure 7.29c*. At the lower end the width increased by *ca* 0·04%, which also agrees with *Figure 7.29c*. On the other hand, at the upper end of the die the width decreased. However, this may have been due to a permanent set caused by the weight of the die while it was hanging from its upper holes.

Example 24. The swelling which occurs with circular dies is illustrated in *Figure 7.32* for specimens made from H 13, 75 mm diameter × 50 mm. Heating was carried out at 1050°C in a salt bath. Cooling in oil resulted in greater deformation than did cooling in a martempering bath. During

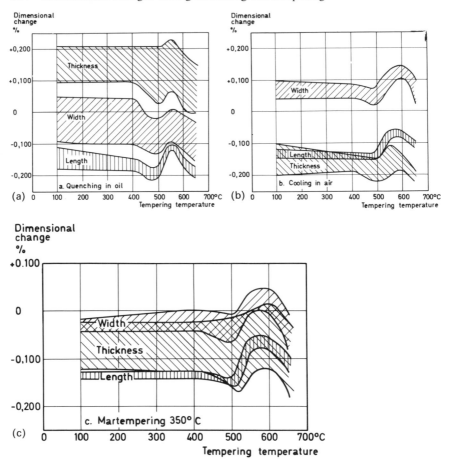

Figure 7.29a–c Changes in dimensions of tools made from steel H 13 after
hardening from 1050°C in different media and then tempering at
temperatures up to 650°C. Specimen dimensions: 25 × 50 × 150 mm.
Longitudinal direction = direction of rolling

tempering of the specimen the values shown in *Figure 7.33* were obtained.

Rings made from H 13 follow by and large the deformation pattern which is based on the effects of thermal stresses. This implies that the more drastic the cooling medium, the greater is the reduction in the diameter.

Example 25. Four rings, diameter 201/217 × 120 mm and two rings diameter 254/216 mm made from H 13 and with 0·4% V (H 11 mod.) were quenched from 1025°C in a martempering bath at 400°C. Tempering was carried out at 560°C, twice for 2 h. The variation in all dimensions was within the scatter range of ±0·05 mm.

Example 26. Six rings made from H 11 mod. with dimensions as shown in *Figure 7.34* were hardened in oil from 1025°C. They were tempered at about 560°C, twice for 2 h. *Figure 7.34* shows the changes in dimension as a function of diameter. The majority of the dimensional changes are within

Dimensional change scale ⊢——⊣ = 0,05 mm

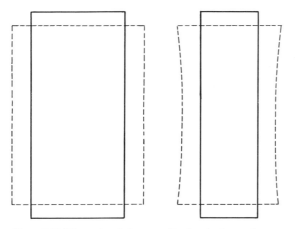

Figure 7.30 Dimensional changes after hardening and tempering of a die block made from steel H 13. Air hardening from 1050°C. Temering at 600°C twice for 2 h. Hardness 415 HB. Size 150 × 250 × 550 mm

the range $-0{\cdot}10{\pm}0{\cdot}03\%$ in the case of the outer diameter, and within $-0{\cdot}20{\pm}0{\cdot}03\%$ for the inner diameter. Increases as well as reductions in height were observed.

Example 27. If a martempering bath is not available and air cooling is inadvisable, the steel may be hardened by first cooling it in oil for some seconds and then letting it cool in air. This produces about the same dimensional stability as that achieved with martempering. When using the latter treatment, however, it is possible for the temperature of the bath to influence the change in dimensions as will be shown in the example.

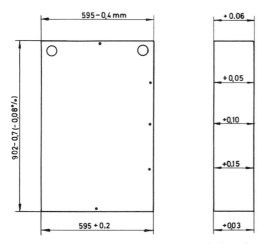

Figure 7.31 Dimensional changes after hardening of a die-casting die made from steel H 13. Size 902 × 595 × 205 mm

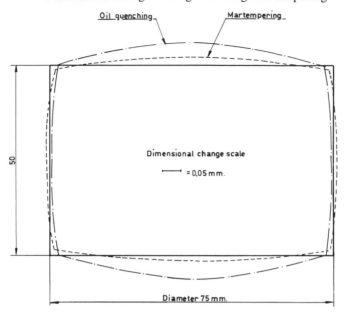

Figure 7.32 Deformation after martempering and oil quenching from 1050°C of a die made from steel H 13, diameter 75 × 50 mm

In this case plates made from BH 10A (UHB QRO 45) were used of size 50 × 75 × 100 mm, and were hardened from 1050°C. Regardless of the method of cooling employed, the length decreased, whereas the width varied around the 0-value. After hardening, the plates were in all cases thicker at the centre than at the edges. In *Table 7.6* the variations in thickness after tempering at 550°C provide a measure of the suitability of the coolant. Cooling in a martempering bath at 350°C gives the smallest change in dimensions.

Figure 7.35 shows the values measured after tempering at different temperatures for the specimen which was quenched in a martempering bath at 350°C.

Table 7.6 Effect of the coolant on change in thickness during hardening of UHB QRO 45. Tempering temperature 550°C ($t = 50$ mm)

| Cooling medium | Thickness variation mm | | |
	Variation limits		Range of variation
Martempering bath 350°C	−0·01	−0·02	0·01
Air	±0·00	+0·07	0·07
Martempering bath 250°C	±0·00	+0·14	0·14
Oil 30 s, followed by air	−0·01	+0·14	0·15
Oil 60 s, followed by air	−0·06	+0·12	0·18
Martempering bath 200°C	−0·01	+0·20	0·21
oil	−0·04	+0·33	0·37

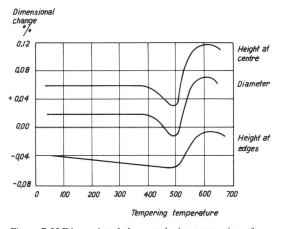

Figure 7.33 Dimensional changes during tempering of
H 13 after martempering from 1050°C. Specimen
dimensions: diameter 75 mm × 50 mm

7.3.5 High-speed steels

As regards deformation during hardening and tempering high-speed steels
exhibit some similarity to the high-alloy chromium steels. Because of the
high solution temperatures involved, high-speed steels have a relatively
high retained austenite content after hardening. This means that we must
expect shrinkage in all dimensions, except possibly in the length
dimension, when hardening from normal temperatures. When tempering
at around 600°C, all dimensions increase quite considerably.

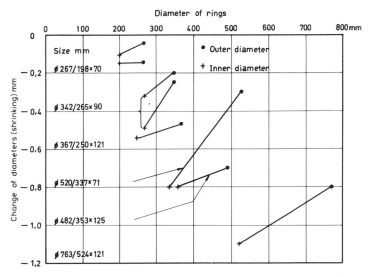

Figure 7.34 Dimensional changes after hardening of rings made from steel
H 11, modified. Hardening temperature 1050°C, quenching in oil,
tempering at 560°C, twice for 2 h

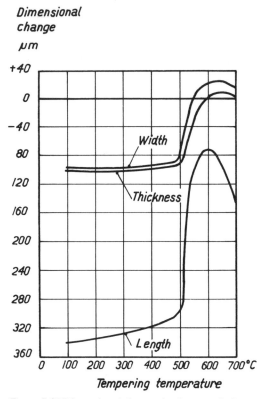

Figure 7.35 Dimensional changes in plates made from
B H 10A (U H B Q R O 45) shown as functions of
tempering temperature. Martempering 1050–350°C.
Specimen dimensions: 50 × 75 × 150 mm

Example 28. Test plates made from M 2 size 25 × 70 × 100 mm were
pre-heated at 850°C and austenitized at 1210°C for 6 min, then quenched
in oil, in martempering bath at 400°C, and cooled in air, respectively. The
specimens were tempered for 1 h at ten different temperatures up to
650°C. As a check, one series was tempered at 550°C only. This tempering
operation was carried out twice with intervening checks on dimensions.
The average values for the changes in dimensions are shown for all
specimens in *Figure 7.36.* During hardening a reduction in volume
occurred, due to the high retained austenite content, and the
thickness—taken as a percentage—decreased the most, the width coming
next. On quenching in oil there was also a reduction in the length, whereas
on martempering the length remained almost unaltered. On cooling in air
there was a very slight increase in length. On tempering up to 500°C there
was hardly any change in dimensions at all, but at higher temperatures
there was a sudden increase in all dimensions. The specimens which were
tempered only at 550°C increased in size during the first tempering
operation by roughly the same amount as the specimens which were
tempered continuously up to 550°C. During the second tempering of the
first-mentioned specimens, the dimensions increased to a value which

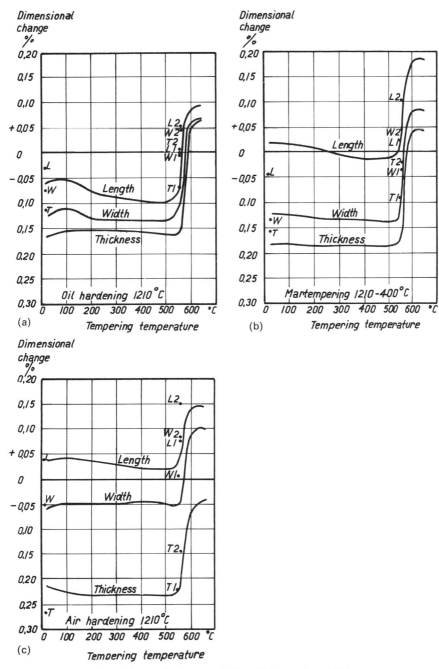

Figure 7.36 Dimensional changes in plates made from high-speed steel M 2 after hardening from 1210 °C in different cooling media followed by tempering. Specimen dimensions: 25 × 70 × 100 mm. The curves were plotted from measurements taken on the same specimens as they were tempered successively at 100, 200, 300 °C, etc. The letters L, W and T at the points indicated denote the changes in length, width and thickness of the specimens tempered at 550 °C only. The digits 1 and 2 after these letters refer to the actual dimensions after one and two tempering operations respectively. The longitudinal direction = direction of rolling

corresponds roughly to tempering at 570°C for 1 h. Tempering for 2 h at 550°C is equivalent to 1 h at 570°C (*see Figure 6.62*). After hardening in oil the specimen had become visibly deformed, which was *not* the case after martempering or air hardening (*see Figure 7.37*).

Example 29. A series of tests, exactly similar to the one described in *Example 28* was carried out with steel M 42. Both steels behave in roughly the same way, but the variations between the dimensional changes in the different measured directions are considerably greater in the case of M 42 (*see Figure 7.38*). The increases in size after tempering in the range around 500°C are also considerably greater for this steel than for M 2. *Figure 7.39* provides sketches in which the scale shows the dimensional changes.

It is extremely important to be familiar with the large increases in size occurring in hardened high-speed steels after tempering at about 550°C because the majority of high-speed steels are tempered precisely in this range. Concerning the changes in diameter of round dies, we can work on the basis of an average value for width and thickness for the plates in *Examples 28* and *29*.

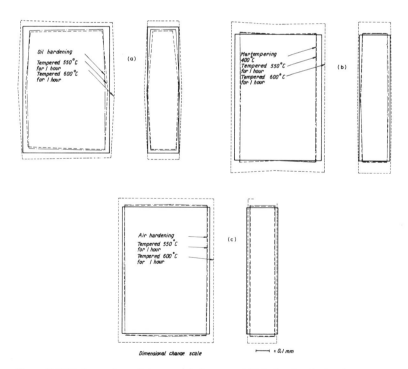

Figure 7.37 Deformation of plates made from steel M 2 after hardening from 1210°C in different cooling media, followed by tempering at 550 and 600°C, 1 h. Specimen dimensions: 25 × 70 × 100 mm

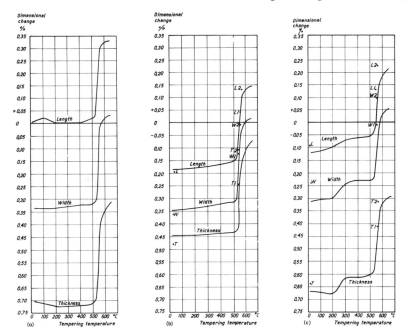

Figure 7.38 Dimensional changes in plates made from steel M 42 after hardening in different cooling media, followed by tempering. Specimen dimensions: 25 × 70 × 100 mm. The curves were plotted from measurements taken on the same specimens as they were tempered successively at 100, 200, 300 °C, etc. The letters L, W and T at the points indicated denote the change in length, width and thickness for specimens tempered at 550 °C only. The digits 1 and 2 after these letters refer to the actual dimensions after one and two tempering operations respectively. Hardening temperature 1200 °C. (a) Quenching in oil; (b) Martempering at 400 °C; (c) Cooling in air

7.4 Dimensional changes during case hardening

During recent decades considerable interest has been devoted particularly by gear manufacturers to coping with the deformations which occur during case hardening. One of the reasons for this work is that it is desirable to avoid the expensive post-hardening grinding operations. The dimensional changes which occur during case hardening are governed by a wide variety of factors, and of these by way of introduction we can mention:

1. *The hardenability of the steel.* As the hardenability increases and the thickness of the material decreases, the volume increase becomes larger.
2. *Type of steel.* Cr–Ni steels, Cr–Ni–Mo steels and to some extent Cr–Mn steels behave in a similar manner. On the other hand Cr–Mo steels can exhibit some variations in behaviour as regards change in shape.

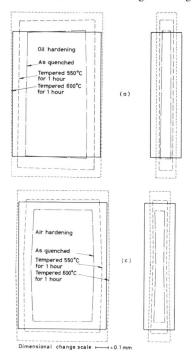
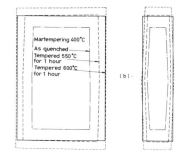

Figure 7.39 Deformation of plates made from steel M 42 after hardening from 1200 °C for 6 min in different cooling media, followed by tempering at 550 and 600 °C for 1 h. Specimen dimensions: 25 × 70 × 100 mm

3. *Depth of case hardening.* This factor is extremely difficult to assess, because a considerable influence on the structure, properties and thickness of the case-hardened layer is exercised both by the carbon content in the surface layer and by the hardening temperature. However, it is clearly obvious that the depth of case hardening influences the dimensional changes.

4. *Method of hardening.* In principle direct hardening causes the least dimensional variations and double hardening causes the greatest changes. The deformation increases with the number of heating and cooling operations.

5. *Material dimensions.* A clear tendency towards shrinkage in the diameter of gears has been observed in connexion with small or moderate dimensions or material thicknesses. As the dimensions of gears and rings increase, the shrinkage decreases, and at a certain dimension there is an increase in the diameter.

Example 30. Rings with outer diameter 100 mm and inner diameter 50 and 75 mm, and height 50 mm were fabricated from steel BS 655M13 (3·25% Ni–Cr steel), BS 637M17 (En 352) and from BS 805M20 (En 362). The rings were gas carburized at 940 °C to a depth of about 1 mm. Prior to quenching which was carried out in oil and martempering bath, respectively, the temperature was lowered to 830 °C. After tempering at 170 °C the dimensions of the rings were checked. It was noted that there had been a reduction in the outer and inner diameters and an increase in the height of all the rings. No significant difference was observed in the changes in dimensions between the various grades. Prior to the finish machining, the rings had been stress-relieved or normalized, respectively. Untreated rings were also included in the test. It was not possible to detect whether any influence was exerted by the different types of pre-treatment. Quenching in oil and in martempering bath also gave quite similar results.

The dimensional changes which are shown for steel En 352 in *Figure 7.40* and which comprise the average value for 18 rings of each dimension, are also representative for the other grades. Throughout, the dimensional changes were greater in the case of the thin-walled rings the volume of which had also increased more than that of the thick-walled rings. The increase in volume is shown in the bar diagram in *Figure 7.41*. It is probable that the smaller increase in volume on the part of the thick-walled rings was due to unsatisfactory through-hardening. The somewhat reduced hardenability of steel En 362 is also reflected in the results from the martempering.

Example 31. Rings made from En 352 with dimensions larger than those employed in *Example 30,* were case hardened in the same way as the latter, by martempering and tempered at 170 °C and 300 °C. The dimensional changes occurring are shown in *Figure 7.42. Examples 30* and *31* show clearly that the reduction in diameter decreases as the thickness of the rings increases and there is a critical dimension above which the diameter increases. This situation has also been confirmed by making measurements on gears having dimensions similar to those employed in *Example 31* (*see Example 36*). By tempering at temperatures exceeding 170 °C the dimensions can be affected in the manner illustrated in *Figure 7.42*.

Dim. ⌀ 100/75x50 mm.

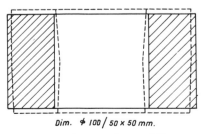

Dim. ⌀ 100/50 x 50 mm.

Dimensional change scale: ⊢⎯⊣ = 0.05 mm

Figure 7.40 Changes in dimensions after case hardening of rings made from B S 637M17, En 352. Case-hardening depth 1·0 mm

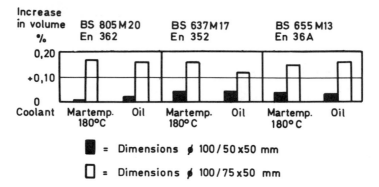

Figure 7.41 Changes in volume after case hardening of rings made from
different case-hardening steels

Example 32. The influence of the depth of case hardening was examined
using rings made from En 352 size 100/50 × 50 mm. The rings were pack
carburized at 900 °C to a depth of 1·0 and 0·5 mm respectively. The rings
were subsequently hardened in oil from 820 °C and tempered at 170 °C.
One ring was hardened without being carburized. The dimensional
changes are shown in *Figure 7.43*. As the depth of case hardening
increases, a greater increase in volume is encountered during hardening,
due to the formation of a larger quantity of martensite. Undoubtedly, the
change in diameter is governed by the depth of case hardening.

Example 33. Bushings made from En 352 were to be produced with
tolerances for which turning was sufficient. It was desirable that the
bushings should be case hardened without the necessity of subsequent
grinding to adjust the dimensions. A test batch of 10 bushings with
dimensions as shown in *Figure 7.44* was produced and check
measurements were made. The bushings were gas carburized at 940 °C to a
depth of 1 mm. After the temperature had dropped to 830 °C, they were
martempered at 180 °C and then tempered at 180 °C.

The dimensional variations between the various points (A–C and D–F
respectively) were equal (±0·01 mm) and the different bushings showed
similar scatter. The values in *Table 7.7* were obtained.

Table 7.7 Changes in dimensions occurring during case hardening of bushings

Measurement	Prioor to hardening mm	After hardening mm	Change%
Outer diameter	55·29	55·22	−0·13
Inner diameter	44·08	43·93	−0·34
Length	103·55	103·49	−0·06

All dimensions decreased but the relatively large reduction in the inner
diameters resulted in an increase of 0·43% in volume. This increase in
volume is more than twice the amount occurring with the thin-walled rings
mentioned in *Example 30* and this can be ascribed to the efficient
through-hardening and to the comparatively large case depth of these
thin-walled bushings.

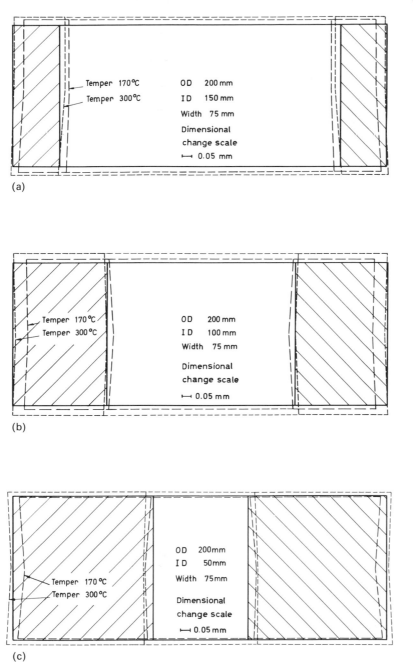

(a)

(b)

(c)

Figure 7.42 Deformation occurring during case hardening of rings made from BS 637M17, En 352

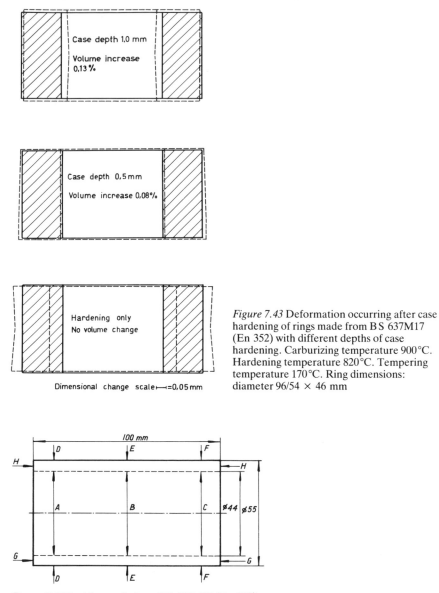

Figure 7.43 Deformation occurring after case hardening of rings made from B S 637M17 (En 352) with different depths of case hardening. Carburizing temperature 900°C. Hardening temperature 820°C. Tempering temperature 170°C. Ring dimensions: diameter 96/54 × 46 mm

Figure 7.44 Bushing made from B S 637M17 (En 352)

To avoid shrinkage, *hardening on mandrels* is carried out, whereby the mandrel is made from a tool steel which is hardened and tempered at 400°C at least. When the quenching of gears or sleeves with inner splines is being carried out, the mandrels are made cylindrical so that the cooling medium can pass between the mandrel and the component which is to be hardened. For components with cylindrical holes it is advisable to provide the mandrel with grooves. To facilitate extraction of the mandrel, it can be Sulfinuz treated prior to hardening.

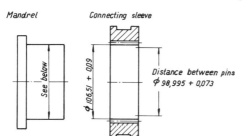

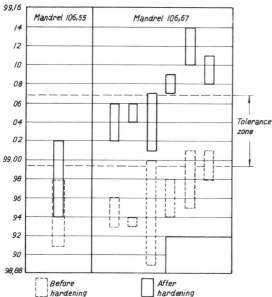

Figure 7.45 Connecting sleeve made from BS 655M13
En 36 A with mandrel of O 1 together with diagram
showing dimensions measured before and after case
hardening. Case depth 0·7 mm

Example 34. A connecting sleeve with dimensions as shown in
Figure 7.45 was made from BS 655M13 and was pack-carburized to a
depth of 0·7 mm. Hardening took place from 820°C. Directly before the
oil quenching, the sleeve was placed on a mandrel, the diameter of which
lay just in the centre of the tolerance range (d = 106·55 mm). During a
trial hardening of one sleeve, the inner diameter increased but only
negligibly. Since the inner diameter of the sleeves was at the lower end of
the tolerance zone, the remaining sleeves were hardened on a mandrel
having a diameter of 106·67 mm. After hardening three of the sleeves were
wholly within the tolerance range, whereas three were on the plus side. A
suitable mandrel diameter would have been 106·00 mm.

Example 35. Prior to the hardening of the splined sleeve as shown in
Figure 7.46 it was found that the outer diameter was on the low side and in
part below the tolerance range. Hardening was carried out in the same way

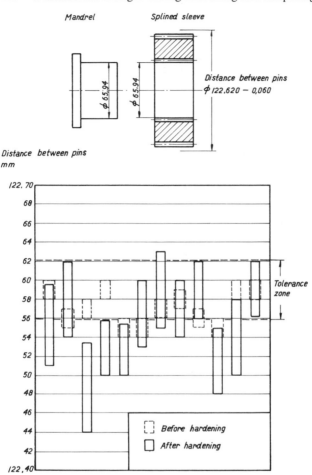

Figure 7.46 Splined sleeve of BS 655M13, En 36 A with mandrel
of O 1 together with diagram showing dimensions measured
before and after case hardening. Case depth 0·7 mm

as in *Example 34*. The diameter of the mandrel was the same as the inner
diameter of the sleeve. The dimensions of the sleeves changed as shown in
Figure 7.46. The amount of shrinkage was greater than desired but the
sleeves could still be used. When all the spline diameters were measured it
was found that the largest dimension occurred in two measuring directions
at right-angles to each other and that the smallest dimension was displaced
at an angle of 45° to this latter dimension. During hardening the sleeves
had tended to adopt a square shape. This change is due to segregations
occurring in ingots that have a square cross-section.

Example 36. During the hardening of gears, as shown in *Figure 7.47*,
carburization took place in gas at 930°C. After the temperature had
dropped to 830°C, the gears were quenched in oil and tempered at 180°C.
The depth of case hardening was about 1 mm. The cross-section of the

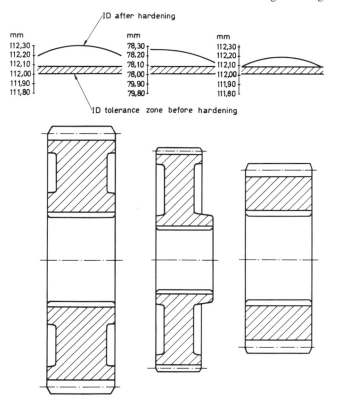

Figure 7.47 Deformation after case hardening of gears made from steel corresponding to En 352 (courtesy Atlas Copco MCT, Avosverken)

steel in these gears was in all cases greater than that in the largest ring described in *Example 31*. There is very considerable similarity between the largest ring and the smallest gear. This also applies to the deformation of the hole. As the thickness of material in the gears increased, the hole diameter increased during hardening. At that part of the intermediate gear, where the hub is broader, the increase in diameter was negligible. In this case we can regard the hub as a separate ring which—considering its dimensions—should shrink, but where shrinkage is prevented by the larger portion of the gear, the hole in which increases in size during hardening.

In each gear the diameter of the hole increased so much that the wheel could not be used. In order to rescue the unhardened gears eight throughgoing holes were drilled in the larger wheels and corresponding counterboring was made in the smallest gear, as shown in *Figure 7.48*. During subsequent case hardening, we obtained the hole dimensions as shown in the diagram which meant that most of the gears could be used. This investigation, undertaken by Atlas Copco at Avosverken, shows clearly the influence of steel bulk on the dimensional changes occurring during case hardening.

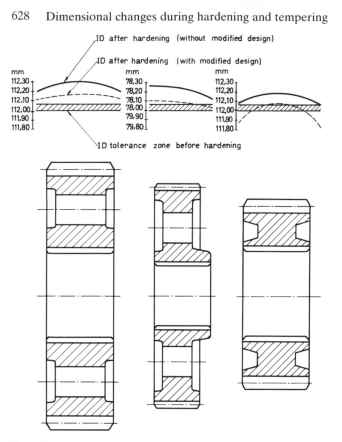

Figure 7.48 Gears of modified design made from steel corresponding to En 352, showing deformation after case hardening (courtesy Atlas Copco MCT, Avosverken)

7.5 Dimensional changes during nitriding

Nitriding is the heat treatment which results in the least dimensional changes and gives the highest hardness values. Since the treatment is carried out at a temperature around only 500°C, no phase transformations take place in the base material. As a result of nitrogen pick-up in the surface layer an increase in volume does occur which occasions a swelling of the surface layer, besides setting up a stress condition between the surface layer and the core.

The amount of swelling ranges from 0·005 to 0·02 mm, depending on the length of the nitriding treatment. If the nitriding takes less than 60 h we can expect that the swelling will be 0·01–0·015 mm per side. With components which have fairly heavy sections, e.g. more than 50 mm, the dimensional changes are restricted mainly to the swelling as mentioned above. When the section size, compared with the thickness of the nitrided layer, is reduced there occurs a dimensional change, the magnitude of which is determined by the ratio of the cross-sectional areas of the nitrided layer and the core.

The following approximate equation can be employed to calculate the increase in length:

$$\Delta l = k \frac{N}{K} \%$$

where Δl — increase in length%

k = constant, the magnitude of which depends on the section and the grade of material

N = cross-sectional area of the nitrided layer

K = cross-sectional area of the core

Cross-sectional area of the nitrided layer + cross-sectional area of the core = total cross-sectional area. This equation applies mainly to N/K values less than 0·7.

The following values (*Table 7.8*) for the constant k were obtained by experiment.

Table 7.8

Section	BS 905M39 (En 41 B)	BS 817M40 (En 24 and En 29 B)
Circular	0·3	0·15
Rectangular	0·4	0·2

To provide a guide for calculating the dimensional variations occurring during nitriding, some general examples will be given first, followed by details of several cases where both calculations and measurements were made.

Example 37. A straight-edge, size 10 × 20 × 1000 mm made from BS 905M39 (En 41 B) is to be nitrided for 60 h. The depth of nitriding is estimated at 0·52 mm and the swelling of the surface layer at 0·01 mm.

Area of nitrided layer N = 30 mm²
Area of core K = 170 mm²

$$\Delta l = 0.4 \times \frac{30}{170} = 0.071\%$$

The change in length is $\dfrac{0.071 \times 1000}{100} + 2 \times 0.01 = 0.73$ mm

Example 38. A ring of En 29 B size 150 × 134 × 10 mm is to be nitrided for 30 h. Depth of nitriding is 0·35 mm. Swelling of the surface layer is 0·005 mm. When calculating the areas we take as a basis the longitudinal cross-sectional area of the ring which is 10 × 8 mm:

Area of nitrided layer N = 12 mm²
Area of core K = 68 mm²

$$\Delta d = \frac{0 \cdot 2 \times 12}{68} = 0 \cdot 035\%$$

The change in diameter is calculated on the basis of the mean diameter 142 mm.

$$\frac{0 \cdot 035 \times 142}{100} = 0 \cdot 05 \text{ mm}$$

The outer diameter increases by $0 \cdot 05 + 2 \times 0 \cdot 005 = 0 \cdot 06$ mm.
The inner diameter increases by $0 \cdot 05 - 2 \times 0 \cdot 005 = 0 \cdot 04$ mm.

Example 39. A number of rings size outer diameter = 368 mm, inner diameter = 348 mm and height = 181 mm which were to be nitrided could not be ground after nitriding. The rings were made from BS 817M40 (En 24). Prior to final machining the rings were stress relieved and then nitrided for 30 h. With the aid of the equation presented previously, the increase in the diameters was found to be as follows:

$OD = 0 \cdot 094$ mm
$ID\ \ = 0 \cdot 074$ mm

These values were taken as a basis for dimensioning the rings prior to nitriding.

The increase in diameter of eight rings was measured after nitriding, the following average values being obtained:

$OD = 0 \cdot 087$ mm
$ID\ \ = 0 \cdot 067$ mm

In addition to this there was some ovality which when calculted for all rings amounted to $\pm 0 \cdot 03$ mm. The rings could be used for the intended purpose without grinding.

Example 40. When a certain definite relationship exists between the outer and inner diameters, the inner diameter will shrink. To gain a better understanding of the effect of this, calculations were made for rings made from En 41 B with a width of 50 mm, outer diameter 70 mm and with varying sizes for the inner diameter. A depth of nitriding of 0·50 mm was stipulated causing a surface swelling of 0·01 mm. The results are shown in *Figure 7.49.* As indicated by the graph, the inner diameter shrinks in rings having a wall thickness greater than 15 mm.

Example 41. Practical tests were made by Hubbard and Robinson (*Figure 7.50*), using the same steels as in *Example 38.* The nitriding period during these tests was 72 h and the nitriding temperature was 525 °C which gives both a greater depth of nitriding and more swelling than that shown in *Example 40.* Bearing this in mind, the agreement between the two diagrams is extremely close.

During the nitriding of unsymmetrical bodies it is possible for irregular deformation to occur which may be due on the one hand to the fact that on one side a body which is otherwise symmetrical has a larger surface area, e.g. it incorporates a U-groove. The face having the largest surface area will expand more during nitriding. This circumstance is well illustrated by the following example:

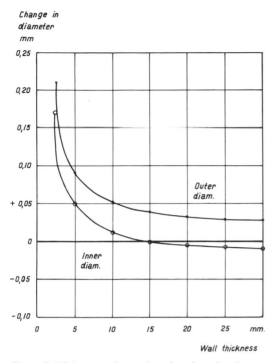

Figure 7.49 Diameter changes as a function of wall
thickness of rings 40 mm in height, 70 mm outer
diameter and of varying wall thickness, after nitriding
for 60 h at 510°C. Steel BS 905M39, En 41 B

Example 42. During the test the specimen shown in *Figure 7.51* was
used. In all cases the surfaces of the slot were insulated so that no swelling
could take place there. The other plane surfaces were also insulated.
Nitriding was performed for 30 h and 60 h. The material comprising the
test bodies was BS 905M39 (En 41 B). Results are shown in *Table 7.9.*

Table 7.9 Dimensional changes during the nitriding of US Navy test specimens

Specimen number	Nitriding of	Nitriding period h	Slot width mm Before nitriding	After nitriding	Dimensional change mm
1	Inside and outside	30	5·02	4·98	−0·04
2	Inside	30	5·03	5·35	+0·32
3	Outside	30	5·04	4·65	−0·39
4	Outside	60	5·02	4·45	−0·57

Example 43. The cylinders shown in *Figure 7.52* were made from a
Cr–Mn case-hardening steel. Previously, during case hardening, the
deformation produced was relatively large and grinding became expensive.
Hence tests were made employing gas nitriding for 30 h at 510°C. The
surface hardness was *ca* 625 HV and at a depth of 0·4 mm the hardness of

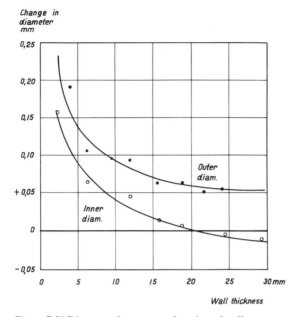

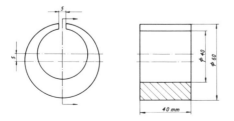

Figure 7.50 Diameter changes as a function of wall thickness of rings 50 mm in height, 70 mm outer diameter and of varying wall thickness, after nitriding for 72 h at 525 °C. Steel corresponds to B S 905M39, En 41 B (after Hubbard and Robinson)

the nitrided case was 400 H V. Measurements made at 20 points on three cylinders showed that all dimensions had increased during nitriding. The maximum increase, viz. 0·11 mm, was observed on the outer diameter (see the arrow in *Figure 7.52*). The inner diameter followed next with an increase of 0·08 mm. The diameters in 90° angle increased in size by up to 0·04 mm. The largest measured diameter increases were due to a larger dimensional increase taking place in the intervening sections where the wall thickness is least on the one side and where there are milled recesses on the other. Both of these factors co-operate in the sense of making the linear increase the largest in this section, with the result that the diameter measurement in the 90° angle increases the most.

Example 44. When making sealing rings from B S 905M39 (En 41 B) it was found after check measurements that the rings, after lapping of the plane surfaces, were concave on one side and convex on the other. This

Figure 7.51 U S Navy distortion-test specimen, dimensions somewhat modified, for use in investigating dimensional changes

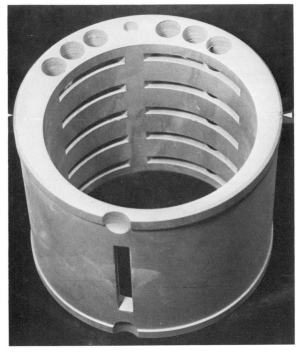

Figure 7.52 Nitrided cylinder for compressed-air engine

change in shape had occurred because lapping was carried out down to different depths on the plane surfaces. This mistake was rectified by fine-grinding the rings prior to nitriding, after which only very slight lapping was required.

7.6 Ageing

By ageing of tool steels we mean in the present context the change that a hardened and usually tempered steel undergoes at room temperature, or in certain cases at somewhat higher temperatures, changes that can result in altered dimensions. The principles underlying these changes have been discussed in Sections 1.4 and 7.2. The ageing which takes place within the temperature range concerned here involves decomposition of the martensite to ε-carbide whereby the volume is reduced, and also the transformation of retained austenite to bainite which is accompanied by an increase in volume. Some form of stress relief also takes place during ageing and this can result in warping. The grade of steel, and its heat treatment and the temperature at which ageing takes place, all play their part in governing which of the two effects mentioned above will predominate during ageing.

In carbon steel which has aged at room temperature the decomposition of martensite is the dominant factor if the steel has not been tempered

$\frac{\Delta L}{L} \times 10^6$

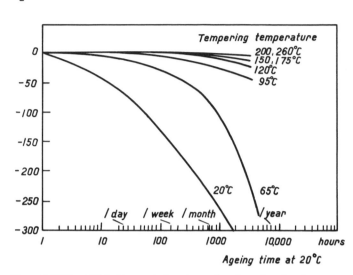

Figure 7.53 Steel W 1. Change in length as a function of ageing time at
20°C after specimens were hardened and tempered at different
temperatures (after Averbach *et al*.[5])

prior to ageing. As the tempering temperature increases prior to ageing,
this effect is reduced. As shown in *Figure 7.53* after tempering at 200°C
relatively high stability has been achieved.

In certain cases conditions are more complicated in the alloyed steels. To
gain some insight into the ageing process, this phenomenon will be
discussed below for steel W1 and O1. The data have been derived from
American sources[5, 6, 7]. For the sake of comparison we have also included
the values for steel L 3 having the composition 1·0% C, 1·5% Cr and
0·20% V.

After the test specimens had been hardened they were tempered, usually
for 1 h at 150°C, they were subzero cooled or subjected to a combination
of both treatments. The test specimens were then kept at room
temperature and were measured after certain intervals of time. The results
obtained with steel W 1 are shown in *Table 7.10*.

Regardless of the treatment employed after hardening, a reduction in
length occurs during ageing. After subzero-cooling alone the contraction is
greater because of the larger quantity of martensite which decomposes
during ageing, coupled with the fact that the effect of any retained
austenite transformation is almost completely absent. This effect is
enhanced, the lower the subzero-cooling temperature employed. The
smallest change observed in the tests took place after tempering at 150°C.
According to *Figure 7.53* the changes are even less after tempering at 200
and 260°C. The conclusion to be drawn from these tests is that tempering
at 180–200°C is adequate to stabilize steel W 1. *Table 7.11* shows the
values obtained during the corresponding investigations made on steel
O 1.

Table 7.10 Steel W1. Change in length occurring as a result of ageing at 20 °C
Pre-treatment: Hardening from 785 °C in water followed by tempering and subzero-cooling as shown

Treatment temperatures °C	Hard- ness HRC	Change in length $\frac{\Delta L}{L} \cdot 10^{2*}$ after ageing at 20°C				
		1 week	1 month	3 months	1 year	3 years
—	66	−90	−175	−265	−405	—
150	65	—	−3	6	−8	−12
−86	67	−110	−205	−310	−480	—
−196	66·5	−120	−240	−350	−525	—
150, −86	65·5	−4	−8	−10	−14	—
150, −196	65·5	−6	−10	−14	−17	—
150, −86, 150	65·5	−5	−8	−10	−14	—
150, −196, 150	65·5	−5	−8	−11	−16	—

* 0·0001%

The results are fundamentally similar to those obtained with steel W 1. Tests made on steel L 3 also exhibit largely similar results and the values obtained with this steel after hardening and tempering at different temperatures, and after hardening, subzero-cooling and tempering respectively, are therefore likely to be applicable to steel O 1 (*see Figures 7.54* and *7.55*).

After hardening and tempering at 150–200 °C the effects of ageing can be ignored, according to these experiments. If tempering is undertaken after subzero-cooling to −196 °C, it must on the other hand be performed at 260 °C in order to prevent any reduction in volume during ageing.

Experiments involving martempering of steel O 1 showed that an increase in volume occurs during ageing if tempering is undertaken at 120 °C and that the changes in dimensions can be ignored after tempering at 150–200 °C.

'Artificial ageing', i.e. heating to 100–150 °C, e.g. for 24 h, does not result in any improvement as regards dimensional stability at room temperature for these steels if they have already been tempered in the normal temperature range, i.e. 180–200 °C. This fact has been confirmed

Table 7.11 Steel O 1. Change in length occurring as a result of ageing at 20 °C
Pre-treatment: Hardening in oil followed by tempering and subzero-cooling as shown

Treatment temperatures °C	Hard- ness HRC	Change in length $\frac{\Delta L}{L} \cdot 10^{2*}$ after ageing at 20°C				
		1 week	1 month	3 months	1 year	3 years
—	64	−14	−30	−48	−87	—
150	63	−1	−1	−2	−3	−1
−86	66	−36	−71	−103	−185	—
−196	66	−50	−91	−128	−118	—
150, −86	62·5	−3	−5	−7	−8	—
150, −196	63	−5	−8	−10	−13	—
150, −86, 150	62	−2	−5	−6	−8	—

* 0·0001%

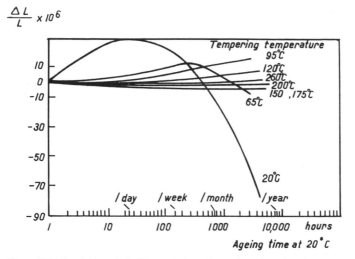

Figure 7.54 Steel A I S I L 3. Change in length as a function of ageing time at 20 °C after specimens were hardened and tempered at different temperatures (after Averbach *et al.*[5])

by an extensive investigation on steel S S 2092, 30 mm diameter × 100 mm. After hardening and tempering at 170 °C for 1 h the specimens were aged at temperatures between 130–170 °C for 6–96 h. After this ageing process quite large dimensional changes were observed. Subsequently both the aged and the un-aged test bars were kept at room temperature for 42 months, during which time two measurements were

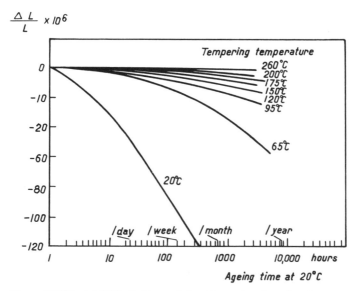

Figure 7.55 Steel A I S I L 3. Change in length as a function of ageing time at 20 °C after hardening, subzero treatment and subsequent tempering at different temperatures (after Averbach *et al.*[5])

made at definite intervals. Except for some few values all the dimensional changes measured during this 'natural ageing' were less than ± 0.003 mm. With regard to the majority of measuring gauges such dimensional changes are of very little practical importance.

In all the cases mentioned above only one tempering operation was performed, but since this took place after the steel had (after hardening) cooled to room temperature, double tempering was not required. In those cases where stringent requirements are imposed regarding dimensional stability, to be on the safe side a double tempering should always be carried out. It is then also important that the tool should be allowed to cool slowly to room temperature so that no thermal stresses are created.

Since measuring appliances can be subjected to forces which may involve the transformation of retained austenite to martensite it is most advisable to undertake subzero-cooling between the tempering operations.

With high-alloy steels such as D 6 and M 2 the conditions prevailing during ageing are much more complicated. If extremely high dimensional stability is required attention must be devoted mainly to ensuring a well-defined heat treatment process, which is modified on the basis of tests made to establish susceptibility to ageing.

7.7 Designing for heat treatment

A very considerable responsibility rests on the heat treater, particularly during the hardening of tools. He has to undertake a treatment which costs about £1 per kilogram, whereas the kilogram price of the tool can be 10–50 times this. The risks of cracking are great even if the heat-treatment personnel are experts at their job. Their work will be greatly facilitated if the shape of the tool is such as to facilitate heat treatment. There are two simple rules which the designer should bear in mind when he designs a die or machine component which is to be heat treated, and these rules can be stressed as follows:

1. Large radii (fillets) and smooth changes of section in the material.
2. As uniform a distribution of the material as possible.

The main causes of quenching cracks are stresses which are set up as a result of temperature differences in different parts of the material, also the transformation stresses during formation of martensite[8].

It is extremely difficult to avoid harmful stresses in steel during hardening. At sharp re-entrant corners, stress concentrations prevail whereby the stresses can exceed the tensile strength of the steel and result in cracking. *Figures 7.56a–d* show some examples of tools which cracked because of their unsuitable shape.

(*a*) The main reason is that there is no fillet radius. The tool was made from the through-hardening steel D 2. Tensile stresses were set up in the surface layer during hardening and this resulted in cracking.

(*b*) The plate was made from a shallow-hardening steel. The material cooled most rapidly at the corners and martensite formed first in these areas. When martensite formation then started in the sides, the compressive stresses around the holes were so great that fracture

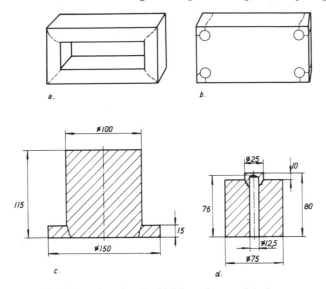

Figure 7.56 Examples of tools which have fractured during hardening because of unsuitable design

occurred. If there had been more material around the holes martensite formation would have occurred there later. Furthermore the stresses would not have been so great across a larger cross-sectional area. By insulating these holes it is possible to reduce the cooling rate and the degree of through hardening.

(*c*) The die was made from steel O 1 and hardened in oil. It cracked at the change of section between the cylindrical parts because of dissimilar material thickness and the small fillet radius. During quenching the thin plate became martensitic first and expanded, so that the material in the upper part was subjected to tensile stresses. Cooling was not so efficient in the fillet and the dimensions of the upper part were too large to admit of martensite formation. The upper part shrank during the cooling and the tensile stresses increased at the fillet where the structure was inhomogeneous and could not accommodate the stresses. In this case not only was the design incorrect; the choice of material was also dubious. It might have been possible to rescue the component by quenching it in a martempering bath with high cooling intensity. The chances of a successful outcome would have improved had the die been made from S S 2550.

(*d*) This component was also made from O 1. It was hardened in oil and cracked at the sudden change of section. Martensite formation occurred first in the smaller, cylindrical part, which caused the surrounding material to become deformed. During the martensite formation in the larger cylindrical section, the latter increased in volume, which created tensile stresses in the central parts. Furthermore the tensile stresses increased as a result of thermal shrinkage of the smaller, cylindrical part. This die would probably

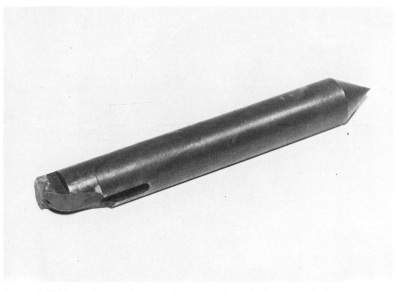

Figure 7.57 Mandrel made from steel W 1 which cracked through the keyway
owing to absence of fillet radius at bottom of groove

not have fractured, if the hole had gone right through or if the fillet
radius had been at least 5 mm. Martempering is recommended for
this die.

Figure 7.57 shows a mandrel made from steel W 1 which cracked in the
keyway which has no bottom radius. *Figure 7.58* shows a part of a die made
from SS 2550 which cracked because of sudden changes of section.

Figure 7.58 Tool made from steel SS 2550 which cracked during hardening
owing to excessively non-uniform sections

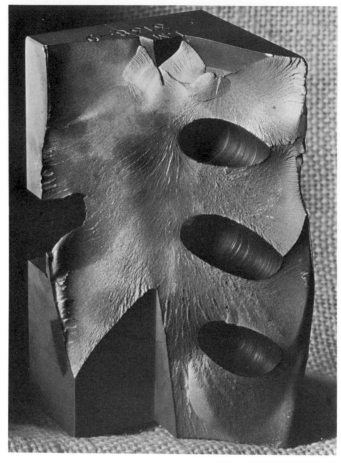

(a)

Figure 7.59a, b Cracked tool made from steel O 1. The die has been quenched in oil, 1 ×

During the oil-hardening of through-hardening steels, tensile stresses are created in the surface. Very often hardness cracks are formed in such dies as a result of sharp punch marks. The die in *Figure 7.59* is a typical example. It was made from steel O 1 and hardened by oil quenching, which was resorted to because this method of hardening would give rise to a smaller increase in length than if quenching in a martempering bath were used. It was possible to solve the problem in either of the following two ways:

1. Keep the sharp punch marks, quench in a martempering bath and grind to the required dimensions afterwards.
2. Engrave by means of an engraving machine, which gives smooth transitions, and quench in oil. The client selected this solution.

(b)

References

1. FREHSER, J., and LOWITZER, O., 'The Process of Dimensional Change During the Heat Treatment of Tools Steels', *Stahl u. Eisen,* **77**, No. 18, 1221–1233 (1957) (in German)
2. LEMENT, B. S., *Distortion in Tool Steel,* ASM (1959)
3. FREHSER, J., 'Anisotropic Dimensional Changes During the Heat Treatment of Ledeburitic Chrome Tool Steels', *Archiv für das Eisenhüttenwesen,* **24**, No. 11/12, 483–495 (1953) (in German)
4. BROWN, R, L., and COHEN, M., 'Stress Relaxation of Hardened Steel', *Met. Prog.,* 66–71 (February 1962)
5. AVERBACH, B. L., COHEN, M., and FLETCHER, S. G., 'The Dimensional Stability of Steel, Part III. Decomposition of Martensite and Austenite at Room Temperature', *Trans. ASM,* **40**, 728 (1948)
6. LEMENT, B. S., and AVERBACH, B. L., *Summary Report I. Measurement and Control of the Dimensional Behaviour of Metals,* Massachusetts Institute of Technology (September 1955)
7. LEMENT, B. S., AVERBACH, B. L., and COHEN, M., 'The Dimensional Stability of Steel, Part IV. Tool Steel', *Trans. ASM,* **41**, 1061 (1949)
8. THELNING, K-E., 'Why Does Steel Crack During Hardening?', *Härterei-Techn. Mitt.,* **25**, No. 7, 271–281 (1970) (in German)

Tables

8.1 Weight tables for steel bars

Table 8.1 Round and square bars. Metric units
Weight in kg/m. Specific gravity 7·85

Size mm	⊘	▨	Size mm	⊘	▨	Size mm	⊘	▨
5	0·15	0·20	47	13·6	17·3	89	48·8	62·2
6	0·22	0·28	48	14·2	18·1	90	49·9	63·6
7	0·30	0·38	49	14·8	18·9	91	51·1	65·0
8	0·39	0·50	50	15·4	19·6	92	52·2	66·4
9	0·50	0·64	51	16·0	20·4	93	53·3	67·9
10	0·62	0·79	52	16·7	21·2	94	54·5	69·4
11	0·75	0·95	53	17·3	22·1	95	55·6	70·9
12	0·89	1·13	54	18·0	22·9	96	56·8	72·4
13	1·04	1·33	55	18·7	23·8	97	58·0	73·9
14	1·21	1·54	56	19·3	24·6	98	59·2	75·4
15	1·39	1·77	57	20·0	25·5	99	60·4	76·9
16	1·58	2·01	58	20·7	26·4	100	61·7	78·5
17	1·78	2·27	59	21·5	27·3	105	68·9	86·6
18	2·00	2·54	60	22·2	28·3	110	74·6	95·0
19	2·23	2·83	61	22·9	29·2	115	81·5	104
20	2·47	3·14	62	23·7	30·2	120	88·8	113
21	2·72	3·46	63	24·5	31·2	125	96·3	123
22	2·98	3·80	64	25·3	32·2	130	104	133
23	3·26	4·15	65	26·1	33·2	135	112	143
24	3·55	4·52	66	26·9	34·2	140	121	154
25	3·85	4·91	67	27·7	35·2	145	130	165
26	4·17	5·31	68	28·5	36·3	150	139	177
27	4·49	5·72	69	29·4	37·4	155	148	189
28	4·83	6·15	70	30·2	38·5	160	158	201
29	5·19	6·60	71	31·1	39·6	165	168	214
30	5·55	7·07	72	32·0	40·7	170	178	227
31	5·92	7·54	73	32·8	41·8	175	189	240
32	6·31	8·04	74	33·8	43·0	180	200	254
33	6·71	8·55	75	34·7	44·2	185	211	269
34	7·13	9·07	76	35·6	45·3	190	223	283
35	7·55	9·62	77	36·6	46·5	195	234	299
36	7·99	10·2	78	37·5	47·8	200	247	314
37	8·44	10·8	79	38·5	49·0	210	272	346
38	8·90	11·3	80	39·5	50·2	220	298	380
39	9·38	11·9	81	40·5	51·5	230	326	415
40	9·86	12·6	82	41·5	52·8	240	355	452
41	10·4	13·2	83	42·5	54·1	250	386	491
42	10·9	13·9	84	43·5	55·4	260	417	531
43	11·4	14·5	85	44·5	56·7	270	449	572
44	11·9	15·2	86	45·6	58·1	280	483	615
45	12·5	15·9	87	46·7	59·4	290	518	660
46	13·1	16·6	88	47·7	60·8	300	555	707

Molybdenum High Speed Steels weigh approximately 3% more.
High Speed Steels of 18–4–1 type weigh approximately 10% more.

Table 8.2 Hexagonal and octagonal bars. Metric units
Weight in kg/m. Specific gravity 7·85

Size mm	⬡	⯃	Size mm	⬡	⯃	Size mm	⬡	⯃	Size mm	⬡	⯃
5	0·17	0·16	15	1·53	1·46	25	4·25	4·06	35	8·33	7·97
6	0·25	0·23	16	1·74	1·66	26	4·60	4·40	37	9·31	8·90
7	0·33	0·32	17	1·97	1·88	27	4·96	4·74	39	10·3	9·89
8	0·44	0·42	18	2·20	2·11	28	5·33	5·09	41	11·4	10·9
9	0·55	0·53	19	2·45	2·35	29	5·72	5·47	43	12·6	12·0
10	0·68	0·65	20	2·72	2·60	30	6·12	5·85	45	13·8	13·2
11	0·82	0·79	21	3·00	2·87	31	6·53	6·25	47	15·0	14·4
12	0·98	0·94	22	3·29	3·15	32	6·96	6·66	50	17·0	16·3
13	1·15	1·10	23	3·60	3·44	33	7·41	7·08	53	19·1	18·3
14	1·33	1·27	24	3·92	3·74	34	7·86	7·52	56	21·3	20·4

Table 8.3 Flat bars, width 10 to 40 mm. Metric units
Weight in kg/m. Specific gravity 7·85

Thick-ness mm	10	11	12	13	14	15	16	17	18	19	20	25	30	35	40
5	0·39	0·43	0·47	0·51	0·55	0·59	0·63	0·67	0·71	0·75	0·79	0·98	1·18	1·37	1·57
6	0·47	0·52	0·57	0·61	0·66	0·71	0·75	0·80	0·85	0·89	0·94	1·18	1·41	1·65	1·88
7	0·55	0·60	0·66	0·71	0·77	0·82	0·88	0·93	0·99	1·04	1·10	1·37	1·65	1·92	2·20
8	0·63	0·69	0·75	0·82	0·88	0·94	1·00	1·07	1·13	1·19	1·26	1·57	1·88	2·20	2·51
9	0·71	0·78	0·85	0·92	0·99	1·06	1·13	1·20	1·27	1·34	1·41	1·77	2·12	2·47	2·83
10	0·79	0·86	0·94	1·02	1·10	1·18	1·26	1·33	1·41	1·49	1·57	1·96	2·36	2·75	3·14
11	—	0·95	1·04	1·12	1·21	1·30	1·38	1·47	1·55	1·64	1·73	2·16	2·59	3·02	3·45
12	—	—	1·13	1·22	1·32	1·41	1·51	1·60	1·70	1·79	1·88	2·36	2·83	3·30	3·77
13	—	—	—	1·33	1·43	1·53	1·63	1·73	1·84	1·94	2·04	2·55	3·06	3·57	4·08
14	—	—	—	—	1·54	1·65	1·76	1·87	1·98	2·09	2·20	2·75	3·30	3·85	4·40
15	—	—	—	—	—	1·77	1·88	2·00	2·12	2·24	2·36	2·94	3·53	4·12	4·71
16	—	—	—	—	—	—	2·01	2·14	2·26	2·39	2·51	3·14	3·77	4·40	5·02
17	—	—	—	—	—	—	—	2·27	2·40	2·54	2·67	3·34	4·00	4·67	5·34
18	—	—	—	—	—	—	—	—	2·54	2·68	2·83	3·53	4·24	4·95	5·65
19	—	—	—	—	—	—	—	—	—	2·83	2·98	3·73	4·47	5·22	5·97
20	—	—	—	—	—	—	—	—	—	—	3·14	3·93	4·71	5·30	6·28
25	—	—	—	—	—	—	—	—	—	—	—	4·91	5·89	6·87	7·85
30	—	—	—	—	—	—	—	—	—	—	—	—	7·07	8·24	9·42
35	—	—	—	—	—	—	—	—	—	—	—	—	—	9·62	11·0
40	—	—	—	—	—	—	—	—	—	—	—	—	—	—	12·6

Molybdenum High Speed Steels weigh approximately 3% more.
High Speed Steels of 18–4–1 type weigh approximately 10% more.

Table 8.4 Flat bars, width 45 to 130 mm. Metric units
Weight in kg/m. Specific gravity 7·85

Thick-ness mm	45	50	55	60	65	70	75	80	85	90	95	100	110	120	130
5	1·77	1·96	2·16	2·36	2·55	2·75	2·94	3·14	3·34	3·53	3·73	3·93	4·32	4·71	5·10
6	2·12	2·36	2·59	2·83	3·06	3·30	3·53	3·77	4·00	4·24	4·47	4·71	5·18	5·65	6·12
7	2·47	2·75	3·02	3·30	3·57	3·85	4·12	4·40	4·67	4·95	5·22	5·50	6·04	6·59	7·14
8	2·83	3·14	3·45	3·77	4·08	4·40	4·71	5·02	5·34	5·65	5·97	6·28	6·91	7·54	8·16
9	3·18	3·53	3·89	4·24	4·59	4·95	5·30	5·65	6·01	6·36	6·71	7·07	7·77	8·48	9·18
10	3·53	3·93	4·32	4·71	5·10	5·50	5·89	6·28	6·67	7·07	7·46	7·85	8·64	9·42	10·2
11	3·89	4·32	4·75	5·18	5·61	6·04	6·48	6·91	7·34	7·77	8·20	8·64	9·50	10·4	11·2
12	4·24	4·71	5·18	5·65	6·12	6·59	7·07	7·54	8·01	8·48	8·95	9·42	10·4	11·3	12·3
13	4·59	5·10	5·61	6·12	6·63	7·14	7·65	8·16	8·67	9·18	9·69	10·2	11·2	12·5	13·3
14	4·95	5·50	6·04	6·59	7·14	7·69	8·24	8·79	9·34	9·89	10·4	11·0	12·1	13·2	14·3
15	5·30	5·89	6·48	7·07	7·65	8·24	8·83	9·42	10·0	10·6	11·2	11·8	13·0	14·1	15·3
16	5·65	6·28	6·91	7·54	8·16	8·79	9·42	10·1	10·7	11·3	11·9	12·6	13·8	15·1	16·3
17	6·01	6·67	7·34	8·01	8·67	9·34	10·0	10·7	11·3	12·0	12·7	13·3	14·7	16·0	17·4
18	6·36	7·07	7·77	8·48	9·18	9·89	10·6	11·3	12·0	12·7	13·4	14·1	15·5	17·0	18·4
19	6·71	7·46	8·20	8·95	9·69	10·4	11·2	11·9	12·7	13·4	14·2	14·9	16·4	17·9	19·4
20	7·07	7·85	8·64	9·42	10·2	11·0	11·8	12·6	13·4	14·1	14·9	15·7	17·3	18·8	20·4
25	8·83	9·81	10·8	11·8	12·8	13·7	14·7	15·7	16·7	17·7	18·6	19·6	21·6	23·6	25·5
30	10·6	11·8	13·0	14·1	15·3	16·5	17·7	18·8	20·0	21·2	22·4	23·6	25·9	28·3	30·6
35	12·4	13·7	15·1	16·5	17·9	19·2	20·6	22·0	23·4	24·7	26·1	27·5	30·2	33·0	35·7
40	14·1	15·7	17·3	18·8	20·4	22·0	23·6	25·1	26·7	28·3	29·8	31·4	34·5	37·7	40·8
45	15·9	17·7	19·4	21·2	23·0	24·7	26·5	28·3	30·0	31·8	33·6	35·3	38·9	42·4	45·9
50	—	19·6	21·6	23·6	25·5	27·5	29·4	31·4	33·4	35·3	37·3	39·3	43·2	47·1	51·0
55	—	—	23·8	25·9	28·1	30·2	32·4	34·5	36·7	38·9	41·0	43·2	47·5	51·8	56·1
60	—	—	—	28·3	30·6	33·0	35·3	37·7	40·0	42·4	44·8	47·1	51·8	56·5	61·2
65	—	—	—	—	33·2	35·7	38·3	40·8	43·3	45·9	48·5	51·0	56·1	61·2	66·3
70	—	—	—	—	—	38·5	41·2	44·0	46·7	49·5	52·2	55·0	60·5	65·9	71·4
75	—	—	—	—	—	—	44·2	47·1	50·0	53·0	55·9	58·9	64·8	70·7	76·5
80	—	—	—	—	—	—	—	50·2	53·4	56·5	59·7	62·8	69·1	75·4	81·6
90	—	—	—	—	—	—	—	—	—	63·6	67·1	70·7	77·7	84·8	91·9
100	—	—	—	—	—	—	—	—	—	—	—	78·5	86·4	94·2	102
125	—	—	—	—	—	—	—	—	—	—	—	—	—	—	128
150	—	—	—	—	—	—	—	—	—	—	—	—	—	—	—
175	—	—	—	—	—	—	—	—	—	—	—	—	—	—	—
200	—	—	—	—	—	—	—	—	—	—	—	—	—	—	—
250	—	—	—	—	—	—	—	—	—	—	—	—	—	—	—

Molybdenum High Speed Steels weigh approximately 3% more.
High Speed Steels of 18–4–1 type weigh approximately 10% more.

Table 8.5 Flat bars, width 140 to 350 mm. Metric units
Weight in kg/m. Specific gravity 7·85

Thick-ness mm	140	150	160	170	180	190	Width mm 200	210	220	230	240	250	275	300	350
5	5·50	5·89	6·28	6·67	7·07	7·46	7·85	8·24	8·64	9·03	9·42	9·81	10·8	11·8	13·7
6	6·59	7·07	7·54	8·01	8·48	8·95	9·42	9·89	10·4	10·8	11·3	11·8	13·0	14·1	16·5
7	7·69	8·24	8·79	9·34	9·89	10·4	11·0	11·5	12·1	12·6	13·2	13·7	15·1	16·5	19·2
8	8·79	9·42	10·1	10·7	11·3	11·9	12·6	13·2	13·8	14·4	15·1	15·7	17·3	18·8	22·0
9	9·89	10·6	11·3	12·0	12·7	13·4	14·1	14·8	15·5	16·3	17·0	17·7	19·4	21·2	24·7
10	11·0	11·8	12·6	13·4	14·1	14·9	15·7	16·5	17·3	18·1	18·8	19·6	21·6	23·6	27·5
11	12·1	13·0	13·8	14·7	15·5	16·4	17·3	18·1	19·0	19·9	20·7	21·6	23·7	25·9	30·2
12	13·2	14·1	15·1	16·0	17·0	17·9	18·8	19·8	20·7	21·7	22·6	23·6	25·9	28·3	33·0
13	14·3	15·3	16·3	17·4	18·4	19·4	20·4	21·4	22·5	23·5	24·5	25·5	28·1	30·6	35·7
14	15·4	16·5	17·6	18·7	19·8	20·9	22·0	23·1	24·2	25·3	26·4	27·5	30·2	33·0	38·5
15	16·5	17·7	18·8	20·0	21·2	22·4	23·6	24·7	25·9	27·1	28·3	29·4	32·4	35·3	41·2
16	17·6	18·8	20·1	21·4	22·6	23·9	25·1	26·4	27·6	28·9	30·1	31·4	34·5	37·7	44·0
17	18·7	20·0	21·4	22·7	24·0	25·4	26·7	28·0	29·4	30·7	32·0	33·4	36·7	40·0	46·7
18	19·8	21·2	22·6	24·0	25·4	26·9	28·3	29·7	31·1	32·5	33·9	35·3	38·9	42·4	49·5
19	20·9	22·4	23·9	25·4	26·9	28·3	29·8	31·3	32·8	34·3	35·8	37·3	41·0	44·8	52·2
20	22·0	23·6	25·1	26·7	28·3	29·8	31·4	33·0	34·5	36·1	37·7	39·3	43·2	47·1	55·0
25	27·5	29·4	31·4	33·4	35·3	37·3	39·3	41·2	43·2	45·1	47·1	49·1	54·0	58·9	68·7
30	33·0	35·3	37·7	40·0	42·4	44·8	47·1	49·5	51·8	54·2	56·5	58·9	64·8	70·7	82·4
35	38·5	41·2	44·0	46·7	49·5	52·2	55·0	57·7	60·5	63·2	65·9	68·7	75·6	82·4	96·2
40	44·0	47·1	50·2	53·4	56·5	59·7	62·8	65·9	69·1	72·2	75·4	78·5	86·4	94·2	110
45	49·5	53·0	56·5	60·1	63·6	67·1	70·7	74·2	77·7	81·3	84·8	88·3	97·1	106	124
50	55·0	58·9	62·8	66·7	70·7	74·6	78·5	82·4	86·4	90·3	94·2	98·1	108	118	137
55	60·5	64·8	69·1	73·4	77·7	82·0	86·4	90·7	95·0	99·3	104	108	119	130	151
60	65·9	70·7	75·4	80·1	84·8	89·5	94·2	98·9	104	108	113	118	130	141	165
65	71·4	76·5	81·6	86·7	91·9	97·0	102	107	112	117	123	128	140	153	179
70	76·9	82·4	87·9	93·4	98·9	104	110	115	121	126	132	137	151	165	192
75	82·4	88·3	94·2	100	106	112	118	124	130	135	141	147	162	177	206
80	87·9	94·2	101	107	113	119	126	132	138	144	151	157	173	188	220
90	98·9	106	113	120	127	134	141	148	155	163	170	177	194	212	247
100	110	118	126	134	141	149	157	165	173	181	188	196	216	236	275
125	137	147	157	167	177	186	196	206	216	226	236	245	270	294	343
150	—	177	188	200	212	224	236	247	259	271	283	294	324	353	412
175	—	—	—	—	247	261	275	288	302	316	330	343	378	412	481
200	—	—	—	—	—	—	314	330	345	361	377	393	432	471	550
250	—	—	—	—	—	—	—	—	—	—	—	491	540	589	687

Molybdenum High Speed Steels weigh approximately 3% more.
High Speed Steels of 18–4–1 type weigh approximately 10% more.

Table 8.6 Round and square bars. Inch units
Weight in lb/ft

Size	○	□	Size	○	□	Size	○	□
$\frac{3}{16}$	0·09	0·12	$\frac{13}{16}$	1·76	2·25	$2\frac{1}{4}$	13·5	17·2
$\frac{13}{64}$	0·11	0·14	$\frac{53}{64}$	1·83	2·33	$2\frac{5}{16}$	14·3	18·2
$\frac{7}{32}$	0·13	0·16	$\frac{27}{32}$	1·90	2·42	$2\frac{3}{8}$	15·1	19·2
$\frac{15}{64}$	0·15	0·19	$\frac{55}{64}$	1·97	2·51	$2\frac{7}{16}$	15·9	20·2
$\frac{1}{4}$	0·17	0·21	$\frac{7}{8}$	2·04	2·60	$2\frac{1}{2}$	16·7	21·3
$\frac{17}{64}$	0·19	0·24	$\frac{57}{64}$	2·12	2·70	$2\frac{9}{16}$	17·5	22·3
$\frac{9}{32}$	0·21	0·27	$\frac{29}{32}$	2·19	2·79	$2\frac{5}{8}$	18·4	23·4
$\frac{19}{64}$	0·24	0·30	$\frac{59}{64}$	2·27	2·89	$2\frac{11}{16}$	19·3	24·6
$\frac{5}{16}$	0·26	0·33	$\frac{15}{16}$	2·35	2·99	$2\frac{3}{4}$	20·2	25·7
$\frac{21}{64}$	0·29	0·37	$\frac{61}{64}$	2·43	3·09	$2\frac{13}{16}$	21·1	26·9
$\frac{11}{32}$	0·32	0·40	$\frac{31}{32}$	2·51	3·19	$2\frac{7}{8}$	22·1	28·1
$\frac{23}{64}$	0·35	0·44	$\frac{63}{64}$	2·59	3·29	$2\frac{15}{16}$	23·0	29·3
$\frac{3}{8}$	0·38	0·48	1	2·67	3·40	3	24·0	30·6
$\frac{25}{64}$	0·41	0·52	$1\frac{1}{32}$	2·84	3·62	$3\frac{1}{8}$	26·1	33·2
$\frac{13}{32}$	0·44	0·56	$1\frac{1}{16}$	3·02	3·84	$3\frac{1}{4}$	28·2	35·9
$\frac{27}{64}$	0·48	0·61	$1\frac{3}{32}$	3·19	4·07	$3\frac{3}{8}$	30·4	38·7
$\frac{7}{16}$	0·51	0·65	$1\frac{1}{8}$	3·38	4·30	$3\frac{1}{2}$	32·7	41·7
$\frac{29}{64}$	0·55	0·70	$1\frac{5}{32}$	3·57	4·55	$3\frac{5}{8}$	35·1	44·7
$\frac{15}{32}$	0·59	0·75	$1\frac{3}{16}$	3·77	4·80	$3\frac{3}{4}$	37·6	47·8
$\frac{31}{64}$	0·63	0·80	$1\frac{7}{32}$	3·97	5·05	$3\frac{7}{8}$	40·1	51·1
$\frac{1}{2}$	0·67	0·85	$1\frac{1}{4}$	4·17	5·31	4	42·7	54·4
$\frac{33}{64}$	0·71	0·90	$1\frac{9}{32}$	4·38	5·58	$4\frac{1}{4}$	48·2	61·4
$\frac{17}{32}$	0·75	0·96	$1\frac{5}{16}$	4·60	5·86	$4\frac{1}{2}$	54·1	68·9
$\frac{35}{64}$	0·80	1·02	$1\frac{11}{32}$	4·82	6·14	$4\frac{3}{4}$	60·3	76·7
$\frac{9}{16}$	0·85	1·08	$1\frac{3}{8}$	5·05	6·43	5	66·8	85·0
$\frac{37}{64}$	0·89	1·14	$1\frac{13}{32}$	5·28	6·72	$5\frac{1}{4}$	73·6	92·7
$\frac{19}{32}$	0·94	1·20	$1\frac{7}{16}$	5·52	7·03	$5\frac{1}{2}$	80·8	103
$\frac{39}{64}$	0·99	1·26	$1\frac{15}{32}$	5·76	7·34	$5\frac{3}{4}$	88·3	112
$\frac{5}{8}$	1·04	1·33	$1\frac{1}{2}$	6·01	7·65	6	96·1	122
$\frac{41}{64}$	1·10	1·40	$1\frac{9}{16}$	6·52	8·30	$6\frac{1}{2}$	113	144
$\frac{21}{32}$	1·15	1·46	$1\frac{5}{8}$	7·05	8·98	7	131	167
$\frac{43}{64}$	1·21	1·54	$1\frac{11}{16}$	7·60	9·68	$7\frac{1}{2}$	150	191
$\frac{11}{16}$	1·26	1·61	$1\frac{3}{4}$	8·18	10·4	8	171	218
$\frac{45}{64}$	1·32	1·68	$1\frac{13}{16}$	8·77	11·2	9	216	275
$\frac{23}{32}$	1·38	1·76	$1\frac{7}{8}$	9·39	12·0	10	267	340
$\frac{47}{64}$	1·44	1·83	$1\frac{15}{16}$	10·0	12·8	11	323	411
$\frac{3}{4}$	1·50	1·91	2	10·7	13·6	12	385	490
$\frac{49}{64}$	1·57	1·99	$2\frac{1}{16}$	11·4	14·5	13	451	574
$\frac{25}{32}$	1·63	2·08	$2\frac{1}{8}$	12·1	15·4	14	523	666
$\frac{51}{64}$	1·70	2·16	$2\frac{3}{16}$	12·8	16·3	15	601	765

Molybdenum High Speed Steels weigh approximately 3% more.
High Speed Steels of 18–4–1 type weigh approximately 10% more.

Table 8.7 Hexagonal and octagonal bars. Inch units
Weight in lb/ft

Size	Hex	Oct	Size	Hex	Oct	Size	Hex	Oct	Size	Hex	Oct
$\frac{3}{16}$	0·10	0·10	$\frac{1}{2}$	0·74	0·70	$\frac{13}{16}$	1·94	1·86	$1\frac{1}{4}$	4·60	4·40
$\frac{7}{32}$	0·14	0·13	$\frac{17}{32}$	0·83	0·79	$\frac{27}{32}$	2·10	2·01	$1\frac{5}{16}$	5·07	4·85
$\frac{1}{4}$	0·18	0·18	$\frac{9}{16}$	0·93	0·89	$\frac{7}{8}$	2·25	2·16	$1\frac{3}{8}$	5·57	5·32
$\frac{9}{32}$	0·23	0·22	$\frac{19}{32}$	1·04	0·99	$\frac{29}{32}$	2·42	2·31	$1\frac{1}{2}$	6·63	6·34
$\frac{5}{16}$	0·29	0·28	$\frac{5}{8}$	1·15	1·10	$\frac{15}{16}$	2·59	2·48	$1\frac{5}{8}$	7·77	7·44
$\frac{11}{32}$	0·35	0·33	$\frac{21}{32}$	1·27	1·21	$\frac{31}{32}$	2·76	2·64	$1\frac{3}{4}$	9·01	8·63
$\frac{3}{8}$	0·41	0·40	$\frac{11}{16}$	1·39	1·33	1	2·94	2·82	$1\frac{7}{8}$	10·4	9·90
$\frac{13}{32}$	0·49	0·46	$\frac{23}{32}$	1·52	1·46	$1\frac{1}{16}$	3·32	3·18	2	11·8	11·3
$\frac{7}{16}$	0·56	0·54	$\frac{3}{4}$	1·66	1·58	$1\frac{1}{8}$	3·73	3·56	$2\frac{1}{8}$	13·3	12·7
$\frac{15}{32}$	0·65	0·62	$\frac{25}{32}$	1·80	1·72	$1\frac{3}{16}$	4·15	3·97	$2\frac{1}{4}$	14·9	14·3

Table 8.8 Flat bars, width ⅜ to 1½ in. Inch units
Weight in lb/ft

Thick-ness	$\frac{3}{8}$	$\frac{7}{16}$	$\frac{1}{2}$	$\frac{9}{16}$	$\frac{5}{8}$	$\frac{11}{16}$	$\frac{3}{4}$	$\frac{13}{16}$	$\frac{7}{8}$	$\frac{15}{16}$	1	$1\frac{1}{8}$	$1\frac{1}{4}$	$1\frac{3}{8}$	$1\frac{1}{2}$
$\frac{3}{16}$	0·24	0·28	0·32	0·36	0·40	0·44	0·48	0·52	0·56	0·60	0·64	0·72	0·79	0·88	0·96
$\frac{7}{32}$	0·28	0·33	0·37	0·42	0·47	0·51	0·56	0·61	0·65	0·70	0·74	0·84	0·93	1·02	1·12
$\frac{1}{4}$	0·32	0·37	0·43	0·48	0·53	0·59	0·64	0·69	0·74	0·80	0·85	0·96	1·06	1·17	1·28
$\frac{9}{32}$	0·36	0·42	0·48	0·54	0·60	0·66	0·72	0·78	0·84	0·90	0·96	1·08	1·20	1·31	1·44
$\frac{5}{16}$	0·40	0·46	0·53	0·60	0·67	0·73	0·80	0·87	0·93	1·00	1·06	1·20	1·33	1·46	1·60
$\frac{3}{8}$	0·48	0·56	0·64	0·72	0·80	0·88	0·96	1·04	1·12	1·20	1·28	1·33	1·59	1·75	1·91
$\frac{7}{16}$	—	0·65	0·74	0·84	0·93	1·02	1·12	1·21	1·30	1·39	1·49	1·67	1·86	2·05	2·33
$\frac{1}{2}$	—	—	0·85	0·96	1·07	1·17	1·28	1·38	1·49	1·60	1·70	1·91	2·13	2·34	2·55
$\frac{9}{16}$	—	—	—	1·08	1·20	1·31	1·44	1·56	1·67	1·79	1·91	2·15	2·39	2·63	2·87
$\frac{5}{8}$	—	—	—	—	1·33	1·46	1·60	1·73	1·86	1·99	2·12	2·39	2·66	2·92	3·19
$\frac{11}{16}$	—	—	—	—	—	1·61	1·76	1·90	2·04	2·19	2·34	2·63	2·92	3·22	3·51
$\frac{3}{4}$	—	—	—	—	—	—	1·91	2·07	2·23	2·39	2·55	2·86	3·19	3·50	3·83
$\frac{13}{16}$	—	—	—	—	—	—	—	2·25	2·41	2·59	2·76	3·11	3·45	3·80	4·14
$\frac{7}{8}$	—	—	—	—	—	—	—	—	2·60	2·79	2·98	3·34	3·72	4·09	4·46
$\frac{15}{16}$	—	—	—	—	—	—	—	—	—	2·99	3·19	3·59	3·98	4·38	4·78
1	—	—	—	—	—	—	—	—	—	—	3·40	3·82	4·25	4·68	5·10
$1\frac{1}{8}$	—	—	—	—	—	—	—	—	—	—	—	4·30	4·78	5·27	5·74
$1\frac{1}{4}$	—	—	—	—	—	—	—	—	—	—	—	—	5·31	5·85	6·38
$1\frac{3}{8}$	—	—	—	—	—	—	—	—	—	—	—	—	—	6·43	7·02
$1\frac{1}{2}$	—	—	—	—	—	—	—	—	—	—	—	—	—	—	7·65

Molybdenum High Speed Steels weigh approximately 3% more.
High Speed Steels of 18–4–1 type weigh approximately 10% more.

Table 8.9 Flat bars, width 1⅝ to 5 in. Inch units
Weight in lb/ft

Thick-ness	1⅝	1¾	2	2¼	2½	2¾	3	3¼	3½	3¾	4	4¼	4½	4¾	5
								Width							
$\frac{3}{16}$	1·04	1·12	1·28	1·44	1·59	1·75	1·91	2·07	2·23	2·39	2·55	2·71	2·87	3·03	3·19
$\frac{7}{32}$	1·21	1·30	1·49	1·68	1·86	2·05	2·23	2·42	2·60	2·79	2·98	3·16	3·35	3·54	3·72
$\frac{1}{4}$	1·38	1·49	1·70	1·92	2·12	2·34	2·55	2·76	2·98	3·19	3·40	3·61	3·83	4·04	4·25
$\frac{9}{32}$	1·56	1·68	1·92	2·15	2·39	2·63	2·87	3·11	3·35	3·59	3·83	4·07	4·31	4·55	4·79
$\frac{5}{16}$	1·73	1·86	2·12	2·39	2·65	2·92	3·19	3·45	3·72	3·99	4·25	4·52	4·78	5·05	5·31
$\frac{3}{8}$	2·08	2·23	2·55	2·87	3·19	3·51	3·83	4·15	4·47	4·78	5·10	5·42	5·74	6·06	6·38
$\frac{7}{16}$	2·42	2·60	2·98	3·35	3·72	4·09	4·46	4·83	5·20	5·58	5·95	6·32	6·70	7·07	7·44
$\frac{1}{2}$	2·76	2·98	3·40	3·83	4·25	4·67	5·10	5·53	5·95	6·38	6·80	7·22	7·65	8·08	8·50
$\frac{9}{16}$	3·11	3·35	3·83	4·30	4·78	5·26	5·74	6·22	6·70	7·17	7·65	8·13	8·61	9·09	9·57
$\frac{5}{8}$	3·46	3·72	4·25	4·78	5·31	5·84	6·38	6·91	7·44	7·97	8·50	9·03	9·57	10·1	10·6
$\frac{11}{16}$	3·80	4·09	4·67	5·26	5·84	6·43	7·02	7·60	8·18	8·76	9·35	9·93	10·5	11·1	11·7
$\frac{3}{4}$	4·15	4·47	5·10	5·75	6·38	7·02	7·65	8·29	8·93	9·57	10·2	10·8	11·5	12·1	12·8
$\frac{13}{16}$	4·49	4·84	5·53	6·21	6·90	7·60	8·29	8·98	9·67	10·4	11·1	11·7	12·4	13·1	13·8
$\frac{7}{8}$	4·84	5·20	5·95	6·69	7·44	8·18	8·93	9·67	10·4	11·2	11·9	12·7	13·4	14·1	14·9
$\frac{15}{16}$	5·18	5·58	6·38	7·18	7·97	8·77	9·57	10·4	11·2	12·0	12·8	13·6	14·3	15·1	15·9
1	5·53	5·95	6·80	7·65	8·50	9·35	10·2	11·1	11·9	12·8	13·6	14·5	15·3	16·2	17·0
1⅛	6·22	6·70	7·65	8·61	9·57	10·5	11·5	12·4	13·4	14·3	15·3	16·3	17·2	18·2	19·1
1¼	6·91	7·44	8·50	9·56	10·6	11·7	12·8	13·8	14·9	15·9	17·0	18·1	19·1	20·2	21·3
1⅜	7·60	8·18	9·35	10·5	11·7	12·9	14·0	15·2	16·4	17·5	18·7	19·9	21·0	22·2	23·4
1½	8·29	8·93	10·2	11·5	12·8	14·0	15·3	16·6	17·9	19·1	20·4	21·7	23·0	24·2	25·5
1⅝	8·98	9·67	11·1	12·4	13·8	15·2	16·6	18·0	19·3	20·7	22·1	23·5	24·9	26·3	27·6
1¾	—	10·4	11·9	13·4	14·9	16·4	17·9	19·3	20·8	22·3	23·8	25·3	26·8	28·3	29·8
2	—	—	13·6	15·3	17·0	18·7	20·4	22·1	23·8	25·5	27·2	28·9	30·6	32·3	34·0
2¼	—	—	—	17·2	19·1	21·0	23·0	24·9	26·8	28·7	30·6	32·5	34·4	36·3	38·3
2½	—	—	—	—	21·3	23·4	25·5	27·6	29·8	31·9	34·0	36·1	38·3	40·4	42·5
2¾	—	—	—	—	—	25·7	38·0	30·4	32·7	35·1	37·4	39·7	42·1	44·4	46·8
3	—	—	—	—	—	—	30·6	33·2	35·7	38·3	40·8	43·4	45·9	48·5	51·0
3½	—	—	—	—	—	—	—	—	41·7	44·6	47·6	50·6	53·6	56·5	59·5
4	—	—	—	—	—	—	—	—	—	—	54·4	57·8	61·2	64·6	68·0
4½	—	—	—	—	—	—	—	—	—	—	—	—	68·9	72·7	76·5
5	—	—	—	—	—	—	—	—	—	—	—	—	—	—	85·0
6	—	—	—	—	—	—	—	—	—	—	—	—	—	—	—
7	—	—	—	—	—	—	—	—	—	—	—	—	—	—	—
8	—	—	—	—	—	—	—	—	—	—	—	—	—	—	—
10	—	—	—	—	—	—	—	—	—	—	—	—	—	—	—

Molybdenum High Speed Steels weigh approximately 3% more.
High Speed Steels of 18–4–1 type weigh approximately 10% more.

Table 8.10 Flat bars, width 5½ to 15 in. Inch units
Weight in lb/ft

Thick-ness	5½	6	6½	7	7½	8	8½	9 (Width)	9½	10	11	12	13	14	15
$\frac{3}{16}$	3·51	3·83	4·14	4·47	4·79	5·11	5·43	5·75	6·07	6·39	7·02	7·65	8·29	8·93	9·57
$\frac{7}{32}$	4·09	4·47	4·84	5·21	5·58	5·95	6·32	6·70	7·07	7·44	8·18	8·93	9·67	10·4	11·2
$\frac{1}{4}$	4·67	5·10	5·53	5·96	6·38	6·80	7·22	7·64	8·06	8·48	9·32	10·2	11·0	11·9	12·7
$\frac{9}{32}$	5 26	5 74	6 22	6·70	7·18	7·65	8·13	8·61	9·09	9·57	10·5	11·5	12·5	13·4	14·4
$\frac{5}{16}$	5·84	6·38	6·90	7·44	7·97	8·50	9·03	9·56	10·1	10·6	11·7	12·7	13·8	14·9	15·9
$\frac{3}{8}$	7·02	7·65	8·29	8·92	9·56	10·2	10·8	11·5	12·1	12·8	14·0	15·3	16·6	17·9	19·1
$\frac{7}{16}$	8·18	8·93	9·67	10·4	11·2	11·9	12·7	13·4	14·2	14·9	16·4	17·9	19·4	20·9	22·4
$\frac{1}{2}$	9·35	10·2	11·1	11·9	12·8	13·6	14·5	15·3	16·2	17·0	18·7	20·4	22·1	23·8	25·5
$\frac{9}{16}$	10·5	11·5	12·4	13·4	14·4	15·3	16·3	17·2	18·2	19·1	21·0	22·9	24·8	26·7	28·6
$\frac{5}{8}$	11·7	12·8	13·8	14·9	16·0	17·0	18·1	19·1	20·2	21·2	23·3	25·4	27·5	29·6	31·7
$\frac{11}{16}$	12·9	14·0	15·2	16·4	17·6	18·7	19·9	21·0	22·2	23·3	25·6	27·9	30·2	32·5	34·8
$\frac{3}{4}$	14·0	15·3	16·6	17·9	19·2	20·4	21·7	22·9	24·2	25·4	27·9	30·4	32·9	35·4	37·9
$\frac{13}{16}$	15·2	16·6	18·0	19·3	20·7	22·1	23·5	24·9	26·3	27·7	30·5	33·2	36·0	38·8	41·6
$\frac{7}{8}$	16·4	17·9	19·3	20·8	22·3	23·8	25·3	26·8	28·3	29·8	32·7	35·7	38·7	41·7	44·7
$\frac{15}{16}$	17·5	19·1	20·7	22·3	23·9	25·5	27·0	28·7	30·2	31·9	35·0	38·1	41·2	44·3	47·4
1	18·7	20·4	22·1	23·8	25·5	27·2	28·9	30·6	32·3	34·0	37·4	40·8	44·2	47·6	51·0
$1\frac{1}{8}$	21·0	23·0	24·9	26·8	28·9	30·6	32·5	34·4	36·3	38·2	42·0	45·8	49·6	53·4	57·2
$1\frac{1}{4}$	23·4	25·5	27·6	29·8	31·9	34·0	36·1	38·2	40·4	42·5	46·7	51·0	55·2	59·4	63·7
$1\frac{3}{8}$	25·7	28·1	30·4	32·7	35·1	37·4	39·7	42·1	44·4	46·8	51·4	56·1	60·8	65·4	70·1
$1\frac{1}{2}$	28·1	30·6	33·2	35·7	38·3	40·8	43·4	45·9	48·5	51·0	56·1	61·2	66·3	71·4	76·5
$1\frac{5}{8}$	30·4	33·2	35·9	38·7	41·4	44·2	47·0	49·7	52·5	55·2	60·8	66·3	71·8	77·3	82·8
$1\frac{3}{4}$	32·7	35·7	38·7	41·7	44·6	47·6	50·6	53·6	56·6	59·5	65·5	71·4	77·4	83·3	89·3
2	37·4	40·8	44·2	47·6	51·0	54·4	57·8	61·2	64·6	68·0	74·8	81·6	88·4	95·2	102
$2\frac{1}{4}$	42·1	45·9	49·7	53·6	57·4	61·2	65·1	68·9	72·7	76·5	84·2	91·9	99·5	107	115
$2\frac{1}{2}$	46·8	51·0	55·3	59·5	63·8	68·0	72·3	76·5	80·8	85·0	93·5	102	111	119	128
$2\frac{3}{4}$	51·4	56·1	60·8	65·4	70·1	74·8	79·4	84·1	88·8	93·5	103	112	121	131	140
3	56·1	61·2	66·3	71·4	76·5	81·6	86·7	91·8	96·9	102	112	122	133	143	153
$3\frac{1}{2}$	65·5	71·4	77·4	83·3	89·3	95·2	101	107	113	119	131	143	155	167	179
4	74·8	81·6	88·4	95·2	102	109	116	122	129	136	150	163	177	190	204
$4\frac{1}{2}$	84·2	91·8	99·4	107	115	122	130	138	145	153	168	184	199	214	230
5	93·6	102	111	119	127	136	145	153	162	170	187	204	222	238	256
6	—	122	133	143	153	163	173	184	194	204	224	244	266	286	306
7	—	—	—	167	179	190	202	214	226	238	262	286	310	334	358
8	—	—	—	—	—	218	232	244	258	272	300	326	354	380	408
10	—	—	—	—	—	—	—	—	340	374	408	444	476	512	

Molybdenum High Speed Steels weigh approximately 3% more.
High Speed Steels of 18–4–1 type weigh approximately 10% more.

8.2 Conversion tables for temperature

Table 8.11a Conversion table for °Celsius (Centigrade) and °Fahrenheit (−250° to 81°)
$$°C = \tfrac{5}{9}\,(°F -32) \qquad °F = 32 + \tfrac{9}{5}\!\cdot\!°C$$
The known number of degrees Centigrade or Fahrenheit is found in the line of figures marked
O, whereafter the desired number of degrees is read off in the left or right column of figures

C	O	F	C	O	F	C	O	F	C	O	F
−157	−250	−418	−60·6	−77	−107	−31·1	−24	−11·2	−1·7	29	84·2
−154	−245	−409	−60·0	−76	−105	−30·6	−23	−9·4	−1·1	30	86·0
−151	−240	−400	−59·4	−75	−103	−30·0	−22	−7·6	−0·6	31	87·8
−148	−235	−391	−58·9	−74	−101	−29·4	−21	−5·8	0	32	89·6
−146	−230	−382	−58·3	−73	−99·4	−28·9	−20	−4·0	0·6	33	91·4
−143	−225	−373	−57·8	−72	−97·6	−28·3	−19	−2·2	1·1	34	93·2
−140	−220	−364	−57·2	−71	−95·8	−27·8	−18	−0·4	1·7	35	95·0
−137	−215	−355	−56·7	−70	−94·0	−27·2	−17	1·4	2·2	36	96·8
−134	−210	−346	−56·1	−69	−92·2	−26·7	−16	3·2	2·8	37	98·6
−132	−205	−337	−55·6	−68	−90·4	−26·1	−15	5·0	3·3	38	100·4
−129	−200	−328	−55·0	−67	−88·6	−25·6	−14	6·8	3·9	39	102·2
−126	−195	−319	−54·4	−66	−86·8	−25·0	−13	8·6	4·4	40	104·0
−123	−190	−310	−53·9	−65	−85·0	−24·4	−12	10·4	5·0	41	105·8
−121	−185	−301	−53·3	−64	−83·2	−23·9	−11	12·2	5·6	42	107·6
−118	−180	−292	−52·8	−63	−81·4	−23·3	−10	14·0	6·1	43	109·4
−115	−175	−283	−52·2	−62	−79·6	−22·8	−9	15·8	6·7	44	111·2
−112	−170	−274	−51·7	−61	−77·8	−22·2	−8	17·6	7·2	45	113·0
−109	−165	−265	−51·1	−60	−76·0	−21·7	−7	19·4	7·8	46	114·8
−107	−160	−256	−50·6	−59	−74·2	−21·1	−6	21·2	8·3	47	116·6
−104	−155	−247	−50·0	−58	−72·4	−20·6	−5	23·0	8·9	48	118·4
−101	−150	−238	−49·4	−57	−70·6	−20·0	−4	24·8	9·4	49	120·2
−98	−145	−229	−48·9	−56	−68·8	−19·4	−3	26·6	10·0	50	122·0
−96	−140	−220	−48·3	−55	−67·0	−18·9	−2	28·4	10·6	51	123·8
−93	−135	−211	−47·8	−54	−65·2	−18·3	−1	30·2	11·1	52	125·6
−90	−130	−202	−47·2	−53	−63·4	−17·8	0	32	11·7	53	127·4
−87	−125	−193	−46·7	−52	−61·6	−17·2	1	33·8	12·2	54	129·2
−84	−120	−184	−46·1	−51	−59·8	−16·7	2	35·6	12·8	55	131·0
−82	−115	−175	−45·6	−50	−58·0	−16·1	3	37·4	13·3	56	132·8
−79	−110	−166	−45·0	−49	−56·2	−15·6	4	39·2	13·9	57	134·6
−76	−105	−157	−44·4	−48	−54·4	−15·0	5	41·0	14·4	58	136·4
−73·3	−100	−148	−43·9	−47	−52·6	−14·4	6	42·8	15·0	59	138·2
−72·8	−99	−146	−43·3	−46	−50·8	−13·9	7	44·6	15·6	60	140·0
−72·2	−98	−144	−42·8	−45	−49·0	−13·3	8	46·4	16·1	61	141·8
−71·7	−97	−143	−42·2	−44	−47·2	−12·8	9	48·2	16·7	62	143·6
−71·1	−96	−141	−41·7	−43	−45·4	−12·2	10	50·0	17·2	63	145·4
−70·6	−95	−139	−41·1	−42	−43·6	−11·7	11	51·8	17·8	64	147·2
−70·0	−94	−137	−40·6	−41	−41·8	−11·1	12	53·6	18·3	65	149·0
−69·4	−93	−135	−40·0	−40	−40·0	−10·6	13	55·4	18·9	66	150·8
−68·9	−92	−134	−39·4	−39	−38·2	−10·0	14	57·2	19·4	67	152·6
−68·3	−91	−132	−38·9	−38	−36·4	−9·4	15	59·0	20·0	68	154·4
−67·8	−90	−130	−38·3	−37	−34·6	−8·9	16	60·8	20·6	69	156·2
−67·2	−89	−128	−37·8	−36	−32·8	−8·3	17	62·6	21·1	70	158·0
−66·7	−88	−126	−37·2	−35	−31·0	−7·8	18	64·4	21·7	71	159·8
−66·1	−87	−125	−36·7	−34	−29·2	−7·2	19	66·2	22·2	72	161·6
−65·6	−86	−123	−36·1	−33	−27·4	−6·7	20	68·0	22·8	73	163·4
−65·0	−85	−121	−35·6	−32	−25·6	−6·1	21	69·8	23·3	74	165·2
−64·4	−84	−119	−35·0	−31	−23·8	−5·6	22	71·6	23·9	75	167·0
−63·9	−83	−117	−34·4	−30	−22·0	−5·0	23	73·4	24·4	76	168·8
−63·3	−82	−116	−33·9	−29	−20·2	−4·4	24	75·2	25·0	77	170·6
−62·8	−81	−114	−33·3	−28	−18·4	−3·9	25	77·0	25·6	78	172·4
−62·2	−80	−112	−32·8	−27	−16·6	−3·3	26	78·8	26·1	79	174·2
−61·7	−79	−110	−32·2	−26	−14·8	−2·8	27	80·6	26·7	80	176·0
−61·1	−78	−108	−31·7	−25	−13·0	−2·2	28	82·4	27·2	81	177·8

Table 8.11b Conversion tables for °Celsius (Centigrade) and °Fahrenheit (82° to 1065°)
$°C = \frac{5}{9}(°F - 32)$ $°F = 32 + \frac{9}{5} \cdot °C$
The known number of degrees Centigrade or Fahrenheit is found in the line of figures marked O, whereafter the desired number of degrees is read off in the left or right column of figures

C	O	F	C	O	F	C	O	F	C	O	F
27·8	82	179·6	135	275	527	282	540	1004	429	805	1481
28·3	83	181·4	138	280	536	285	545	1013	432	810	1490
28·9	84	183·2	141	285	545	288	550	1022	435	815	149a
29·4	85	185·0	143	290	554	291	555	1031	438	820	1508
30·0	86	186·8	146	295	563	293	560	1040	441	825	1517
30·6	87	188·6	149	300	572	296	565	1049	443	830	1526
31·1	88	190·4	152	305	581	299	570	1058	446	835	1535
31·7	89	192·2	154	310	590	302	575	1067	499	840	1544
32·2	90	194·0	157	315	599	304	580	1076	452	845	1553
32·8	91	195·8	160	320	608	307	585	1085	454	850	1562
33·3	92	197·6	163	325	617	310	590	1094	457	855	1571
33·9	93	199·4	166	330	626	313	595	1103	460	860	1580
34·4	94	201·2	168	335	635	316	600	1112	463	865	1589
35·0	95	203·0	171	340	644	318	605	1121	466	870	1598
35·6	96	204·8	174	345	653	321	610	1130	468	875	1607
36·1	97	206·6	177	350	662	324	615	1139	471	880	1616
36·7	98	208·4	179	355	671	327	620	1148	474	885	1625
37·2	99	210·2	182	360	680	329	625	1157	477	890	1634
37·8	100	212	185	365	689	332	630	1166	479	895	1643
41	105	221	188	370	698	335	635	1175	482	900	1652
43	110	230	191	375	707	338	640	1184	485	905	1661
46	115	239	193	380	716	341	645	1193	488	910	1670
49	120	248	196	385	725	343	650	1202	491	915	1679
52	125	257	199	390	734	346	655	1211	493	920	1688
54	130	266	202	395	743	349	660	1220	496	925	1697
57	135	275	204	400	752	352	665	1229	499	930	1706
60	140	284	207	405	761	354	670	1238	502	935	1715
63	145	293	210	410	770	357	675	1247	504	940	1724
66	150	302	213	415	779	360	680	1256	507	945	1733
68	155	311	216	420	788	363	685	1265	510	950	1742
71	160	320	218	425	797	366	690	1274	513	955	1751
74	165	329	221	430	806	368	695	1283	516	960	1760
77	170	338	224	435	815	371	700	1292	518	965	1769
79	175	347	227	440	824	374	705	1301	521	970	1778
82	180	356	229	445	833	377	710	1310	524	975	1787
85	185	365	232	450	842	379	715	1319	527	980	1796
88	190	374	235	455	851	382	720	1328	529	985	1805
91	195	383	238	460	860	385	725	1337	532	990	1814
93	200	392	241	465	869	388	730	1346	535	995	1823
96	205	401	243	470	878	391	735	1355	538	1000	1832
99	210	410	246	475	887	393	740	1364	541	1005	1841
102	215	419	249	480	896	396	745	1373	543	1010	1850
104	220	428	252	485	905	399	750	1382	546	1015	1859
107	225	437	254	490	914	402	755	1391	549	1020	1868
110	230	446	257	495	923	404	760	1400	552	1025	1877
113	235	455	260	500	932	407	765	1409	554	1030	1886
116	240	464	263	505	941	410	770	1418	557	1035	1895
118	245	473	266	510	950	413	775	1427	560	1040	1904
121	250	482	268	515	959	416	780	1436	563	1045	1913
124	255	491	271	520	968	418	785	1445	566	1050	1922
127	260	500	274	525	977	421	790	1454	568	1055	1931
129	265	509	277	530	986	424	795	1463	571	1060	1940
132	270	518	279	535	995	427	800	1472	574	1065	1949

Table 8.11c Conversion tables for °Celsius (Centigrade) and °Fahrenheit (1070° to 2125°)
°C = ⁵⁄₉ (°F −32)　　°F = 32+⁹⁄₅·°C
The known number of degrees Centigrade or Fahrenheit is found in the line of figures marked O, whereafter the desired number of degrees is read off in the left or right column of figures

C	O	F	C	O	F	C	O	F	C	O	F
577	1070	1958	724	1335	2435	871	1600	2912	1018	1865	3389
579	1075	1967	727	1340	2444	874	1605	2921	1021	1870	3398
582	1080	1976	729	1345	2453	877	1610	2930	1024	1875	3407
585	1085	1985	732	1350	2462	879	1615	2939	1027	1880	3416
588	1090	1994	735	1355	2471	882	1620	2948	1029	1885	3425
591	1095	2003	738	1360	2480	885	1625	2957	1032	1890	3434
593	1100	2012	741	1365	2489	888	1630	2966	1035	1895	3443
596	1105	2021	743	1370	2498	891	1635	2975	1038	1900	3452
599	1110	2030	746	1375	2507	893	1640	2984	1041	1905	3461
602	1115	2039	749	1380	2516	896	1645	2993	1043	1910	3470
604	1120	2048	752	1385	2525	899	1650	3002	1046	1915	3479
607	1125	2057	754	1390	2534	902	1655	3011	1049	1920	3488
610	1130	2066	757	1395	2543	904	1660	3020	1052	1925	3497
613	1135	2075	760	1400	2552	907	1665	3029	1054	1930	3506
616	1140	2084	763	1405	2561	910	1670	3038	1057	1935	3515
618	1145	2093	766	1410	2570	913	1675	3047	1060	1940	3524
621	1150	2102	768	1415	2579	916	1680	3056	1063	1945	3533
624	1155	2111	771	1420	2588	918	1685	3065	1066	1950	3542
627	1160	2120	774	1425	2597	921	1690	3074	1068	1955	3551
629	1165	2129	777	1430	2606	924	1695	3083	1071	1960	3560
632	1170	2138	779	1435	2615	927	1700	3092	1074	1965	3569
635	1175	2147	782	1440	2624	929	1705	3101	1077	1970	3578
638	1180	2156	785	1445	2633	932	1710	3110	1079	1975	3587
641	1185	2165	788	1450	2642	935	1715	3119	1082	1980	3596
643	1190	2174	791	1455	2651	938	1720	3128	1085	1985	3605
646	1195	2183	793	1460	2660	941	1725	3137	1088	1990	3614
649	1200	2192	796	1465	2669	943	1730	3146	1091	1995	3623
652	1205	2201	799	1470	2678	946	1735	3155	1093	2000	3632
654	1210	2210	802	1475	2687	949	1740	3164	1096	2005	3641
657	1215	2219	804	1480	2696	952	1745	3173	1099	2010	3650
660	1220	2228	807	1485	2705	954	1750	3182	1102	2015	3659
663	1225	2237	810	1490	2714	957	1755	3191	1104	2020	3668
666	1230	2246	813	1495	2723	960	1760	3200	1107	2025	3677
668	1235	2255	816	1500	2732	963	1765	3209	1110	2030	3686
671	1240	2264	818	1505	2741	966	1770	3218	1113	2035	3695
674	1245	2273	821	1510	2750	968	1775	3227	1116	2040	3704
677	1250	2282	824	1515	2759	971	1780	3236	1118	2045	3713
679	1255	2291	827	1520	2768	974	1785	3245	1121	2050	3722
682	1260	2300	829	1525	2777	977	1790	3254	1124	2055	3731
685	1265	2309	832	1530	2786	979	1795	3263	1127	2060	3740
688	1270	2318	835	1535	2795	982	1800	3272	1129	2065	3749
691	1275	2327	838	1540	2804	985	1805	3281	1132	2070	3758
693	1280	2336	841	1545	2813	988	1810	3290	1135	2075	3767
696	1285	2345	843	1550	2822	991	1815	3299	1138	2080	3776
699	1290	2354	846	1555	2831	993	1820	3308	1141	2085	3785
702	1295	2363	849	1560	2840	996	1825	3317	1143	2090	3794
704	1300	2372	852	1565	2849	999	1830	3326	1146	2095	3803
707	1305	2381	854	1570	2858	1002	1835	3335	1149	2100	3812
710	1310	2390	857	1575	2867	1004	1840	3344	1152	2105	3821
713	1315	2399	860	1580	2876	1007	1845	3353	1154	2110	3830
716	1320	2408	863	1585	2885	1010	1850	3362	1157	2115	3839
718	1325	2417	866	1590	2894	1013	1855	3371	1160	2120	3848
721	1330	2426	868	1595	2903	1016	1860	3380	1163	2125	3857

Table 8.11d Conversion table for °Celsius (Centigrade) and °Fahrenheit (2130° to 3200°)

$$°C = \tfrac{5}{9}(°F - 32) \qquad °F = 32 + \tfrac{9}{5} \cdot °C$$

The known number of degrees Centigrade or Fahrenheit is found in the line of figures marked O, whereafter the desired number of degrees is read off in the left or right column of figures.

C	O	F	C	O	F	C	O	F	C	O	F
1166	2130	3866	1313	2395	4343	1460	2660	4820	1607	2925	5297
1168	2135	3875	1316	2400	4352	1463	2665	4829	1610	2930	5306
1171	2140	3884	1318	2405	4361	1466	2670	4838	1613	2935	5315
1174	2145	3893	1321	2410	4370	1468	2675	4847	1616	2940	5324
1177	2150	3902	1324	2415	4379	1471	2680	4856	1618	2945	5333
1179	2155	3911	1327	2420	4388	1474	2685	4865	1621	2950	5342
1182	2160	3920	1329	2425	4397	1477	2690	4874	1624	2955	5351
1185	2165	3929	1332	2430	4406	1479	2695	4883	1627	2960	5360
1188	2170	3938	1335	2435	4415	1482	2700	4892	1629	2965	5369
1191	2175	3947	1338	2440	4424	1485	2705	4901	1632	2970	5378
1193	2180	3956	1341	2445	4433	1488	2710	4910	1635	2975	5387
1196	2185	3965	1343	2450	4442	1491	2715	4919	1638	2980	5396
1199	2190	3974	1346	2455	4451	1493	2720	4928	1641	2985	5405
1202	2195	3983	1349	2460	4460	1496	2725	4937	1643	2990	5414
1204	2200	3992	1352	2465	4469	1499	2730	4946	1646	2995	5423
1207	2205	4001	1354	2470	4478	1502	2735	4955	1649	3000	5432
1210	2210	4010	1357	2475	4487	1504	2740	4964	1652	3005	5441
1213	2215	4019	1360	2480	4496	1507	2745	4973	1654	3010	5450
1216	2220	4028	1363	2485	4505	1510	2750	4982	1657	3015	5459
1218	2225	4037	1366	2490	4514	1513	2755	4991	1660	3020	5468
1221	2230	4046	1368	2495	4523	1516	2760	5000	1663	3025	5477
1224	2235	4055	1371	2500	4532	1518	2765	5009	1666	3030	5486
1227	2240	4064	1374	2505	4541	1521	2770	5018	1668	3035	5495
1229	2245	4073	1377	2510	4550	1524	2775	5027	1671	3040	5504
1232	2250	4082	1379	2515	4559	1527	2780	5036	1674	3045	5513
1235	2255	4091	1382	2520	2568	1529	2785	5045	1677	3050	5522
1238	2260	4100	1385	2525	4577	1532	2790	5054	1679	3055	5531
1241	2265	4109	1388	2530	4586	1535	2795	5063	1682	3060	5540
1243	2270	4118	1391	2535	4595	1538	2800	5072	1685	3065	5549
1246	2275	4127	1393	2540	4604	1541	2805	5081	1688	3070	5558
1249	2280	4136	1396	2545	4613	1543	2810	5090	1691	3075	5567
1252	2285	4145	1399	2550	4622	1546	2815	5099	1693	3080	5576
1254	2290	4154	1402	2555	4631	1549	2820	5108	1696	3085	5585
1257	2295	4163	1404	2560	4640	1552	2825	5117	1699	3090	5594
1260	2300	4172	1407	2565	4649	1554	2830	5126	1702	3095	5603
1263	2305	4181	1410	2570	4658	1557	2835	5135	1704	3100	5612
1266	2310	4190	1413	2575	4667	1560	2840	5144	1707	3105	5621
1268	2315	4199	1416	2580	4676	1563	2845	5153	1710	3110	5630
1271	2320	4208	1418	2585	4685	1566	2850	5162	1713	3115	5639
1274	2325	4217	1421	2590	4694	1568	2855	5171	1716	3120	5648
1277	2330	4226	1424	2595	4703	1571	2860	5180	1718	3125	5657
1279	2335	4235	1427	2600	4712	1574	2865	5189	1721	3130	5666
1282	2340	4244	1429	2605	4721	1577	2870	5198	1724	3135	5675
1285	2345	4253	1432	2610	4730	1579	2875	5207	1727	3140	5684
1288	2350	4262	1435	2615	4739	1582	2880	5216	1729	3145	5693
1291	2355	4271	1438	2620	4748	1585	2885	5225	1732	3150	5702
1293	2360	4280	1441	2625	4757	1588	2890	5234	1735	3155	5711
1296	2365	4289	1443	2630	4766	1591	2895	5243	1738	3160	5720
1299	2370	4298	1446	2635	4775	1593	2900	5252	1741	3165	5729
1302	2375	4307	1449	2640	4782	1596	2905	5261	1743	3170	5738
1304	2380	4316	1452	2645	4793	1599	2910	5270	1749	3180	5756
1307	2385	4325	1454	2650	4802	1602	2915	5279	1754	3190	5774
1310	2390	4334	1457	2655	4811	1604	2920	5288	1760	3200	5792

8.3 Conversion tables for size

Table 8.12a Inches to millimetres
(1 in = 25·40 mm)

in	0	$\frac{1}{16}$	$\frac{1}{8}$	$\frac{3}{16}$	$\frac{1}{4}$	$\frac{5}{16}$	$\frac{3}{8}$	$\frac{7}{16}$
0	0·00	1·59	3·18	4·76	6·35	7·94	9·53	11·11
1	25·40	26·99	28·58	30·16	31·75	33·34	34·93	36·51
2	50·80	52·39	53·98	55·56	57·15	58·74	60·33	61·91
3	76·20	77·79	79·38	80·96	82·55	84·14	85·73	87·31
4	101·60	103·19	104·78	106·36	107·95	109·54	111·13	112·71
5	127·00	128·59	130·18	131·76	133·35	134·94	136·53	138·11
6	152·40	153·99	155·58	157·16	158·75	160·34	161·93	163·51
7	177·80	179·39	180·98	182·56	184·15	185·74	187·33	188·91
8	203·20	204·79	206·38	207·96	209·55	211·14	212·73	214·31
9	228·60	230·19	231·78	233·36	234·95	236·54	238·13	239·71
10	254·00	255·59	257·18	258·76	260·35	261·94	263·53	265·11
11	279·40	280·99	282·58	284·16	285·75	287·34	288·93	290·51
12	304·80	306·39	307·98	309·56	311·15	312·74	314·33	315·91
13	330·20	331·79	333·38	334·96	336·55	338·14	339·73	341·31
14	355·60	357·19	358·78	360·36	361·95	363·54	365·13	366·71
15	381·00	382·59	384·18	385·76	387·35	388·94	390·53	392·11
16	406·40	407·99	409·58	411·16	412·75	414·34	415·93	417·51
17	431·80	433·39	434·98	436·56	438·15	439·74	441·33	442·91
18	457·20	458·79	460·38	461·96	463·55	465·14	466·73	468·31
19	482·60	484·19	485·78	487·36	488·95	490·54	492·13	493·71
20	508·00	509·59	511·18	512·76	514·35	515·94	517·53	519·11
21	533·40	534·99	536·58	538·16	539·75	541·34	542·93	544·51
22	558·80	560·39	561·98	563·56	565·15	566·74	568·33	569·91
23	584·20	585·79	587·38	588·96	590·55	592·14	593·73	595·31
24	609·60	611·19	612·78	614·36	615·95	617·54	619·13	620·71
25	635·00	636·59	638·18	639·76	641·35	642·94	644·53	646·11
26	660·40	661·99	663·58	665·16	666·75	668·34	669·93	671·51
27	685·80	687·39	688·98	690·56	692·15	693·74	695·33	696·91
28	711·20	712·79	714·38	715·96	717·55	719·14	720·73	722·31
29	736·60	738·19	739·78	741·36	742·95	744·54	746·13	747·71
30	762·00	763·59	765·18	766·76	768·35	769·94	771·53	773·11
31	787·40	788·99	790·58	792·16	793·75	795·34	796·93	798·51
32	812·80	814·39	815·98	817·56	819·15	820·74	822·33	823·91
33	838·20	839·79	841·38	842·96	844·55	846·14	847·73	849·31
34	863·60	865·19	866·78	868·36	869·95	871·54	873·13	874·71
35	889·00	890·59	892·18	893·76	895·35	896·94	898·53	900·11
36	914·40	915·99	917·58	919·16	920·75	922·34	923·93	925·51
37	939·80	941·39	942·98	944·56	946·15	947·74	949·33	950·91
38	965·20	966·79	968·38	969·96	971·55	973·14	974·73	976·31
39	990·60	992·19	993·78	995·36	996·95	998·54	1 000·13	1 001·71
40	1 016·00	1 017·59	1 019·18	1 020·76	1 022·35	1 023·94	1 025·53	1 027·11

Table 8.12b Inches to millimetres
(1 in = 25·40 mm)

in	$\frac{1}{2}$	$\frac{9}{16}$	$\frac{5}{8}$	$\frac{11}{16}$	$\frac{3}{4}$	$\frac{13}{16}$	$\frac{7}{8}$	$\frac{15}{16}$
0	12·70	14·29	15·88	17·46	19·05	20·64	22·23	23·81
1	38·10	39·69	41·28	42·86	44·45	46·04	47·63	49·21
2	63·50	65·09	66·68	68·26	69·85	71·44	73·03	74·61
3	88·90	89·49	92·08	93·66	95·25	96·84	98·43	100·01
4	114·30	115 89	117·48	119·06	120·65	122·24	123·83	125·41
5	139·70	141·29	142·88	144·46	146·05	147·64	149·23	150·81
6	165·10	166·69	168·28	169·86	171·45	173·04	174·63	176·21
7	190·50	192·09	193·68	195·26	196·85	198·44	200·03	201·61
8	215·90	217·49	219·08	220·66	222·25	223·84	225·43	227·01
9	241·30	242·89	244·48	246·06	247·65	249·24	250·83	252·41
10	266·70	268·29	269·88	271·46	273·05	274·64	276·23	277·81
11	292·10	293·69	295·28	296·86	298·45	300·04	301·63	303·21
12	317·50	319·09	320·68	322·26	323·85	325·44	327·03	328·61
13	342·90	344·49	346·08	347·66	349·25	350·84	352·43	354·01
14	368·30	369·89	371·48	373·06	374·65	376·24	377·83	379·41
15	393·70	395·29	396·88	398·46	400·05	401·64	403·23	404·81
16	419·10	420·69	422·28	423·86	425·45	427·04	428·63	430·21
17	444·50	446·09	447·68	449·26	450·85	452·44	454·03	455·61
18	469·90	471·49	473·08	474·66	476·25	477·84	479·43	481·01
19	495·30	496·89	498·48	500·06	501·65	503·24	504·83	506·41
20	520·70	522·29	523·88	525·46	527·05	528·64	530·23	531·81
21	546·10	547·69	549·28	550·86	552·45	554·04	555·63	557·21
22	571·50	573·09	574·68	576·26	577·85	579·44	581·03	582·61
23	596·90	598·49	600·08	601·66	603·25	604·84	606·43	608·01
24	622·30	623·89	625·48	627·06	628·65	630·24	631·83	633·41
25	647·70	649·29	650·88	652·46	654·05	655·64	657·23	658·81
26	673·10	674·69	676·28	677·86	679·45	681·04	682·63	684·21
27	698·50	700·09	701·68	703·26	704·85	706·44	708·03	709·61
28	723·90	725·49	727·08	728·66	730·25	731·84	733·43	735·01
29	749·30	750·89	752·48	754·06	755·65	757·24	758·83	760·41
30	774·70	776·29	777·88	779·46	781·05	782·64	784·23	785·81
31	800·10	801·69	803·28	804·86	806·45	808·04	809·63	811·21
32	825·50	827·09	828·68	830·26	831·85	833·44	835·03	836·61
33	850·90	852·49	854·08	855·66	857·25	858·84	860·43	862·01
34	876·30	877·89	879·48	881·06	882·65	884·24	885·83	887·41
35	901·70	903·29	904·88	906·46	908·05	909·64	911·23	912·81
36	927·10	928·69	930·28	931·86	933·45	935·04	936·63	938·21
37	952·50	954·09	955·68	957·26	958·85	960·44	962·03	963·61
38	977·90	979·49	981·08	982·66	984·25	985·84	987·43	989·01
39	1 003·30	1 004·89	1 006·48	1 008·06	1 009·65	1 011·24	1 012·83	1 014·41
40	1 028·70	1 030·29	1 031·88	1 033·46	1 035·05	1 036·64	1 038·23	1 039·81

Table 8.13 Decimals of an inch to millimetres

in	mm	in	mm	in	mm	in	mm	in	mm
0·001	0·025 4	0·041	1·041 4	0·081	2·057 4	0·26	6·604	0·66	16·764
0·002	0·050 8	0·042	1·066 8	0·082	2·082 8	0·27	6·858	0·67	17·018
0·003	0·076 2	0·043	1·092 2	0·083	2·108 2	0·28	7·112	0·68	17·272
0·004	0·101 6	0·044	1·117 6	0·084	2·133 6	0·29	7·366	0·69	17·526
0·005	0·127 0	0·045	1·143 0	0·085	2·159 0	0·30	7·620	0·70	17·780
0·006	0·152 4	0·046	1·168 4	0·086	2·184 4	0·31	7·874	0·71	18·034
0·007	0·177 8	0·047	1·193 8	0·087	2·209 8	0·32	8·128	0·72	18·288
0·008	0·203 2	0·048	1·219 2	0·088	2·235 2	0·33	8·382	0·73	18·542
0·009	0·228 6	0·049	1·244 6	0·089	2·260 6	0·34	8·636	0·74	18·796
0·010	0·254 0	0·050	1·270 0	0·090	2·286 0	0·35	8·890	0·75	19·050
0·011	0·279 4	0·051	1·295 4	0·091	2·311 4	0·36	9·144	0·76	19·304
0·012	0·304 8	0·052	1·320 8	0·092	2·336 8	0·37	9·398	0·77	19·558
0·013	0·330 2	0·053	1·346 2	0·093	2·362 2	0·38	9·652	0·78	19·812
0·014	0·355 6	0·054	1·371 6	0·094	2·387 6	0·39	9·906	0·79	20·066
0·015	0·381 0	0·055	1·397 0	0·095	2·413 0	0·40	10·160	0·80	20·320
0·016	0·406 4	0·056	1·422 4	0·096	2·438 4	0·41	10·414	0·81	20·574
0·017	0·431 8	0·057	1·447 8	0·097	2·463 8	0·42	10·668	0·82	20·828
0·018	0·457 2	0·058	1·473 2	0·098	2·489 2	0·43	10·922	0·83	21·082
0·019	0·482 6	0·059	1·498 6	0·099	2·514 6	0·44	11·176	0·84	21·336
0·020	0·508 0	0·060	1·524 0	0·100	2·540 0	0·45	11·430	0·85	21·590
0·021	0·533 4	0·061	1·549 4			0·46	11·684	0·86	21·844
0·022	0·558 8	0·062	1·574 8			0·47	11·938	0·87	22·098
0·023	0·584 2	0·063	1·600 2			0·48	12·192	0·88	22·352
0·024	0·609 6	0·064	1·625 6			0·49	12·446	0·89	22·606
0·025	0·635 0	0·065	1·651 0			0·50	12·700	0·90	22·860
0·026	0·660 4	0·066	1·676 4	0·11	2·794	0·51	12·954	0·91	23·114
0·027	0·685 8	0·067	1·701 8	0·12	3·048	0·52	13·208	0·92	23·368
0·028	0·711 2	0·068	1·727 2	0·13	3·302	0·53	13·462	0·93	23·622
0·029	0·736 6	0·069	1·752 6	0·14	3·556	0·54	13·716	0·94	23·876
0·030	0·762 0	0·070	1·778 0	0·15	3·810	0·55	13·970	0·95	24·130
0·031	0·787 4	0·071	1·803 4	0·16	4·064	0·56	14·224	0·96	24·384
0·032	0·812 8	0·072	1·828 8	0·17	4·318	0·57	14·478	0·97	24·638
0·033	0·838 2	0·073	1·854 2	0·18	4·572	0·58	14·732	0·98	24·892
0·034	0·863 6	0·074	1·879 6	0·19	4·826	0·59	14·986	0·99	25·146
0·035	0·889 0	0·075	1·905 0	0·20	5·080	0·60	15·240	1·00	25·400
0·036	0·914 4	0·076	1·930 4	0·21	5·334	0·61	15·494		
0·037	0·939 8	0·077	1·955 8	0·22	5·588	0·62	15·748		
0·038	0·965 2	0·078	1·981 2	0·23	5·842	0·63	16·002		
0·039	0·990 6	0·079	2·006 6	0·24	6·096	0·64	16·256		
0·040	1·016 0	0·080	2·032 0	0·25	6·350	0·65	16·510		

8.4 Conversion tables for weight (mass)

Table 8.14 Pounds to kilograms
(1 lb = 0·454 kg)

lb	0	1	2	3	4	5	6	7	8	9
0	0·000	0·454	0·907	1·361	1·814	2·268	2·722	3·175	3·629	4·082
10	4·536	4·990	5·443	5·897	6·350	6·804	7·257	7·711	8·165	8·618
20	9·072	9·525	9·979	10·433	10·886	11·340	11·793	12·247	12·701	13·154
30	13·608	14·061	14·515	14·969	15·422	15·876	16·329	16·783	17·237	17·690
40	18·144	18·597	19·051	19·504	19·958	20·412	20·865	21·319	21·772	22·226
50	22·680	23·133	23·587	24·040	24·494	24·948	25·401	25·855	26·308	26·762
60	27·216	27·669	28·123	28·576	29·030	29·484	29·937	30·391	30·844	31·298
70	31·752	32·205	32·659	33·112	33·566	34·019	34·473	34·927	35·380	35·834
80	36·287	36·741	37·195	37·648	38·102	38·555	39·009	39·463	39·916	40·370
90	40·823	41·277	41·731	42·184	42·638	43·091	43·545	43·999	44·452	44·906
100	45·359	45·813	46·266	46·720	47·174	47·627	48·081	48·534	48·988	49·442
110	49·895	50·349	50·802	51·256	51·710	52·163	52·617	53·070	53·524	53·978
120	54·431	54·885	55·338	55·792	56·246	56·699	57·153	57·606	58·060	58·513
130	58·967	59·421	59·874	60·328	60·781	61·235	61·689	62·142	62·596	63·049
140	63·503	63·957	64·410	64·864	65·317	65·771	66·225	66·678	67·132	67·585
150	68·039	68·493	68·946	69·400	69·853	70·307	70·761	71·214	71·668	72·121
160	72·575	73·028	73·482	73·936	74·389	74·843	75·296	75·750	76·204	76·657
170	77·111	77·564	78·018	78·472	78·925	79·379	79·832	80·286	80·740	81·193
180	81·647	82·100	82·554	83·008	83·461	83·915	84·368	84·822	85·275	85·729
190	86·183	86·636	87·090	87·543	87·997	88·451	88·904	89·358	89·811	90·265
200	90·719	91·172	91·626	92·079	92·533	92·987	93·440	93·894	94·347	94·801
210	95·255	95·708	96·162	96·615	97·069	97·522	97·976	98·430	98·883	99·337
220	99·790	100·244	100·698	101·151	101·605	102·058	102·512	102·966	103·419	103·873
230	104·326	104·780	105·234	105·687	106·141	106·594	107·048	107·502	107·955	108·409
240	108·862	109·316	109·770	110·223	110·677	111·130	111·584	112·037	112·491	112·945
250	113·398	113·852	114·305	114·759	115·213	115·666	116·120	116·573	117·027	117·481
260	117·934	118·388	118·841	119·295	119·749	120·202	120·656	121·109	121·563	122·017
270	122·470	122·924	123·377	123·831	124·284	124·738	125·192	125·645	126·099	126·552
280	127·006	127·460	127·913	128·367	128·820	129·274	129·728	130·181	130·635	131·088
290	131·542	131·996	132·449	132·903	133·356	133·810	134·264	134·717	135·171	135·624
300	136·078	136·531	136·985	137·439	137·892	138·346	138·799	139·253	139·707	140·160

Table 8.15 British units to kilograms

	1 ounce (oz)	=	0·028	kg
16 ounces	= 1 pound (lb)	=	0·454	kg
28 pounds	= 1 quarter (qr)	=	12·70	kg
4 quarters	= 1 hundredweight (cwt)	=	50·80	kg
20 hundredweights	= 1 ton	=	1016·05	kg

British units	kg	British units	kg	British units	kg
1 oz	0·028	19 lb	8·618	1 ton	1 016·05
2 oz	0·057	20 lb	9·072	2 ton	2 032·10
3 oz	0·085	21 lb	9·525	3 ton	3 048·15
4 oz	0·113	22 lb	9·979	4 ton	4 064·20
5 oz	0·142	23 lb	10·433	5 ton	5 080·25
6 oz	0·170	24 lb	10·886	6 ton	6 096·30
7 oz	0·198	25 lb	11·340	7 ton	7 112·35
8 oz	0·227	26 lb	11·793	8 ton	8 128·40
9 oz	0·255	27 lb	12·247	9 ton	9 144·45
10 oz	0·283	28 lb	12·70	10 ton	10 160·50
11 oz	0·312			11 ton	11 176·55
12 oz	0·340	1 qr	12·70	12 ton	12 192·60
13 oz	0·369	2 qr	25·40	13 ton	13 208·65
14 oz	0·397	3 qr	38·10	14 ton	14 224·70
15 oz	0·425	4 qr	50·80	15 ton	15 240·75
16 oz	0·454	1 cwt	50·80	16 ton	16 256·80
		2 cwt	101·60	17 ton	17 272·85
1 lb	0·454	3 cwt	152·41	18 ton	18 288·90
2 lb	0·907	4 cwt	203·21	19 ton	19 304·95
3 lb	1·361	5 cwt	254·01	20 ton	20 321·00
4 lb	1·814	6 cwt	304·81	21 ton	21 337·05
5 lb	2·268	7 cwt	355·61	22 ton	22 353·10
6 lb	2·722	8 cwt	406·42	23 ton	23 369·15
7 lb	3·175	9 cwt	457·22	24 ton	24 385·20
8 lb	3·629	10 cwt	508·02	25 ton	25 401·25
9 lb	4·082	11 cwt	558·82		
10 lb	4·536	12 cwt	609·62		
11 lb	4·989	13 cwt	660·43		
12 lb	5·443	14 cwt	711·23		
13 lb	5·897	15 cwt	762·03		
14 lb	6·350	16 cwt	812·83		
15 lb	6·804	17 cwt	863·64		
16 lb	7·257	18 cwt	914·44		
17 lb	7·711	19 cwt	965·24		
18 lb	8·165	20 cwt	1 016·05		

8.5 Conversion tables for stress (pressure)

Table 8.16 Tons per square inch to kiloponds per square millimetre

tonf/ in²	kp/ mm²	tonf/ in²	kp/ mm²	tonf/ in²	kp/ mm²	tonf/ in²	kp/ mm²	tonf/ in²	kp/ mm²
1	1·6	31	48·8	61	96·1	91	143	121	191
2	3·2	32	50·4	62	97·6	92	145	122	192
3	4·7	33	52·0	63	99·2	93	147	123	194
4	6·3	34	53·5	64	101	94	148	124	195
5	7·9	35	55·1	65	102	95	150	125	197
6	9·5	36	56·7	66	104	96	151	126	198
7	11·0	37	58·3	67	105	97	153	127	200
8	12·6	38	59·8	68	107	98	154	128	202
9	14·2	39	61·4	69	109	99	156	129	203
10	15·7	40	63·0	70	110	100	158	130	205
11	17·3	41	64·6	71	112	101	159	131	206
12	18·9	42	66·1	72	113	102	161	132	208
13	20·5	43	67·7	73	115	103	162	133	210
14	22·1	44	69·3	74	116	104	164	134	211
15	23·6	45	70·9	75	118	105	165	135	213
16	25·2	46	72·4	76	120	106	167	136	214
17	26·8	47	74·0	77	121	107	169	137	216
18	28·4	48	75·6	78	123	108	170	138	217
19	29·9	49	77·2	79	124	109	172	139	219
20	31·5	50	78·7	80	126	110	173	140	221
21	33·1	51	80·3	81	128	111	175	141	222
22	34·7	52	81·9	82	129	112	176	142	224
23	36·2	53	83·5	83	131	113	178	143	225
24	37·8	54	85·0	84	132	114	180	144	227
25	39·4	55	86·6	85	134	115	181	145	228
26	41·0	56	88·2	86	135	116	183	146	230
27	42·5	57	89·8	87	137	117	184	147	232
28	44·1	58	91·3	88	139	118	186	148	233
29	45·7	59	92·9	89	140	119	187	149	235
30	47·3	60	94·5	90	142	120	189	150	236

Table 8.17 Tons per square inch to newtons per square millimetre

tonf/in^2	N/mm^2	tonf/in^2	N/mm^2	tonf/in^2	N/mm^2	tonf/in^2	N/mm^2	tonf/in^2	N/mm^2
1	15	31	479	61	942	91	1405	121	1868
2	31	32	494	62	958	92	1421	122	1884
3	46	33	510	63	973	93	1436	123	1899
4	62	34	525	64	988	94	1452	124	1915
5	77	35	541	65	1004	95	1467	125	1930
6	93	36	556	66	1019	96	1483	126	1945
7	108	37	571	67	1035	97	1498	127	1961
8	124	38	587	68	1050	98	1514	128	1976
9	139	39	602	69	1066	99	1529	129	1992
10	154	40	618	70	1081	100	1544	130	2007
11	170	41	633	71	1097	101	1559	131	2023
12	185	42	649	72	1112	102	1575	132	2038
13	201	43	664	73	1127	103	1590	133	2054
14	216	44	680	74	1143	104	1606	134	2069
15	232	45	695	75	1158	105	1621	135	2084
16	247	46	710	76	1174	106	1637	136	2100
17	263	47	726	77	1189	107	1652	137	2115
18	278	48	741	78	1205	108	1668	138	2131
19	293	49	757	79	1220	109	1683	139	2146
20	309	50	772	80	1236	110	1698	140	2162
21	324	51	788	81	1251	111	1714	141	2177
22	340	52	803	82	1266	112	1729	142	2192
23	355	53	819	83	1282	113	1745	143	2208
24	371	54	834	84	1297	114	1760	144	2223
25	386	55	849	85	1313	115	1776	145	2239
26	402	56	865	86	1328	116	1791	146	2254
27	417	57	880	87	1344	117	1806	147	2270
28	432	58	896	88	1359	118	1822	148	2285
29	448	59	911	89	1375	119	1837	149	2301
30	463	60	927	90	1390	120	1853	150	2316

Table 8.18 Kiloponds per square millimetre to newtons per square millimetre

kp/mm²	N/mm²	kp/mm²	N/mm²	kp/mm²	N/mm²	kp/mm²	N/mm²	kp/mm²	N/mm²	kp/mm²	N/mm²
1	10	51	500	101	990	151	1480	201	1970	251	2460
2	20	52	510	102	1000	152	1490	202	1980	252	2470
3	30	53	520	103	1010	153	1500	203	1990	253	2480
4	40	54	530	104	1020	154	1510	204	2000	254	2490
5	50	55	540	105	1030	155	1520	205	2010	255	2500
6	60	56	550	106	1040	156	1530	206	2020	256	2510
7	70	57	560	107	1050	157	1540	207	2030	257	2520
8	80	58	570	108	1060	158	1550	208	2040	258	2530
9	90	59	580	109	1070	159	1560	209	2050	259	2540
10	100	60	590	110	1080	160	1570	210	2060	260	2550
11	110	61	600	111	1090	161	1580	211	2070	261	2560
12	120	62	610	112	1100	162	1590	212	2080	262	2570
13	130	63	620	113	1110	163	1600	213	2090	263	2580
14	140	64	630	114	1120	164	1610	214	2100	264	2590
15	150	65	640	115	1130	165	1620	215	2110	265	2600
16	160	66	650	116	1140	166	1630	216	2120	266	2610
17	170	67	660	117	1150	167	1640	217	2130	267	2620
18	180	68	670	118	1160	168	1650	218	2140	268	2630
19	190	69	680	119	1170	169	1660	219	2150	269	2640
20	200	70	690	120	1180	170	1670	220	2160	270	2650
21	210	71	700	121	1190	171	1680	221	2170	271	2660
22	220	72	710	122	1200	172	1690	222	2180	272	2670
23	230	73	720	123	1210	173	1700	223	2190	273	2680
24	240	74	730	124	1220	174	1710	224	2200	274	2690
25	250	75	740	125	1230	175	1720	225	2210	275	2700
26	260	76	750	126	1240	176	1730	226	2220	276	2710
27	260	77	760	127	1250	177	1740	227	2230	277	2720
28	270	78	770	128	1260	178	1750	228	2240	278	2730
29	280	79	770	129	1270	179	1760	229	2250	279	2740
30	290	80	780	130	1280	180	1770	230	2260	280	2750
31	300	81	790	131	1290	181	1780	231	2270	281	2760
32	310	82	800	132	1290	182	1790	232	2280	282	2770
33	320	83	810	133	1300	183	1800	233	2290	283	2780
34	330	84	820	134	1310	184	1810	234	2300	284	2790
35	340	85	830	135	1320	185	1810	235	2310	285	2800
36	350	86	840	136	1330	186	1820	236	2320	286	2810
37	360	87	850	137	1340	187	1830	237	2320	287	2820
38	370	88	860	138	1350	188	1840	238	2330	288	2830
39	380	89	870	139	1360	189	1850	239	2340	289	2840
40	390	90	880	140	1370	190	1860	240	2350	290	2840
41	400	91	890	141	1380	191	1870	241	2360	291	2850
42	410	92	900	142	1390	192	1880	242	2370	292	2860
43	420	93	910	143	1400	193	1890	243	2380	293	2870
44	430	94	920	144	1410	194	1900	244	2390	294	2880
45	440	95	930	145	1420	195	1910	245	2400	295	2890
46	450	96	940	146	1430	196	1920	246	2410	296	2900
47	460	97	950	147	1440	197	1930	247	2420	297	2910
48	470	98	960	148	1450	198	1940	248	2430	298	2920
49	480	99	970	149	1460	199	1950	249	2440	299	2930
50	490	100	980	150	1470	200	1960	250	2450	300	2940

Table 8.19 Pounds per square inch to kiloponds per square millimetre
(1000 lb per sq. inch = 0·703 kp per mm²)

lb per sq. inch	0	1000	2000	3000	4000	5000	6000	7000	8000	9000
>0	0·0	0·7	1·4	2·1	2·8	3·5	4·2	4·9	5·6	6·3
10 000	7·0	7·7	8·4	9·1	9·8	10·6	11·3	12·0	12·7	13·4
20 000	14·1	14·8	15·5	16·2	16·9	17·6	18·3	19·0	19·7	20·4
30 000	21·1	21·8	22·5	23·2	23·9	24·6	25·3	26·0	26·7	27·4
40 000	28·1	28·8	29·5	30·2	30·9	31·6	32·3	33·0	33·8	34·5
50 000	35·2	35·9	36·6	37·3	38·0	38·7	39·4	40·1	40·8	41·5
60 000	42·2	42·9	43·6	44·3	45·0	45·7	46·4	47·1	47·8	48·5
70 000	49·2	49·9	50·6	51·3	52·0	52·7	53·4	54·1	54·8	55·5
80 000	56·3	57·0	57·7	58·4	59·1	59·8	60·5	61·2	61·9	62·6
90 000	63·3	64·0	64·7	65·4	66·1	66·8	67·5	68·2	68·9	69·6
100 000	70·3	71·0	71·7	72·4	73·1	73·8	74·5	75·2	75·9	76·6
110 000	77·3	78·0	78·7	79·5	80·2	80·9	81·6	82·3	83·0	83·7
120 000	84·4	85·1	85·8	86·5	87·2	87·9	88·6	89·3	90·0	90·7
130 000	91·4	92·1	92·8	93·5	94·2	94·9	95·6	96·3	97·0	97·7
140 000	98·4	99·1	99·8	100·5	101·2	102·0	102·7	103·4	104·1	104·8
150 000	105·5	106·2	106·9	107·6	108·3	109·0	109·7	110·4	111·1	111·8
160 000	112·5	113·2	113·9	114·6	115·3	116·0	116·7	117·4	118·1	118·8
170 000	119·5	120·2	120·9	121·6	122·3	123·0	123·7	124·4	125·2	125·9
180 000	126·6	127·3	128·0	128·7	129·4	130·1	130·8	131·5	132·2	132·9
190 000	133·6	134·3	135·0	135·7	136·4	137·1	137·8	138·5	139·2	139·9
200 000	140·6	141·3	142·0	142·7	143·4	144·1	144·8	145·5	146·2	146·9
210 000	147·6	148·4	149·1	149·8	150·5	151·2	151·9	152·6	153·3	154·0
220 000	154·7	155·4	156·1	156·8	157·5	158·2	158·9	159·6	160·3	161·0
230 000	161·7	162·4	163·1	163·8	164·5	165·2	165·9	166·6	167·3	168·0
240 000	168·7	169·4	170·1	170·9	171·6	172·3	173·0	173·7	174·4	175·1
250 000	175·8	176·5	177·2	177·9	178·6	179·3	180·0	180·7	181·4	182·1
260 000	182·8	183·5	184·2	184·9	185·6	186·3	187·0	187·7	188·4	189·1
270 000	189·8	190·5	191·2	191·9	192·6	193·3	194·1	194·8	195·5	196·2
280 000	196·9	197·6	198·3	199·0	199·7	200·4	201·1	201·8	202·5	203·2
290 000	203·9	204·6	205·3	206·0	206·7	207·4	208·1	208·8	209·5	210·2
300 000	210·9	211·6	212·3	213·0	213·7	214·4	215·1	215·8	216·5	217·2
310 000	217·9	218·6	219·3	220·0	220·7	221·4	222·2	222·8	223·6	224·3
320 000	225·0	225·7	226·4	227·1	227·8	228·5	229·2	229·9	230·6	231·3
330 000	232·0	232·7	233·4	234·1	234·8	235·5	236·2	236·9	237·6	238·3
340 000	239·0	239·7	240·4	241·1	241·8	242·5	243·2	243·9	244·7	245·4
350 000	246·1	246·8	247·5	248·2	248·9	249·6	250·3	251·0	251·7	252·4
360 000	253·1	253·8	254·5	255·2	255·9	256·6	257·3	258·0	258·7	259·4
370 000	260·1	260·8	261·5	262·2	262·9	263·6	264·3	265·0	265·7	266·4
380 000	267·2	267·9	268·6	269·3	270·0	270·7	271·4	272·1	272·8	273·5
390 000	274·2	274·9	275·6	276·3	277·0	277·7	278·4	279·1	279·8	280·5
400 000	281·2	281·9	282·6	283·3	284·0	284·7	285·4	286·1	286·8	287·5

8.6 Conversion tables for energy

Table 8.20 Kilopond metres to footpounds

kp m	ft lb	kp m	ft lb	kp m	ft lb
0·0	0·00	12·0	86·80	24·0	173·6
0·5	3·62	12·5	90·41	24·5	177·2
1·0	7·23	13·0	94·03	25·0	180·8
1·5	10·85	13·5	97·65	25·5	184·4
2·0	14·47	14·0	101·3	26·0	188·1
2·5	18·08	14·5	104·9	26·5	191·7
3·0	21·70	15·0	108·5	27·0	195·3
3·5	25·32	15·5	112·1	27·5	198·9
4·0	28·93	16·0	115·7	28·0	202·5
4·5	32·55	16·5	119·3	28·5	206·1
5·0	36·17	17·0	123·0	29·0	209·8
5·5	39·78	17·5	126·6	29·5	213·4
6·0	43·40	18·0	130·2	30·0	217·0
6·5	47·01	18·5	133·8	30·5	220·6
7·0	50·63	19·0	137·4	31·0	224·2
7·5	54·25	19·5	141·0	31·5	227·8
8·0	57·86	20·0	144·7	32·0	231·5
8·5	61·48	20·5	148·3	32·5	235·1
9·0	65·10	21·0	151·9	33·0	238·7
9·5	68·71	21·5	155·5	33·5	242·3
10·0	72·33	22·0	159·1	34·0	245·9
10·5	75·95	22·5	162·7	34·5	249·5
11·0	79·56	23·0	166·4	35·0	253·2
11·5	83·18	23·5	170·0	35·5	256·8
12·0	86·80	24·0	173·6	36·0	260·4

Table 8.21 Footpounds to kilopond metres

ft lb	kp m	ft lb	kp m	ft lb	kp m	ft lb	kp m	ft lb	kp m
1	0·14	31	4·29	61	8·43	91	12·6	121	16·7
2	0·28	32	4·42	62	8·57	92	12·7	122	16·9
3	0·41	33	4·56	63	8·71	93	12·9	123	17·0
4	0·55	34	4·70	64	8·85	94	13·0	124	17·1
5	0·69	35	4·84	65	8·99	95	13·1	125	17·3
6	0·83	36	4·98	66	9·12	96	13·3	126	17·4
7	0·97	37	5·12	67	9·26	97	13·4	127	17·6
8	1·11	38	5·25	68	9·40	98	13·6	128	17·7
9	1·24	39	5·39	69	9·54	99	13·7	129	17·8
10	1·38	40	5·53	70	9·68	100	13·8	130	18·0
11	1·52	41	5·67	71	9·82	101	14·0	131	18·1
12	1·66	42	5·81	72	9·95	102	14·1	132	18·3
13	1·80	43	5·94	73	10·1	103	14·2	133	18·4
14	1·94	44	6·08	74	10·2	104	14·4	134	18·5
15	2·07	45	6·22	75	10·4	105	14·5	135	18·7
16	2·21	46	6·36	76	10·5	106	14·7	136	18·8
17	2·35	47	6·50	77	10·7	107	14·8	137	18·9
18	2·49	48	6·64	78	10·8	108	14·9	138	19·1
19	2·63	49	6·77	79	10·9	109	15·1	139	19·2
20	2·77	50	6·91	80	11·1	110	15·2	140	19·4
21	2·90	51	7·05	81	11·2	111	15·4	141	19·5
22	3·04	52	7·19	82	11·3	112	15·5	142	19·6
23	3·18	53	7·33	83	11·5	113	15·6	143	19·8
24	3·32	54	7·47	84	11·6	114	15·8	144	19·9
25	3·46	55	7·60	85	11·8	115	15·9	145	20·1
26	3·59	56	7·74	86	11·9	116	16·0	146	20·2
27	3·73	57	7·88	87	12·0	117	16·2	147	20·3
28	3·87	58	8·02	88	12·2	118	16·3	148	20·5
29	4·01	59	8·16	89	12·3	119	16·5	149	20·6
30	4·15	60	8·30	90	12·4	120	16·6	150	20·7

Table 8.22 Kilopond metres to joules

kp m	J	kp m	J	kp m	J	kp m	J	kp m	J
0·1	1	5·1	50	10·1	99	15·1	148	20·1	197
0·2	2	5·2	51	10·2	100	15·2	149	20·2	198
0·3	3	5·3	52	10·3	101	15·3	150	20·3	199
0·4	4	5·4	53	10·4	102	15·4	151	20·4	200
0·5	5	5·5	54	10·5	103	15·5	152	20·5	201
0·6	6	5·6	55	10·6	104	15·6	153	20·6	202
0·7	7	5·7	56	10·7	105	15·7	154	20·7	203
0·8	8	5·8	57	10·8	106	15·8	155	20·8	204
0·9	9	5·9	58	10·9	107	15·9	156	20·9	205
1·0	10	6·0	59	11·0	108	16·0	157	21·0	206
1·1	11	6·1	60	11·1	109	16·1	158	21·1	207
1·2	12	6·2	61	11·2	110	16·2	159	21·2	208
1·3	13	6·3	62	11·3	111	16·3	160	21·3	209
1·4	14	6·4	63	11·4	112	16·4	161	21·4	210
1·5	15	6·5	64	11·5	113	16·5	162	21·5	211
1·6	16	6·6	65	11·6	114	16·6	163	21·6	212
1·7	17	6·7	66	11·7	115	16·7	164	21·7	213
1·8	18	6·8	67	11·8	116	16·8	165	21·8	214
1·9	19	6·9	68	11·9	117	16·9	166	21·9	215
2·0	20	7·0	69	12·0	118	17·0	167	22·0	216
2·1	21	7·1	70	12·1	119	17·1	168	22·1	217
2·2	22	7·2	71	12·2	120	17·2	169	22·2	218
2·3	23	7·3	72	12·3	121	17·3	170	22·3	219
2·4	24	7·4	73	12·4	122	17·4	171	22·4	220
2·5	25	7·5	74	12·5	123	17·5	172	22·5	221
2·6	26	7·6	75	12·6	124	17·6	173	22·6	222
2·7	26	7·7	76	12·7	125	17·7	174	22·7	223
2·8	27	7·8	77	12·8	126	17·8	175	22·8	224
2·9	28	7·9	77	12·9	127	17·9	176	22·9	225
3·0	29	8·0	78	13·0	128	18·0	177	23·0	226
3·1	30	8·1	79	13·1	129	18·1	178	23·1	227
3·2	31	8·2	80	13·2	129	18·2	179	23·2	228
3·3	32	8·3	81	13·3	130	18·3	180	23·3	229
3·4	33	8·4	82	13·4	131	18·4	181	23·4	230
3·5	34	8·5	83	13·5	132	18·5	181	23·5	231
3·6	35	8·6	84	13·6	133	18·6	182	23·6	232
3·7	36	8·7	85	13·7	134	18·7	183	23·7	232
3·8	37	8·8	86	13·8	135	18·8	184	23·8	233
3·9	38	8·9	87	13·9	136	18·9	185	23·9	234
4·0	39	9·0	88	14·0	137	19·0	186	24·0	235
4·1	40	9·1	89	14·1	138	19·1	187	24·1	236
4·2	41	9·2	90	14·2	139	19·2	188	24·2	237
4·3	42	9·3	91	14·3	140	19·3	189	24·3	238
4·4	43	9·4	92	14·4	141	19·4	190	24·4	239
4·5	44	9·5	93	14·5	142	19·5	191	24·5	240
4·6	45	9·6	94	14·6	143	19·6	192	24·6	241
4·7	46	9·7	95	14·7	144	19·7	193	24·7	242
4·8	47	9·8	96	14·8	145	19·8	194	24·8	243
4·9	48	9·9	97	14·9	146	19·9	195	24·9	244
5·0	49	10·0	98	15·0	147	20·0	196	25·0	245

Table 8.23 Footpounds to joules

ft lb	J	ft lb	J	ft lb	J	ft lb	J	ft lb	J
1	1·4	31	42·0	61	82·7	91	123	121	164
2	2·7	32	43·4	62	84·1	92	125	122	166
3	4·1	33	44·7	63	85·4	93	126	123	167
4	5·4	34	46·1	64	86·8	94	127	124	168
5	6·8	35	47·5	65	88·1	95	129	125	170
6	8·1	36	48·8	66	89·5	96	130	126	171
7	9·5	37	50·2	67	90·8	97	132	127	173
8	10·8	38	51·5	68	92·2	98	133	128	174
9	12·2	39	52·9	69	93·6	99	134	129	175
10	13·6	40	54·2	70	94·9	100	135	130	177
11	14·9	41	55·6	71	96·3	101	137	131	178
12	16·3	42	56·9	72	97·6	102	138	132	180
13	17·6	43	58·3	73	99·0	103	139	133	181
14	19·0	44	59·7	74	100	104	141	134	182
15	20·3	45	61·0	75	102	105	142	135	183
16	21·7	46	62·4	76	103	106	144	136	184
17	23·0	47	63·7	77	104	107	145	137	185
18	24·4	48	65·1	78	106	108	146	138	187
19	25·8	49	66·4	79	107	109	148	139	188
20	27·1	50	67·8	80	108	110	149	140	190
21	28·5	51	69·1	81	110	111	151	141	191
22	29·8	52	70·5	82	111	112	152	142	192
23	31·2	53	71·9	83	113	113	153	143	194
24	32·5	54	73·2	84	114	114	155	144	195
25	33·9	55	74·6	85	115	115	156	145	197
26	35·3	56	75·9	86	117	116	157	146	198
27	36·6	57	77·3	87	118	117	159	147	199
28	38·0	58	78·6	88	119	118	160	148	201
29	39·3	59	80·0	89	121	119	162	149	202
30	40·7	60	81·3	90	122	120	163	150	203

8.7 Conversion table for fracture toughness units

Table 8.24 Fracture toughness units
$$MN\,m^{-3/2} - KSI\sqrt{in} - N\,mm^{-3/2} - kp\,mm^{-3/2}$$

	$MN\,m^{-\frac{3}{2}}$	$KSI\,\sqrt{in}$	$N\,mm^{-\frac{3}{2}}$	$kp\,mm^{-\frac{3}{2}}$
$MN\,m^{-\frac{3}{2}}$	1	0·910 1	31·623	3·223 5
$KSI\,\sqrt{in}$	1·098 8	1	34·747	3·542 0
$N\,mm^{-\frac{3}{2}}$	0·031 623	0·028 780	1	0·101 94
$kp\,mm^{-\frac{3}{2}}$	0·310 22	0·282 33	9·806 7	1

$$1\,hbar\,\sqrt{cm} = 1\,MN\,m^{-\frac{3}{2}}$$

8.8 Conversions for some common units

Table 8.25 Common SI units to British units

Length

1 km	=	0·621 371 mile
1 m	=	1·093 61 yd
1 mm	=	0·039 370 1 in
1 μm	=	39·370 1 μin

Area

1 km²	=	247·105 acres
1 m²	=	1·195 99 yd²
1 mm²	=	0·001 550 00 in²

Volume

1 m³	=	1·307 95 yd³
1 dm³	=	0·035 314 7 ft³
1 cm³	=	0·061 023 7 in³
1 l (litre)	=	0·220 0 gal
1 l (litre)*	=	0·264 2 U S gal

Velocity

1 km/h	=	0·621 371 mile/h
1 m/s	=	3·280 84 ft/s

Acceleration

1 m/s²	=	3·280 84 ft/s²

Mass

1 kg	=	2·204 62 lb
1 g	=	0·035 274 0 oz
	=	15·432 4 gr (grain)

Density

1 kg/m³	=	0·062 428 0 lb/ft³

Force

1 N	=	0·224 809 lbf
	=	7·233 01 pdl

Torque

1 N m	=	0·737 562 lbf ft
	=	23·730 4 pdl ft

Pressure, stress

1 N/m²	=	0·000 145 038 lbf/in²
1 kN/m²	=	20·885 4 lbf/ft²
	=	0·295 300 inHg
1 MN/m²	=	0·064 749 0 tonf/in²
	=	145·038 lbf/in²

Viscosity (dynamic viscosity)

1 N s/m²	=	0·020 885 4 lbf s/ft²

Viscosity, kinematic

1 m²/s	=	10·763 9 ft²/s

Energy

1 J	=	0·737 562 ft lbf
1 kJ	=	0·277 778 W h

* The US gallon has been included for convenience.

Table 8.26 Common non-S I metric units to British units

Length

1 Å   =   0·003 937 01 μin

Area

1 ha	=	2·471 05 acres
1 decare	=	0·247 105 acre
1 a (are, 100 m^2)	=	119·599 yd^2

Acceleration

1 Gal (1 cm/s^2)	=	0·032 808 4 ft/s^2
1 mGal or mgal	=	393·701 μin/s^2

Mass

1 t (tonne or metric ton)	=	0·984 207 ton
1 metric carat	=	3·086 47 gr

Force

1 kgf or kp (kilopond)	=	2·204 62 lbf
1 dyn	=	0·224 809 × 10^{-5} lbf

Pressure

1 atm (standard atmosphere)	=	14·695 9 lbf/in^2
1 at (technical atmosphere)	=	14·223 3 lbf/in^2
1 bar	=	14·503 8 lbf/in^2
1 hpz (hectopieze)	=	14·503 8 lbf/in^2
1 mmHg (conventional) ⎫	= ⎧	0·039 370 1 inHg (conventional)
1 torr ⎬	⎨	0·019 336 8 lbf/in^2
1 mbar ⎭	=	0·014 503 8 lbf/in^2

Viscosity (dynamic viscosity)

1 cP (centipoise)   =   2·088 54 × 10^{-5} lbf s/ft^2

Viscosity, kinematic

1 cSt (centistoke)   =   5·580 01 in^2/h

Energy (work, heat)

1 cal$_{IT}$	=	0·003 968 32 Btu
1 erg	=	0·737 562 × 10^{-7} ft lbf

Table 8.27 Non-SI metric units to SI units

Length

1 Å (angstrom)	=	10^{-10} m

Area

1 ha (hectare)	=	10^4 m^2
1 decare	=	1000 m^2
1 a (are)	=	100 m^2

Acceleration

1 Gal (galileo)	=	0·01 m/s^2 = 1 cm/s^2
1 mGal or mgal	=	0·01 mm/s^2

Mass

1 t (tonne or metric ton)	=	1000 kg
1 q (quintal)	=	100 kg
1 metric carat	=	0·2 g

Force

1 sn (sthene)	=	1000 N
1 kgf or kp (kilopond)	=	9·806 65 N
1 dyn	=	10^{-5} N

Pressure, stress

1 Mbar	=	100 GN/m^2
1 kbar	=	100 M N/m^2
1 atm	=	101·325 kN/m^2 [X CGPM, 1954]
1 bar	=	10^5 N/m^2
1 pz (pieze)	=	10^3 N/m^2
1 hpz	=	10^5 N/m^2
1 mmHg (conventional)	=	133·322 N/m^2
1 torr	=	133·322 N/m^2
1 mbar	=	100 N/m^2 = 1 hN/m^2
1 kgf/cm^2 (at)	=	98·066 5 kN/m^2
1 kgf/m^2	=	9·806 65 N/m^2
1 mmH$_2$O (conventional)	=	9·806 65 N/m^2
1 millitorr	=	0·133 322 N/m^2
1 μHg, i.e. μmHg (conventional)	=	0·133 322 N/m^2
1 μbar	=	0·1 N/m^2

Viscosity (dynamic viscosity)

1 cP (centipoise)	=	0·001 N s/m^2

8.9 Elements

Table 8.28 Atomic number and atomic weight

Symbol	Name	Atomic number	Atomic weight	Symbol	Name	Atomic number	Atomic weight
A or Ar	Argon	18	39·948	Mg	Magnesium	12	24·312
Ac	Actinium	89	—	Mn	Manganese	25	54·9380
Ag	Silver	47	107·870	Mo	Molybdenum	42	95·94
Al	Aluminium	13	26·9815	N	Nitrogen	7	14·0067
Am	Americium	95	—	Na	Sodium	11	22·9898
As	Arsenic	33	74·9216	Nb	Niobium	41	92·906
At	Astatine	85	—	Nd	Neodymium	60	144·24
Au	Gold	79	196·967	Ne	Neon	10	20·183
B	Boron	5	10·811 ±0·003	Ni	Nickel	28	58·71
				No	Nobelium	102	—
Ba	Barium	56	137·34	Np	Neptunium	93	—
Be	Beryllium	4	9·0122	O	Oxygen	8	15·9994 ±0·0001
Bi	Bismuth	83	208·980				
Bk	Berkelium	97	—	Os	Osmium	76	190·2
Br	Bromine	35	79·909	P	Phosphorus	15	30·9738
C	Carbon	6	12·01115 ±0·00005	Pa	Protoactinium	91	—
				Pb	Lead	82	207·19
Ca	Calcium	20	40·08	Pd	Palladium	46	106·4
Cd	Cadmium	48	112·40	Pm	Promethium	61	—
Ce	Cerium	58	140·12	Po	Polonium	84	—
Cf	Californium	98	—	Pr	Praseodymium	59	140·907
Cl	Chlorine	17	35·453	Pt	Platinum	78	195·09
Cm	Curium	96	—	Pu	Plutonium	94	—
Co	Cobalt	27	58·9332	Ra	Radium	88	—
Cr	Chromium	24	51·996	Rb	Rubidium	37	85·47
Cs	Caesium	55	132·905	Re	Rhenium	75	186·2
Cu	Copper	29	63·54	Rh	Rhodium	45	102·905
Dy	Dysprosium	66	162·50	Rn	Radon	86	—
Er	Erbium	68	167·26	Ru	Ruthenium	44	101·07
Es	Einsteinium	99	—	S	Sulphur	16	32·064 ±0·003
Eu	Europium	63	151·96				
F	Fluorine	9	18·9984	Sb	Antimony	51	121·75
Fe	Iron	26	55·847	Sc	Scandium	21	44·956
Fm	Fermium	100	—	Se	Selenium	34	78·96
Fr	Francium	87	—	Si	Silicon	14	28·086 ±0·001
Ga	Gallium	31	69·72				
Gd	Gadolinium	64	157·25	Sm	Samarium	62	150·35
Ge	Germanium	32	72·59	Sn	Tin	50	118·69
H	Hydrogen	1	1·00797 ±0·00001	Sr	Strontium	38	87·62
				Ta	Tantalum	73	180·948
He	Helium	2	4·0026	Tb	Terbium	65	158·924
Hf	Hafnium	72	178·49	Tc	Technetium	43	—
Hg	Mercury	80	200·59	Te	Tellurium	52	127·60
Ho	Holmium	67	164·930	Th	Thorium	90	232·038
I	Iodine	53	126·9044	Ti	Titanium	22	47·90
In	Indium	49	114·82	Tl	Thallium	81	204·37
Ir	Iridium	77	192·2	Tm	Thulium	69	168·934
K	Potassium	19	39·102	U	Uranium	92	238·03
Kr	Krypton	36	83·80	V	Vanadium	23	50·942
La	Lanthanum	57	138·91	W	Tungsten	74	183·85
Li	Lithium	3	6·939	Xe	Xenon	54	131·30
Lu	Lutetium	71	174·97	Y	Yttrium	39	88·905
Lw	Lawrencium	103	—	Yb	Ytterbium	70	173·04
Md	Mendeleevium	101	—	Zn	Zinc	30	65·37
				Zr	Zirconium	40	91·22

NOTES
1. The above atomic weights are based on the exact number 12 for the carbon isotope 12, as agreed between the International Unions of Pure and Applied Physics and of Pure and Applied Chemistry, 1961.
2. The values given normally indicate the mean atomic weight of the mixture of isotopes found in nature. Particular attention is drawn to the values for hydrogen, boron, carbon, oxygen, silicon and sulphur, where the deviation shown is due to variation in relative concentration of isotopes.

Index